Optical Communication

Optical Communication

Edited by **Vladimir Latinovic**

Clanrye International

New Jersey

Published by Clanrye International,
55 Van Reypen Street,
Jersey City, NJ 07306, USA
www.clanryeinternational.com

Optical Communication
Edited by Vladimir Latinovic

International Standard Book Number: 978-1-63240-401-5 (Hardback)

Printed in China.

Contents

	Preface	**IX**
Section 1	**Optical Communication: General Concepts**	**1**
Chapter 1	**New Quantum Cipher Optical Communication: Y-00** K. Harasawa	**3**
Chapter 2	**Numerical Technique for Electromagnetic Field Computation Including High Contrast Composite Material** Hiroshi Maeda	**25**
Chapter 3	**Effects of Core and Cladding on Optical Guidance Properties of Holey Fibers** Jeong Kim	**39**
Chapter 4	**3D ICS with Optical Interconnections** Edward G. Kostsov, Sergey V. Piskunov and Mike B. Ostapkevich	**55**
Section 2	**Optical Communication: Optical Components**	**83**
Chapter 5	**Optical Phase-Conjugation of Picosecond Four-Wave Mixing Signals in SOAs** Narottam Das, Hitoshi Kawaguchi and Mohammad Razaghi	**85**
Chapter 6	**Photonic Components for Analog Fiber Links** Alejandro García Juárez, Ignacio Enrique Zaldívar Huerta, Jorge Rodríguez Asomoza and María del Rocío Gómez Colín	**113**
Chapter 7	**Chaotic Dynamics of Semiconductor Lasers for Secure Optical Communication** Muzahim I. Azawe	**151**

Section 3 Optical Communication: Optical Systems 167

Chapter 8 **Terrestrial Coherent Free-Space Optical Communication Systems** 169
Mingbo Niu, Julian Cheng and Jonathan F. Holzman

Chapter 9 **Realization of Mixed WDM Transmission System** 201
Vjaceslavs Bobrovs and Girts Ivanovs

Chapter 10 **Next Generation Optical Wireless Communication Systems Using Fiber Direct Coupled Optical Antennas** 233
Koichi Takahashi

Section 4 Optical Communication: Optical Networks 265

Chapter 11 **Subwavelength and Nanometer Diameter Optical Polymer Fibers as Building Blocks for Miniaturized Photonics Integration** 267
Xiaobo Xing, Huaqing Yu, Debin Zhu, Jiapeng Zheng, Huang Chen, Wei Chen and Jiye Cai

Chapter 12 **Efficient Forward Error Correction Decoder Design for High-Speed Optical Networking** 299
Bo Yuan, Li Li and Zhongfeng Wang

Chapter 13 **Rate-Adaptive Free-Space Optical Links Over Atmospheric Turbulence and Misalignment Fading Channels** 321
Antonio García-Zambrana, Carmen Castillo-Vázquez and Beatriz Castillo-Vázquez

Section 5 Optical Communication: Signal Processing 341

Chapter 14 **All-Optical Signal Processing for High Spectral Efficiency (SE) Optical Communication** 343
Y. Ben Ezra, B.I. Lembrikov, Avi Zadok, Ran Halifa and D. Brodeski

Chapter 15 **Stable Higher-Charged Vortex Solitons in Optically-Induced Lattices** 367
Liangwei Dong and Huijun Li

Section 6 Optical Communication: MIMO Systems 391

Chapter 16 **Performance Analysis of SISO and MIMO FSO Communication Systems Over Turbulent Channels** 393
Kostas Peppas, Hector E. Nistazakis, Vasiliki D. Assimakopoulos and George S. Tombras

Chapter 17 **Multiple-Input Multiple-Output (MIMO) Optical Wireless Communications** 417
Oswaldo González

Permissions

List of Contributors

Preface

It is often said that books are a boon to mankind. They document every progress and pass on the knowledge from one generation to the other. They play a crucial role in our lives. Thus I was both excited and nervous while editing this book. I was pleased by the thought of being able to make a mark but I was also nervous to do it right because the future of students depends upon it. Hence, I took a few months to research further into the discipline, revise my knowledge and also explore some more aspects. Post this process, I begun with the editing of this book.

Optical communication has a significant presence in the telecommunications and data networking sectors. Optical communication is an essential part of telecommunication systems, data processing and networking. It can be performed visually or by using electronic devices. It consists of a transmitter that encodes any information into an optical signal, a channel that carries the signal to its desired destination, and a receiver that reproduces that information from the received optical signal. This book deals with the basic concepts, components, systems, networks, signal processing, etc. In the past few decades, optical mechanism and other improved signal processing functions have been considered in-depth for optical communications systems. The book also explains MIMO system thoroughly. We intend to help our readers with the compilation of researches and theories mentioned in this book.

I thank my publisher with all my heart for considering me worthy of this unparalleled opportunity and for showing unwavering faith in my skills. I would also like to thank the editorial team who worked closely with me at every step and contributed immensely towards the successful completion of this book. Last but not the least, I wish to thank my friends and colleagues for their support.

Editor

Section 1

Optical Communication: General Concepts

Chapter 1

New Quantum Cipher Optical Communication: Y-00

K. Harasawa

Additional information is available at the end of the chapter

1. Introduction

1.1. Introduction of Y-00 (overview of network security)

Data volume handled by individuals and companies on the internet is significantly increasing at present. The dissemination of cloud computing causes a lot of important information to flow on networks. And such information is stored in data centers and servers. Meanwhile, cyber terrorism and other crimes that aim at such important information are also on the increase and their techniques have been advanced. To respond to these threats, advanced security measures are implemented in Layer 2 (data link layer) and higher layers of the Open System Interconnection (OSI) reference model. However, safety measures of Layer 1 (physical layer) that forms a transmission path have not been established although Layer 1 is an open area. In such a network broadly two issues exist.

a. For the security of Layer 2 and higher layers cryptography pursuing mathematical complexity is used for decryption calculation. And the basis of safety greatly depends on the performance of computer used by a eavesdropper for decryption. (The safety deteriorates with the increase in performance of computer.)
b. Security hole shifts to Layer 1 and Layer 1 becomes relatively vulnerable when the safety of Layer 2 and higher layers is strengthened.

For this reason, physical security measures are required in Layer 1 to improve safety. Especially in communication lines requiring high safety, measures for constant monitoring through a dedicated optical fiber path of special line route are presented. But such measures require very high running costs and therefore can be realized only for special purposes. Researches on quantum cryptography for the safety of transmission paths have been made all over the world by above-mentioned background [1]. The quantum cryptography currently studied mainly in Japan and Europe is normally the Quantum Key Distribution

(QKD) system using photon transmission. The mainstream of this system is generally called BB84 that was proposed by C.H.Bennett and G.Brassard in 1984 [2]. On the other hand, Y-00 is a new quantum cipher system published by H.P.Yuen(Professor of Northwestern University in US) in 2000 [3]. BB84 uses photon transmission for the QKD system. But Y-00 is the stream cipher system that uses quantum noise existing in continuous light of a laser diode (LD) to directly encrypt data. Y-00, H.P.Yuen and O.Hirota(Professor of Tamagawa University in Japan) made theoretical verification for safety in reports exceeding the Shannon Limit for cryptography in their respective papers for the final version [4]. However, implementation conditions are limited when practical use and dissemination are considered with the current optical communication technology. As one of the Y-00 features, it has been proved that Y-00 is stronger than the current cipher though the ultimate safety (unconditional safety) cannot be achieved by the current technology [5]. Data cannot be completely deciphered by a ciphertext only attack in the currently installed equipment [6]. Also safety can be ensured by the amount of time up to decryption on the fast correlation attack. Y-00 is a physical cipher system and the safety does not depend on calculation amount like general mathematical cryptography. It information hidden by physical phenomena (quantum effect) must be extracted. No shortcut exists because physical working time for extracting this information becomes the basis of safety. Furthermore, the amount of time can be ensured. Astronomical time of this amount of time can be realized also for the currently installed equipment.

1.2. Concept of Y-00

The QKD system that controls photon and transmits data with key information put on a single photon can theoretically show the safety effect provided that One-Time-Pad can be achieved as described in section 1.1. (The absolute condition required by One Time Pad is key distribution that achieves unconditional safety.) To achieve this unconditional safety by single-photon transmission, ideal devices and transmission path are necessary. And various conditions for implementation are really added. These conditions become a loop hole. Therefore that disables absolutely secure key distribution and deteriorates safety level to allow eavesdropping [7,8]. For this reason, the practical use of the QKD system is difficult while maintaining safety in the current optical communication system. Adaptability to large-capacity optical communication networks handling a great deal of information and construction of a new special transmission system at the time of introduction of a security system require much costs.Therefore, applications that are available be limited. Realizing a system by reducing these limiting conditions as much as possible is also an important task. Research and development for Y-00 have been placing the highest priority on the application to existing optical communication systems and possibility that made based on current optical communication technologies. As a result, the WDM transmission system can be shared and no further special infrastructure needs to be constructed. Because Y-00 effectively uses the existing optical modulation system, it can follow up technology trends of normal optical transceiver modules. And it also enables high-speeding, integration, and power saving of the equipment.

2. Principle and outline of Y-00

2.1. Y-00 started from two directions (phase modulation and intensity modulation)

As described above, research of Y-00 was started by H.P.Yuen and other people at the Northwestern University in the United States. Also research and development on implementation are being made at NuCrypt Limited Liability Company (member company of Northwestern University) [9-11]. In this research, phase modulation on coherent light is used as base and encryption is performed by using phase fluctuation (quantum fluctuation) of light. The phase modulation angle of the light becomes multiple value densely. And logic "1" and logic "0" (binary) are distinguished in the 180 degree opposite phase combination. Only one combination is selected from multiple binary combinations (called bases) for each bit from the key information in synchronization with the receiver. The multi-value phase information on the same circumference is closely arranged like overlapping by quantum fluctuation. Eavesdroppers who have no key information must accurately detect phase information. Normal receivers can predict a set of selected base values by using the common key information and therefore can distinguish information. Several-thousands of phase angle values to be used for modulation are set and Non-Linear Feedback Shift Register (NLFSR) and Advanced Encryption Standard (AES) are used for sorting the base to enhance the safety. They use phase modulation for the Y-00 base to apply it to optical space transmission and achieve highly secure communication between aircraft and the ground. In Japan, O.Hirota of Tamagawa University (in Japan) who had made research together with H.P.Yuen started research of Y-00 on the basis of the widely spread intensity modulation of light [12]. Implementation is made by Hitachi Information & Communication Engineering , Ltd. and its application to existing optical fiber networks is considered [13-18].

2.2. Safety of Y-00

Figure1 shows the idea of theoretical safety of Y-00. Theoretical research on ultimate safety of Y-00 is in progress. But if implementation and practicality are considered by present technologies, safety is restricted. However, this restriction can ensure nearly ultimate safety though it is limited by providing essential conditions (astronomical amount of time) for decryption by eavesdroppers through a quantum physical phenomenon. The quantum physical phenomenon means quantum noise (quantum fluctuation) which is an absolute phenomenon that cannot be removed theoretically. Because this phenomenon is completely random, it does not correlate with measured data and the phenomenon cannot be copied. If this phenomenon can diffuse the effect over the signal area, completely decipher is impossible. But differentiation in receiving conditions between normal receiver and eavesdropper is difficult. This will generate trade-off between receiving sensitivity of normal receivers and safety. Therefore ultimate safety cannot be pursued with technical conditions aimed at practical use. Eavesdroppers still cannot avoid physical phenomena of these conditions and acquire correct data. Therefore, direct decryption (cipher text only attack) is impossible. Furthermore, eavesdroppers will attempt to seek for data correlation to get the initial key from acquired sample data like fast correlation attack that makes exhaustive search. In the exhaustive search, phenomena cannot be copied correctly due to effects of physical phenomena. It is disabling parallel processing [19]. Therefore, eavesdropper must be stacked in serial to find correlation of data. In addition, the sample data volume required for decryption is an order of 1E20~1E30 bytes or more by implementing the safety enhancing measures described later. That memory capacity to store the sample data are 1E20~1E30 bytes or more and the sample data acquisition time is several tens of millions to several hundred millions of years (at 10Gbps transmission) even if effects of physical phenomena cannot be diffused over the signal area as described above. This is limited and that is considered to be indecipherable safety. Because it does not depend on computer's performance unlike the present cryptography that pursues mathematical complexity.

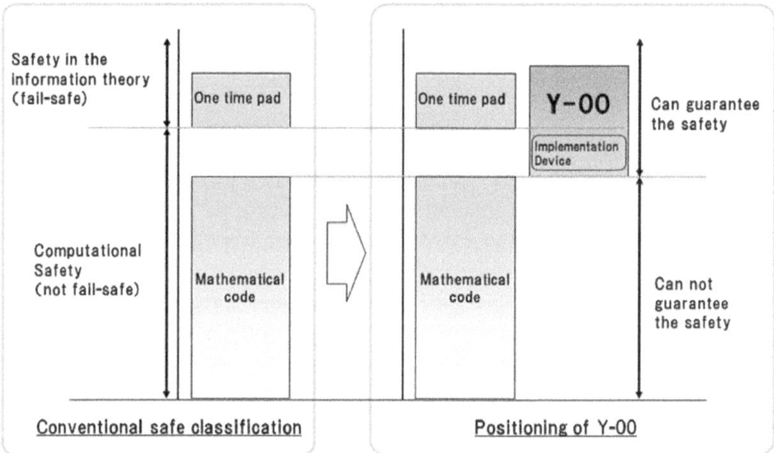

Figure 1. Image of the safety of Y-00

2.3. Idea of physical safety

This section mainly describes an example of optical intensity modulation that is generally used in optical communication. A semiconductor laser diode used in general optical communication uses high-output continuous light (coherent light). The generation probability of each photon that forms this coherent light in the phase and optical power directions is unstable during measurement due to quantum phenomenon, and the value of the probability is not uniquely determined. The distribution of this undefined value is handled as a part of shot noise that is generated during the photoelectric conversion in optical communication, which is an element that degrades the receiving sensitivity. However, effects of other classical noise (including thermal noise of receiving amplifier) generated in the receiver become dominant to the receiving sensitivity of optical communication using the normal optical intensity modulation system. Normal receivers of the devised optical modulation system in Y-00 can maintain little quantum noise effect as in the case of conventional optical communication system. Also Y-00 produces an effect of significantly degrading the receiving sensitivity (to an unreceivable level) for eavesdroppers. In other words, is established safety by great difference of Signal-to-Noise-ratio (S/N) between at normal receiver and eavesdropper. In this case, safety will be improved by as close as possible to 0.5 the error rate of eavesdroppers. Measured value of quantum noise that affects optical communication varies at completely random against the phase value and power value of light as described in section 2.2. Therefore, the phase modulation system and intensity modulation system used in the existing optical communication can be applied to the Y-00 encryption system that uses this quantum noise. The following describes basic idea to establish safety.

1. Multi-value modulation

The normal optical communication performs a multiple value transfer to increase transmission capacity. However in Y-00, only a single bit out of multiplexed values is used for transmitting information and other values are dummy information for eavesdroppers. It is important for multi-values to establish safety that the quantum noise distribution sufficiently overlaps between adjacent levels (both in optical intensity and phase). The number of values becomes several thousands or more depending on conditions.

2. Encryption and decryption

In the encryption by a transmitter, a combination of binary data (1-bit "1" or "0" level value) is selected for each bit of transmit data from multiple signal values created under conditions (1) using the initial key by the multi-value selection information. This selected binary combination is called base in the same way as phase modulation. The amplitude of this base (between two values) determines the receiving sensitivity of normal receivers in the case of intensity modulation. Therefore, 1/2 (180 degrees for phase modulation) of the maximum signal amplitude is the best value for normal receivers to obtain the optimum receiving sensitivity (Figure2). Decryption process of receiver is essential to distinguish the base that varies in each bit by the best threshold value at signal reception. The amplitude of "1" and "0" levels during reception is estimated based on the multi-value selection information generated by the initial key shared with the transmitter and the threshold value is momentarily moved to the best point to distinguish "1" level and "0" level of the signal. Synchronization of multi-level selection information is critical at this time between the transmitter and receiver. This information is changed at random in each bit between transmitter and receiver (Figure2) [17,18,20-22].

3. Tapping

People other than those who are engaged in optical communication believe in many cases that optical fiber does not be able to tapping unlike electric wires. The principle of optical fiber transmission is well known. Light travels in an optical fiber while repeating reflection using the refraction of light generated by junction of the core and clad in the optical fiber.

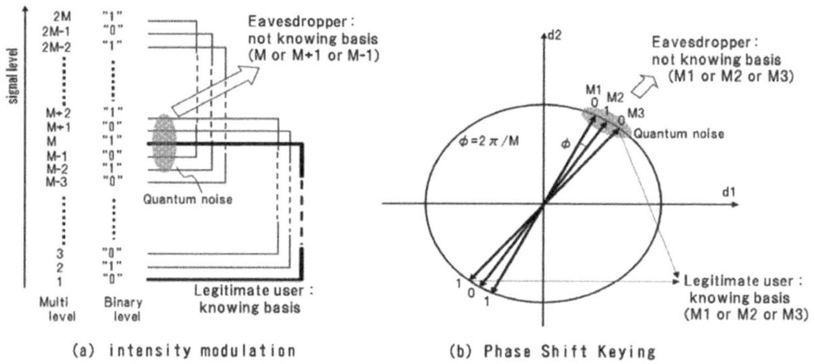

Figure 2. Signal basis and quantum noise

This reflectance varies by bending the optical fiber. If the optical fiber is bent at a sharp angle in particular, the refractive index of the core and clad extremely changes. Therefore optical signal not be able to total reflection. Part of the optical signal will leak out for that. Signal monitoring equipment that uses this principle has been commercialized and used as a measuring instrument. Tapping data from optical fibers has become relatively easy at present due to the technical advance (including high-speed, high-sensitivity detector and low-noise optical amplifier) in optical communication as shown in this example. Measures for improving safety to independently protect transmission paths have become imperative for background mentioned above (Figure3) [22,23].

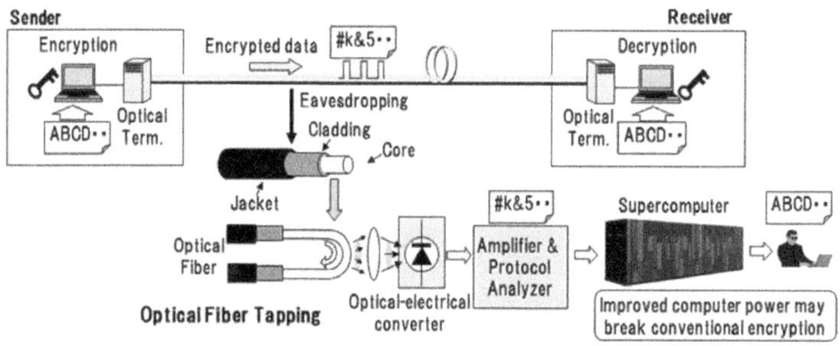

Figure 3. Eavesdropping from optical fiber

3. Implementation of Y-00

3.1. Basic configuration

The Y-00 encryption transmission equipment shares the initial key (Seed Key) between transmitter and normal receiver and performs synchronization processing on each side.

And it configures a pair of transceivers between transmitter and receiver. Figure4 shows the basic configuration of the transmitter of the transceiver. A running key that actually makes encryption is generated from the Seed Key shared by the transmitter and receiver. And the base information (a pair of combination) is selected by the running key from multiple values. Furthermore, the signal level to be used actually is determined from this base information. Input data is converted to a multi-value level by the code modulator and is then output from the subsequent electrical/optical (E/O) converter as a Y-00 encryption optical signal. Figure5 shows the basic configuration of the receiver of the transceiver. The Y-00 encryption optical signal sent from the transmitter is converted to an electrical signal (voltage value) by the O/E converter. And base information is created by the Seed Key in the same procedure as the transmitter. A threshold value that allows the best reception signal to be distinguished is selected from this base information. And a value of 0 or 1 is distinguished by the decoder concurrently with the decoding processing to restore the previous data. This synchronized work between transmitter and receiver performs all processing in each bit of the data transmission rate.

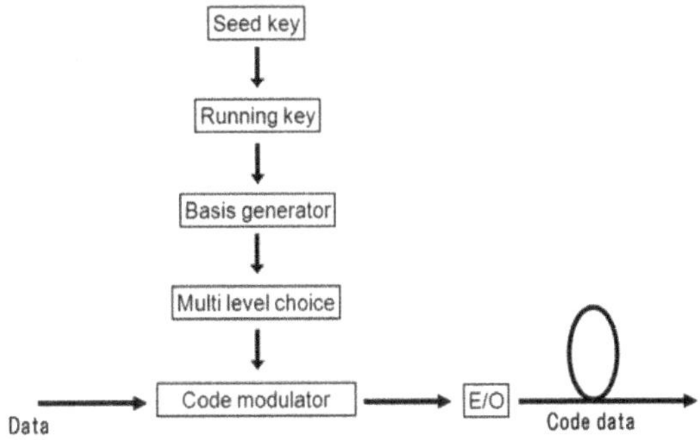

Figure 4. Structure of transmitter of Y-00

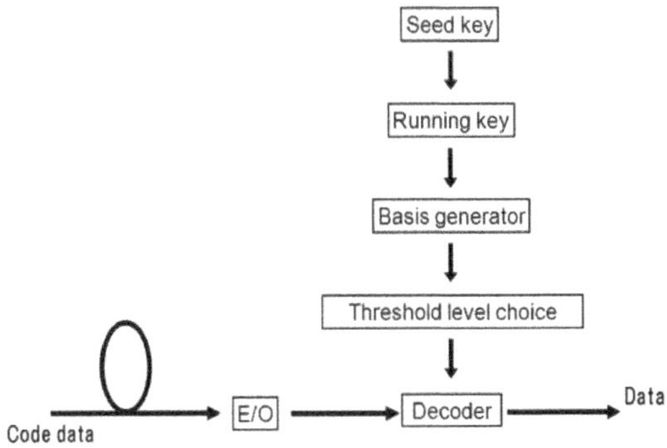

Figure 5. Structure of receiver of Y-00

3.2. Enhancing safety

The probability of eavesdropper's signal level detection error increases with the increase in noise distribution range as described about the safety of Y-00 in the previous section. This effect makes it difficult to extract the correlation of signal level samples acquired by a eavesdropper. Therefore more samples are required. This determines irreducible absolute amount of time necessary to obtain the number of samples required for decryption. This time is for the safety of the people it is possible to secure almost forever (finite strictly) if it is an astronomical value (hundreds of millions of years). Quantum noise contained in coherent light is Poisson distribution dependent on the average optical power as is well known. But this quantum noise is effective between adjacent multi-value levels but it cannot affect entire signal area. For this reason, Y-00 enhances safety using various methods [5,17,18]. This quantum noise effect diffusion method is called Randomization. The following introduces several typical methods.

1. Overlap Selection Keying (OSK)

OSK is a method that allocates logic information (1 or 0) to "High" or "Low" level of light at random to make it difficult for eavesdroppers to decipher signal information "1" or "0" even if they can detect the reception level [16]. Logic of the signal is randomly allocated to a "1" and "0" for each 1bit in order to achieve the OSK. This operation makes it difficult to distinguish whether the signal level detected by tapping is positive logic "1" or negative logic "0". This method is the same as general stream cipher, but is different in purpose and effect. In the basic Y-00, adjacent signal levels are replaced with bit information "1" and "0" alternately. But eavesdroppers can predict the code by focusing on "1" or "0" every other level. Therefore, the safety level is equivalent to the case when the number of values (number of bases) is reduced to 1/2. To solve this problem, changing bit information on a bit basis can maintain degree of difficulty of decryption [12,15,17,24].

2. Keyed Deliberate Signal Randomization (KDSR)

Uniforming the conditional probability of multi-value signal detection in Y-00 is important in terms of cryptographic theory. KDSR is used to perform the uniforming [14]. It is a correction technology that does not directly enhance safety but spuriously evenly expands the quantum noise distribution that is effective for eavesdroppers in detecting the multi-value signal level. This method produces effects equivalent to the case where S/N deterioration effect (overlap of quantum noise between adjacent multi-value signal levels) of eavesdroppers is diffused to a wide area as described in the previous section. Figure6 illustrates this mechanism.

Y-00 of optical intensity modulation is causing a level detection error by the quantum noise distribution overlaps of the adjacent signal level as described in the previous section. It is preferable to uniform the entire multi-value signal level. But it degrades the receiving sensitivity of normal receivers making communication difficult. KDSR slightly fluctuates the selected signal level by shifting a part of multi-value level selections conditions at random to solve this problem. Figure 6 (a) shows this state. The k's true value M is diffused to a range of M±2 by diffusion using KDSR in this example. Furthermore, the noise effect can be

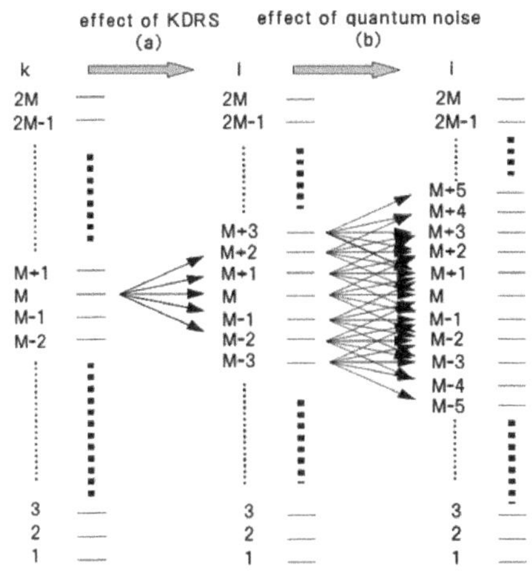

Figure 6. Mechanism of the spread by the random shifter.

diffused to a range of M±5 with the effect of quantum noise distribution (b). The quantum noise distribution function P(k|l) to the individual signal level k diffused by effects of KDSR is evenly arranged, where the l signal level measurement error probability is P(l|i) as the effect of quantum noise at this time. At this time, the effect (a) of KDSR is P(k|l). In addition, the quantum noise effect on the l that are distributed with the effect of (a) is P(l|i). Therefore, the conditional probability P(k|i) of error of true signal level k is shown by the following expression.

$$P(k\mid i)=\sum_{i} P(k\mid l)P(l\mid i) \tag{1}$$

With respect to effects on the receiving sensitivity of normal receivers at that time, the signal level degradation P_{KDS} is shown by the following expression. Conditions are described below. Also ±n is quantum noise diffusion effect by KDSR, 2M is number of multi-values and P_{2M} is full signal amplitude.

$$P_{KDSR}=2\left(\frac{P_{2M}}{2M}\right)|n| \tag{2}$$

For example, if 2M=4096 and n=±3, the level of effects on normal receivers deteriorates to about 1/683 of the signal's full amplitude power. This effect is slight for normal receivers. The following describes KDSR in terms of quantum noise distribution. Figure7 illustrates the range of effects of signal level "i" on the adjacent level when KDSR is not applied. The quantum noise effect is exerted to the reference level ±2 in this example. For this reason, the base selection information error probability is biased and therefore eavesdroppers can estimate a part of the base selection information more easily. KDSR is applied to the base selection information that determines base selection as shown in Figure6 to diffuse noise effects on multi-value levels as shown in Figure8 to solve this problem. Thus the bias in the probability distribution of base selection information error is reduced. And making it very difficult to estimate the base selection information [17,25].

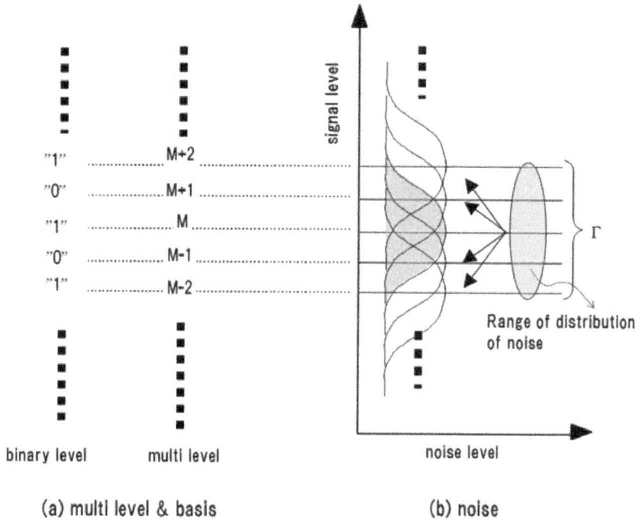

Figure 7. The spread of the noise (KDSR nothing)

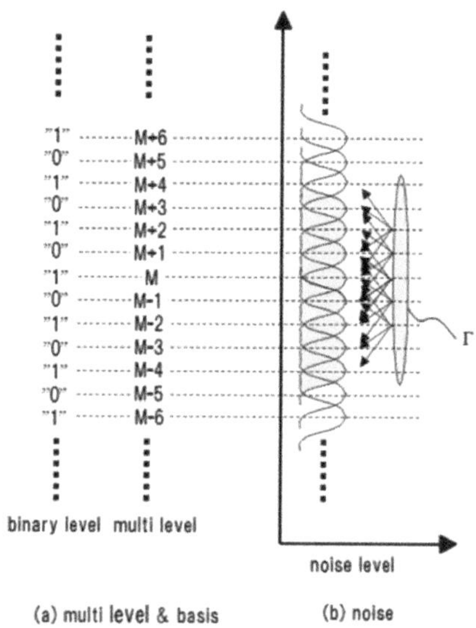

Figure 8. The spread of the noise (KDR)

3. Irregular mapping

Bit error positions become uneven due to effects of the quantum noise distribution in the basic model. Therefore, fast correlation attack may be enabled if the key length is short in the basic array (alternate arrangement of "1" and "0") of bit information of adjacent multi-value levels determined by the base information [25]. However, bit error positions must be uniform to disable such fast correlation attack. Irregular mapping has been developed in order to provide immunity against fast correlation attack even when the short key length [26]. This method disables eavesdroppers to decrypt Y-00 cipher except for complete Brute Force Attack. Figure9 shows the concept of irregular mapping. Synchronization is established between transmitter and receiver. Then bit information is arranged irregularly in the mapping of the multi-value level corresponding to the base. Bit error positions are evenly diffused because the arrangement of the bit information of adjacent multi-value levels is irregular even if the quantum noise distribution effect range is physically the same. This effect disables the fast correlation attack that uses non-uniformity of bit error rate for decryption when the multi-value signal is returned to bit information [17].

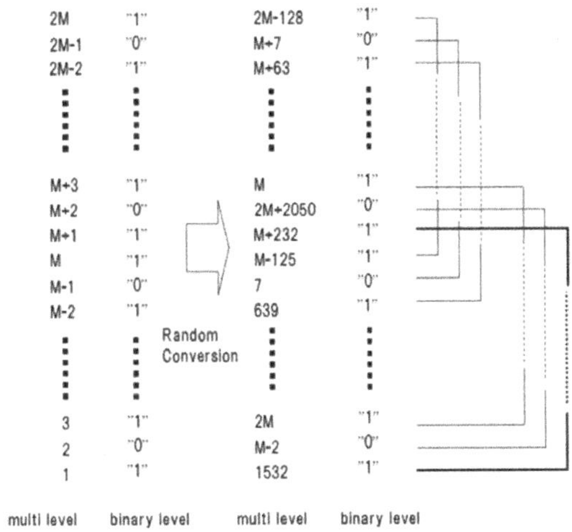

Figure 9. Irregular mapping

3.3. Y-00 encryption circuit

Figure10 shows the configuration of the encryption and modulation circuit that is actually mounted in the Y-00 transmitter. Clock is extracted from input plaintext data for self-synchronization by the Clock Data Recovery (CDR) function. The information for synchronization processing and control is added to the original signal in the Y-00 transmission. Therefore, clock frequency is converted in the FIFO circuit and change the data rate. And frame processing is performed by the framer then information required for synchronization is added. OSK processing is added to this signal to generate a main signal to be the original signal as described in section 3.1. On the other hand, multi-value level selection signal is generated as follows. Running Key is generated from the Seed Key as a first. Then a base selection signal to be the original signal is generated using the randomly mapped base configuration information. This selection signal generates a multi-level selection signal level after processing by KDSR. The multi-value level selection signal that is the same as the main signal is weighted by each driver circuit and added to determine the multi-value level and generate an encoding signal for encryption of Y-00. The operating principle of this final-stage processing is the same as that of the Digital to Analog (D/A) converter. By driving the optical external modulator using the Y-00 signal generated, the Y-00 encryption signal becomes an optical signal with valid quantum noise effect. [17,18].

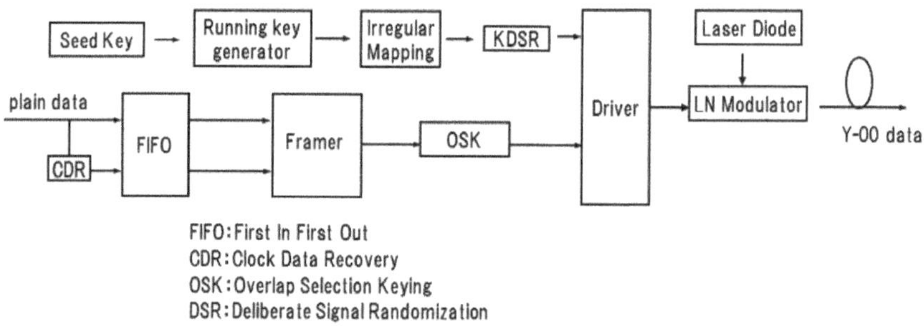

Figure 10. The configuration of the coding of the Y-00 transmitter

3.4. Decoder circuit

Figure11 shows the configuration of the decoder circuit in the Y-00 receiver. The Seed Key information and irregular mapping information are also provided in the receiver as common information. The basic circuit configuration of the decoder is the same as the encryption circuit of the transmitter. However, the receiver does not perform the decoding of KDSR. Therefore, from the decoder outputs a decoding signal (threshold value selection information) of Y-00. The threshold value controller distinguishes the signal on a bit by bit to generate a threshold value level with the best level and timing. Optimum adjustment is made for the threshold level and timing by the Automatic Gain Control (AGC) amplifier and the threshold value controller. Furthermore, the decoder establishes bit synchronization and key synchronization for encryption of Y-00 [17,18].

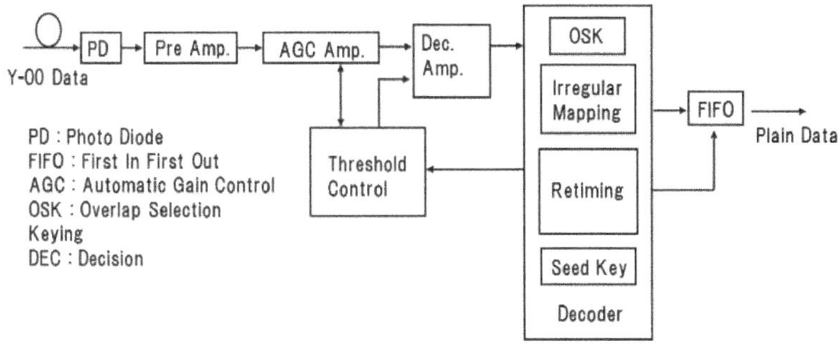

Figure 11. The configuration of the decoding circuit of the Y-00 receiver

4. Transmission experiment and performance evaluation trial production result

4.1. Basic characteristics

To verified the adaptability to existing optical communication networks as Y-00 encryption equipment. Prototype equipment was produced and evaluated based on the content in section 3. The prototype targeted standard specifications of OC-48 (Optical Carrier: SONET standard) optical communication as IEEE standard considering practical use. Table1 shows evaluation results [17]. The transmission rate of original plaintext data is 2.48832 Gbps conforming to OC-48 and the average optical output power is 0 dBm. This is achieved transmission distance of 50 km without relay.

A receiving sensitivity of -15.3 dBm was obtained at a bit error rate (BER) of 1E-12 with a transmission rate of 2.48832 Gbps and an average optical output power of 0 dBm.

Item	unit	result
data rate	Gbps	2.48832
transmission distance	km (w/o amp.)	50
output optical power	dBm (ave.)	0
number of basis (number of levels)	- (-)	2048 (4096)

Table 1. Major characteristics

4.2. Transmission experiment

Low-delay real-time transmission of encrypted uncompressed full-specification High-Definition-Television (HDTV) moving picture data was performed using the prototype Y-00 transmission equipment. Figure12 shows the BER in the back-to-back transmission. The minimum receiving sensitivity is -15.3 dBm when BER=1E-12. The average input power of the

Y-00 transmission equipment (receiver) is approx -10 dBm and the margin of 5.3 dB. It enabling 40 km transmission when considering the optical fiber loss (approx. 0.25 dB/km). Figure13 shows the transmission system in the experiment. The transmitter converts the OC-48 optical signal from 1.5 Gbps moving picture data of signal source by the High Definition Serial Digital Interface to Synchronous Digital Hierarchy (HD-SDI/SDH) converter. And then encrypts the signal by the Y-00 cipher transmission equipment (transmitter). The encrypted optical signal is transmitted through a 40 km single mode fiber (SMF). The receiver decodes encryption signal. And then restores the original moving picture data by the HD-SDI/SDH converter. The latency of the transmission system shown in Figure13 is approx 500 μs. It is achieving secure real-time high-definition moving picture transmission that hardly shows visible delay in monitor images before and after transmission. This result has verified that the Y-00 cipher transmission equipment is applicable to medical sector and financial system networks which require real-time response [27,28].

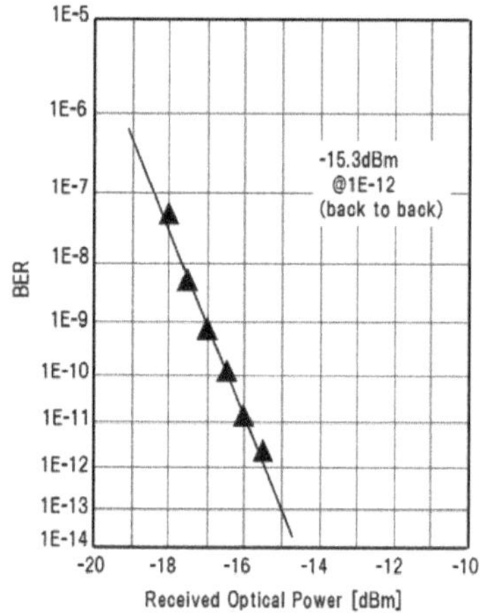

Figure 12. Receiver sensitivity of the regular receiver

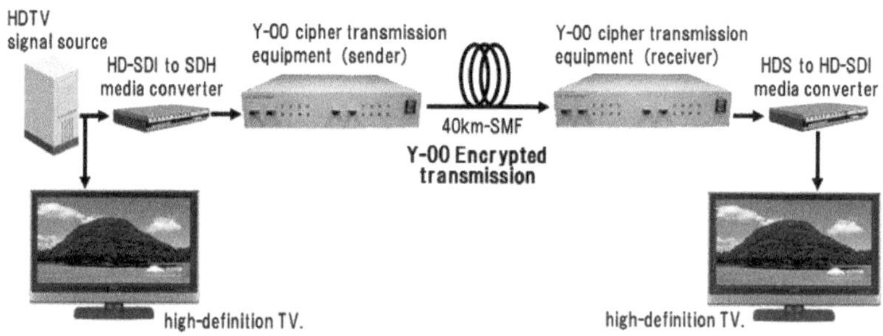

Figure 13. Real-time HDTV transmission experiment

4.3. Field test using a commercial line

We made a transmission experiment using an existing commercial line optical fiber to obtain further prospect of practical use. Figure14 shows the system of the transmission experiment that was actually made. The distance of each transmission span is 48 km and the average span attenuation is 14.5 dB. We made a transmission experiment of total distance 192 km with relay at three location using optical fiber amplifiers (EDFA). Figure15 shows the result of receiving sensitivity measured at the reception end. Figure16 shows waveforms of encrypted and decoded signals. We verified a receiving sensitivity of -18.4 dBm and -19.4 dBm respectively at a BER of 1E-12 in 192 km bidirectional transmission. Also we verified adaptability to optical amplifier repeater transmission. In addition, we have confirmed that the encryption of Y-00 can be applied in Fiber Channel (FC) and Gigabit Ether (GbE). The measured latency value of the transmission system was 1.29 ms in total including the delay of fiber length. Furthermore, we made WDM transmission experiment multiplexing optical output signals from two opposed Y-00 units and verified error-free transmission at each wavelength [17,18,21].

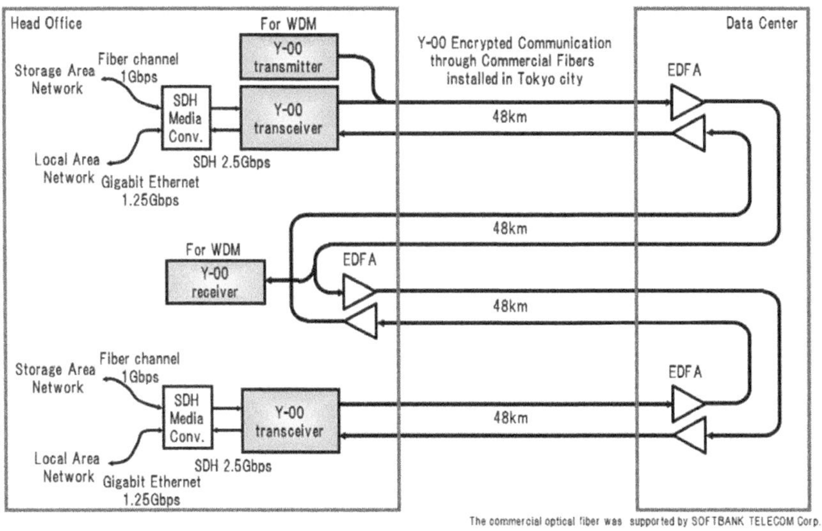

Figure 14. 192 km relay Y-00 encrypted transmission through commercial fibers

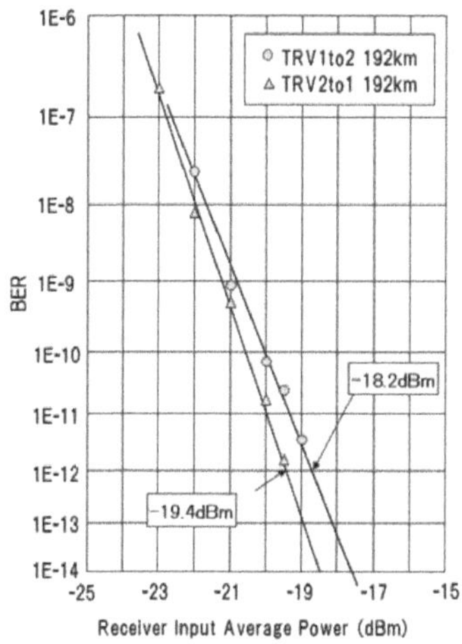

Figure 15. Received optical power sensitivity (192 km)

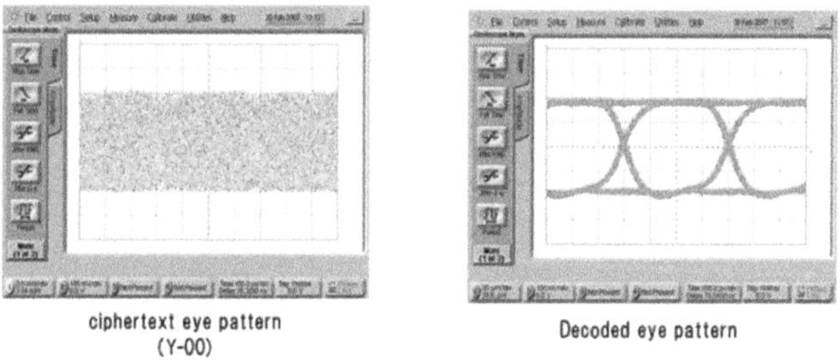

Figure 16. Y-00 transmission wave pattern

4.4. Application to 10 Gbps transmission

This section describes a trial toward large-volume transmission that is the trend of optical communication. Y-00 encryption transmission equipment for 10 Gbps transmission based on optical intensity modulation has been developed in japan [30,31]. The design concept of this equipment is the same as the above-mentioned 2.4 Gbps transmission equipment except that dedicated high-speed devices have been developed to realize the equipment. This section describes the result of 360 km transmission experiment using optical fibers (for experiment laid in Tamagawa University) installed in the field. Figure17 shows the configuration of the transmission system. The 360 km transmission path contains nine EDFA for relay at intervals of 40 km using standard single mode fibers (SMF). Dispersion values are adjusted by the dispersion compensating fiber (DCF) and the tunable dispersion compensator (TDC) to set the residual dispersion to +1170 ps/nm. The optical interface conforms to OC-192. The optical output power is -1.7 dBm in the back-to-back transmission and the full-amplitude extinction ratio is 2.5 dB. Furthermore, the encryption contains various types of randomization for enhance safety. The transmission path is also provided with an optical preamplifier and an optical bandpass filter in the receiver to ensure the S/N for normal receivers.

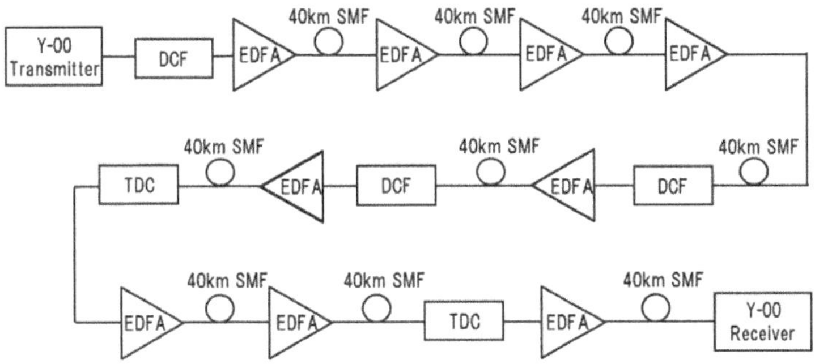

Figure 17. 360 km Y-00 transmission system

Figure18 shows transmission waveforms (eye diagram). They are encrypted waveforms with no eye-opening at each transmission distance. Figure 19 shows characteristic of normal receiver (back-to-back, 40 km, 60 km and 80 km of non repeater transmittion and optical amplifier repeater transmission of 360 km) and the BER of eavesdropper. The minimum receiving sensitivity is -12.2 dBm (BER=1E-12) as shown in Figure 19. And the BER of 360-km transmission is 5.0 × 1E-7. Furthermore, we obtained results that satisfy receiving sensitivity -4 dBm at a BER of 5.0 × 1E-5 which is the target specification considering code error correction under all conditions. We evaluated adjacent signal detection of multi-value signal in the back-to-back transmission to evaluate tapping capability. And obtained a satisfactory result of eavesdropper's BER larger than 0.4. This evaluation has proved that the Y-00 transmission equipment is sufficiently applicable to 10 Gbps transmission. Thus we could obtain prospects for high-speed transmission [29,30,31].

5. Conclusion (future prospects and possibilities)

Based on the Yuen 2000 protocol (Y-00) theory as the research result of H.P.Yuen and O.Hirota, we have developed the Y-00 encryption transmission equipment using quantum noise effects and have verified the practicality of the equipment. We verified the safety and adaptability to existing systems based on trial production results of the equipment and obtained prospects for practical use. The results show the high completeness of the equipment. Hitachi Information & Communication Engineering has been engaged in the development of prototype equipment and is further improving the reliability of the Y-00 encryption transmission equipment for the productization (Figure20). Trial production results show that the Y-00 system can achieve long-haul, large-capacity, high-speed real-

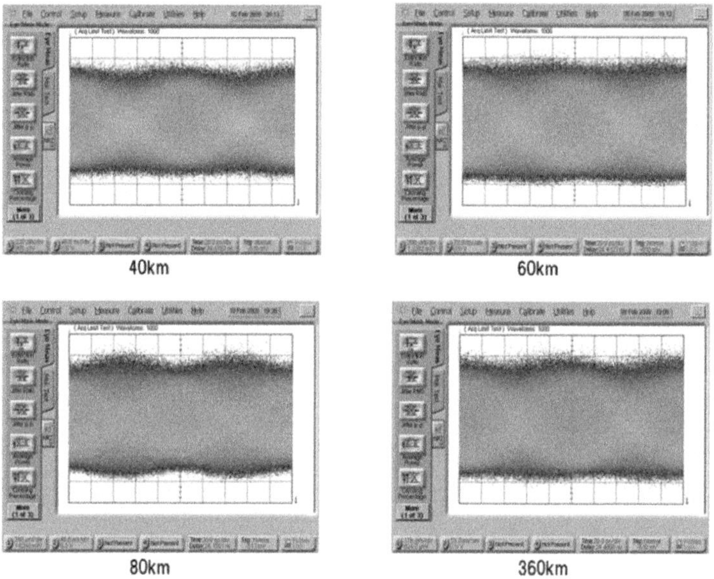

Figure 18. 10 Gbps Y-00 transmission wave form

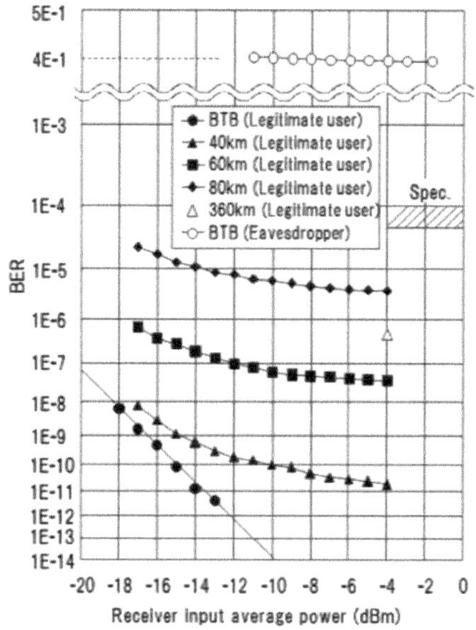

Figure 19. Bit error rate (10 Gbps)

time transmission with low latency. Thus application of the Y-00 system to various fields can be expected. The Y-00 system is also applicable to uncompressed high-definition image transmission in particular, which extends the range of use. Since conventional optical communication technologies that are being developed at present can be used technically, development trends (such as large capacity, downsizing, and power-saving) can be maintained in common. Furthermore, existing optical communication infrastructures are available and allowing co-existence and combined use with current systems and reducing initial costs. Thus we can expect the use of the Y-00 system in wide applications.

Figure 20. Y-00 encryption transmission equipment for the productization

Author details

K. Harasawa
Hitachi Information & Communication Engineering, Ltd., Japan

Acknowledgement

I would like to thank Prof. Osamu Hirota who provided carefully considered feedback and valuable comments. Special thanks also go to Prof. Kiichi Yamashita, Mr.Makoto Honda, Mr. Shigeto Akutus and Mr.Yoshifumi Doi whose opinions and information have helped me very much throughout the production of this study.

6. References

[1] K. Kitayama, M. Sasaki, S. Araki, M. Tsubokawa, A. Tomita, K. Inoue, K. Harasawa, Y. Nagasako, A. Takada "Security in Photonic Networks: Threats and Security Enhancement", IEEE/OSA Journal of Lightwave Technology, vol. 29, no. 21, p. 3210-3222, 2011.

[2] C. H. Bennett, G. Brassard, "Quantum Cryptography: Public Key Distribution and Coin Tossing", Proceedings of IEEE International Conference on Computers Systems and Signal Processing, Bangalore India, p175-179, December 1984.
[3] H. P. Yuen, "A new quantum cryptography," Report in Northwestern University, 2000.
[4] O. Hirota, T. Iwakoshi, F. Futami, K. Harasawa, " Getting around the Shannon limit of cryptography", SPIE, Newsroom, 10. 1117/2. 1201008. 003069, 2010. http://spie. org/documents/Newsroom/Imported/003069/003069_10. pdf
[5] O. Hirota, "Practical security analysis of a stream cipher by the Yuen 2000 protocol" Physical Review, A 76, 032307, 2007.
[6] K. Kato, O. Hirota, "Randomization techniques for the intensity modulation based quantum stream cipher and progress of experiment", SPIE conference on Quantum Communications and Quantum Imaging IX, Proceedings vol. 8163, 2011.
[7] I. Gerhardt, Q. Liu, A. L. Linares, J. Skaar, C. Kurtsiefer, V. Makarov, "Full-field implementation of a perfect eavesdropper on a quantum cryptography system", Nature Communications, vol. 2 349 (14 June 2011)
[8] I. Gerhardt, Q. Liu, A. L. Linares, V. Scarani, J. Skaar, V. Makarov, C. Kurtsiefer, "Experimentally faking the violation of Bells inequalities", Physical Review Letters, 107, 170404, 2011.
[9] E. Corndorf, C. Liang, G. S. Kanter, P. Kumar, H. P. Yuen, "Quantum-noise randomized data encryption for wavelength-division-multiplexed fiber-optic networks" Physical Review A, vol. 71, 062326, 2005.
[10] G. S. Kanter, E. Corndorf, C. Liang, V. S. Grigoryan, P. Kumar, "Exploiting quantum and classical noises for securing high-speed optical communication networks", Fluctuation and Noise in Photonics and Quantum Optics III, edited by P. R. Hemmer, J. R. Gea-Banacloche, P. Heszler, Sr., M. S. Zubairy, Proceedings of SPIE, vol. 5842, 2005.
[11] C. Liang, G. S. Kanter, E. Corndorf, P. Kumar, "Quantum Noise Protected Data Encryption in a WDM Network", IEEE Photonics Technology Letters, vol. 17, No. 7, JULY 2005.
[12] O. Hirota, "Optical Communication Network and Quantum Cryptography", The Transactions of the IEICE B, No. 4, p478-486, 2004.
[13] K. Harasawa, M. Honda, S. Iwata, N. Kanazawa, T. Kanamaru, O. Hirota, "Basic experiment of quantum cryptography based on optical communications" , Proceedings of the Society Conference of IEICE, Communication, B-10-34, 2004.
[14] T. Hosoi, K. Harasawa, M. Honda, S. Akutsu, Y. Kobayashi, O. Hirota, "Field Transmission Experiment of 2. 5G Y-00 Transmitter/Receiver" , Proceedings of the IEICE General Conference, B-10-80, Communication (2), 419, 2007.
[15] M. Fuse, S. Furusawa, T. Ikushima, O. Hirota, "Development of an ultra high－secure and 1 Gbps optical transmission system using quantum noise diffusion cryptography", ECOC 2005. 31st European Conference on Optical Communication, Proceedings vol. 3, p555-556, 2005.
[16] M. Shimizu, T. Uno, K. Sako, K. Ohhata, K. Yamashita, K. Harasawa, S. Hirota, "Modulator driver LSI for 10 Gb/s quantum stream cipher optical transceiver using

Yuen-2000 protocol (Y-00)", Proceedings of the Society Conference of IEICE, C-12-8, Electronics (2), 63, 2007 ·

[17] K. Harasawa, O. Hirota, K. Yamashita, M. Honda, S. Akutsu, T. Hosoi, Y. Doi, K. Ohhata, T. Katayama, T. Shimizu, "Consideration of the Implementation Circuit of Randomization for Physical Cipher by Yuen 2000 protocol", The Transactions of the IEICE C, vol. J91-C, No8, p1-10, 2008.

[18] K. Harasawa, O. Hirota, K. Yamashita, M. Honda, K. Ohhata, S. Akutsu, and Y. Doi,, "Quantum encryption communication over a 192 Km, 2. 5 Gbit/sec line with optical tranceivers employing Yuen-2000 protocol based intensity modulation", IEEE/OSA. Journal of Light Wave Technology, vol-29, No. 3, p316-323, 2011.

[19] O. Hirota, K. Kurosawa, "Immunity against correlation attack on quantum stream cipher by Yuen 2000 protocol", Quantum Information Processing, vol. 6, No-2, p81-91, 2007.

[20] O. Hirota, T. Shimizu, T. Katayama, K. Harasawa, "10 Gbps quantum stream cipher by Y-00 for super HDTV transmission with provable security", Quantum Communications and Quantum Imaging V, Proceedings of SPIE, vol. 6710, Sep. 25, 2007.

[21] T. Hosoi, K. Harasawa, S. Akutsu, M. Honda, Y. Kobayashi, O. Hirota, "Field Transmission Experiment of 2. 5G Y-00 Transmitter/Receiver", Proceedings of the General Conference of IEICE, Communication (2), 419, B-10-80, 2007 ·

[22] S Etemad A. Agarwal, T. Banwell, G. Crescenzo, J. Jackel, R. Menendez, P. Toliver, "An Overlay Photonic Layer Security Approach Scalable to 100 Gb/s", Communications Magazine, IEEE, vol. 46, Issue 8, p32-39, 2008.

[23] S. Kay. Miller, "Fiber Optic Networks Vulnerable to Attack", Information Security Magazine, November 15, 2006.

[24] O. Hirota, M. Sohma, M. Fuse, K. Kato, "Quantum stream cipher by Yuen 2000 protocol: Design and experiment by intensity modulation scheme", Physical Review A -72, 022335, 2005

[25] S. Donnet, A. Thangaraj, M. Bloch, J. Cussey, J. Merolla, Security of Y-00 under heterodyne measurement and fast correlation attack, Physics Letters A, 356, p406-410, 2006.

[26] T. Shimizu, O. Hirota, "Randomization of running key for quantum stream cipher Y-00", Technical report of IEICE, OCS, PN, CS, 2007.

[27] Y. Doi, S. Akutsu, T. Hosoi, M. Honda, Harasawa, O. Hirota, T. Katayama, "Hi-Vision Transmission of Y-00 Quantum Cryptography Transceiver", Proceedings of the IEICE General Conference, Communication(2), 328, B-10-45, 328, 2008

[28] Nature Photonics Technology Conference Report, p11, 23-25 October 2007 Tokyo, Japan http://www. natureasia. com/en/events/photonics/2007_photon_conf_report. pdf (accessed 20 April 2012)

[29] S. Akutsu, Y. Doi, T. Hosoi, M. Honda, K. Harasawa, O. Hirota, T. Katayama, "Field Relay Transmission experiment of Y-00 Quantum Cryptography Transceiver", Proceedings of the Society Conference of IEICE, Communication (2), 223, 2007-08-29

[30] Y. Doi, S,Akutsu, M. Honda, K. Harasawa, O. Hirota, S. Kawanishi, O. Kenichi, K. Yamashita, "Field Transmission Experiments of 10 Gbit/s Stream Cipher by Quantum

Noise for Optical Network", Proceedings of the IEICE General Conference, Communication(2) 2010, 369, 2010.

[31] O. Hirota, K. Ohhata, M. Honda, S. Akutsu, Y. Doi, K. Harasawa, and K. Yamashita, "Experiments of 10 Gbit/sec quantum stream cipher applicable to optical Ethernet and optical satellite link", SPIE conference on quantum communication and quantum imaging VII; Proceedings of SPIE, vol. 7465, 2009.

Chapter 2

Numerical Technique for Electromagnetic Field Computation Including High Contrast Composite Material

Hiroshi Maeda

Additional information is available at the end of the chapter

1. Introduction

Periodic structure plays important role with its sensitivity for frequency in microwave communication and optical integrated circuit system [1][2]. Also, in optical fiber communication[3], such periodic structure is used for supporting optical wave to be guided in core of the fiber. In such application, a defect in the structure generally works as a cavity or a waveguide by making use of its high selectivity. However, by lacking periodicity in the structure with a defect, general mathematical approaches have often difficulties, due to inconvenience in description of the problem. In such cases, computational simulation technique for electromagnetic wave propagation and scattering is very important and effective. By the recent development of computers, it is possible to model a large scale periodic structure with defect.

In simulation of electromagnetic phenomena, finite difference time domain (FDTD) method [4] [5] is most widely used. However, in numerical analysis of the wave behavior near boundary, where the dielectric constant is quite different on each side, a special care is required. Supposing in lossless dielectric medium, it is well known that wavelength of the electromagnetic wave changes due to the dielectric constant. Because of this compression of wavelength in high dielectric constant medium, grid size of space becomes rather coarse compared with material with lower dielectric constant. Therefore, accuracy of finite difference approximation deteriorate in most computation with uniform grid size. In numerical analysis of periodic structure such as photonic crystal(PC), the dielectric constant is generally quite high compared with its background medium.

A constrained interpolation profile (CIP) method[6][7] is payed much attention because of its accurate simulation result compared with conventional FDTD method. In this paper, on scattering problem by a dielectric cylinder with high contrast with its background air, performance of CIP method is compared with analytical approximated method using Hertz potential [8] and conventional FDTD simulation [9]. As a measure of accuracy of CIP and FDTD simulation, an normalized cross correlation function is defined and compared with it, by setting analytical approximated result as a reference. Consequently, results of CIP method showed better correlation than that of FDTD method. As applications of CIP method, analysis of electromagnetic field propagation in Y-shaped branching waveguide and Mach-Zehnder interferometer in two dimensional photonic crystal structure were demonstrated. Both of analysis results showed reasonable behaviour. Especially for asymmetrical Mach-Zehnder interferometer, the measurement result by microwave model and the numerical result of CIP corresponded to each other. Complicated output characteristics of asymmetric Mach-Zhender interferometer was interpreted very well by refering to the electric field profile obtained by CIP method.

2. Formulation of electromagnetic field propagation by CIP method

Applications of CIP method to a problem of electromagnetic wave scattering is reported in Ref.[7]. Outline of the CIP method and its application to solve two dimensional Maxwell's equation is briefly explained in following subsections, based on the reference.

2.1. CIP method and advection equation

We treat wave propagation with velocity u in isotropic and uniform space. The wave form is expressed by $f(x,t)$ as a function of space x and time t. When the function $f(x,t)$ satisfies following equation

$$\frac{\partial f}{\partial t} \pm u\frac{\partial f}{\partial x} = 0, \tag{1}$$

Eq.(1) is called an advection equation. This equation holds for arbitrary wave function f with variable $x \mp ut$, including electromagnetic wave propagation. This equation means that each point on the wave form moves to $+\Delta x$ or $-\Delta x$ from current point with velocity u after passage of Δt. This property is important to calculate electromagnetic field both in uniform space and on boundary.

2.2. Interpolation by cubic polynomial

In CIP method, wave function is approximated by following cubic polynomial,

$$F_i(x) = a_i(x - x_i)^3 + b_i(x - x_i)^2 + c_i(x - x_i) + d_i, \tag{2}$$

where a_i, b_i, c_i and d_i are unknown coefficients to be determined by making use of function value at each discrete point. Suppose that data at discrete points $F_i(x_i)$ and $F_i(x_{i-1})$ are known with its derivatives $dF_i(x_i)/dx$ and $dF_i(x_{i-1})/dx$. Substituting x_i and x_{i-1} into Eq.(2) and its derivative function, we obtain

$$F_i(x_i) = d_i = f_i, \tag{3}$$

$$\frac{dF_i(x_i)}{dx} = c_i = g_i, \tag{4}$$

$$\begin{aligned} F_i(x_{i-1}) &= -a_i\Delta x^3 + b_i\Delta x^2 - c_i\Delta x + d_i \\ &= f_{i-1}, \end{aligned} \tag{5}$$

$$\frac{dF_i(x_{i-1})}{dx} = 3a_i\Delta x^2 - 2b_i\Delta x + c_i = g_{i-1}, \tag{6}$$

where $x_i - x_{i-1} = \Delta x$ is used. According to property of advection Eq.(1), the latest values of $F_i^{n+1} = f_i^{n+1}$ and its derivative $dF_i^{n+1}/dx = g_i^{n+1}$ are obtained by

$$f_i^{n+1} = a_i\xi^3 + b_i\xi^2 + g_i^n\xi + f_i^n \tag{7}$$

$$g_i^{n+1} = 3a_i\xi^2 + 2b_i\xi + g_i^n \tag{8}$$

where, $\xi = -u\Delta t$. By solving Eq.(3) to Eq.(6) with renewal of coefficients a_i, b_i, c_i and d_i, we have updated value of the function f and its derivative g at each discrete point from Eq.(7) and Eq.(8).

2.3. Conversion of Maxwell's equation into advection equation

Maxwell's curl equations for electric field vector $\boldsymbol{E}$ and magnetic field vector $\boldsymbol{H}$ in lossless, isotropic and nonconductive material are given as follows,

$$\nabla \times \boldsymbol{H} = \varepsilon \frac{\partial \boldsymbol{E}}{\partial t} \tag{9}$$

$$\nabla \times \boldsymbol{E} = -\mu \frac{\partial \boldsymbol{H}}{\partial t} \tag{10}$$

where ε is permittivity and μ is permeability.

Assuming two dimensional uniform space along with z axis (i.e. $\partial/\partial z = 0$), Maxwell equations are decomposed into two sets of polarization. We treat E-wave which includes (H_x, H_y, E_z) as the component. For E-wave, Maxwell's equations are reduced to followings:

$$\frac{\partial E_z}{\partial y} = -\mu \frac{\partial H_y}{\partial t} \tag{11}$$

$$\frac{\partial E_z}{\partial x} = \mu \frac{\partial H_y}{\partial t} \tag{12}$$

$$\frac{\partial H_y}{\partial x} - \frac{\partial H_x}{\partial y} = \varepsilon \frac{\partial E_z}{\partial t} \tag{13}$$

From Eqs.(11) to (13), we obtain

$$\frac{\partial}{\partial x}\mathbf{AW} + \frac{\partial}{\partial \mathbf{y}}\mathbf{BW} + \frac{\partial}{\partial \mathbf{t}}\mathbf{CW} = \mathbf{0}, \tag{14}$$

where

$$\boldsymbol{W} = \begin{pmatrix} H_x \\ H_y \\ E_z \end{pmatrix}, \tag{15}$$

$$\boldsymbol{A} = \begin{pmatrix} 0\,1\,0 \\ 0\,0\,1 \\ 0\,0\,0 \end{pmatrix}, \tag{16}$$

$$\boldsymbol{B} = \begin{pmatrix} -1\,0\,0 \\ 0\;\;0\,0 \\ 0\;\;0\,1 \end{pmatrix}, \tag{17}$$

$$\boldsymbol{C} = \begin{pmatrix} 0 & 0 & -\varepsilon \\ 0 & -\mu & 0 \\ \mu & 0 & 0 \end{pmatrix}. \tag{18}$$

Split step procedure is used for Eq.(14), then we have

$$\frac{\partial}{\partial x}\boldsymbol{AW} + \frac{\partial}{\partial t}\boldsymbol{CW} = 0, \quad (\boldsymbol{W}^n \to \boldsymbol{W}^*) \tag{19}$$

$$\frac{\partial}{\partial y}\boldsymbol{BW} + \frac{\partial}{\partial t}\boldsymbol{CW} = 0, \quad \left(\boldsymbol{W}^* \to \boldsymbol{W}^{n+1}\right) \tag{20}$$

where superscripts n and $n+1$ denote time step, and $\boldsymbol{W}^*$ means middle value of EM components. These equations are converged into the advection equation. For plane wave with component (E_z, H_y), Eq.(14) can be written down as

$$\begin{cases} \dfrac{\partial H_y}{\partial x} - \varepsilon \dfrac{\partial E_z}{\partial t} = 0, \\ \dfrac{\partial E_z}{\partial x} - \mu \dfrac{\partial H_y}{\partial t} = 0. \end{cases} \tag{21}$$

By using wave impedance $Z = \sqrt{\mu/\varepsilon}$ and velocity of light $c = 1/\sqrt{\varepsilon\mu}$, above equations can be rewritten as

$$\begin{cases} c\dfrac{\partial \tilde{H}_y}{\partial x} - \dfrac{\partial E_z}{\partial t} = 0, \\ c\dfrac{\partial E_z}{\partial x} - \dfrac{\partial \tilde{H}_y}{\partial t} = 0, \end{cases} \tag{22}$$

where, $\tilde{H}_y = ZH_y$. By addition and subtraction of these equations, we obtain

$$\begin{cases} c\dfrac{\partial(E_z + \tilde{H}_y)}{\partial x} - \dfrac{\partial(E_z + \tilde{H}_y)}{\partial t} = 0, \\ c\dfrac{\partial(E_z - \tilde{H}_y)}{\partial x} + \dfrac{\partial(E_z - \tilde{H}_y)}{\partial t} = 0, \end{cases} \tag{23}$$

for propagation along with x axis. In the same manner, we have advection equations

$$\begin{cases} c\dfrac{\partial(E_z + \tilde{H}_x)}{\partial y} + \dfrac{\partial(E_z + \tilde{H}_x)}{\partial t} = 0, \\ c\dfrac{\partial(E_z - \tilde{H}_x)}{\partial y} - \dfrac{\partial(E_z - \tilde{H}_x)}{\partial t} = 0, \end{cases} \tag{24}$$

for y-direction. Solving these equation with adequate Δx, Δy, and Δt, it is possible to renew electromagnetic field at each discrete point. Similarly, three dimensional analysis can be performed.

3. Numerical examples

3.1. Scattering of plane wave by a dielectric rod with high contrast permittivity to background

In this section, scattered wave by a dielectric cylinder, which is obtained by CIP and FDTD method, is compared with analytical approximated solution [8]. Model of a dielectric

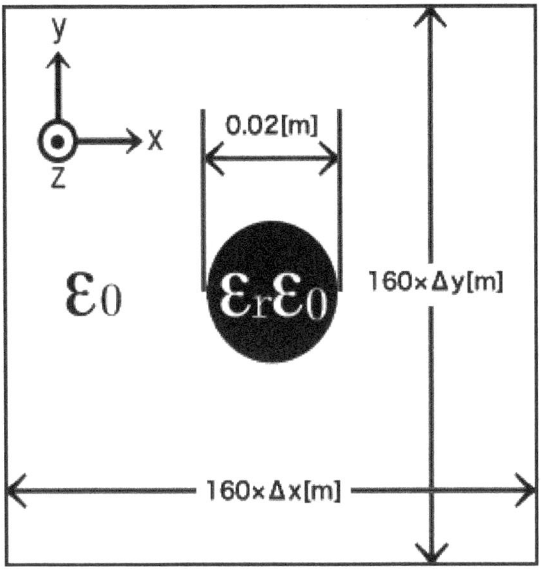

Figure 1. Illustration of the analysis region with coordinate system. A black circle indicates a dielectric cylinder with dielectric constant ε_r.

cylinder and the coordinate system is illustrated in Fig.1. Two dimensional analysis region is $80.0 \times 80.0\ mm^2$ and diameter of the cylinder is $20.0\ mm$. Plane wave of E-polarization with frequency $f = 10.0\ GHz$ is given as continuous incident wave.

In analytical approximated approach, wave function is expanded into summation of Bessel and Hankel functions as basis function. Electromagnetic field is determined so that boundary condition on surface of a dielectric cylinder is satisfied. In this demonstration, convergence of electromagnetic field is confirmed by increasing number of truncation of basis function.

Simulation for the structure was demonstrated by setting $\Delta x = \Delta z = 0.5mm$ and $\Delta t = 0.17ps$, both for CIP and FDTD method. As examples of total electric field intensity near a cylinder, the profiles are shown in Fig.2 and Fig.3, where relative dielectric constant ε_r of the cylinder is chosen to be 25.0, and 36.0, respectively [9]. The background medium is air and its relative dielectric constant $\varepsilon_r = 1.0$ is supposed. These high values of ε is needed for constructing photonic crystal structure, which requires high contrast with respect to the background medium.

From these field profiles, we can evaluate normalized cross correlation function defined as,

$$\eta = \frac{\sum_{i,j} \left| E_{z,Ana}(x_i, y_j) E_{z,\{CIP|FDTD\}}(x_i, y_j) \right|}{\sum_{i,j} \left\{ E_{z,Ana}(x_i, y_j) \right\}^2}, \tag{25}$$

where (x_i, y_j) is discrete point in analyzing region, $E_{z,Ana}(x_i, y_j)$ is E_z field by analytical approximated solution, $E_{z,\{CIP|FDTD\}}(x_i, y_j)$ is that by CIP and FDTD method, respectively.

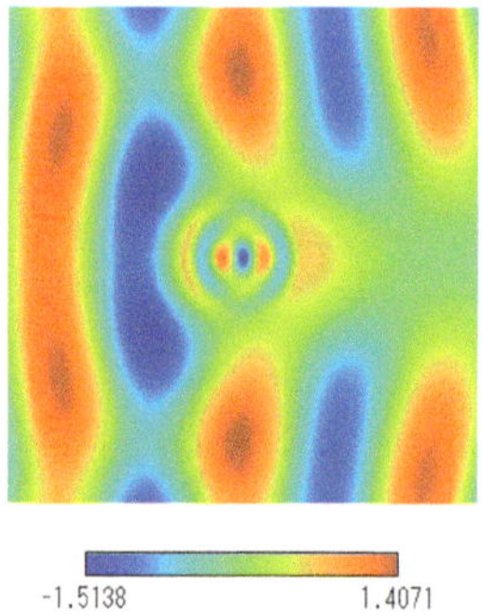

(a) Analytical approximated method

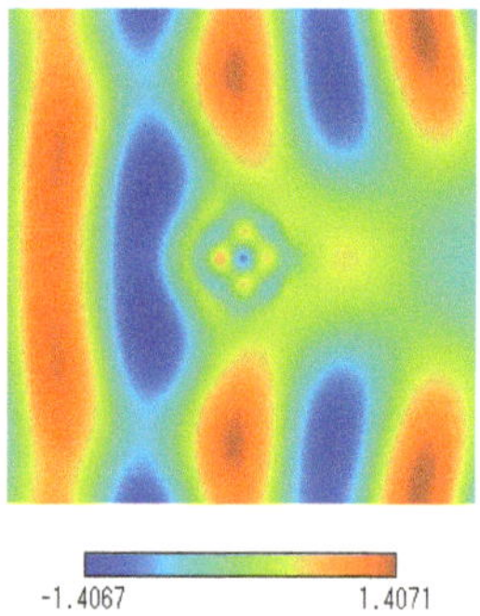

(b) CIP method

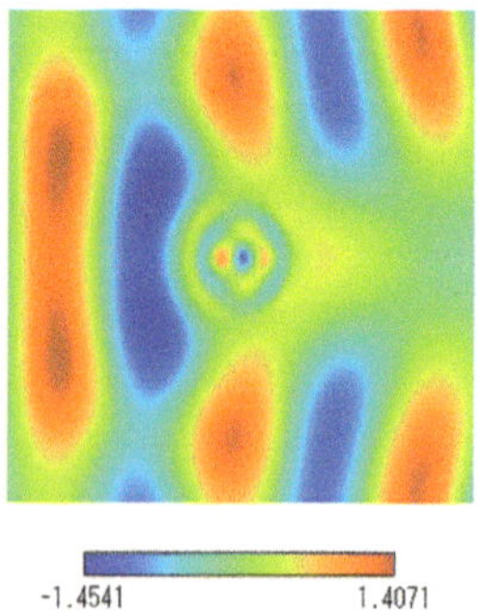

(c) FDTD method

Figure 2. Field profile of E_z by (a) analytical approximated solution, (b) CIP method and (c) FDTD method, respectively, for a case that ε_r of the cylinder is 25.0.

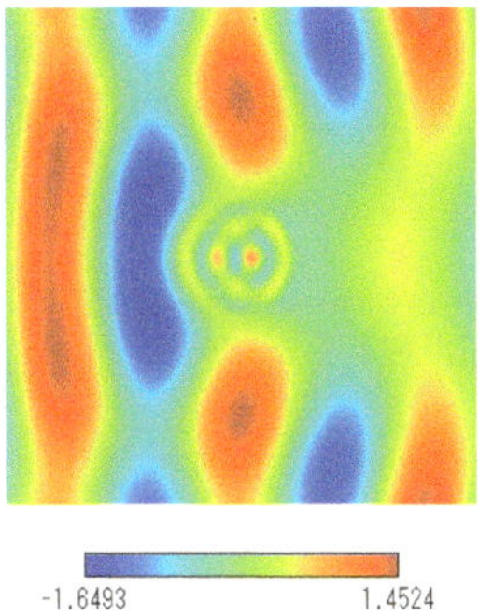

(a) Analytical approximated method

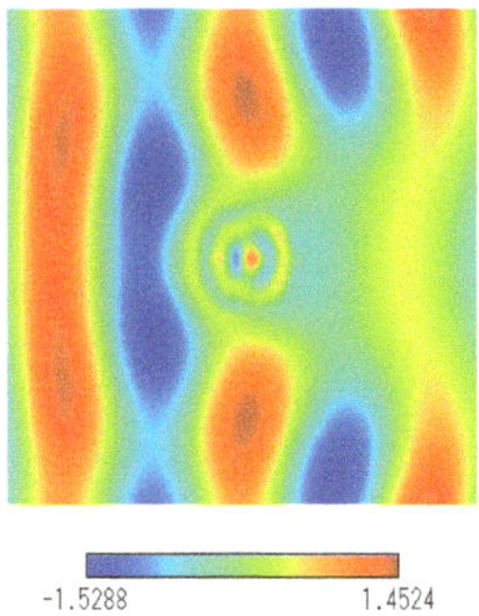

(b) CIP method

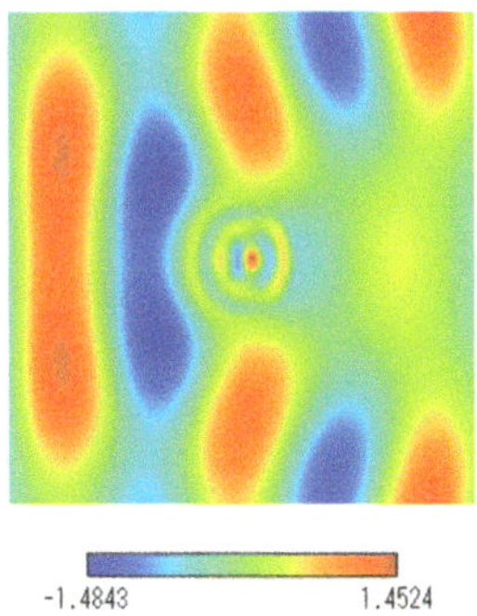

(c) FDTD method

Figure 3. Field profile of E_z by (a) analytical approximated solution, (b) CIP method and (c) FDTD method, respectively, for $\varepsilon_r = 36.0$.

ε_r	4.0	16.0	25.0	36.0
FDTD(%)	86.55	81.19	83.67	87.73
CIP(%)	98.04	96.69	96.44	98.17

Table 1. Values of normalized cross correlation.

Using Eq.(25), we obtain Table.1. From Table.1, values of normalized cross correlation by CIP method is always better than that by FDTD method for typical four kinds of relative dielectric constant ε_r of the cylinder.

3.2. Y-shaped branch waveguide in two-dimensional photonic crystal with triangular lattice

In Fig.4, two dimensional scale model of Y-shaped branch waveguide is illustrated. From port 1, continuous electric field with Gaussian profile along with vertical axis are given. The electromagnetic field is composed of unique electric field which is oriented to vertical direction to the paper surface and magnetic fields which are parallel to L_0 and W_0 axes. Therefore, the incident wave is E-polalization. Spot size $2w_0$ of the Gaussian profile corresponds to waveguide full width $\sqrt{3}P$, where $P = 26.5mm$ is lattice period. Dielectric

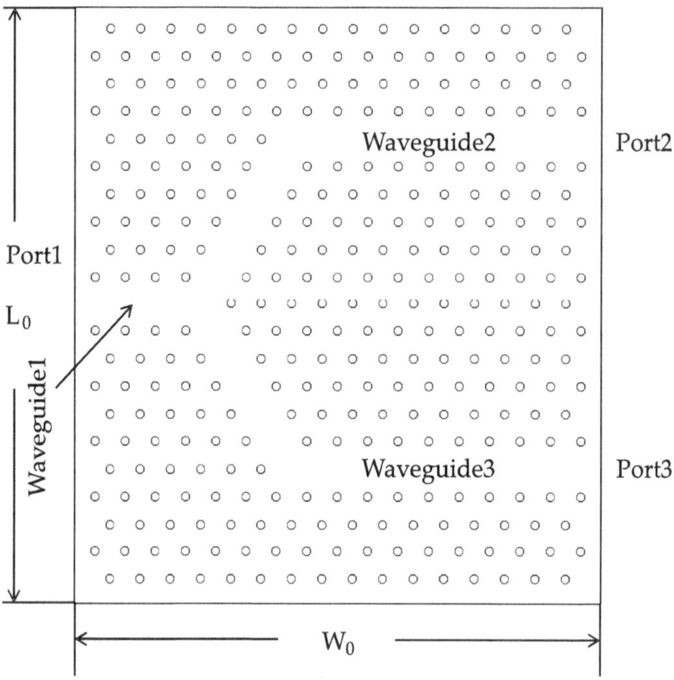

Figure 4. Illustration of scale model of Y-shaped branch waveguide, where $L_0 = 485.5mm$ and $W_0 = 477.0mm$.

rods have same parameters with previous numerical examples. Example of the total electric field profile with frequency $f = 4.0GHz$ by CIP method is shown in Fig.5, where parameters were set as follows; space discretization $\Delta = 0.5mm$, time step $\Delta t = 2.5 \times 10^{-13} sec$, total cell numbers 648×659, cells on diameter of the cylinder 10, respectively. From the Fig.6, it is found that incident wave is equally divided by the branch circuit.

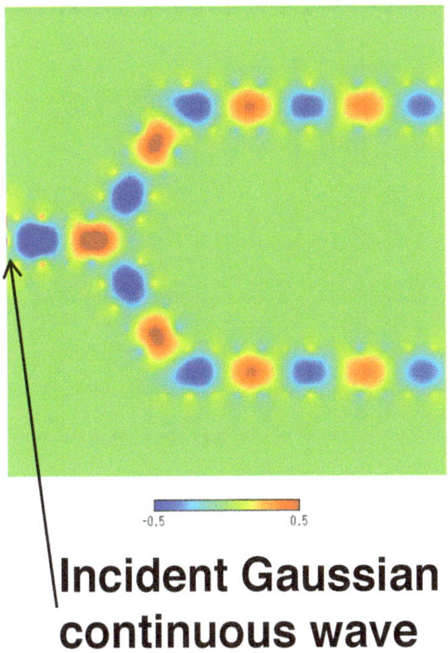

Figure 5. Electric field profile of E-polarised wave in Y-shaped branch waveguide.

In Fig.7, maximum electric field profile at output after filtering is shown. Two pairs of rods were inserted simultaneously in each output waveguide for filtering different frequencies. Length of the cavity is $2.5P$ and another one is $3.0P$, respectively, where P is lattice period. As expected, frequencies with half-wavelength-along-waveguide which corresponding to each cavity length are filtered and obtained in each port.

3.3. Mach-Zehnder interferometer in two dimensional photonic crystal with triangular lattice

In Fig.8, Mach-Zehnder(MZ) type interferometers are depicted. The structure is situated in two dimensional photonic crystal with triangular lattice. To investigate interference at combining point, asymmetrical structure with different arm lengths are compared with symmetrical one. At first, electric field profile in symmetrical MZ structure is shown in Fig.9 The output characteristics of the asymmetric structure was investigated by experiment using a model in same microwave frequency range. The measurement results are shown in Fig.10.

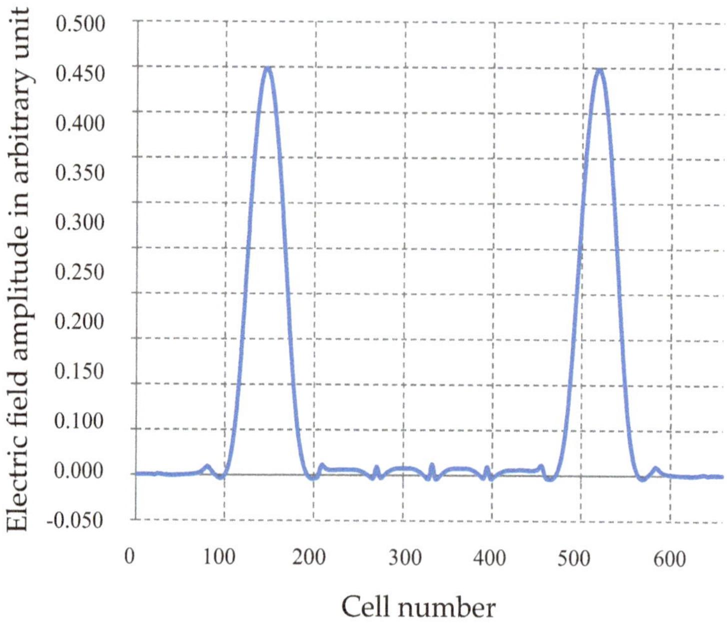

Figure 6. Maximum electric field profile at output port. Two identical peaks locates on center of each output waveguide.

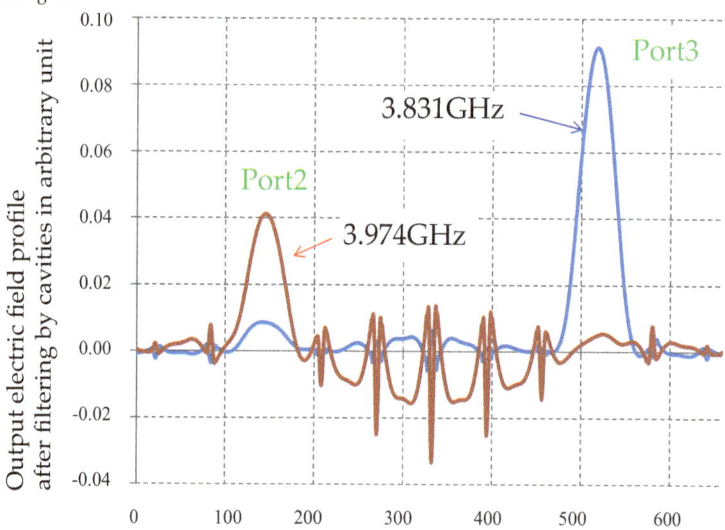

Figure 7. Electric field profile at output after filtering. Fields with maximum amplitude in each output port were depicted.

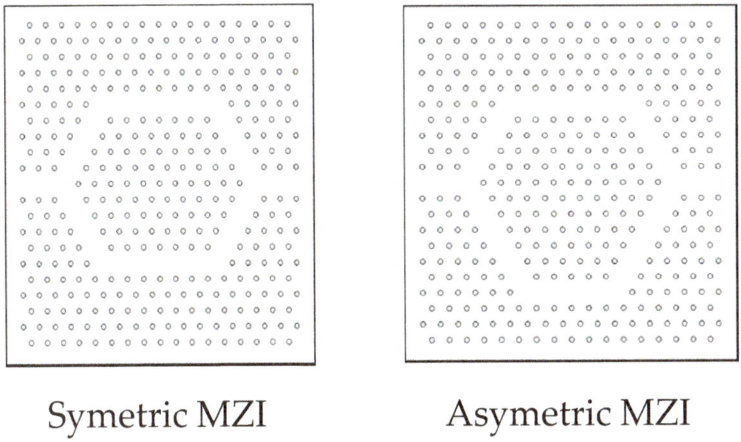

Figure 8. Mach-Zehnder interferometer(MZI) in two dimensional photonic crystal structure.

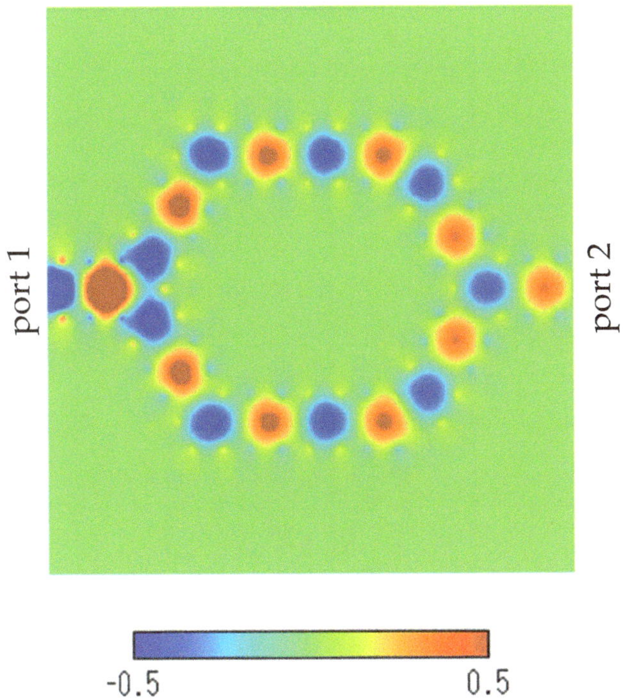

Figure 9. Electric field profile in symmetrical MZ structure.

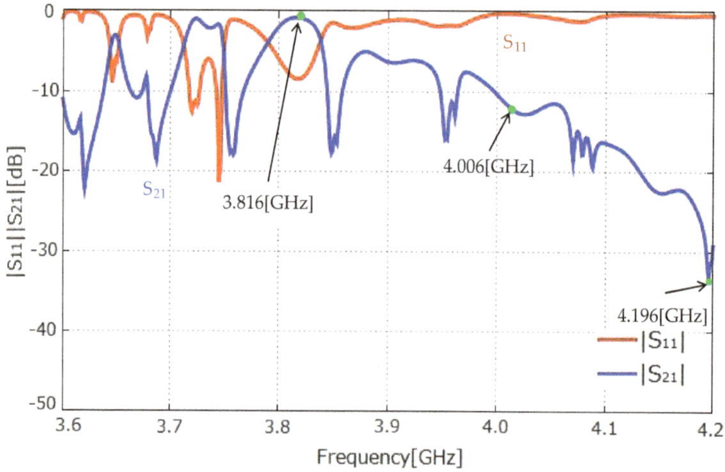

Figure 10. Transmission(S_{21}) and reflection(S_{11}) characteristics of asymmetric MZ interferometer by experiment.

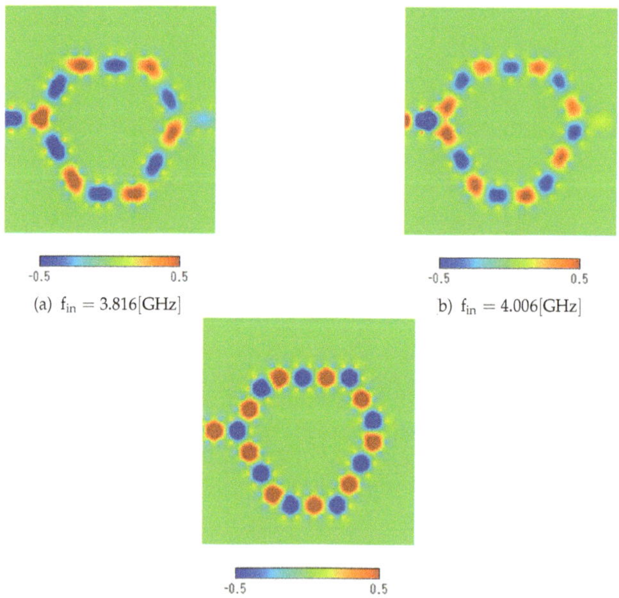

(a) $f_{in} = 3.816$[GHz]

b) $f_{in} = 4.006$[GHz]

(c) $f_{in} = 4.196$[GHz]

Figure 11. Electric field profile for typical input frequency f_{in}.

In Fig.10, it is found that the asymmetric MZ interferometer (MZI) shows maximum and minimum transmission at frequencies 3.816GHz and 4.196GHz, which are indicated by arrows and dots in the figure. Measurement of symmetrical MZI showed almost flat and relatively high transmission characteristics over same frequency range, comparing with result of asymmetrical one. Therefore, simulations by CIP method were demonstrated for these two frequencies and additional one frequency 4.006GHz as middle level of output.

The electric field profiles by CIP method along with MZI are shown for these three frequencies in Fig.11. The electric field profile at output port is also indicated in Fig.12. From these figures, it is found that electric field propagated along with two arms comes to combining point with relatively small phase difference in Fig.11(a) and (b), while electric field become extinct for the two fields comes to combining point with out of phase in Fig.11(c) . In Fig.12(c), the maximum electric field is quite small compared with output with small phase difference in MZI as shown in Fig.12(a) and (b). From Fig.10 to 12, complicated output characteristics were qlearly interpreted by numerical results by CIP method. This is because CIP method provides more precise results of electromagnetic wave scattering compared with FDTD method, as we saw in subsection 3.1. From these results, it was shown that superiority and significance of the CIP method for designing photonic crystal structure which is composed of periodic structure with high contrast of material constant.

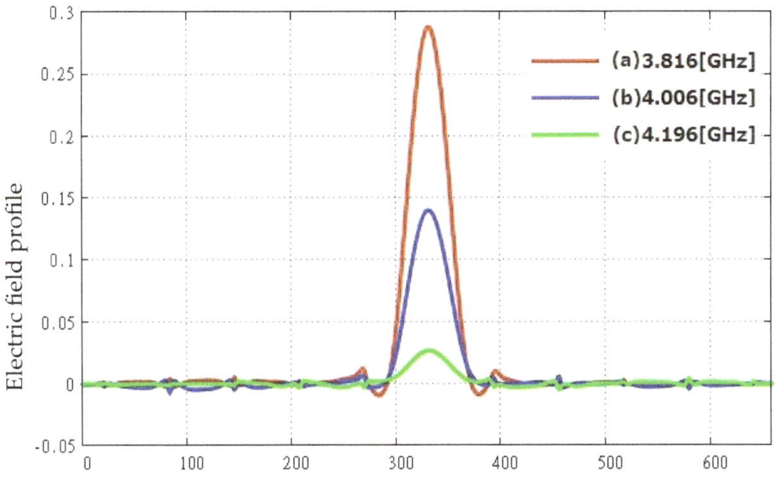

Figure 12. Maximum electric field profile at output port for typical three frequencies.

4. Conclusion

By using CIP method, numerical analysis of scattered electromagnetic field by a dielectric cylinder was demonstrated. Referring to result of analytical approximated approach, result of CIP method showed good accuracy in comparison with result by conventional FDTD method. As examples of designing of photonic crystal structure with high-contrast material profile, Y-shaped branch waveguide and Mach-Zehnder interferometer in photonic crystal structure with triangular lattice were numerically demonstrated. The CIP method showed reasonable results of branch circuit and the filtering characteristics by using cavity. Also for Mach-Zehnder interferometer, numerical results of electric field profile by CIP method implemented experimental result of the structure for typical frequencies.

For example, designing of filtering device for microwave communication or guiding device for optical fiber communication system, the CIP method is expected to show superior performance in accuracy compared with conventional FDTD method. Application of CIP method to design electromagnetic or optical signal processing devices with some defects in periodic structure can be designed by CIP method.

Acknowledgment

Author would like to express his thanks to Mr. H. Terashima, Ms. Y. Zhang, Mr. T. Maruyama, and Mr. S. Hironaka for their engagement as master course research in Graduate School of Fukuoka Institute of Technology, Fukuoka, Japan from 2010 to 2012.

Author details

Hiroshi Maeda
Dept. of Information and Communication Eng., Fukuoka Institute of Technology, Japan

5. References

[1] Yasumoto, K. ed. (2006). *Electromagnetic Theory and Applications for Photnic Crystals*, CRC Press, 0-84933677-5, New York.

[2] Noda, S. & Baba, T. ed. (2003). *Roadmap on Photonic Crystals*, Kluwer Academic Publishers, 1-4020-7464-6, Boston.

[3] Bjarklev, A., Broeng, J. & Bjarklev, A. S. (2003). *Photonic Crystal Fibres*, Kluwer Academic Publishers, 1-4020-7610-X, Boston.

[4] Yee, K. S. "Numerical solution of initial boundary value problems involving Maxwell's equations in isotropic media, "*IEEE Transactions on Antennas and Propagation*, Vol.AP-14, No.4, May 1966, pp.302-307.

[5] Taflove, A. ed.(1998) *Advances in Computational Electrodynamics - The Finite-Difference Time-Domain Method*, Artech House Publishers, 0-89006-834-8, Boston.

[6] Yabe, T., Feng X. & Utsumi T. "The constrained interpolation profile method for multiphase analysis, "*Journal of Computational Physics*, Vol.169, 2001, pp.556-593.

[7] Okubo, K. & Takeuchi, N. "Analysis of an electromagnetic field created by line current using constrained interpolation profile method, "*IEEE Transactions on Antennas and Propagation*, Vol. 55, No. 1, Jan. 2007, pp.111-119.

[8] Ishimaru, A. (1991). *Electromagnetic Wave Propagation, Radiation, and Scattering*, Chap. 11, Prentice-Hall International, 0-13-273871-6, New Jersey.

[9] Maeda, H., Terashima, H., Maruyama, T. & Zhang, Y. (2011). *Numerical Analysis of Electromagnetic Wave Scattered by High-Contrast Dielectric Cylinder Using CIP Method, Proc. of The 1st International Workshop on Information Networking and Wireless Communications (INWC 2011), in conjunction with The 14th International Conference on Network-Based Information Systems (NBiS 2011)*, pp. 491-496.

Chapter 3

Effects of Core and Cladding on Optical Guidance Properties of Holey Fibers

Jeong Kim

Additional information is available at the end of the chapter

1. Introduction

Optical fibers work as one of the most important parts for data transmission medium and light-signal processing component in optical communications. Microstructure holey optical fibers (MHOFs) from very new advanced technology have attracted considerable attention in research communities during recent years for the purpose of enhancing the performance of existing optical communication system [1-5]. Basically, the MHOFs can be made of a single material, in contrast to all other types of conventional optical fibers, which are normally manufactured with two or more materials. Only by properly designing the geometrical structures of the MHOF, numerous novel properties, such as single mode operation over the entire wavelength range of interest, zero or anomalous group-velocity dispersion, and large effective area, can be obtained. Moreover, various combinations of material composition in the MHOF may offer more opportunities for new and reliable fiber-optic devices.

Since the invention of the MHOFs, lots of applications have been explored and realized recently. Some of the novel applications include second-harmonic generation with broader bandwidth compared to some conventional fibers, polarization-preserving capability with elliptical air holes, fiber lasers, nonlinear-optic devices, and even highly sensitive fiber sensors. This special optical fiber, which is different from the conventional step-index fiber or graded-index fiber, can be fabricated by changing multiple parameters such as the diameter of each small air hole, its shape and location coordinates, the spacing between the adjacent holes, and the number of air-hole cladding layers. Because of the structural uniqueness, the MHOF can be designed to provide small or nearly constant dispersion and highly nonlinearity with single-mode operation characteristic over a wide wavelength range in optical communications.

In this chapter, following a brief review about the principle of light-guiding of the conventional optical fiber, general features of the MHOF will be introduced in section 3.

Showing the arrangement method of small-hole cladding on a transverse plane, modelling of the MHOF without and with an air core is addressed. For the electromagnetic analysis of the proposed microstructures with complicated geometries, the utilizations of a finite difference method (FDM) and a finite-difference time-domain (FDTD) technique are considered in the investigation of optical guidance properties, such as normalized propagation constant, mode field distribution, effective area, and chromatic dispersion. Employing the two numerical methods provides cross-verification and additive confidence in the accuracy of the results. Based on the FDTD and the FDM techniques, how the cladding structure of the HMOF affects optical guidance properties are analyzed in section 4. Then the investigation is extended to find out the effects of the size and shape of the core in the holey fiber on optical propagation properties in section 5.

2. Principle of optical guidance

Normally, optical fibers as cylindrical dielectric waveguides consist of core and cladding regions with at least two or more materials. Each region can be made of silica glass or other transparent materials such as plastic. Although practical optical fibers have several or a few layers of cylindrical cladding regions with cross-sectional ring shapes uniform along the light-travelling direction, these cladding regions can be functionally combined together and represented by the average cladding refractive index. The refractive index difference between the core and cladding can provide light confinement and hence guidance of light signal along the fiber. Upon travelling through the fiber, the optical signal is both attenuated and distorted. These effects impose a limit on how far the signal can travel in a fiber before it degrades beyond an acceptable level. The wavelength (λ) of operation is an important consideration in fiber-optic communication systems. Practical fiber optic systems operate in a wavelength range of 0.8 μm to 1.6 μm, where glass attenuation is low. In particular, the silica glass fiber exhibits local minimums in the spectral attenuation curve at 1.3-μm and 1.55-μm regions, which are often referred to as the transmission windows. Many commercial fiber-optic communication systems are designed for operation at these two wavelengths.

Optical fibers, as any other electromagnetic waveguide, can support discrete modes of propagation [6]. Assuming the cladding is just air instead of the multiple cylindrical cladding layers with respective material parameters, a simplified step-index optical fiber can be considered as shown in Figure 1(a). For more specific values, the simplified step-index waveguide has a core radius (a) of 1.0 μm, core refractive index (n_1) of 1.45, and cladding refractive index of 1.0. Then total internal reflection due to the difference between the core refractive index and the average cladding index can contribute to optical light guidance in the fiber. Analytical approach to calculate guided light propagation characteristics involves the method of separation of variables and the boundary conditions. For circularly cylindrical waveguides, the radial dependence of electromagnetic fields is governed by the Bessel differential equation, hence the solutions are described in terms of the Bessel and the modified Bessel functions. Figure 1(b) compares the effective refractive index ($\bar{\beta}$) of the fundamental mode obtained from the exact analytical solution and a numerical technique. The numerical approach will be described in the following section.

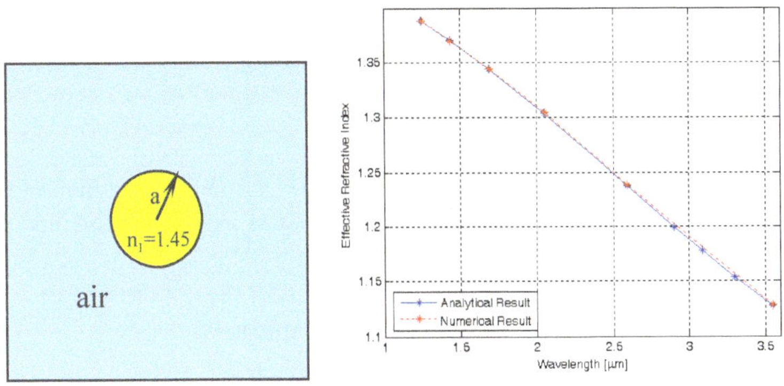

(a) (b)

Figure 1. (a) Simplified step-index optical fiber rod, (b) comparison of normalized propagation constants of the fundamental mode in the fiber of (a)

Based on the analytical and numerical results of the normalized propagation constants, it is reasonably found that the dielectric rod waveguide supports the first four modes at λ = 1.3 μm, which agrees well with the results from the Gloge's mode chart for this waveguide. And transverse electric field distributions, Ex and Ey, can be obtained by using the numerical method. Figure 2 illustrates the normalized field distributions of the first two modes for the optical rod waveguide at the operating wavelength of 1.3 μm. Here, it is noted that the side length of each square cell for the numerical analysis is 0.1 μm.

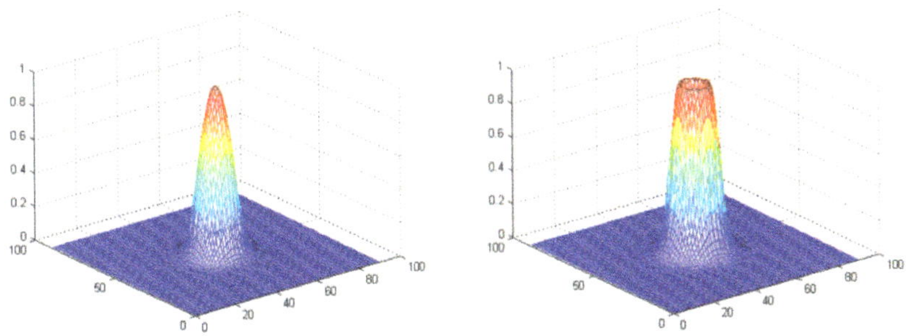

(a) (b)

Figure 2. Normalized field distributions of the first two modes for the optical rod fiber

Once the electric field distribution has been determined, the effective core area for the fundamental mode can be obtained using the following expression [7]:

$$A_{eff} = \frac{\left(\int_{-\infty}^{\infty} \int_{-\infty}^{\infty} |\mathbf{E}(x,y)|^2 \, dxdy \right)^2}{\int_{-\infty}^{\infty} \int_{-\infty}^{\infty} |\mathbf{E}(x,y)|^4 \, dxdy} \tag{1}$$

where $\mathbf{E}(x,y)$ is the electric field on a transverse plane. For the optical waveguide of Figure 1(a), the results of the effective area are 1.7925 μm^2, 2.0624 μm^2, and 2.2148 μm^2 at the wavelengths of 0.8 μm, 1.3 μm, and 1.55 μm, respectively.

3. FDTD and FDM algorithms for the MHOF analysis

Optical light is electromagnetic wave in nature, and thus its propagation properties are governed by the laws of electrodynamics which are collectively known as Maxwell's equations. It is known that guiding of light signals in the MHOF with photonic bandgap structures relies on constructive interference effect due to the periodic arrangement of identical air holes. On the other hand, when the air holes become random in size, location, or both, the phenomenon of total internal reflection takes place and the index difference between the core and cladding provides light confinement and hence guidance of light along the fiber. Although the propagation characteristics of complicated structures like arbitrary MHOFs cannot be calculated easily using analytical methods, there are ways to solve electromagnetic problems numerically.

In this section, two numerical techniques of finite-difference time-domain (FDTD) and finite difference method (FDM) are addressed considering extension to the analysis of holey optical fibers with arbitrary air-hole distributions. Each of these techniques has certain advantages. Using the FDTD method, the continuous electromagnetic field in a finite volume of space is sampled at distinct points in a space lattice and at equally spaced sampling points in time. The sampled data at the points are used for numerical calculations of allowed modes, without generating spurious mode solutions, in a given waveguide. Despite being an effective technique for calculation of propagation constants of guided modes, the FDTD method is not well suited for the evaluation of individual mode field distributions. This is because the source is an impulse function in the time domain covering an infinite spectrum, thus field-distribution solutions are superposition of all possible modes. To alleviate this problem with propagation constants available from FDTD, individual mode field distributions are obtained using the FDM, which can quickly and conveniently provide individual mode field solutions.

The FDTD has gained considerable popularity in recent years, because this method provides robust solutions, based on Maxwell's equations [8], and can readily accommodate complex-valued material properties. An arbitrary material object can be approximated by building up unit cells for which field component positions are disposed with the desired values of permittivity and permeability. Once the geometry of the object is specified in the numerical simulation region, source condition is modelled somewhere in the region. Initially, it is assumed that all fields within the calculation domain are identically zero. Then, an incident wave is enforced to enter the numerical calculation region.

Using the MKS system of units, let us first consider Maxwell's curl equations expressed as:

$$\nabla \times \mathbf{E} = -\mu \frac{d\mathbf{H}}{dt} \tag{2}$$

$$\nabla \times \mathbf{H} = \varepsilon \frac{d\mathbf{E}}{dt} \tag{3}$$

where ε is the electrical permittivity constant in F/m and μ is the magnetic permeability constant in H/m. Expanding the curl expressions and equating the like components, the system of six coupled partial differential equations are formed for the FDTD analysis of electromagnetic wave interactions with general three-dimensional objects. It should be noted that the electric and magnetic field components (Ex, Ey, Ez, Hx, Hy, and Hz) are inter-related. That is, Maxwell's equations do not directly yield electric and magnetic field values, but rather relate the rate of change between electric and magnetic field values.

Adopting central finite difference approximation for space and time derivatives with accuracy to the second order, the following approximations as representative examples in a three dimensional (3D) FDTD formulation can be developed:

$$E_x^{n+1}(i+\tfrac{1}{2},j,k) = E_x^n(i+\tfrac{1}{2},j,k) + \frac{\Delta t}{\varepsilon_0 \varepsilon_r (i+\tfrac{1}{2},j,k)} \left\{ \left[\frac{H_z^{n+1/2}(i+\tfrac{1}{2},j+\tfrac{1}{2},k) - H_z^{n+1/2}(i+\tfrac{1}{2},j-\tfrac{1}{2},k)}{\Delta y} \right] - \left[\frac{H_y^{n+1/2}(i+\tfrac{1}{2},j,k+\tfrac{1}{2}) - H_y^{n+1/2}(i+\tfrac{1}{2},j,k-\tfrac{1}{2})}{\Delta z} \right] \right\} \tag{4}$$

$$H_y^{n+1/2}(i+\tfrac{1}{2},j,k+\tfrac{1}{2}) = H_y^{n-1/2}(i+\tfrac{1}{2},j,k+\tfrac{1}{2}) + \frac{\Delta t}{\mu} \left\{ \left[\frac{E_z^n(i+1,j,k+\tfrac{1}{2}) - E_z^n(i,j,k+\tfrac{1}{2})}{\Delta x} \right] - \left[\frac{E_x^n(i+\tfrac{1}{2},j,k+1) - E_x^n(i+\tfrac{1}{2},j,k)}{\Delta z} \right] \right\} \tag{5}$$

$$H_z^{n+1/2}(i+\tfrac{1}{2},j+\tfrac{1}{2},k) = H_z^{n-1/2}(i+\tfrac{1}{2},j+\tfrac{1}{2},k) + \frac{\Delta t}{\mu} \left\{ \left[\frac{E_x^n(i+\tfrac{1}{2},j+1,k) - E_x^n(i+\tfrac{1}{2},j,k)}{\Delta y} \right] - \left[\frac{E_y^n(i+1,j+\tfrac{1}{2},k) - E_y^n(i,j+\tfrac{1}{2},k)}{\Delta x} \right] \right\} \tag{6}$$

where i, j, k, and n are integers for Δx, Δy, Δz, and Δt, respectively, as the space and time increments [9].

Since optical fibers such as MHOFs generally have no variations along the direction of propagation and variations of material properties are limited to the transverse directions as shown in Figure 3, the 3D FDTD formulation can be simplified to the compact two dimensional (2D) FDTD algorithm [10]. By using phasor notation with the axial propagation constant (β), the first-order partial derivatives with respect to z are replaced with -jβ, because the z-dependence of fields is as exp(-jβz). And two adjacent fields required for the first-order derivatives in the discretized space region can be represented by a field at the mid point between them. Based on these two facts, the following formulation as an example is obtained:

$$H_y^{n+1/2}(i+\tfrac{1}{2},j) = H_y^{n-1/2}(i+\tfrac{1}{2},j) + \frac{\Delta t}{\mu}\left\{\left[\frac{E_z^n(i+1,j)-E_z^n(i,j)}{\Delta x}\right] + j\beta\cdot E_x^n(i+\tfrac{1}{2},j)\right\} \quad (7)$$

The resulting 2D algorithm takes advantage of significant reduction in the required computer memory allocation and running time. Thus, for computer-calculation of arbitrary waveguides that are uniform along the direction of wave propagation, only modelling of the cross sections of waveguides is sufficient.

Along with this efficient algorithm, infinite media in the 2D space for an arbitrary electromagnetic object need to be modelled carefully, because computer memory is limited in the calculation region even with advanced current technology. In order to model regions extending to infinity, a perfectly matched layer (PML) as a highly effective absorbing boundary condition (ABC) is designed at the outer lattice boundary of a calculation domain. Ideally, the absorbing medium is only as thick as a few lattice cells, highly absorbing, reflectionless to all impinging waves, and effective over the full range of operating wavelengths.

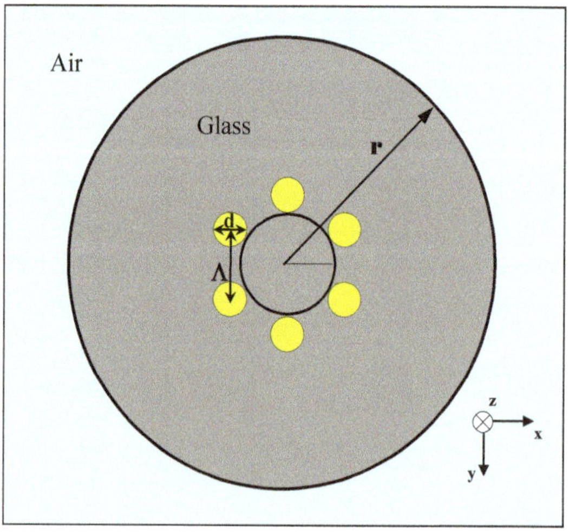

Figure 3. Schematic of a cross section for an MHOF with one layer of air holes in a hexagonal arrangement

Similarly to the development of the FDTD algorithm, the FDM formulation can be derived from the coupled Maxwell's equations [8]. For continuous waves in linear and isotropic media, combining Eqs (2) and (3) results in the following vectorial wave equation:

$$\nabla \times \nabla \times \mathbf{E} - n^2 k_0^2 \mathbf{E} = 0 \tag{8}$$

where n is the refractive index and k_0 is the propagation constant in free space. Many waveguiding devices, like optical fibers, can be viewed as z-invariant, or piecewise z-invariant structures. For those structures, the refractive index n(x,y,z) varies slowly along the propagation direction z, which is valid for most photonic guided-wave devices. By using the vector identity of $\nabla \times \nabla \times = \nabla(\nabla \cdot) - \nabla^2$, Eq (8) can be written as

$$\nabla^2 \mathbf{E} + n^2 k_0^2 \mathbf{E} = \nabla(\nabla \cdot \mathbf{E}) \tag{9}$$

Also with the reasonable assumption of negligible time dependency along the z-axis, the FDM formulation as in Eq (9) can be implemented by replacing spatial derivatives with finite difference approximations. Here, it is noted that the transverse component of (9) is

$$\nabla^2 \mathbf{E}_t + n^2 k_0^2 \mathbf{E}_t = \nabla_t \left(\nabla_t \cdot \mathbf{E}_t + \frac{\partial E_z}{\partial z} \right) \tag{10}$$

where the subscript "t" stands for the transverse components. Since the longitudinal component may be readily obtained by application of the following zero divergence (Gauss's law) constraint:

$$\nabla \cdot \left(n^2 \mathbf{E} \right) = 0, \tag{11}$$

the transverse components are sufficient to describe the full-vectorial natures of the electromagnetic field in an optical waveguide.

For the initial investigation of guidance properties of MHOFs, the optical fiber shown in Figure 3 is computer-analyzed. Generally, the MHOF geometry can be described with two parameters, pitch length (Λ) and diameter (d), as indicated in Figure 3. Here, the pitch length is the distance between centers of two closest air holes with the cylindrical shape. For the MHOF of Figure 3, each small air hole has a diameter of 1.4 μm, constituting a hexagon with Λ = 1.7 μm. The glass portion surrounding the six air holes of the yellow regions has a refractive index of 1.45. The outer radius (r) of the holey fiber is assumed to be 10 μm. Also, the outside region of the MHOF is air.

Once the cross section of a holey fiber is defined in a proper calculation domain, the FDTD simulation can be undertaken with several specified parameters, such as τ in defining a Gaussian source, Δt for stable simulation, the total number (n_{tot}) of time steps for sampling data in the time domain, and reasonable values of β. Here, in order to avoid numerical divergence and ensure stability of the FDTD algorithm, an appropriate Δt needs to be selected to satisfy the following stability condition:

$$\Delta t \leq \frac{1}{c_M \left[\frac{1}{\Delta x^2} + \frac{1}{\Delta y^2} + \frac{1}{\Delta z^2} \right]^{1/2}} \tag{12}$$

where c_M is the maximum wave phase velocity within a given numerical model. Summarizing the mechanism behind the FDTD analysis, the computer simulation proceeds by the following steps:

- Choose appropriate parameter values (τ, Δt, n_{tot}, and β)
- Data sampling of a field component in the time domain
- Take the Fourier transform of the time data
- Get spectral data of a field component
- Pick mode frequencies associated with the β value
- Collect β and mode frequency data
- Make a plot of mode index versus wavelength

Figure 4 illustrates the characteristic curves, which are obtained from the FDTD calculation, for the first three lower-order modes of the MHOF defined in Figure 3.

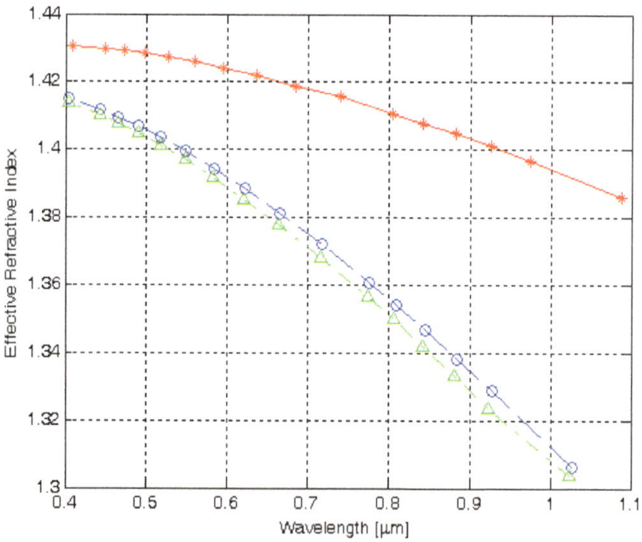

Figure 4. Effective refractive index versus wavelength for the first three modes in an MHOF with one layer of air holes

The red curve with the star symbols plots the normalized propagation constant for the first mode versus wavelength, while the blue and green curves show the normalized propagation constants for the second and the third modes, respectively. The results indicate that the HMOF with a single hexagonal air-hole cladding layer supports multimode guiding.

4. Effects of cladding structure on optical guidance properties

Based on the FDTD and the FDM techniques, important optical guidance properties of the fundamental mode, such as normalized propagation constant, chromatic dispersion, field distributions, and effective area, have been investigated for the HMOF with different structural parameter values. In this section, how the cladding structure of the HMOF affects optical guidance properties is first analyzed. Then in the next section, the investigation is extended to find out the effects of the core in the holey fiber design on optical propagation properties.

For the case when the MHOF has Λ = 2.0 μm, d = 1.2 μm, and r = 12.0 μm with three hexagonal cladding layers of air holes as in Figure 5, the effective refractive index characteristics are computer-calculated by the FDTD analysis approach. Assuming that the core index, which is the same as the surrounding medium of the small air holes, is 1.45 and the geometry of the MHOF is uniform along the z-axis, Figure 6 shows the results for the effective refractive index of the fundamental mode obtained from the FDTD method without accounting for the material dispersion (D_{mat}). In this analysis, a Gaussian pulse source, which excites the MHOF at one end, is used with the parameter value of $\tau = 0.25\text{x}10^{-6}$. The expression describing the source is given as

$$S_G = \exp\left[-\frac{(x^2+y^2)}{2\tau^2}\right] \tag{13}$$

where x and y are perpendicular coordinates in the transverse plane on a cross section of the holey fiber. In Figure 6, the 3D plot of the source is also shown inset.

In order to see the effect of the thickness of the dielectric layer surrounding the air holes, the outer radius (r) of the holey fiber is reduced from 12 μm to 9 μm. As shown in Figure 6, no significant change on the propagation constant occurs. The propagation constants for the 12-μm and 9-μm radius MHOFs are almost the same except for the longest wavelength. The reason is that the field of the fundamental mode is nearly entirely confined to a region of 9-μm radius, which is about 6 times the operating wavelength of λ = 1.55 μm. It should be also noted that both of the MHOFs allow only a single mode over the wavelength range from 0.9 μm to 1.8 μm.

Now, with the same parameter values (Λ = 2.0 μm and d = 1.2 μm) as in Figure 5, the MHOF is numerically analyzed by changing the number of hexagonal air-hole cladding layers. Since it is demonstrated that increasing the outer radius (r) about 6 times larger than the operating wavelength doesn't affect the propagation characteristics significantly, the dielectric glass material surrounding the air holes has been assumed to have the enough thickness. Here, however, it is assumed that the high-index material of the MHOF is pure fused silica (100 mol % SiO_2) with refractive index n_1, while the low-index region is air. That

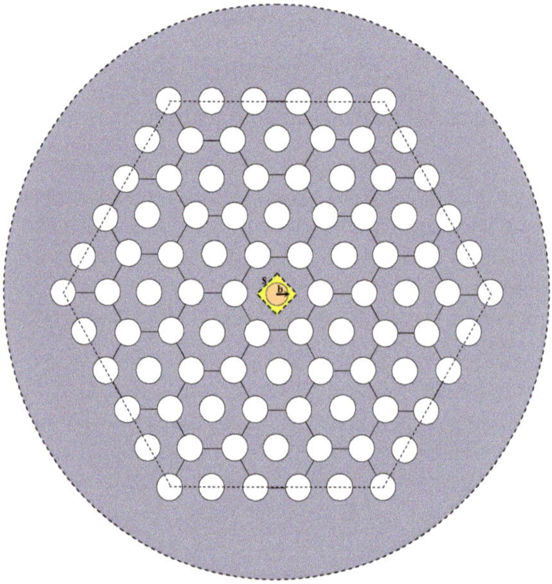

Figure 5. Transverse structure of the MHOF (The air holes in the fifth layer from the core are linked by the dashed line, forming a hexagonal ring.)

is, in order to account for the material dispersion effect of the glass [7], the wavelength dependence of the glass refractive index is directly incorporated in the effective index calculation through the following widely-utilized Sellmeier's equation:

$$n_1(\lambda) = \left[1 + \sum_{j=1}^{3} \frac{A_j \lambda^2}{\lambda^2 - \lambda_j^2}\right]^{1/2} \tag{14}$$

where material constants, A_j and λ_j, for the pure fused silica are given as ($A_1 = 0.6961663$, $\lambda_1 = 0.0684043$), ($A_2 = 0.4079426$, $\lambda_2 = 0.1162414$), and ($A_3 = 0.8974794$, $\lambda_3 = 9.8961610$).

Continuing numerical analysis for the optical guidance properties of the step-index rod fiber case, chromatic dispersion is calculated using the FDM analysis and compared to the exact analytical results. With the effective refractive index results in section 2, the chromatic dispersion (D_{ch}) can be calculated by using the following relationship:

$$D_{\mathrm{ch}} = -\frac{\lambda}{c} \frac{d^2 \overline{\beta}}{d\lambda^2} \tag{15}$$

where c is the speed of light in free space [6]. The comparison of the chromatic dispersion results is shown in Figure 7(a) for the dielectric rod waveguide. Here, the blue curve is from the analytical solution, while the red curve is obtained for the FDM analysis. It is noticed

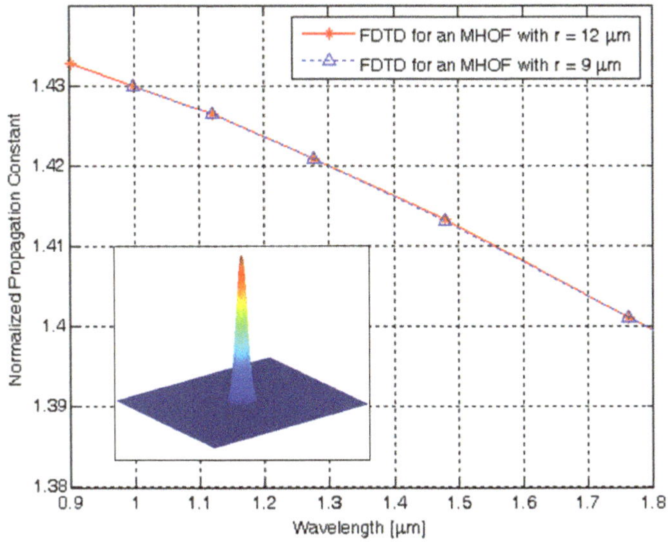

Figure 6. Normalized propagation constants of 3-layer MHOFs with different radii (Inset: Gaussian source for the FDTD analysis)

that the two results agree well. This provides additive confidence that the FDM analysis can accurately predict chromatic dispersion in holey fibers.

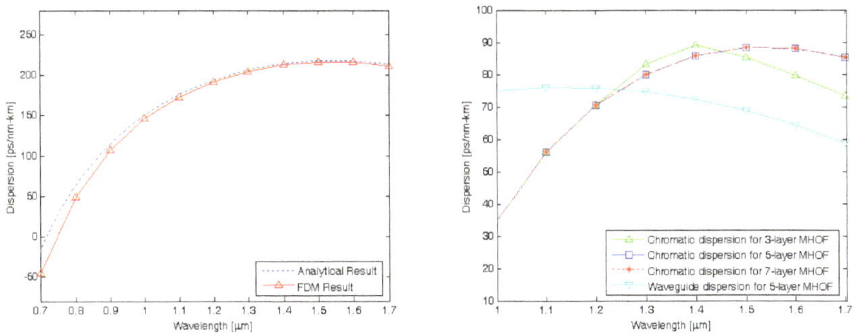

(a) (b)

Figure 7. Dispersion results: (a) for the dielectric rod waveguide of Figure 1(a), (b) for the MHOFs with $\Lambda = 2.0$ μm, d = 1.2 μm, and different numbers of air-hole cladding layers

Again for the MHOF shown in Figure 6, initially, the number of air-hole layers is chosen to be 5. The background glass material (n_1) is assumed to be pure fused silica with enough extension. The blue curve in Figure 7(b) shows the chromatic dispersion as a function of wavelength for the fundamental mode in the MHOF with five hexagonal cladding layers of air holes. The cyan (light-blue) curve illustrates the waveguide dispersion for the same holey fiber without material dispersion included. Compared with the chromatic dispersion result, the waveguide dispersion varies rather gradually from about 78.985 ps/nm·km to 59.537 ps/nm·km over the wavelength range between 1.0 μm and 1.7 μm. It is noticed that the five-layer MHOF provides positive chromatic and waveguide dispersions over the same wavelength range.

When the number of hexagonal air-hole cladding layers is changed from 5 to 3, it is noticed that the chromatic dispersion curves for the both cases are very close at the shorter wavelength range but deviation occurs from 1.2 μm. However, with the increased number of air-hole layers to 7, the D_{ch} dispersion changes very close to the case of the five-layer MHOF. And adding more air-hole cladding layers like nine or eleven layers does not significantly change the chromatic dispersion result from the five-layer holey fiber in the operation wavelength of interest for general optical communications. This behaviour is reasonably attributed to the fact that when the operating wavelength is shorter, electromagnetic fields are more confined to the core region and only the core region has a major impact on the optical guidance properties. By comparison, when the operating wavelength is longer, fields spread more to the cladding region and the index profile of the cladding region has more influence on the effective refractive index.

Incorporating the material dispersion for the MHOF with five layers of air holes, the field solution of the fundamental mode is obtained by using the finite difference technique. For instance, Figure 8 illustrates the normalized field pattern for the Ex electric component of the MHOF at the operating wavelength of 1.55 μm. Figure 8(a) demonstrates the top view with the colorbar scale in the right side and Figure 8(b) does the 3D view. As expected, it is observed that most of the energy is confined within the core region. Also for the fundamental mode of the MHOF with three, seven or nine cladding layers of air holes, about the same field shape is maintained at the same wavelength.

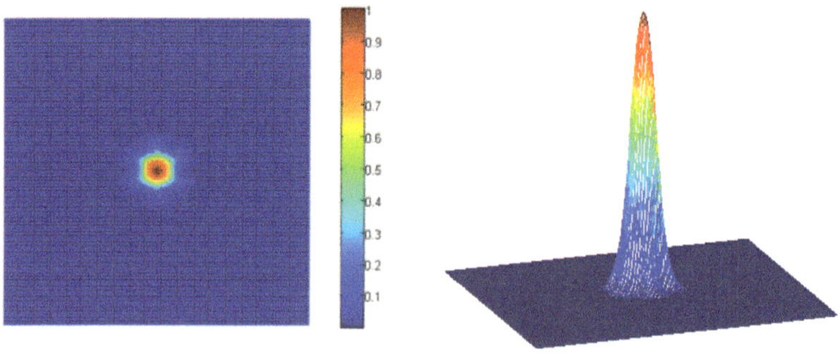

(a) (b)

Figure 8. Normalized field pattern for the Ex electric component of the fundamental mode at λ = 1.55 μm in the MHOF with five layers of air holes and the parameter values of Λ = 2.0 μm and d = 1.2 μm: (a) top view with the colorbar scale; (b) 3D view

Furthermore, similarly to the results of optical guidance properties for the normalised propagation constant, chromatic dispersion, and field distribution, those for the effective core area also tend to change very little when the number of hexagonal air-hole cladding layers becomes larger than five. In other words, adding more air-hole layers to the cladding beyond a certain point will not significantly affect the propagation constant, dispersion, and, in fact, all guidance properties of the MHOF waveguide.

5. Effects of the central air core on optical guidance properties

For the MHOF defined with the geometrical parameters of Λ = 2.0 μm and d = 1.2 μm, and five cladding layers of air holes, how the core structure affects optical guidance properties is also investigated. By sizing the central air-hole radius (b) as shown in Figure 5, the MHOF with an air core is analyzed to obtain its propagation characteristics, based on FDM and FDTD methods for the full-vectorial analysis.

Getting the normalised propagation constant of the given optical fiber and computing its second derivative with respect to λ, dispersion characteristics can be evaluated. Figure 9 shows chromatic dispersion variations for the MHOF, as the radius of the central air hole is sized from 0 μm to 0.4 μm. This result indicates that the dispersion flatness can be achieved around 1.3-μm wavelength by designing the MHOF with b = 0.25 μm. Specifically, the chromatic dispersion (ps/nm·km) is -0.7724, 0.2635, -0.4532 at wavelengths of 1.2, 1.3, and 1.4 μm, respectively. This near-zero ultra-flattened dispersion is highly desirable in long-haul fiber-optic communication systems.

When the circular hole at the center is changed to the square air hole with the side length (s) of 0.4432 μm, which has the area equal to that of the former, the chromatic dispersion versus wavelength for the fundamental mode in the MHOF is also depicted as the green dashed line with the diamond symbols in Figure 9. As noticed, the chromatic dispersion curves for the both cases are close at the shorter wavelength range around 1.0 μm, compared to the gap between two curves at the wavelength range around 1.7 μm.

The effective area for the fundamental mode is also evaluated by using the field distribution results, as shown in Figure 10. The red dashed line represents the effective area for the MHOF at the operating wavelength of 1.3 μm and the blue solid line at 1.55 μm, as the radius of the core hole (b) is sized from 0 to 0.6 μm. It is noted that the effective core area tends to increase with the size of the central air hole at both the operation wavelengths. The reason is that with bigger core hole the electromagnetic fields spread further into the cladding region. Thus sizing the core hole provides a decent way to control the effective area. Also, notice that the effective areas at the longer wavelength of 1.55 μm generally result in larger values for each different hole radius, compared to those at 1.3 μm. While the effective area is 6.7926 μm^2 for the MHOF with s = 0.4432 μm at λ = 1.55 μm, as indicated by the dotted line, those with b = 0.25 μm at λ = 1.3 μm and 1.55 μm are 6.691 μm^2 and 7.145 μm^2, respectively.

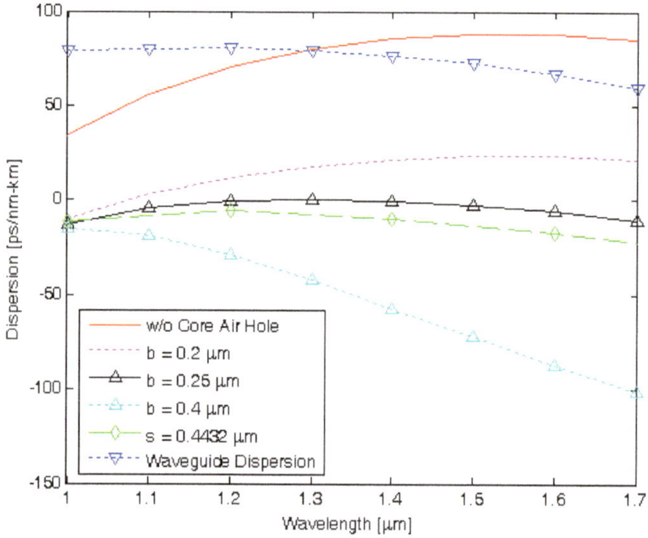

Figure 9. Chromatic and waveguide dispersions for the MHOF with the five layers of air holes, the parameter values of $\Lambda = 2.0$ μm and $d = 1.2$ μm, and different central core air holes

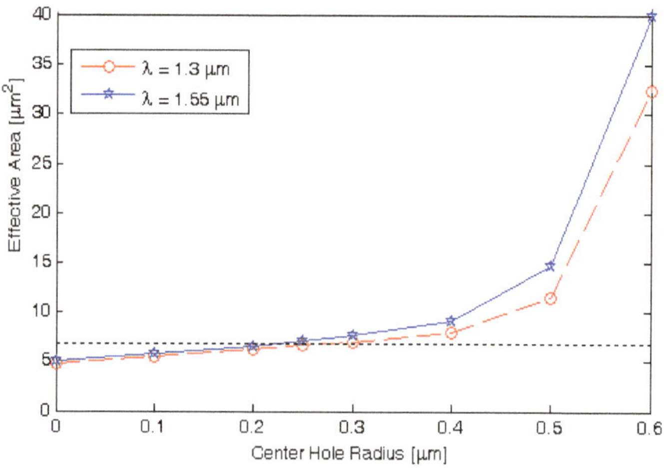

Figure 10. Effective area for the fundamental mode versus the radius of the central air hole for the same MHOF as given in Figure 9

6. Conclusions

For the investigation of optical guidance properties of MHOFs, two numerical techniques of the FDTD and FDM are employed, considering extension to the analysis of holey optical fibers with random hole distributions in terms of material, shape, size, or location [11]. In this investigation, it is found that the outer radius (r) of the MHOF with Λ = 2.0 μm, d = 1.2 μm, and the three hexagonal air-hole cladding layers doesn't affect the propagation characteristics significantly, if its thickness is about 6 times larger than the operating wavelength. And the 3-layer MHOF supports only a single mode over the wavelength range from 0.9 μm to 1.8 μm, whereas the MHOF with Λ = 1.7 μm, d = 1.4 μm, and the single hexagonal air-hole cladding layer multimode guiding.

In the case of the MHOF with Λ = 2.0 μm, d = 1.2 μm, and r = 12.0 μm, the guidance characteristics, such as normalized propagation constant, chromatic dispersion, mode field distribution, and effective area tend to change very little when the number of hexagonal air-hole cladding layers becomes larger than five. In other words, adding more hexagonal air-hole layers to the cladding beyond a certain point will not significantly affect the propagation constant, dispersion, and, in fact, all guidance properties of the MHOF waveguide.

By comparison, the size of the central core air hole has fair effect on the optical guidance properties of MHOFs. The MHOF with Λ = 2.0 μm, d = 1.2 μm, b = 0.25 μm, and the five hexagonal cladding layers provides -4.2647 ps/nm·km of chromatic dispersion and 7.145 μm^2 of effective area at λ = 1.55 μm. Similarly, the MHOF with the square air core of the equal area shows close results with -15.4055 ps/nm·km of the former and 6.7926 μm^2 of the latter. As a general guideline, when the operating wavelength is shorter, electromagnetic fields are more confined to the core region and only the core region has a major impact on the optical guidance properties. Conversely, when the operating wavelength is longer, fields spread more to the cladding region and the index profile of the cladding region has more influence on these.

Even thought the holey optical fibers have complicated geometries, which make the electromagnetic analysis difficult, the results in this investigation can be utilized as a general guide. And depending on the desired applications, the MHOF with various design parameters are expected to be useful for novel developments related to optical communication in many areas.

Author details

Jeong Kim
Chungnam National University, Department of Electric, Electronic and Communication Engineering Education, Daejeon, South Korea

Acknowledgement

This study was financially supported by research fund of Chungnam National University in 2010 and 2011.

7. References

[1] Knight, J.C.; Broeng, J.; Birks, T.A. & Russell, P.St.J. (1998). Photonic Band Gap Guidance in Optical Fibers. *Science,* Vol.282, (November 1998), pp. 1476-1478, ISSN 1095-9203

[2] Petrovic, J. (2008). Modelling of Long Period Gratings in Photonic Crystal Fibres and Sensors Based on Them, In: *Recent Advances in Modelling and Simmulation,* InTech, G. Petrone & G. Cammarata, (Ed.), pp. 417-432, ISBN 978-3-902613-25-7, Vienna, Austria

[3] Monro, T.M.; Richardson, D.J.; Broderick, N.G.R. & Bennet, P.J. (2000). Modeling Large Air Fraction Holey Optical Fibers. *Journal of Lightwave Technology,* Vol.18, No.1, (January 2000), pp. 50-56

[4] Ranka, J.K.; Windeler, R.S. & Stentz, A.J. (2000). Optical Properties of High-Delta Air Silica Microstructure Optical Fibers. *Optics Letters,* Vol.25, No.11, (June 2000), pp. 796-798

[5] Benabid, F.; Couny, F.; Knight, J.C.; Birks, T.A. & Russell, P.St.J. (2005). Compact, Stable and Efficient All-Fibre Gas Cells Using Hollow-Core Photonic Crystal Fibres. *Nature,* Vol.434, (March 2005), pp. 488-491, ISSN 0028-0836

[6] Keiser, G. (2010). *Optical Fiber Communications,* McGraw-Hill, ISBN 978-007-3380-71-1, New York, U.S.A.

[7] Agrawal, G.P. (2010). *Fiber-Optic Communication Systems,* Wiley, ISBN 978-047-0505-11-3, New York, U.S.A.

[8] Hayt, W.H.Jr. & Buck, J.A. (2011). *Engineering Electromagnetics,* McGraw-Hill, ISBN 978-007-3380-66-7, New York, U.S.A.

[9] Kim, J.I. (2007). Holey Fiber Analysis Employing FDM and FDTD Methods on the Transverse Plane. *Microwave and Optical Technology Letters,* Vol.49, No.7, (July 2007), pp. 1592-1594

[10] Taflove, A. & Hagness, S.C. (2005). *Computational Electrodynamics: the Finite-Difference Time-Domain Method,* Artech House, ISBN 978-158-0538-32-9, Boston, U.S.A.

[11] Pickrell, G.; Kominsky, D.; Stolen, R.; Ellis, F.; Kim, J.; Safaai-Jazi, A. & Wang, A. (2004). Microstructural Analysis of Random Hole Optical Fibers. *Photonics Technology Letters,* Vol.16, No.2, (February 2004), pp. 491-493

Chapter 4

3D ICS with Optical Interconnections

Edward G. Kostsov, Sergey V. Piskunov and Mike B. Ostapkevich

Additional information is available at the end of the chapter

1. Introduction

Today's microelectronics, whose main driving force of development has always been the needs in computing devices, has achieved exceptionally great results over the last decade.

The exponential growth in performance of microelectronic devices, predicted in 1965 by Gordon Moore and formulated as an empirical law, was sufficiently accurate for more than 45 years: the computing power of single-chip microprocessor-based systems increased almost by a factor of four every three years. Simultaneously with the improvement of parameters of ICs the performance of supercomputers has increased. The performance has increased both due to a reduced duration of a cycle and due to pipelining and parallelization of computations. Currently, petaflops computers are built on the basis of 10^5 - 10^6 single-chip processors with a clock frequency of 10 GHz.

At the same time, the analysis of the element base of microelectronics has shown that the main obstacle to improve the performance of computer devices is the problem of connections, both on the surface of individual substrates, and between them. The area occupied by conductors is about 70% of the overall area of a crystal itself in modern VLSI. The total length of conductors exceeds the linear dimensions of a crystal by more than a thousand times. The energy spent on charging the conductors is 60-70% of all energy losses. Placement of the conductors on the surface of a crystal requires significant technical efforts, while the limit value of capacitance of a conductor, which is 10^{-11} F/m., is attainable [1]. On schematic level, there is also a problem of RC propagation delay in the connections. And consequently, the velocity of signal propagation in ICs is much lower than that of light (5-30 times) depending on the degree of integration.

Thus, the main source of increasing the performance of computers at the previous period of their evolution, which is concluded in increasing the performance of a single gate and the degree of integration, is nearly exhausted. Thus, it can be stated that the clock frequency of computers is determined by circuit limitations, rather than by physical and technological ones.

No less complicated problems arise while exchanging information between processor chips. With an increase in their size the number of inter-chip connections increases as well. And this reduces the clock speed by almost an order of magnitude. With decreasing the cycle duration, the above-mentioned connections play an increasingly significant role.

From the stated-above it can be concluded that the solution to the problem of connections and thus further improvement of the performance of computer devices can be obtained by the following ways:

- increasing the degree of parallelization of information processing at all levels down to elementary operations (a maximum depth of parallelization) and increasing the decentralization of functions of storage, control, and data processing, and transition from long logical links within the surface of a chip to the local relations between the neighboring gates, i.e. by transition to a VLSI with a homogeneous structure;
- transition from connecting chips placed in a plane to 3D packs of VLSIs from these chips and to their vertical interconnections.

It follows from general physical considerations that a simple way of introducing the "third dimension" into the structure of communications is using optical links. Such channels have the following advantages:

- A high degree of parallelism of the information transfer from plane to plane makes possible to use highly parallel algorithms for processing (down to elementary operations), and thus to create high-performance computing.
- Capability of optical synchronization allowing delivering a synchronization signal to any point of a chip practically without delay, e.g. from one light source outside an IC.
- Optical communications channels are free from parasitic effects of the mutual influence because of the neutrality of photons. With increasing the clock frequencies (especially in the gigahertz frequency range), the advantages of optical signals grow as capacitive coupling between electronic conductors with increasing frequency grows as well.

There is always an interest in the use of optics for high-performance computer devices. Recently, a special attention of developers of such devices has been drawn to the development of a new element base, creation of optoelectronic integrated circuits (OEIC), for the organization of inter-IC and inter-processor connections in order to attain high throughput and low power consumption [1-4].

The optical communication bandwidth over short distances within a few millimeters already competes with electric conductors with sufficiently low energy consumption, up to 1 pJ/bit/m or less. The placement of optical communication channels directly on the chip surface can significantly reduce this bandwidth. The reports are already known about the first experimental data on providing the density of the bandwidth 37 Gbit/s/mm2 and higher by using a switched CMOS pair of a vertical-cavity surface-emitting laser (VCSEL) and a CMOS-compatible avalanche photodetector (CMOS-APD) system. The frequency of switching light sources VCSEL, electro-optic modulators MQW based on superlattices GaAs / GaAlAs, InGaAsP, etc. as well as that of their corresponding photodetectors has already reached values of 20 ÷ 50 GHz, which in the short term, can grow up to 70 ÷ 80 GHz. The level of technology of constructing field-programmable smart-pixel arrays and FP-SPA systems based on the use of optical channels in free space, is now able to provide the exchange of information between the two surfaces at a rate exceeding 10 Tbit/cm2.

Currently, however, optical communication channels are used only for the information transfer problems. Their application directly to the microelectronic structures and in the channels of information processing virtually does not evolve. The main reason is that the size of components of gates (modulators and light sources) due to the wave nature of the optical signal is much larger than that of the active elements of modern microelectronics. Entirely optical computers performing massively parallel computations typically contain rather large elements: lenses, shadowgrams, spatial light modulators, etc. and cannot be created using microelectronics technology.

At the same time, it may also be noted that with the development of nanophotonics, the advent of the light sources and receivers with nanometer dimensions (see, e.g. [5-7]), the optical channels could be used not only for the exchange of information between ICs, but also for the information processing, with the 3D logic devices. Thus, the construction of optoelectronic schemes, whose implementation fits well into the existing technology of building semiconductor circuits, is actual.

The *objectives* of this paper are: a) to consider the possibility of creating high performance and manufacturable multi-layer (3D) chips in which a processing of information and its mass exchange between the layers is performed using optical communication channels and whose parameters are compatible with those of microelectronic circuits, b) to represent the algorithmic and computer tools providing the design and research of simulation models of such 3D algorithms and structures with massive parallelism that are oriented to an optoelectronic implementation.

The principles of the gate, where logic signals are represented by the presence or absence of a light flux driven by an electric field are described in the paper. The possibility of the physical implementation of 3D logic circuits based on such elements is analyzed and its advantages are discussed. The methods and means of the algorithmic design of 3D ICs with a homogeneous structure for execution of algorithms with fine-grain parallelism are described. These tools include a formal model of fine-grained computing, Parallel Substitution Algorithm (PSA) and a simulation system (WinALT), which is based on PSA and provides the construction of models of devices with a 3D architecture. Features of the system are demonstrated by constructing models of 3D optoelectronic matrices for parallel data processing. Matrices can be characterized by a high performance, simplicity of cells, homogeneity and simplicity of the topology. Simulation models help in acquiring an objective view of the complexity and performance of the matrices and confirm the reasonability of the transition to 3D VLSI in order to overcome the problem of connections that arises in modern 2D VLSI. The WinALT system is available in Internet. The expediency of transforming WinALT to an online environment for supporting the simulation of 3D computational structures with the users' participation in the network (virtual) community is discussed.

2. The physical basis of implementation of 3D optical logic circuits

2.1. Optoelectronic gate

Let us consider features of constructing a universal gate, which is based on the modulation of a light flux by an electric field [8-11], and when only a single transition of energy (between light and electric signal), the source of a light flux is shared by many gates. The scheme of such a gate is presented in Figure 1. It has the following main components: a light modulator LM controlled by the electric field, EC converter of a light signal to an electric signal (e.g. a photoelectric transducer), energy storage or a load element LE (capacitance for dynamic circuits). The element controls the intensity of the luminous flux transmitted by LM, and with the help of the light flux entering the optical input of EC.

The most optimal physical and technical solution, as follows from [8-11], is obtained when the light modulator is electro-optical, the energy converter is photovoltaic and the energy storage is electrostatic. Such a decision is mainly due to the consideration of energy. In this sense, the use of magneto-optical and acousto-optic light modulators is limited by a high energy consumption and by the complexity of the direct energy conversion (photomagnetic or photoacoustic).

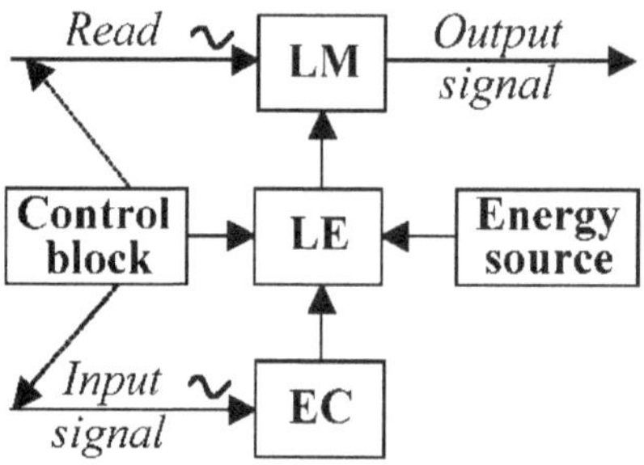

Figure 1. Flowchart of universal gate

The state of a gate is determined by the ability of the light modulator to pass a light flux coming on its input (state "0" or "1"). This state is unambiguously related to the amount of energy stored, which, in turn, is determined by the intensity of the light flux entering the energy converter (entrance gate). According to Figure 1 a gate in the dynamic mode operates as follows:

- supply of energy from the energy source (voltage pulse) to LE at the time t_0 (operation "erasing information");
- supply of a light flux on the input of the transmitter PE at the time $t_0 + \Delta t$ (operation "information recording"), where Δt is the duration of a cycle;
- supply of a light flux on the input of LM at the time $t_0 + 2\Delta t$ (operation "read information").

Assume that the modulator lets the light flux pass when the amount of energy stored in LE, and does not let the light pass otherwise. Let us also assume that when EC receives a light flux composed of the streams $X_1, X_2,..., X_n$, such energy in the load element is released. Then the light flux on the output of the modulator Y (at the moment of reading) will be the Pierce logical function:

$$Y = \overline{X_1 \vee X_2 \vee \ldots \vee X_n}\,.$$

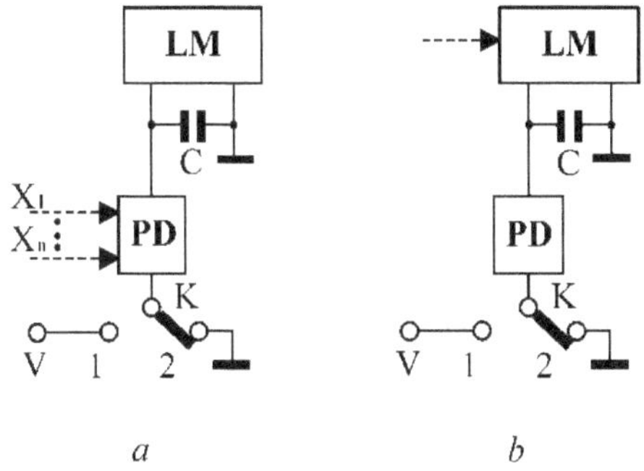

Figure 2. Universal gate with optical connections

This function forms a complete basis of Boolean functions, so the element proposed can be considered universal.

The electric functional scheme of the gate is shown in Figure 2, where K is the key, V is the power supply voltage. The gate implements the Pierce function and it operates by cycles according to the following general description.

cycle 1 - "erasing information", key K is in the position 1, and capacitance C is being charged by the source when the photodetector FP is illuminated.

cycle 2 - "storing" K is in position 2 (see Figure 2, a), the light flux with intensity J1 or J2 (depending on the level of illumination corresponding to signal "1" or "0", i.e. whether there is at least one of Xi not equal to "0") comes to the optical input of the gate (photodetector FP). According to the level of illumination the capacitance C is discharged with the time constants R_TC, or R_CC, and the $R_TC \gg R_CC$, where R_T is dark resistance of FP, and R_C is its resistance under illumination.

cycle 3 - "reading", key K is in position 2 (see Figure 2, b) and the light signal passes to the optical input of the modulator. The light flux corresponding to "0" light is obtained from the output if signal "1" arrived at the input in the second cycle. And vice versa, signal "1" with a high intensity of light is on the output of LM if there was signal "0" in the previous cycle.

The storage time of information received by the element is proportional to the value of R_TC. However, we can create a dynamic memory by inclusion of two gates so that the output of one of them be connected to the input of another and vice versa, as shown in Figure 3. Its storage time is limited only by the time of maintaining the voltage on the power supply.

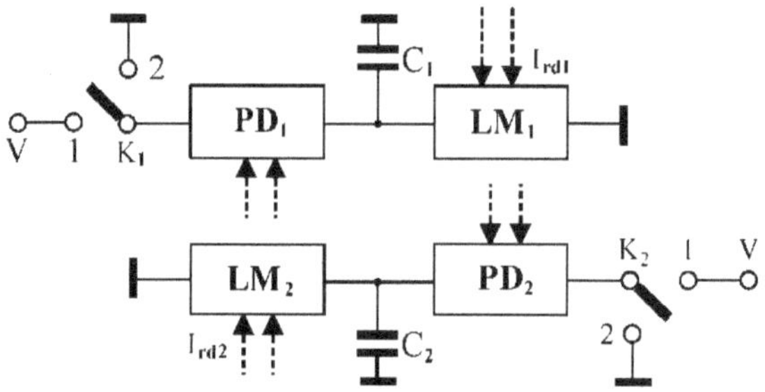

Figure 3. A dynamic memory cell

When constructing digital devices based on these optical gates the interlayer connections are implemented by placing pairs of "modulator - photodetector" in different layers-planes so that the optical output light modulator of one gate is geometrically aligned with the photodetector optical input of another gate.

2.2. A sample of constructing a dynamic memory cell

Such a cell is intended for storing information in the electric form on capacitors and transferring information between cells by pulsed optical signals, while a direct transfer of charge does not take place. The scheme of the cell is depicted in Figure 3.

Assume that a bit of information (e.g. "0") is written in the upper gate 1. The switch K_1 was shot to pin 2 and PD_1 was closed in order to do so.

Cycle 1. "Erasing" of information in lower gate 2 by charging the capacitance C_2 under lit PD_2 and the switch K_2 shot to pin 1 of the power source with voltage *V*.

Cycle 2. The reading of information from element 1 by the light signal I_{rd1}. The switches K_1 and K_2 are in position 2. The output signal from LM_1 is the input writing signal for element 2.

Cycle 3. The "erasing" of information in element 1. The switch K_1 is in position 1 and the capacitance C_1 is connected to the power source with voltage *V* and is charging.

Cycle 4. The reading of information from element 2 by the light signal I_{rd2}. The switches K_1 and K_2 are in position 2. The output signal from LM_2 is the input writing signal for element 1.

The information in a cell is represented by a pair of light signals ("0", "1") or by voltages taken from pins of capacitances. It can be kept as long as needed as a result of iterative repetition of cycles 1-4. If "1" was written to a cell in the initial state, it would have kept a couple of "1", "0".

Similarly, other schemes of optical elements can be built to implement the basic logic functions: repeater, inverter, AND, NOT-AND, OR, NOT-OR, sum by modulo 2, implication. The analysis of transfer characteristics for such elements is presented in [10].

The described principle of implementation of inter-layer communications allows creating functionally flexible devices by replacing electric logical communications by the optical ones and by using 3D structures of gates. A computational device would contain a minimum number of elements and electric connections if the following rules of its creation are followed: parallel electric circuits are used to supply power to elements and parallel logical links are done using optical channels.

An illustration of the functionality of a logic gate with optical connections is presented in [10, 11] using samples constructing 1D and 2D shift registers, switch and matrix processor for parallel image processing. The processor consists of a control block and a program controlled cellular automaton. The control block stores a program and fetches its instructions. The cellular automaton performs information storage and processing [11]. A cell in cellular automaton transforms information by changing its own state and those states of its neighbors. Any transformation is represented as a sequence of elementary transformations. Each elementary transformation is defined by the contents of a command that comes to a cell from the control unit. The transformation of information in a cellular automaton is performed simultaneously by all cells.

The following conclusions can be drawn from the analysis of the functioning of the described devices (primarily the matrix processor), the specific features of their design and comparison with microelectronic devices, capable of performing similar functions.

1. A cell of matrix processor contains significantly less elementary components (estimated by two orders of magnitude) as compared to conventional chip with similar functional capabilities.
2. The total number of cycles spent on execution of logical operations required for image processing is also smaller, at least by a factor of N (where N is the number of lines in the image). The time of the complete image processing does not depend on its size.
   In particular, a very small number of clock cycles is required to execute such operations as selection of a contour image, noise filtering, noise filtering with masking, extension of lines, etc., which are quite complex for electronic circuits. And the number of cycles remains unchanged when image dimensions increase.
3. The area occupied by the connections is 7 - 8% of the total area of a substrate, as shown by the analysis made. The described devices have a higher volume density of placement of elements, the reliability of connections, noise immunity, and ultimate manufacturability. In addition, the design of an optical VLSI is considerably simpler, since there is no need to take a complex configuration of interconnections into consideration.

A breadboard construction of a cell of optic dynamic memory is created in order to demonstrate the possibility of practical implementation of optoelectronic gates [11]. It consists of two optoelectronic logic gates. Lithium niobate crystals are used as optoelectronic modulators.

The study of the dynamic memory model has made possible to draw the following conclusions:

- transfer of information from one element to another can be done without loss of information an unlimited number of times;
- the use of threshold photodetectors allows reducing of the operating voltage and of energy required for switching a gate down to acceptable levels for microelectronic implementation.

3. The Cellular technology constructing 3D logical structures

Efforts to create optoelectronic circuits have given impetus to the development of algorithmic and software tools that aid in solution of the problem of design and study of 3D (multilayer) digital structures, focused on the use of optical interlayer connections.

These tools map a primordial parallel algorithm of a problem solution onto an architecture with a massive spatial-temporal parallelism. The orientation to constructing structures of models that consist of huge sets of rather simple and homogenous computing devices (cells), mainly with local links, and placed in a 3D space is their basic property.

The considered tools include:

- a formal model of fine-grain (cellular) computations, which is called the Algorithm of Parallel Substitutions (PSA) and which serves as the basis of a method for synthesis of parallel architectures;
- a simulation system of parallel computational processes, which is used for the construction, debugging of models of 3D logical structures as well as for the extraction of the characteristics from these models.

A detailed description of the cellular technology is given in [11].

3.1. PSA: A generalized model of fine-grain computations

Conceptually PSA unites in itself a substitutional character of Markov's algorithm [12] and spatial parallelism of a cellular automaton [13] basing on an associative mechanism of application of operations, which is common to both of them. PSA represents a "true parallelism" of computations, when all the allowable operations are executed at each step for all the available data. The main idea of PSA is concluded in the following three statements:

- The processed information is presented in the form of a cellular array, which is a set of cells. Each cell is data (a bit, a character, a number, etc.) with its name (its "location" within array, which is an element of a set of names *M*) in the array. A set of data belongs to a certain finite alphabet *A*.
- The algorithm is defined by a set of parallel substitutions. They have left-hand and right-hand sides (left and right parts). The expression in the left part generates cellular arrays, one for each cell name in the processed cellular array. If processed arrays contain one or more such arrays, then the substitution is applicable. Its execution means that a certain "base" part of the found array generated by its right part is replaced by array generated by its left part for the same cell name. Its execution means that a certain

"base" part of the found array is replaced by the array generated by the right part of the parallel substitution for the same cell name.

- The process of computations is iterative. At each iteration, all the substitutions applicable to the processed cellular array are executed. The computation is over when no substitutions applicable to the array obtained at the previous iteration were found. This array is the result of work of PSA.

Executing the substitution in the form of replacement of one cellular array by another permits representation of such replacement as replacement of one spatial image by another. This is rather essential for a visual construction of computer-aided model of optoelectronic logical structure and visual representation of computational process of such a model, which is distributed in time and space.

A formal description of PSA is presented in [14]. Let us demonstrate the idea of PSA by a simple example. Let us consider PSS Φ_σ for adding many non-negative binary numbers as a sample.

Let $A = \{0, 1\}$, $M = N^2$, $N = \{0, 1, 2, ...\}$. A transformed cellular array is a 2D rectangular table whose cells are numbered according to left coordinate system (i is abscissa, j is ordinata). The name of cell m is a pair of values i, j. The digits of numbers are kept in the strings of the table. The least significant bits occupy the right-most column. PSA Φ_σ consists of two parallel substitutions:

$$\Phi_\sigma = \begin{cases} \Theta_1^\sigma : \{(1,\langle i,j\rangle)(1,\langle i,j+1\rangle)(0,\langle i+1,j\rangle)\} * \{(0,\langle i,j-1\rangle)(0,\langle i+1,j-1\rangle)\} \to \\ \quad \{(0,\langle i,j\rangle)(0,\langle i,j+1\rangle)(1,\langle i+1,j\rangle)\}; \\ \Theta_2^\sigma : \{(1,\langle i,j\rangle)(0,\langle i,j+1\rangle)\} * \{(0,\langle i,j-1\rangle)\} \to \{(0,\langle i,j\rangle)(1,\langle i,j+1\rangle)\}. \end{cases}$$

The left part of the substitution is to the left of the arrow, while its right part is to the right of the arrow. Shift functions are written in angle brackets in the definitions of substitutions Φ_σ. These functions set the location of cells of the left and right parts of a substitution as related to each other. A cell description is confined in parentheses. Cell states are shown to the left of commas. The "base" parts of the left parts of substitutions are at the left of asterisks. When particular values of the pair i, j are substituted into the shift functions, the cellular arrays associated with this name are obtained.

The description of substitutions allows their geometric representation depicted in Figure 4, a. In this case, the left and right parts of commands are defined by templates, and the search of occurrences of the left parts of substitutions is done by shifting their templates above cells of the table along axes. One step of transformation of the source table containing numbers 9, 15, 5 is presented in Figure 4, b. After carrying out four steps the result is placed in the top string, while the rest of them contain zeros.

The expressive capabilities of PSA augment if an alphabet A is extended by introduction of *variable* and *functional* symbols. An alphabet A serves as a domain for variables symbols and as a range for functional symbols. In the case of graphic representation of substitutions,

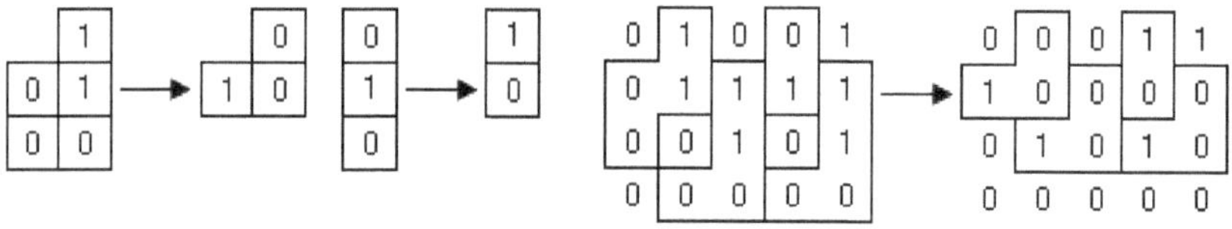

Figure 4. Cellular adder: a – graphic representation of PSA commands, b – one step of execution of algorithm

variable symbols can be written into cells of templates of the left part of substitution, while the functional symbols can be in the cells of the right part. Such a substitution is called *functional* substitution. When an alteration of a state is performed for a cell located in the processed array under a template cell of the right part of a certain command containing a functional symbol, it is not specific data from the template cell, but a result of evaluation of a certain function that is written into this cell. The states of cells in the processed array that are below the cells of a template of the left part can be arguments of such functions. Using the functional substitutions helps: a) in reducing PSA notation to a considerably more concise form in theoretical issues, b) in practical issues to represent rather complicated devices such as ALU by a single cell.

3.2. A simulation system of 3D digital structures

3.2.1. Overview of the system

A physical and technical rationale for the construction of multilayer optoelectronic structures is given In Section 2. Evidently the construction of real 3D structures requires a large amount of work associated in the first place with selection of an optimal kind of structure from the standpoint of technical parameters among a big number of possible variants. A manual solution of such a task is virtually impossible. Thus, a computer-aided tool must be available for the research and development of models of 3D structures. Such an instrument was built and it is called WinALT [15]. The user's interface of the system coincides with the standard user's interface in Windows applications. A simulation model is represented by a project that contains a number of sub-windows. Each sub-window can hold graphic or text objects of a model. Creation and editing of graphic objects are carried out by means of toolbars, menus and dialog windows. The system is freely distributed. System's site [16] contains a section "installation", which includes manuals on installation, uninstallation and system's distributive package. An open architecture of the system enables the user to participate in extension of the system functionality.

3.2.2. Adequacy of the system language to the problem domain

The WinALT was developed simultaneously with computer-aided models of parallel algorithms and structures under their strong influence. This influence has manifested itself in the WinALT system by a wide employment of visualization tools both for supporting the construction of parallel algorithm descriptions (graphic representation of objects corresponding to cellular arrays, the left and right parts of substitutions) and for their simulation (capability to view the dynamics of application of each substitution). What is more, a special attention was given to the development of tools that provide the visualization of 3D (multilayer) objects and that allow viewing the transformation of cell states in any layer of a 3D cellular array. This is due to the orientation of the development of computer-aided models to fine-grain architectures that are promising for the implementation in the form of multilayer VLSI. The simulation language contains three parts.

The first part of the language is designed to describe parallel computations in the form of parallel substitutions. It is fully based on PSA.

The objective of the second part of the language is the description of sequential computations. This part is essentially based on Pascal. It provides statements for the description of simulation program structure, control operators, assignment operator and subroutine call by name or by reference. These statements can be used in a model program for the description of sequential control when needed. These means can be used in a model program for the description of sequential control, when it is needed, for the definition of functions that describe cell states and, also, for the construction of such service functions within a model program as menu definition, graph drawing or initial data input.

The third part of the language provides importing libraries into a model program. These are **dll** libraries written in **C/C++** and embedded into the simulation system. This helps in enriching the functionality of simulation tools to suit the user.

Let us describe the first and the third parts in greater detail.

The Description of the Parallel Part of Simulation Language. This part has a clear division into graphic and analytical subparts. Graphic objects are cellular arrays and templates. An image of an object is composed of color cells located along horizontal and vertical axes and, also, along the axis that goes from the user to the screen. The origin in the template is called its *center*.

A color is used to visualize a cell's state. Its state can belong to any cell's data type supported by the library of data formats (see below). A name can be assigned to a cell in addition to a typed value as an additional property. It has to be unique within the scope of one cellular object. This makes possible to implement functional substitutions in WinALT. There is a special neutral **void** state of a cell (depicted by a diagonal cross on a color background).

A parallel substitution in WinALT simulation language is set by a bunch of operators **in-at-do**. The names of cellular arrays and templates can be used as parameters in these operators. In the case of one-block structure of a device (a single cellular array), a parallel substitution is described as follows. The parameter of **in** operator is the name of a processed cellular array. A name of template of the left part of substitution is a parameter in **at** operator, while that of its right part is a parameter in **do** operator. The execution of substitution is done in two phases. During the first one, the center of a template of the left part is moved along the axes in a processed cellular array, and all its occurrences in this array are marked accurate to the empty cells. At the second phase, the states of cells of the processed cellular array in all these occurrences are replaced by the states of cells from the template of the right part also accurate to the empty cells. Taking multi-block structure of the device into consideration means the following. The parameter in operator **in** is a list of names of cellular arrays placed in brackets and separated by commas. The lists in operators **at** and **do** are arranged similarly, but instead of names of cellular arrays the names of templates are used in them. Combining patterns in the list means that the movement of patterns in the images of their corresponding cellular arrays is coordinated by substitution of the same set of coordinates into the centers of all the templates of the operators **at** and **do**. The coordinates of cells of the cellular array that is in the head of the list **in** operator form these sets. Such substitutions are called vector substitutions and allow describing parallel transformations of information in compositions of cellular arrays.

A bunch of operators **in-at-do** and a description of a function serve as an analog of functional substitution. Operator **at** contains the names of templates, in which some cells are named. A function uses these names as input and output variables. The name of a function is used as a parameter in operator **do**. A functional substitution can also be a vector one.

A synchroblock **exhaust-end** (or shortly **ex-end**) is the main structure for definition of an algorithm of a device operation. This block implements an iterative procedure of PSA application for the composite operators describing parallel substitutions that it contains. In addition, there are two more kinds of synchroblocks that were introduced: **clock – end** (**cl – end**) and **change – end** (**ch – end**). The first one executes its substitutions a number of times specified as its parameter. The second one executes its body only once.

Remark. The combination of parallel and sequential parts of the language is attained by the possibility to use operators from the sequential part in synchroblocks. Let us also note that nested synchroblocks can be in WinALT simulation language. The described capabilities allow constructing any parallel-sequential compositions of synchronous transformations of cellular arrays.

Model program. The structure of a model program is quite conventional. It consists of a list of libraries imported to a program (using operators **use**, **import** and **include**), declarations of constants, variables and cellular objects, procedures, functions and the main operator block. The main operator block is placed in operator **begin-end** brackets and contains operators of the first and second parts of the language. A program may include comments, which can be placed in braces. A project can contain any number of simulation programs. They are capable of interacting with each other if necessary.

The third part of the language is based upon a set of WinALT libraries. The functionality of the system is extended by means of external modules. These modules are represented by Windows **dll** files. These external modules contain the interface functions that are used in versatile simulation models. The external modules form several groups that are called libraries. Some of them are briefly described below.

The library of data formats eliminates limitations of a data type that can be represented by cells in a cellular array. The library contains modules for representations of cellular arrays with integer cells (**int8**, **int16**, **int32**, **uint8**, **uint16**, **uint32**), bit cells (**bit**), float cells (**float**) and others. Some external formats are supported by the modules of library, such as **bmp** raster graphics format. The assignment of **default** type for a cellular object means that any of its cells can have any of the above-mentioned formats. The latter can be used for the representation of heterogeneous cellular objects. In GUI, the type of a cellular object can be selected in a combo box within the dialog window of the new object creation.

The library of language functions provides the ability to use such functions in simulation programs as functions of object management (creation, deletion, modification or size alteration), GUI functions (construction of dialog windows and data input based upon them), mathematical functions (**sin**, **cos**, **atan**, **cosh**, **log**, **j0**), console I/O functions (**WriteLn**, **ReadLn**), file I/O functions (**fopen**, **fgets**, **fread**, **feof**, ...) and miscellaneous functions such as **max**, **min**, **null**, **typeof**, **StringLength**, **Time**. Operator **use** activates the modules of this library in a simulation program. A typical module of this library is a **dll** written in **C** or **C++**.

The library of visual modes provides a customizable visualization of a cellular array and its cells. A cell state can be visualized e.g. by color, a directed arrow or by a number or by their certain combination. A 3D cellular array can be shown as a deck of layers or as layers unrolled in a line or in a grid.

New external modules can be added to any of these libraries. Such modules can be created, for example, using Microsoft Visual Studio. Their source texts can be either borrowed from a provided sample, written from the scratch, or taken from an existing library of functions (e.g. ANSI C runtime library).

3.3. Constructing simulation models of a family of optoelectronic matrices in WinALT

3.3.1. Mapping digital schemes onto a customizable cellular automaton

The cellular automaton with Margolus (further CA) neighborhood [13], in which setting cells for execution of a certain set of elementary transformations of information, serves as a logical basis for the construction of a family of optoelectronic matrices. The CA is a double layer automaton. Its first layer functions in the same way as the Margolus automaton, being a rectangular matrix of cells. Let us split the matrix to blocks of 2x2 cells. Let us call it E-partition. Let us split the matrix again to blocks that are shifted as related to the blocks of E-partition by one cell along the vertical and horizontal axes. Let us call it O-partition. Each cell of the first layer can be in one of the three states: "white", "gray" or "black". Let us call this layer *informational*. The layer under the informational one is called *control* layer. Its E-partition coincides with that of the informational layer. This layer keeps the table of settings the control layer of CA. The size of a cell in the table coincides with that of a block. The CA is shown in figures as a sweep of two layers in a plane. The informational layer is on the left side and the control layer is on the right side. The blocks of E-partition are limited by solid lines. The blocks of O-partition are limited by dashed lines. An elementary transformation of information performed in the informational layer is a parallel substitution. Its left and right parts are matrices with 2x2 cells. These matrices are composed of white, gray and black cells. The elementary transformations performed in a CA are enumerated. Setting a CA consists in writing numbers of those elementary transformations that can be performed in the block of informational layer above a cell into that cell. Digital (combinatory) schemes of different kinds (adder, multiplier, etc.) can be implemented by setting a CA. A source combinatory scheme is initially transformed in a way that any of its gates has no more than two inputs and two outputs. All the cells of the informational layer are white in its initial state. The picture of a scheme to be imitated in CA, which is made by gray cells in informational layer of CA is called *image* of digital scheme. The signals transformed in a digital scheme are depicted by black cells in its image. The states of cells in each block of the information layer of a CA form a certain picture called *image* of a block.

The graphic images of commands of parallel substitutions are depicted in Figure 5, a. A number of a command is shown above an arrow. The transformations of information in Margolus cellular automata are performed by alternating E- and O-partitions [13]. The alternation of partitions is substituted by alternation of two groups of shifts of the image of simulated scheme as related to the fixed control layer in a CA. A group of shifts set into correspondence with E-partition is called *E-group*. A group of shifts set into correspondence with O-partition is called *O-group*. The introduction of a setting for CA allows reducing the number of parallel substitution commands that are executed by a single block to two. The command, whose number is specified first in a cell of the settings table of the control layer, can be executed when E-group is active, while the command, whose number is listed second, can be executed when O-group is active. A command is executed if its left part coincides with the image of a block in the informational layer. The execution of a command sets its right part as the image of a block.

Possible variants of signal transmission in the image of a digital scheme (horizontal and diagonal transmissions, branching and crossing of signals) and the indications to the functional elements (AND gate is denoted by the sign '&', OR gate is denoted by '1', addition by modulo 2 is denoted by '=1', half-adder" is depicted at end) are shown in Figure 5, b.

When E-group of shifts is active, the commands are executed for the source disposition of the image of a digital scheme, then for the image shifted by one row of blocks from E-partition up as related to the source image, and then one row of blocks down as related to the source image. The shifts are introduced in order to provide the execution of commands imitating the operation of gates and crossing of signals transmission channels in the image of a digital scheme. When O-group of shifts is turned on, commands are executed for the image obtained by shifting the source one by one row down and one column to the left.

The horizontal transmission of a signal is simulated by commands 1, 2. The diagonal transmission is performed by commands 3, 4. The signal branching is done by 5, 6 and signal crossing - by 8, 10, 15. The operation of AND gate is simulated by the two sets of commands: 9, 16, 17 and 12, 16, 17, because the result on the output of the gate can be recorded either in the right top cell or in the right bottom cell. of the block. Similarly, the simulation of OR gate is done by commands 7, 8, 9 and 10, 11, 12 and the simulation of addition - by modulo two is done by commands 7, 8, 18 and 10, 11, 18. A block of E-partition that does not contain white cells can simulate a gate operation not only with one output, but also with two outputs. Thus, the operation of half-adder is simulated by commands 7, 8, 12. The capability to simulate digital schemes with two inputs and two outputs allows performing both the signal transfer, e.g. by one of the diagonals, and a logical transformation of this signal and another signal. For example, a signal transition from the right bottom cell into the right top cell of a block along with performing AND operation with signals from the left column of cells in the block with writing the result to the right bottom cell of the block is simulated by commands 8, 15, 16.

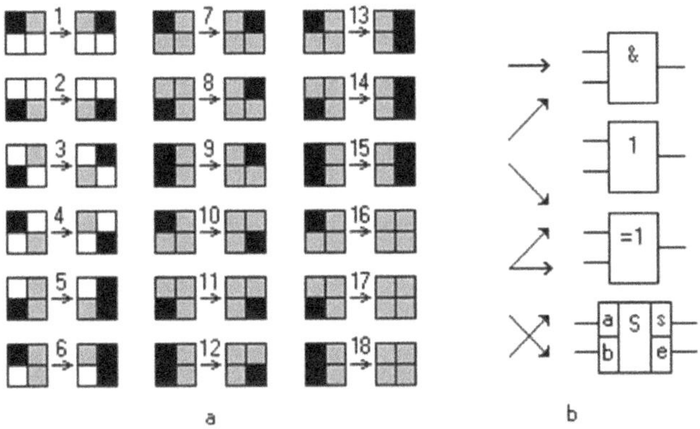

Figure 5. Graphic representation of parallel substitution commands, a - images of commands, b - symbols of signals and functional elements

Let us demonstrate an image of a full adder in CA as a simple sample of what was stated above. It is presented in Figure 6.

Let us comment on this figure. The numbers of commands listed in Figure 5, a are used in the settings table. An image of a digital scheme with mnemonic specification of the functions of its composing blocks except those imitating the transfer of signals under O-partition is given in Figure 6, c in addition to the image in Figure 6, a, b. A sign of operation or an arrow in the image of scheme denotes a cell where result of transformation or data transfer is placed. Such a representation is rather easy to grasp and is introduced solely for the purpose of the reader's convenience. This helps in comparing a combinatory scheme and its image without the table of settings.

Usually, the signal transfers will be omitted when using such a kind of representation of a digital scheme unless that hampers the perception of the image of a scheme. Let us describe the resulting image of a digital scheme. Each of blocks (2, 2) and (3, 3) corresponds to a half-adder, which is a composition of AND gates and of addition by modulo two. Blocks (4, 2) correspond to OR gate.

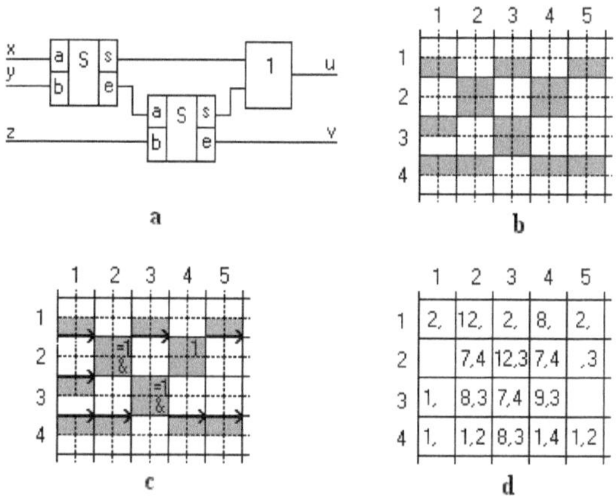

Figure 6. Full adder: a - a combinatory scheme, b - an image of scheme, c - a correspondence of blocks, d - the settings table

Some or all of the gray cells in the first column of CA have to be altered to black ones in order to introduce input signals to a full adder.

A more complicated and realistic sample is presented in Figure 7. The following heuristic criteria are selected for the estimation of quality of a particular mapping. A mapping is optimal if there is at least one such chain from the inputs of a scheme to its outputs, that is composed only of blocks that imitate gates connected with other by corner cells. The image of an eight-bit pyramidal adder [17] transformed into a scheme, in which each gate has two inputs and one or two outputs is presented in Figure 7.

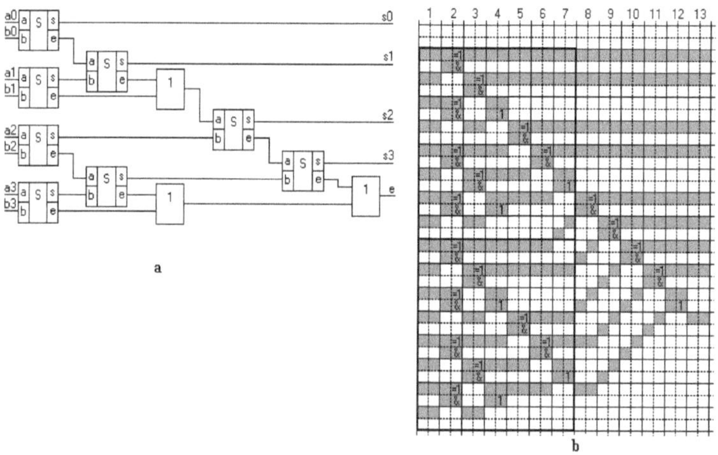

Figure 7. A pyramidal adder: a - the combinatory scheme of a four-bit adder, b - a concise image of eight-bit adder scheme

The image of a four-bit adder, which scheme is depicted in Figure 7, a, is a source fragment for its construction. The images of four-bit adders are contoured by a bold line in Figure 7, b. The rest of the scheme image provides daisy chaining of the selected images of adders. The image of an adder presented in Figure 8, b meets the heuristic criteria. The proposed way of constructing an image of a digital scheme with a specified bit width by connecting its homogeneous fragments of smaller bit width, for each of which the best image is already found, can be used for other kinds of schemes as well.

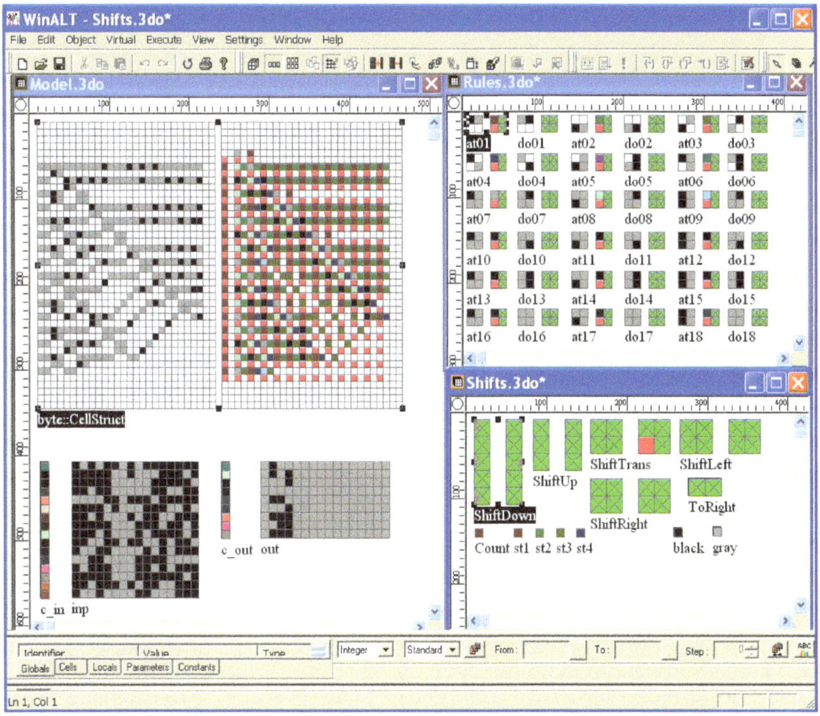

Figure 8. A project window of simulation model of CA that implements an eight-bit adder

From a logic standpoint, a fine-grain structure (a micro-pipeline) is implemented by setting a CA up. This structure simulates a simultaneous operation of many copies of the same digital scheme. Each copy (let us call it a virtual digital scheme) performs a transformation of its own data set. A *stage* of a micro-pipeline forms a vertical column of blocks in E-partition of matrix. Each stage is a "cut" of image of a virtual scheme in a certain phase of transformation of the relevant information. This feature of matrix makes possible to obtain new results in each cycle. A cycle of micro-pipeline operation includes transformation and transfer of data from one stage to another. It means that a cycle includes an execution of the both groups of shifts. A model is constructed in WinALT in order to evaluate the correctness of an eight-bit adder implemented as CA.

3.3.2. CA simulation model implementing eight-bit adder

The screenshot of a window of this project is presented in Figure 8. This window contains sheets with all the graphic objects of the model.

The sweep of a double layer cellular array **byte::CellStruct** is presented at sheet **Model.3do**. The upper layer containing the image of an eight-bit adder is presented at the left. The bottom layer is shown at the right. It contains a computer-aided representation of the settings table of an adder. The division of array layers into blocks is done by setting the cells of the bottom layer with both even coordinates into a state dedicated for this purpose. The numbers of substitution commands are represented by color in the settings table. The numbers of commands are distributed by the block cells as follows: the right and left top cells of block keep the numbers of the first and second commands respectively. The rest of cellular arrays shown in this window are auxiliary. Array **inp** keeps a set of input data. The columns of array **inp** are written in each cycle to the leftmost column of the informational layer of array **byte::CellStruct** using the array **c_in** and become the values of outputs of an adder. Similarly, the results of computations from the very right column of the informational layer of array **byte::CellStruct** are transferred to array **out** that stores the results of pipeline computations. Double layer templates are presented in sheet **Rules.3do**. The templates named **ati** and **doi** are put in correspondence with the left and right sides of substitution command with the number **i**. The upper layer of template **ati** coincides with the left part of the substitution number **i**, while that of template **doi** coincides with the right part of that substitution. The left upper cell of the bottom layer of template **ati** contains number **i** of substitution command. The number is represented by a color. A color of the left bottom cell of the same layer coincides with that of those cells in the bottom layer of array **byte::CellStruct**, which provides its partition into blocks. All the other cells of the both templates are empty (diagonally crossed). The templates used in substitution commands of the procedure imitating clocking of CA are presented in sheet **Shifts.3do**. A cycle of a model execution is divided into the four stages. At each of them parallel substitution commands kept in the settings table are executed:

- above the source image at the first stage;
- above the image shifted two cells up from the source one at the second stage;
- above the image shifted two cells down from the source one at the third stage;
- above the image shifted one cell down and one cell right from the source at the fourth stage (the command interleaving is done in the settings table during this stage: odd and even columns are swapped).

The operation of gates of a scheme is simulated at the first three stages of cycle. The data transfer between gates is simulated at the fourth stage. The cellular image and the settings table return to the initial ("unshifted") state at the end of a cycle.

Let us briefly present main procedures of simulation program. The description of parallel substitution commands is kept in procedure **mainProc**, while the shifts performed at the each stage of a cycle are defined in procedure **shiftImage**. The body of procedure **mainProc** contains operator **in** with parameter **byte::CellStruct** and eighteen bunches of **at-do** operators. The parameters of operators **at** and **do** in the i-th bunch are the templates named **ati** and **doi** respectively. The shifts of image of the digital scheme with respect to settings layer and the interleaving of substitution commands is done in procedure **ShiftImage** by vector functional substitutions. Cellular arrays **byte::CellStruct** and **Count** listed in round brackets in the first operator **in** form a composition, which has the following meaning. The changes in array **byte::CellStruct** happen only when the unicellular array **Count** is in a certain state. For example, if the state of **Count** coincides with that of **st2**, the shift of the image of a digital scheme is done by one row of blocks down from its source placement. This shift is performed by the function **fShift** that uses the values of variables **x** and **y** from template **ShiftDown** in its operator **y:=x**. The operators **at-do** in the second operator **in** alter the state of **Count** in order to set phases of cycle.

3.3.3. A transformation of CA into optoelectronic matrix

Let us list those operations converting optical signals, which are to be carried out in an optoelectronic matrix to implement the above-mentioned CA. The operations have to provide the following for each block of CA: a) a comparison of the image of the left part of a command, whose number is written in the settings table, with the image of block in the image of a scheme, and b) if they coincide, replacement of the image of such block by the image of the right part of a command. Let us demonstrate that an execution of such operations is possible in a four layer matrix (called **S**). The images of left parts of substitution commands are kept in its first layer. The second layer contains the image of a digital scheme. The images of comparison schemes are in the third layer. The images of right parts of substitution commands are in the fourth layer. Matrix **S** is built using the basic gate rather schematically and without details of implementation. All the essential physical elements: modulators, photodetectors, memory cells are considered to have a size equal to one cell. The reasons for that are as follows: 1) matrix **S** built in such way can be easily turned into a data object when it is simulated, 2) the obtained matrix **S** has rather generalized form that makes possible to specify its versatile representations taking into consideration real sizes of its elements later, because these sizes are not known a priori and depend on physical principles of construction of the elements, technology, and the application domain.

The cells of informational layer of CA have three states. But the basic gate is binary. Thus, a transition must be done to binary encoding of the states of matrix **S**. For example, an encoding of white, gray and black cells presented in Figure 9 can be chosen.

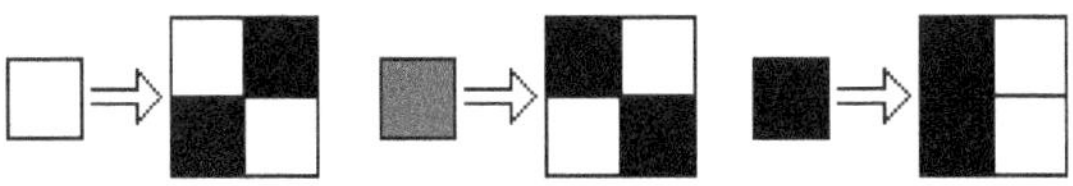

Figure 9. Binary encoding of CA states

A white cell denotes a memory cell containing one, an opened modulator or an opened photodetector. A black cell denotes a memory cell containing zero, a closed modulator or a closed photodetector.

Let us commence the construction of matrix S using its second layer that contains an image of a digital scheme in a binary encoding. An image of each block in CA informational layer is composed of modulator states. Memory cells control the states of modulators (opened, closed). Assume that each cell has two outputs. Modulators are connected to both of them. Let us consider that each memory cell has two inputs, each of which is connected to a photodetector. Let us also assume that the signals from the inputs of a cell appear at its output with a fixed delay. A line of cells is composed of such elements. The first and second cells are modulators. The third cell is a memory cell. The fourth and fifth cells are photodetectors. A memory cell operates in two phases. If it contains one at the first phase (the phase of comparison), it opens the first modulator and closes the second one. And otherwise, it closes the first modulator and opens the second one. At the second phase (phase of recording) if the fourth photodetector is opened, and the fifth one is closed, one is written to memory cell. Otherwise, memory cell resets to zero.

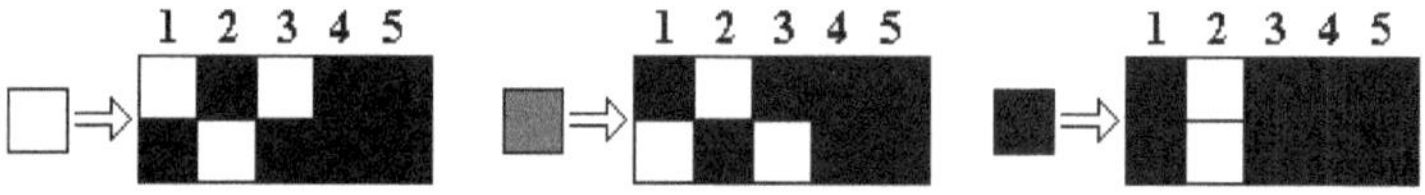

Figure 10. Binary encoding of cell states in the second layer of optoelectronic matrix

An image of the second layer of matrix **S** when it is ready for the phase of comparison is obtained by replacing each cell in the information layer of CA by a group of cells according to the substitution rules in Figure 10. Thus, a block of optoelectronic matrix of size 10×4 cells composed of four lines is put in correspondence to each block of the informational layer of CA.

Remark. Gray cells are also used in the transition to binary encoding, but they denote "void", absence of any hardware.

Let us perform a partition into blocks that have the same sizes and placement as in the rest of matrix layers.

Let us construct the first layer of matrix **S**. Unlike lines of cells form the second layer, the cell number four contains a memory cell and the fifth cell is gray. A memory cell is connected to modulators just as the memory cell occupying the third cell. The basis for setup of the first layer is the settings table of the control layer of CA. The setting of layer starts with encoding of its E-partition, i.e. the cells in columns 3 and 8 of the layer are set in such a way that the image formed by the states of modulators would coincide with the left part of that command, whose number is written first in the cell with exactly the same placement in the settings table of CA. The encoding for O-partition is embedded into the encoding of the layer that was just obtained. For this purpose binary codes set into correspondence to white, gray and black cells of the left part of the second command are written into the fourth column just as it is done for the cells of the third column in Figure 10.

The third layer of the matrix is constructed as follows. Black cells of a single block in the columns 1, 2, 6, 7 denote closed photodetectors. Black cells in the columns 4, 5, 9, 10 denote closed modulators. The rest of columns are empty. They are composed of gray cells. Pairs of photodetectors in rows 1, 2 and 3, 4 of the columns 1, 2 and 6, 7 are connected by OR scheme (parallel). Then all the OR schemes formed by pairs of photodetectors are connected sequentially forming AND scheme. A parallel assembly of all modulators of a block is connected as its load. The scheme obtained in the third layer allows to detect the situation when the images of blocks coded by the states of modulators located in the first and second layer one below another coincide, and in the case of full coincidence to prepare a substitution of the image of block of the second layer by the right part of command kept in the fourth layer.

The construction of the fourth layer can be done in two stages. First, a layer that is exactly the same as the first one for the right parts of commands is built. Then the columns in each block are transposed in the following order: 5, 4, 3, 1, 2, 10, 9, 8, 6, 7.

A fragment of matrix **S** presented in Figure 11 illustrates the procedure described above. The fragment is built for such a block of CA, whose settings table has a cell that contains the numbers of substitution commands 7, 4. The fragment is in such a state when E-shift is done and it appeared that the image of the block in the informational layer of CA coincides with the image of left part of substitution command number 7. A polarized light comes perpendicularly to the first and last layers of the matrix, respectively, from above and from below. The phase of comparison of images of blocks in the layers 1 and 2 of matrix **S** and detection of their coincidence is shown in Figure 11, a. The modulators in columns 4, 5 and 9, 10 are opened. The phase of writing new states of memory cells of the second layer using photodetectors connected to their inputs is depicted in Figure 11, b.

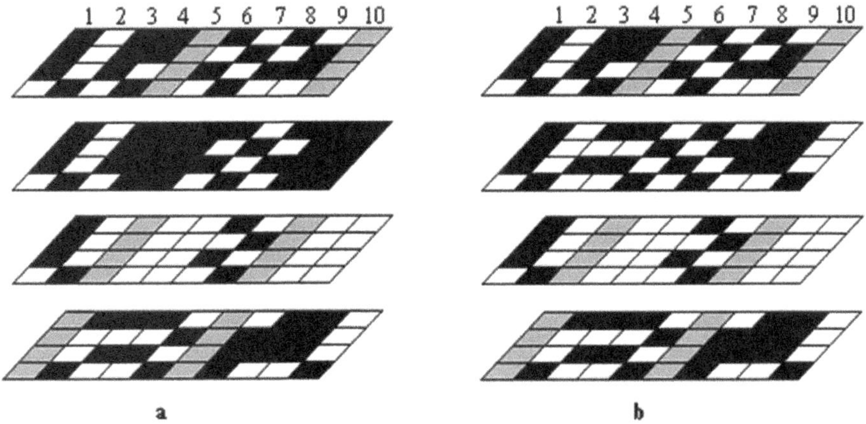

Figure 11. A fragment of the matrix S: a - detection of matching images, b - the phase of setting of new cell states in the second layer.

After a certain fixed delay, memory cells of the second layer set new states for modulators on their outputs. A fixed delay can be implemented, for example, by division of a memory cell into two elements: main and buffer. Modulators are connected to the outputs of the main buffer. Photodetectors are connected to the inputs of a buffer element. One cycle of state transition from a buffer element to a main one serves as a delay. Then the next shift of the image of a digital scheme is performed and the phase of comparison of image blocks can be done again.

It can be stated that CA turns into a simple homogeneous optoelectronic matrix **S**. It is a device that consists of four layers, which contain microscopic passive sources of light (optoelectronic modulators) and light detectors, forming a regular flat structure. The control of light sources and detectors is performed using electronic memory cells. Memory cells form a 2D shift register in order to provide setup of memory cells for storing the image of a scheme and alternation of commands. The obtained matrix is reconfigurable. It can be setup to imitate any combinatory scheme. The logical operations in the matrix (a comparison of the two codes and writing a code to memory) are preformed optically and their corresponding electronic gates simply do not exist. The proposed matrix has high performance. Indeed, after the pipeline has entered in the steady state, each next result is produced on the outputs of the matrix in each cycle.

3.3.4. The family of optoelectronic matrices

Matrix **S** can serve as a basis for the construction of a family of similar matrices. Let us outline the possible ways of their construction. The circuitry of matrices may have rather different implementations. For example, one can construct a matrix of static memory elements (as in the previous section), but it is also possible to construct a matrix with dynamic memory elements. Another way of binary encoding of white, gray and black cells can be chosen. However, we consider more profound transformations of the original matrix. And here the following options are available.

The first way. A number of layers can be reduced to two in a matrix. Let us first note that the need to alternate substitution commands in the control layer is induced by the necessity to execute commands in O-shifts. A modification of CA is proposed in [18] that eliminates the group of O-shifts and as a result the need to alternate commands in the settings layer. Let us select such a CA as a basis for the construction of a new optoelectronic matrix **S1**. The selection of such CA means that the memory cells in columns 4 and 2 can be erased in the lines of blocks of the first and fourth layers along with their modulators and photodetectors.

Let us introduce an additional limitation. Let us consider that a CA is setup once and only for implementation of a single digital scheme. Selecting a constant setting leads to the fact that there is no need to change the state of modulators in the first and fourth layers of the matrix **S** in the process of its operation. In its turn, this means that the memory cell can be removed, an open modulator can be replaced by a transparent plate, and a closed one by a dark plate. But then the first and fourth layers of the matrix **S** can be removed and the modulators in the second and third layers under dark plates can be masked or removed from the substrate. The execution of such a procedure can be demonstrated in greater details using masks in substitution command 1 (Figure 5) taken as a sample. A set of masks is presented in Figure 12.

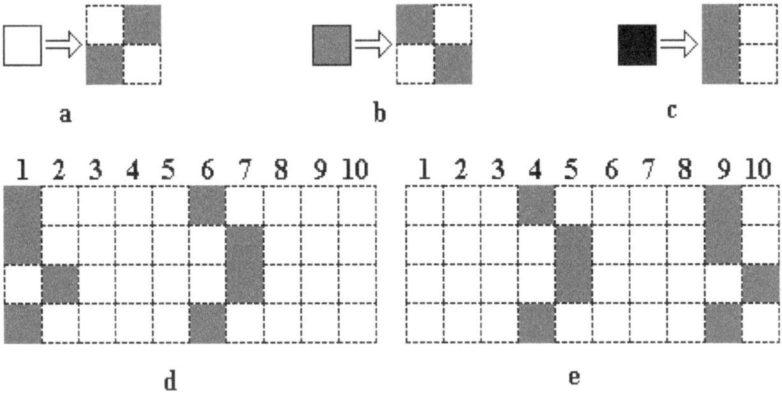

Figure 12. Masks: a - of a white cell, b - of a gray cell, c - of a black cell, d - the left part of substitution command 1, e - the right part of the same command

Masks **d** and **e** are built using masks **a**, **b**, **c**. Mask **d** is superimposed on those blocks of the second layer, which are assigned to execute command 1. The modulators under gray cells are removed. Other elements of a block under white cells remain intact. Mask **e** is applied to such blocks of the third layer, that are located under similar blocks of the second layer, and the same kind of actions is performed. As a result a computer simulation model that implements a full adder for such a matrix was built.

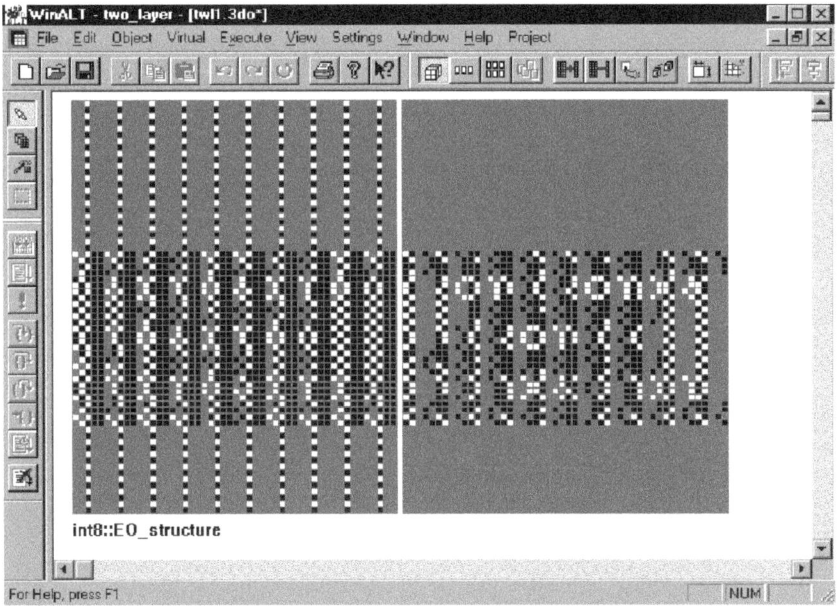

Figure 13. Computer image of a two layer optoelectronic matrix

A certain step of computations performed in a double layer matrix is presented in Figure 13. The obtained model is rather realistic. All the optical and electronic components are shown in Figure 13. By analyzing the screenshot, one can easily draw a conclusion that the first layer of matrix is composed of logically isolated shift registers. Each of their digits has optical inputs and outputs. The second layer composed of isolated single comparison schemes. This way can be considered as one with the minimum fraction of electronic components in the matrix. The double layer matrix and its simulation are presented in details in [18].

However, it is clear that when considering a practical implementation of optoelectronic devices, one has to take into account the achievements of modern microelectronics, where the actual geometric dimensions of the transistors (approximately 30 nm) have become significantly smaller than the sources of light, which can not be less than 0.3 microns due to the physical limitations. They become a weak element in terms of achieving high-density packaging of elements in a chip. Therefore, when constructing such a family of matrices, it is useful to consider the maximum use of electronic components. "Optical" components keep only their main function, which is a three-dimensional organization of the logical connections and processing of cellular (logical) information. The electronic components are then used to perform simpler operations that support a computational process, such as the shift and storage of information, including settings, etc.

The second way. One can augment the electronic part of the hardware of matrix **S** and turn it into the matrix **S2** with extended capabilities. Suppose that there are **k** (**k** > 1) four layer matrices and each **i**-th matrix is set to perform operation $\mathbf{O_i}$. Assume that the sizes of the matrices and the location of inputs and outputs of simulated digital circuits are coordinated. Some of the considerations on how that can be obtained can be found in [19]. Let us combine all the electronic components of all these matrices in order to superpose them in a four layer matrix. It can be done the following way. A set of **k** memory cells is placed in layers 1, 2, 3 of the matrix instead of a single memory cell. Let us combine $\mathbf{k_i}$ cells into a 2D shift register set for execution of operation $\mathbf{O_i}$. A decoder (named **D1**) is placed near each set of memory cells in the first and fourth layers. Similarly, two decoders are placed in the second layer (one **D1** and one that is different from **D1** is named **D2**). Decoder **D1** has one group of outputs and two groups of inputs. The states of outputs can be 0 and 1. They are connected to modulators of layer the just the same way as the outputs of memory cells in the source matrix **S**. The first group of inputs of decoder **D1** is for control. The codes of operations $\mathbf{O_i}$ (**i = 1, ..., k**) come to it. The second group of inputs has **k** pairs of inputs. The pair of outputs of the **i**-th memory cell of a set is connected to the **i**-th pair of inputs. If an operation code $\mathbf{O_i}$ comes to the control input of a decoder, the states of the **i**-th pair of the second input group are sent to the outputs of decoder **D1**. Decoder **D2** is also controlled by operation codes. But its input group has two inputs, and it output group has **k** pairs of outputs. The inputs are connected to the outputs of photodetectors of the second layer the same way as the inputs of memory cells in the source matrix **S**. The **i**-th pair of outputs of decode **D2** is connected to the inputs of the **i**-th memory cell of a set. If an operation code $\mathbf{O_i}$ came to the control input of decoder **D2**, the states of inputs are sent to the outputs of **i**-th pair of decoder **D2**. Each set of input data has to be accompanied by an operation code for such a matrix so as to specify what has to be done with these data. The transfer of operation codes from one micro-pipeline tier to another and their delivery to the decoders of the destination tier must be implemented. These actions can be performed for example by using additional shift registers that have to be inserted in the first, second and third layers of the matrix **S2**. As a result a matrix is built, in which one can change the image of a digital scheme and its setup table dynamically. That means that **k** different digital schemes can reside simultaneously in such a matrix. Thus from a logical standpoint the matrix is dynamically reconfigurable in the process of computations done by its pipeline. Matrix **S2** has undoubted advantages in comparison with matrices **S**, **S1**. The delays of its pipeline are virtually eliminated, because its control device would always "find" a data set and an operation to perform and deliver those to the inputs of matrix **S2**.

The list of matrices is not limited by the samples presented above. Let us briefly outline some more possible ways of constructing new modifications of matrices. For example, a dynamically reconfigurable matrix can be built on the basis of a two layer matrix. In order to make it reconfigurable, one has to introduce electronic masking control first of all. Then an additional hardware is to be added just as it was shown in the previous section. A group of E-shifts of a digital image can be eliminated in all the proposed matrices by increasing the vertical sizes of blocks that are compared in layers at least by a factor of three. This technique helps in shortening the duration of a micro-pipeline cycle. An assessment of creating vertical assemblies built, for example, on the basis of two-layer matrices, seems to be rather interesting. Double-layer matrices are interleaved with layers of light sources in such kind of assemblies.

And finally, let us mark one more important advantage of optoelectronic configurable matrices. Their application replaces the design (logical and physical) and production of digital circuitry by a programming of an already existing matrix. Indeed, in order to get a digital circuit in a matrix, one need only to write specially selected long vectors to its layers, using the shift registers, which are formed by memory elements of the matrix layers.

4. Conclusion

Two most pressing problems appeared in constructing computing systems: 1) an increase in the number and length of the connections inside 2D ICs, 2) a massive parallelization of computations.

Using a 3D (multilayer) IC structure is proposed in this chapter in order to solve these problems. Each of the layers, forming such an IC, has a homogeneous cell structure. A data exchange between these cells is performed by the means of local electronic (intralayer) and parallel optical (interlayer) communication channels. Data exchange between layers is combined with their processing. Cellular 3D ICs are oriented to computations with fine grain parallelism.

The physical and computing aspects of creating such ICs were considered.

The physical aspects include constructing an optoelectronic gate that allows combining single-layer ICs in the two-layer structures and an experimental test of its operability. The assessment of potential of project of a customized device, a matrix processor for the image processing from the standpoint of its performance and of its manufacturability permits to draw the following conclusions about the prospects of the chosen direction of constructing optoelectronic circuits.

Despite the fact that the size of one of the components (light modulator) of the optoelectronic logic gate is much larger than the active components of the modern microelectronics because of the wave nature of the optical signal, the performance and manufacturability of specialized devices of this type can greatly exceed similar characteristics of a purely electronic device. This is due to the fact that the number of elementary components in an optoelectronic device is significantly less than that in a purely electronic device with the same functionality. Reducing the number of components results from a local structure of logical connections and from the unidirectional light propagation (in electronic conductors a required direction of motion of the electrons is obtained by using a set of gates). There are virtually no intersections in optical conductors. This greatly facilitates the design of ICs and lets them be cheaper, more reliable and easily manufacturable. High throughput between layers by the means of optical channels gives a hope that an essentially greater performance could be reachable in comparison with purely electronic schemes.

Computer aspects of the work are devoted to the foundations of the algorithmic design of 3D ICs based on the formal model of fine-grain parallel computing, Parallel Substitution Algorithm, and to WinALT simulation system, presented in [15, 16]. A family of optoelectronic matrices was developed using WinALT. Comparison of two graphic images in all matrices, which is a main logical operation, is carried out optically exactly the same way as in devices built on the basis of optoelectronic gate. The functions of information storage are carried out electrically. High homogeneity, simplicity of topology and a low complexity of cells are the features of the matrix. A wide range of matrices was built. They vary by a number of layers, by functionality and by ratio of optical and electronic components. The proposed matrices can serve as an ALU basis in a general purpose CPU. The selection of a particular choice depends primarily on the kind of physical parameters, on which a designer can be oriented to, and on the kind of operations that a matrix has to perform. And this requires interaction between experts in different domains: physics, algorithm, programming, etc. The above suggests the following conclusions about the directions in which algorithmic design tools are to be developed.

The system's site must evolve into a fully functional online resource providing the ability to create a network (virtual) team consisting of specialists in different domains, and let them actively participate in a joint development of a 3D IC using the Internet.

The mapping of only one version of a fine-grain structure, a reconfigurable CA, to an optoelectronic structure is presented in the paper. There are many fine-grain structures and algorithms for different purposes, which may be of interest for optoelectronic implementation.

It is advisable to perform a constant replenishment of the collection of simulation models on the system's site both by its developers and by users. Also a collection of simulation models of optoelectronic implementations of fine grain structures and algorithms must be created. Let us also note the need to constantly replenish the system by new modules that extend its functionality with the emergence of new kinds of fine-grain structures and algorithms.

Constructing realistic optoelectronic structures requires constructing simulation models with huge amount of data and computations. It is expedient to launch such models on supercomputers. Thus, a WinALT subsystem must be developed for parallel execution of these models on clusters and supercomputers.

Author details

Edward G. Kostsov
Institute of Automation and Electrometry, Russian Academy of Sciences, Russia

Sergey V. Piskunov and Mike B. Ostapkevich
Institute of Computational Mathematics and Mathematical Geophysics, Russian Academy of Sciences, Russia

5. References

[1] Jin-Sung Y, Myung-Jae L, Kang-Yeob P, Woo-Young C. 10-Gb/s 850-nm CMOS OEIC Receiver with a Silicon Avalanche Photodetector. IEEE journal of quantum electronics 2012; 48(2) 229-236.
[2] Young I A, Edris M, Liao J T S; Kern A M, Palermo S, Block B A, Reshotko M R, Chang P L D. Optical I/O Technology for Tera-Scale Computting. IEEE Journal of solid-state circuits 2009; 45(1) 235-248.
[3] Beausoleil R G. Large-Scale Integrated Photonics for High-Performance Interconnects. ACM Journal on emerging technologies in computing systems 2011;7(2), Special Issue SI N 6.
[4] Krishnamoorthy A V, Goossen K W, William J et al. Progress in Low-Power Switched Optical Interconnects. IEEE Journal of selected topics in quantum electronics 2011;17(2) 357-376.
[5] Hoshino K, Yamada K, Matsumoto K, Shimoyama I. Creating a nano-sized light source by electrostatic trapping of nanoparticles in a nanogap. J. Micromech. Microeng. 2006;16 1285-1289.
[6] Nozaki K, Baba T. Laser characteristics with ultimate-small modal volume in photonic crystal slab point-shift nanolasers. Appl. Phys. Lett. 2006;88, 211101.
[7] Nozaki K, Kita S, Baba T. Room temperature continuous wave operation and controlled spontaneous emission in ultrasmall photonic crystal nanolaser. Optics express. 2007;15(12) 7506 -7514.

[8] Egorov V M, Kostsov E G. The prospects for constructing high-performance optical digital computers. Optoelectronics, Instrumentation and Data Processing. 1985;1 106-116.

[9] Egorov V M, Kostsov E G. Microelectronic optical digital computing device. Optoelectronics, Instrumentation and Data Processing. 1989;3 61- 68.

[10] Egorov V. M., Kostsov E. G. Integral optical digital computers. Appl.Opt. 1990;29(8) 1178 - 1185.

[11] Kostsov E. G., Piskunov S.V. 3D ICs with optical interconnections. In: Tverdokhleb P.E. (ed.). 3D Laser Information Technologies. Novosibirsk; 2003. p168-242.

[12] Markov A. A. Theory of Algorithms. Proc. Steklov. Moscow-Leningrad: USSR Academy of Sciences Publishing House. 1954;42.

[13] Toffoli T., Margolus N. Cellular automata machines: a new environment for modeling. Cambridge, Massachusetts: MIT Press; 1987.

[14] Achasova S., Bandman O., Markova V., Piskunov S. Parallel Substitution Algorithm. Theory and Application. Singapore: World Sci.; 1994.

[15] Ostapkevich M., Piskunov S. The Construction of Simulation Models of Algorithms and Structures with Fine-Grain Parallelism in WinALT. In: V. Malyshkin (ed.) PaCT 2011. LNCS. 6873. Heidelberg, Springer; 2011. p192-203.

[16] WinALT site. http://winalt.sscc.ru/ (accessed 25 Apr 2012).

[17] Kartsev M.A., Brick V.A. Computing systems and synchronous arithmetic. Moscow: Radio and Communications; 1981.

[18] Kostsov E.G., Piskunov S.V. Computer design of a two-layer computational array with optical interconnections. Optoelectronics Instrumentation and Data Processing C/C of Avtometriia. Allerton Press. 2000;3 3-14.

[19] Kostsov E G, Piskunov S V, Umrikhina E V. Configurable multilayer pipeline electro-optical structures. Optoelectronics Instrumentation and Data Processing C/C of Avtometriia. Allerton Press. 2005;6 62-71.

Section 2

Optical Communication: Optical Components

Chapter 5

Optical Phase-Conjugation of Picosecond Four-Wave Mixing Signals in SOAs

Narottam Das, Hitoshi Kawaguchi and Mohammad Razaghi

Additional information is available at the end of the chapter

1. Introduction

The application of four-wave mixing (FWM) in semiconductor optical amplifiers (SOAs) has been widely demonstrated to all-optical devices, such as wavelength converters (Vahala *et al.*, 1996; Nesset *et al.*, 1998), optical samplers (Kawaguchi & Inoue, 1998), optical multiplexers/ demultiplexers (Kawanishi *et al.*, 1997; Uchiyama *et al.*, 1998; Tomkos *et al.*, 1999; Buxens *et al.*, 2000; Set *et al.*, 1998), and optical phase conjugators (Dijaili *et al.*, 1990; Kikuchi & Matsumura, 1998; Marcenac *et al.*, 1998; Corchia *et al.*, 1999; Tatham *et al.*, 1993; Ducellier *et al.*, 1996), which are expected to be used in future optical communication systems. Kikuchi & Matsumura have demonstrated the transmission of 2-ps optical pulses at 1.55 μm over 40 km of standard fiber by employing midspan optical phase conjugation in SOAs (Kikuchi & Matsumura, 1998). An ideal phase conjugator must reverse the chirp of optical pulses while maintaining the pulse waveform. Kikuchi & Matsumura (Kikuchi & Matsumura, 1998) have shown that the second-order dispersion is entirely compensated by the optical phase-conjugation obtained using SOAs with a continuous wave (cw) input pump wave. All-optical demultiplexing (DEMUX), based on FWM in SOAs, was also demonstrated. When a single pulse of a time-multiplexed signal train (as probe pulses) and a pump pulse are injected simultaneously into an SOA, the gain and refractive index in the SOA are modulated, and an FWM signal pulse is created by the modulations. Thus, we can obtain a demultiplexed signal as the FWM signal by Das *et al.*, (Das *et al.*, 2000). All-optical DEMUX has been experimentally demonstrated up to 200 Gbit/s by Kawanishi *et al.*, (Kawanishi *et al.*, 1997). Many research reports have been published recently on the theoretical investigaton of the characteristics of FWM for short optical pulses in SOAs. Tang and Shore (Tang & Shore, 1999) theoretically examined the dynamical chirping of mixing pulses and showed that all mixing pulses have negative pulse chirp except in the far edges of trailing pulses, indicating that pulse spectra are primarily red-shifted. The demultiplexed signals obtained as FWM signals may still have optical phase-conjugate characteristics, although they may include waveform distortion and additional chirp. If the demultiplexed signal obtained as the FWM signal has optical phase-conjugate characteristics, the demultiplexed signal can be compressed using a dispersive medium. However, such optical phase-conjugate characteristics of FWM signals have not yet been reported to the best of author's knowledge. Another advantage of FWM with short duration optical pump pulses is the high conversion efficiency. In conventional systems, the FWM conversion efficiency in an SOA is limited due to gain saturation. However, high FWM conversion efficiency can be achieved with short duration optical pump and probe pulses as it is possible to reduce gain saturation and hence increase FWM conversion efficiency in an SOA by applying strong pump intensity (Shtaif & Eisenstein, 1995; Shtaif *et al.*, 1995; Kawaguchi & Inoue, 1996a; Kawaguchi & Inoue, 1996b).

In this chapter, we present the detail numerical simulation results of optical phase-conjugate characteristics of picosecond FWM signal pulses generated in SOAs using the FD-BMP (Das *et al.*, 2000; Razaghi *et al.*, 2009). These simulations are based on the nonlinear propagation equation considering the group velocity dispersion, self-phase modulation (SPM), and two-photon absorption (TPA), with the dependencies on the carrier depletion (CD), carrier heating (CH), spectral-hole burning (SHB), and their dispersions, including the recovery times in SOAs (Hong *et al.*, 1996). The main purpose of our simulations is to provide answers to the following questions: 1) how is the nature of the optical phase-conjugate maintained for a short FWM signal pulse? 2) how does the chirp observed in the FWM signals affect the nature of the optical phase-conjugate? For this reason, we have analyzed the system in which the Fourier transform-limited Gaussian optical pulse is linearly chirped by transmission through a fiber (Fiber I) and then injected into an SOA as a probe pulse, together with a pump pulse that has a 1 ~ 10 ps pulsewidth. The FWM signal is generated by the mixing of the pump pulse and the probe pulse, and is selected by an optical narrow band-pass filter. The FWM signal is then transmitted through another fiber (Fiber II) that has the same group velocity dispersion (GVD) as Fiber I and an appropriate length. The simulations are based on the nonlinear propagation equation considering the GVD, SPM, and TPA, with dependencies on CD, CH, SHB, and dispersion of those properties (Das *et al.*, 2000; Hong *et al.*, 1996).

The FD-BPM is useful to obtain the propagation characteristics of single pulse or milti-pulses using the modified nonlinear Schrödinger equation (MNLSE) (Hong *et al.*, 1996 & Das *et al.*, 2000), simply by changing only the combination of input optical pulses. These are: (1) single pulse propagation (Das *et al.*, 2008), (2) FWM characteristics using two input pulses (Das *et al.*, 2000), (3) optical DENUX using several input pulses (Das *et al.*, 2001), (4) optical phase-conjugation using two input pulses with chirp (Das *et al.*, 2001) and (5) optimum time-delayed FWM characteristics between the two input pump and probe pulses (Das *et al.*, 2007).

2. Analytical model

In this section, we briefly discuss the important nonlinear effects in SOAs, mathematical formulation of modified nonlinear Schrödinger equation (MNLSE), finite-difference beam propagation method (FD-BPM) that is used in the simulation, and nonlinear propagation characteristics of solitary optical pulses in SOAs.

2.1. Important nonlinear effects in SOAs

There are several types of "nonlinear effects" occurs in SOAs. Among them, the important four types of "nonlinear effects" are shown in Fig. 1. These are namely: (i) spectral hole-burning (SHB), (ii) carrier heating (CH), (iii) carrier depletion (CD) and (iv) two-photon absorption (TPA).

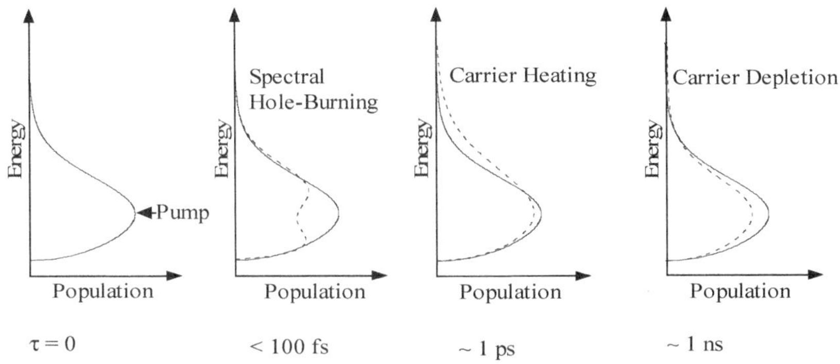

Figure 1. The important nonlinear effects in SOAs: (i) spectral hole-burning (SHB) with a life time of less than 100 fs (i.e., < 100 fs), (ii) carrier heating (CH) with a life time of ~ 1 ps, (iii) carrier depletion (CD) with a life time is ~ 1 ns and (iv) two-photon absorption (TPA).

Figure 1 shows the time-development of the population density in the conduction band after excitation (Das, 2000). The arrow (pump) shown in Fig. 1 is the excitation laser energy. When the life time is less than 100 fs (i.e., < 100 fs), the SHB effect is dominant. The SHB occurs when a narrow-band strong pump beam excites the SOA, which has an inhomogeneous broadening. The SHB arises due to the finite value of intraband carrier-carrier scattering time (~ 50 – 100 fs), which sets the time scale on which a quasi-equilibrium Fermi distribution is established among the carriers in a band. After the life time ~ 1 ps, the SHB effect is relaxed and the CH effect becomes dominant. The process tends to increase the temperature of the carriers beyond the lattice's temperature. The main causes of heating the carriers are (1) the stimulated emission, since it involves the removal of "cold" carriers close to the band edge and (2) the free-carrier absorption, which transfers carriers to high energies within the bands. The "hot"-carriers relax to the lattice temperature through the emission of optical phonons with a relaxation time of ~ 0.5 – 1 ps. The effect of CD remains for about 1 ns. The stimulated electron-hole recombination depletes the carriers, thus reducing the optical gain. The band-to-band relaxation also causes CD, with a relaxation time of ~ 0.2 – 1 ns. For ultrashort optical pumping, the TPA effect also becomes important. An atom makes a transition from its ground state to the excited state by the simultaneous absorption of two laser photons. All these nonlinear effects (mechanisms) are taken into account in the modeling/ simulation and the mathematical formulation of modified nonlinear Schrödinger equation (MNLSE).

2.2. Mathematical Formulation of Modified Nonlinear Schrödinger Equation (MNLSE)

In this subsection, we will briefly explain the theoretical analysis of short optical pulses propagation in SOAs. We start from Maxwell's equations (Agrawal, 1995; Yariv, 1991; Sauter, 1996) and reach the propagation equation of short optical pulses in SOAs, which are governed by the wave equation (Agrawal & Olsson, 1989) in the frequency domain.

$$\nabla^2 \bar{E}(x,y,z,\omega) + \frac{\varepsilon_r}{c^2}\omega^2 \bar{E}(x,y,z,\omega) = 0 \tag{1}$$

where, $\bar{E}(x,y,z,\omega)$ is the electromagnetic field of the pulse in the frequency domain, c is the velocity of light in vacuum and ε_r is the nonlinear dielectric constant which is dependent on the electric field in a complex form. By slowly varying the envelope approximation and integrating the transverse dimensions we arrive at the pulse propagation equation in SOAs (Agrawal & Olsson, 1989; Dienes *et al.*, 1996).

$$\frac{\partial V(\omega,z)}{\partial z} = -i\left\{\frac{\omega}{c}\left[1+\chi_m(\omega)+\Gamma\tilde{\chi}(\omega,N)\right]^{1/2} - \beta_0\right\}V(\omega,z) \tag{2}$$

where, $V(\omega,z)$ is the Fourier-transform of $V(t,z)$ representing pulse envelope, $\chi_m(\omega)$ is the background (mode and material) susceptibility, $\tilde{\chi}(\omega)$ is the complex susceptibility which represents the contribution of the active medium, N is the effective population density, β_0 is the propagation constant. The quantity Γ represents the overlap/ confinement factor of the transverse field distribution of the signal with the active region as defined in (Agrawal & Olsson, 1989).

Using mathematical manipulations (Sauter, 1996; Dienes *et al.*, 1996), including the real part of the instantaneous nonlinear Kerr effect as a single nonlinear index n_2 and by adding the TPA term we obtain the MNLSE for the phenomenological model of semiconductor laser and amplifiers (Hong *et al.*, 1996). The following MNLSE (Hong *et al.*, 1996; Das *et al.*, 2000) is used for the simulation of FWM characteristics and optical phase-conjugation characteristics with input pump and probe pulses in SOAs.

$$\begin{aligned}&\left[\frac{\partial}{\partial z} - \frac{i}{2}\beta_2\frac{\partial^2}{\partial \tau^2} + \frac{\gamma}{2} + \left(\frac{\gamma_{2p}}{2} + ib_2\right)|V(\tau,z)|^2\right]V(\tau,z)\\ &= \left\{\frac{1}{2}g_N(\tau)\left[\frac{1}{f(\tau)} + i\alpha_N\right] + \frac{1}{2}\Delta g_T(\tau)(1+i\alpha_T) - i\frac{1}{2}\frac{\partial g(\tau,\omega)}{\partial \omega}\bigg|_{\omega_0}\frac{\partial}{\partial \tau} - \frac{1}{4}\frac{\partial^2 g(\tau,\omega)}{\partial \omega^2}\bigg|_{\omega_0}\frac{\partial^2}{\partial \tau^2}\right\}V(\tau,z)\end{aligned} \tag{3}$$

Here, we introduce the frame of local time τ (=t - z/v_g), which propagates with a group velocity v_g at the center frequency of an optical pulse. A slowly varying envelope approximation is used in (3), where the temporal variation of the complex envelope function is very slow compared with the cycle of the optical field. In (3), $V(\tau,z)$ is the time domain complex envelope function of an optical pulse, $|V(\tau,z)|^2$ represents the corresponding optical pulse power, and β_2 is the GVD. γ is the linear loss, γ_{2p} is the TPA coefficient, b_2 (= $\omega_0 n_2 / cA$) is the instantaneous SPM term due to the instantaneous nonlinear Kerr effect n_2, ω_0 (= $2\pi f_0$) is the center angular frequency of the pulse, c is the velocity of light in vacuum, A (= wd/Γ) is the effective area (d and w are the thickness and width of the active region, respectively and Γ is the confinement factor) of the active region.

The saturation of the gain due to the CD is given by (Hong *et al.*, 1996)

$$g_N(\tau) = g_0 \exp\left(-\frac{1}{W_s} \int_{-\infty}^{\tau} e^{-s/\tau_s} |V(s)|^2 \, ds \right) \tag{4}$$

where, $g_N(\tau)$ is the saturated gain due to CD, g_0 is the linear gain, W_s is the saturation energy, τ_s is the carrier lifetime.

The SHB function $f(\tau)$ is given by (Hong *et al.*, 1996)

$$f(\tau) = 1 + \frac{1}{\tau_{shb} P_{shb}} \int_{-\infty}^{+\infty} u(s) e^{-s/\tau_{shb}} |V(\tau - s)|^2 \, ds \tag{5}$$

where, $f(\tau)$ is the SHB function, P_{shb} is the SHB saturation power, τ_{shb} is the SHB relaxation time, and α_N and α_T are the line width enhancement factor associated with the gain changes due to the CD and CH.

The resulting gain change due to the CH and TPA is given by (Hong *et al.*, 1996)

$$\begin{aligned} \Delta g_T(\tau) = &-h_1 \int_{-\infty}^{+\infty} u(s) e^{-s/\tau_{ch}} (1 - e^{-s/\tau_{shb}}) |V(\tau - s)|^2 \, ds \\ &- h_2 \int_{-\infty}^{+\infty} u(s) e^{-s/\tau_{ch}} (1 - e^{-s/\tau_{shb}}) |V(\tau - s)|^4 \, ds \end{aligned} \tag{6}$$

where, $\Delta g_T(\tau)$ is the resulting gain change due to the CH and TPA, $u(s)$ is the unit step function, τ_{ch} is the CH relaxation time, h_1 is the contribution of stimulated emission and free-carrier absorption to the CH gain reduction and h_2 is the contribution of TPA.

The dynamically varying slope and curvature of the gain plays a shaping role for pulses in the sub-picosecond range. The first and second order differential net (saturated) gain terms are (Hong *et al.*, 1996),

$$\left. \frac{\partial g(\tau,\omega)}{\partial \omega} \right|_{\omega_0} = A_1 + B_1 \left[g_0 - g(\tau,\omega_0) \right] \tag{7}$$

$$\left. \frac{\partial^2 g(\tau,\omega)}{\partial \omega^2} \right|_{\omega_0} = A_2 + B_2 \left[g_0 - g(\tau,\omega_0) \right] \tag{8}$$

$$g(\tau,\omega_0) = g_N(\tau,\omega_0) / f(\tau) + \Delta g_T(\tau,\omega_0) \quad (9)$$

where, A_1 and A_2 are the slope and curvature of the linear gain at ω_0, respectively, while B_1 and B_2 are constants describing changes in A_1 and A_2 with saturation, as given in (7) and (8).

The gain spectrum of an SOA is approximated by the following second-order Taylor expansion in $\Delta\omega$:

$$g(\tau,\omega) = g(\tau,\omega_0) + \Delta\omega \left.\frac{\partial g(\tau,\omega)}{\partial \omega}\right|_{\omega_0} + \frac{(\Delta\omega)^2}{2} \left.\frac{\partial^2 g(\tau,\omega)}{\partial \omega^2}\right|_{\omega_0} \quad (10)$$

The coefficients $\left.\frac{\partial g(\tau,\omega)}{\partial \omega}\right|_{\omega_0}$ and $\left.\frac{\partial^2 g(\tau,\omega)}{\partial \omega^2}\right|_{\omega_0}$ are related to A_1, B_1, A_2 and B_2 by (7) and (8).

Here we assumed the same values of A_1, B_1, A_2 and B_2 as in (Hong *et al.*, 1996) for an AlGaAs/GaAs bulk SOA.

The time derivative terms in (3) have been replaced by the central-difference approximation in order to simulate this equation by the FD-BPM (Das *et al.*, 2000). In simulation, the parameter of bulk SOAs (AlGaAs/GaAs, double heterostructure) with a wavelength of 0.86 µm (Hong *et al.*, 1996) is used and the SOA length is 350 µm. The input pulse shape is sech^2 and is Fourier transform-limited. The detail parameters are listed in Table 1 (Section 3).

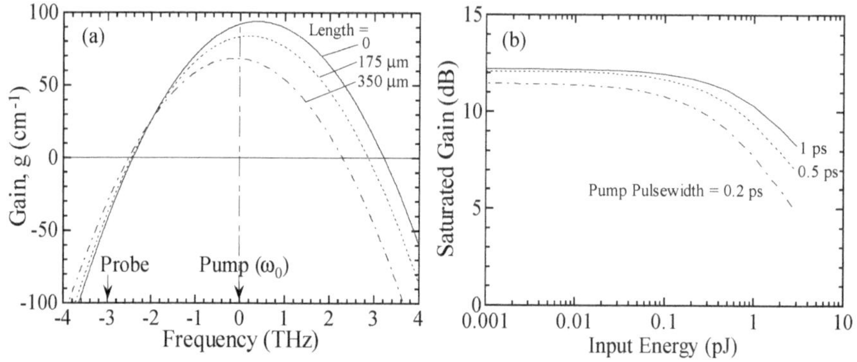

Figure 2. (a) The gain spectra given by the second-order Taylor expansion about the center frequency of the pump pulse ω_0. The solid line shows the unsaturated gain spectrum (length: 0 µm), the dotted and the dashed-dotted lines are a saturated gain spectrum at 175 µm and 350 µm, respectively. Here, the input pump pulse pulsewidth is 1 ps and pulse energy is 1 pJ. (b) Saturated gain versus the input pump pulse energy characteristics of the SOA. The saturation energy decreases with decreasing the input pump pulsewidth. The SOA length is 350 µm. The input pulsewidths are 0.2 ps, 0.5 ps, and 1 ps respectively, and a pulse energy of 1 pJ.

The gain spectra of SOAs are very important for obtaining the propagation and wave mixing (FWM and optical phase-conjugation between the input pump and probe pulses) characteristics of short optical pulses. Figure 2(a) shows the gain spectra given by a second-order Taylor expansion about the pump pulse center frequency ω_0 with derivatives of $g(\tau, \omega)$ by (7) and (8) (Das *et al.*, 2000). In Fig. 2(a), the solid line represents an unsaturated gain spectrum (length: 0 μm), the dotted line represents a saturated gain spectrum at the center position of the SOA (length: 175 μm), and the dashed–dotted line represents a saturated gain spectrum at the output end of the SOA (length: 350 μm), when the pump pulsewidth is 1 ps and input energy is 1 pJ. These gain spectra were calculated using (1), because, the waveforms of optical pulses depend on the propagation distance (i.e., the SOA length). The spectra of these pulses were obtained by Fourier transformation. The "local" gains at the center frequency at z = 0, 175, and 350 μm were obtained from the changes in the pulse intensities at the center frequency at around those positions (Das *et al.*, 2001). The gain at the center frequency in the gain spectrum was approximated by the second-order Taylor expression series. As the pulse propagates in the SOA, the pulse intensity increases due to the gain of the SOA. The increase in pulse intensity reduces the gain, and the center frequency of the gain shifts to lower frequencies. The pump frequency is set to near the gain peak, and linear gain g_0 is 92 cm^{-1} at ω_0. The probe frequency is set -3 THz from for the calculations of FWM characteristics as described below, and the linear gain g_0 is -42 cm^{-1} at this frequency. Although the probe frequency lies outside the gain bandwidth, we selected a detuning of 3 THz in this simulation because the FWM signal must be spectrally separated from the output of the SOA. As will be shown later, even for this large degree of detuning, the FWM signal pulse and the pump pulse spectrally overlap when the pulsewidths become short (<0.5 ps) (Das *et al.*, 2001). The gain bandwidth is about the same as the measured value for an AlGaAs/GaAs bulk SOA (Seki *et al.*, 1981). If an InGaAsP/InP bulk SOA is used we can expect much wider gain bandwidth (Leuthold *et al.*, 2000). With a decrease in the carrier density, the gain decreases and the peak position is shifted to a lower frequency because of the band-filling effect. Figure 2(b) shows the saturated gain versus input pump pulse energy characteristics of the SOA. When the input pump pulsewidth decreases then the small signal gain decreases due to the spectral limit of the gain bandwidth. For the case, when the input pump pulsewidth is short (very narrow, such as 200 fs or lower), the gain saturates at small input pulse energy (Das *et al.*, 2000). This is due to the CH and SHB with the fast response.

Initially, the MNLSE was used by (Hong *et al.*, 1996) for the analysis of "solitary pulse" propagation in an SOA. We used the same MNLSE for the simulation of FWM and optical phase-conjugation characteristics in SOA using the FD-BPM. Here, we have introduced a complex envelope function V(τ, 0) at the input side of the SOA for taking into account the two (pump and probe) or more (multi-pump or prove) pulses.

2.3. Finite-Difference Beam Propagation Method (FD-BPM)

To solve a boundary value problem using the finite-differences method, every derivative term appearing in the equation, as well as in the boundary conditions, is replaced by the central differences approximation. Central differences are usually preferred because they lead to an excellent accuracy (Conte & Boor, 1980). In the modeling, we used the finite-differences (central differences) to solve the MNLSE for this analysis.

Usually, the fast Fourier transformation beam propagation method (FFT-BPM) (Okamoto, 1992; Brigham, 1988) is used for the analysis of the optical pulse propagation in optical fibers by the successive iterations of the Fourier transformation and the inverse Fourier transformation. In the FFT-BPM, the linear propagation term (GVD term) and phase compensation terms (other than GVD, 1st and 2nd order gain spectrum terms) are separated in the nonlinear Schrödinger equation for the individual consideration of the time and frequency domain for the optical pulse propagation. However, in our model, equation (3) includes the dynamic gain change terms, i.e., the 1st and 2nd order gain spectrum terms which are the last two terms of the right-side in equation (3). Therefore, it is not possible to separate equation (3) into the linear propagation term and phase compensation term and it is quite difficult to calculate equation (3) using the FFT-BPM. For this reason, we used the FD-BPM (Chung & Dagli, 1990; Conte & Boor, 1980; Das *et al.*, 2000; Razagi *et al.*, 2009). If we replace the time derivative terms of equation (3) by the below central-difference approximation, equation (11), and integrate equation (3) with the small propagation step Δz, we obtain the tridiagonal simultaneous matrix equation (12)

$$\frac{\partial}{\partial \tau} V_k = \frac{V_{k+1} - V_{k-1}}{2\Delta\tau}, \quad \frac{\partial^2}{\partial \tau^2} V_k = \frac{V_{k+1} - 2V_k + V_{k-1}}{\Delta\tau^2} \tag{11}$$

where, $V_k = V(\tau_k)$, $V_{k+1} = V(\tau_k + \Delta\tau)$, and $V_{k-1} = V(\tau_k - \Delta\tau)$

$$\begin{aligned} -a_k(z+\Delta z)\, V_{k-1}(z+\Delta z) + \{1 - b_k(z+\Delta z)\}\, V_k(z+\Delta z) - c_k(z+\Delta z)\, V_{k+1}(z+\Delta z) \\ = a_k(z)\, V_{k-1}(z) + \{1 + b_k(z)\}\, V_k(z) + c_k(z)\, V_{k+1}(z) \end{aligned} \tag{12}$$

where, $k = 1, 2, 3, \ldots\ldots\ldots\ldots, n$ and

$$a_k(z) = \frac{\Delta z}{2}\left[\frac{i\beta_2}{2\Delta\tau^2} + i\frac{1}{4\Delta\tau}\frac{\partial g(\tau,\omega,z)}{\partial\omega}\Big|_{\omega_0,\tau_k} - \frac{1}{4\Delta\tau^2}\frac{\partial^2 g(\tau,\omega,z)}{\partial\omega^2}\Big|_{\omega_0,\tau_k}\right] \tag{13}$$

$$\begin{aligned} b_k(z) = -\frac{\Delta z}{2}\Bigg[\frac{i\beta_2}{\Delta\tau^2} + \frac{\gamma}{2} + \left(\frac{\gamma_{2p}}{2} + ib_2\right)|V_k(z)|^2 - \frac{1}{2}g_N(\tau_k,\omega_0,z)(1+i\alpha_N) \\ -\frac{1}{2}\Delta g_T(\tau_k,\omega_0,z)(1+i\alpha_T) - \frac{1}{2\Delta\tau^2}\frac{\partial^2 g(\tau,\omega,z)}{\partial\omega^2}\Big|_{\omega_0,\tau_k}\Bigg] \end{aligned} \tag{14}$$

$$c_k(z) = \frac{\Delta z}{2}\left[\frac{i\beta_2}{2\Delta\tau^2} - i\frac{1}{4\Delta\tau}\frac{\partial g(\tau,\omega,z)}{\partial\omega}\Big|_{\omega_0,\tau_k} - \frac{1}{4\Delta\tau^2}\frac{\partial^2 g(\tau,\omega,z)}{\partial\omega^2}\Big|_{\omega_0,\tau_k}\right] \tag{15}$$

where, $\Delta\tau$ is the sampling time and n is the number of sampling. If we know $V_k(z)$, ($k = 1, 2, 3, \ldots\ldots\ldots\ldots, n$) at the position z, we can calculate $V_k(z+\Delta z)$ at the position of $z+\Delta z$

which is the propagation of a step Δz from position z, by using equation (12). It is not possible to directly calculate equation (12) because it is necessary to calculate the left-side terms $a_k(z+\Delta z)$, $b_k(z+\Delta z)$, and $c_k(z+\Delta z)$ of equation (12) from the unknown $V_k(z+\Delta z)$. Therefore, we initially defined $a_k(z+\Delta z) \equiv a_k(z)$, $b_k(z+\Delta z) \equiv b_k(z)$, and $c_k(z+\Delta z) \equiv c_k(z)$ and obtained $V_k^{(0)}(z+\Delta z)$, as the zero-th order approximation of $V_k(z+\Delta z)$ by using equation (12). We then substituted $V_k^{(0)}(z+\Delta z)$ in equation (12) and obtained $V_k^{(1)}(z+\Delta z)$ as the first order approximation of $V_k(z+\Delta z)$ and finally obtained the accurate simulation results by the iteration as used in (Brigham, 1988; Chung & Dagli, 1990; Das *et al.*, 2000; Razagi *et al.*, 2009).

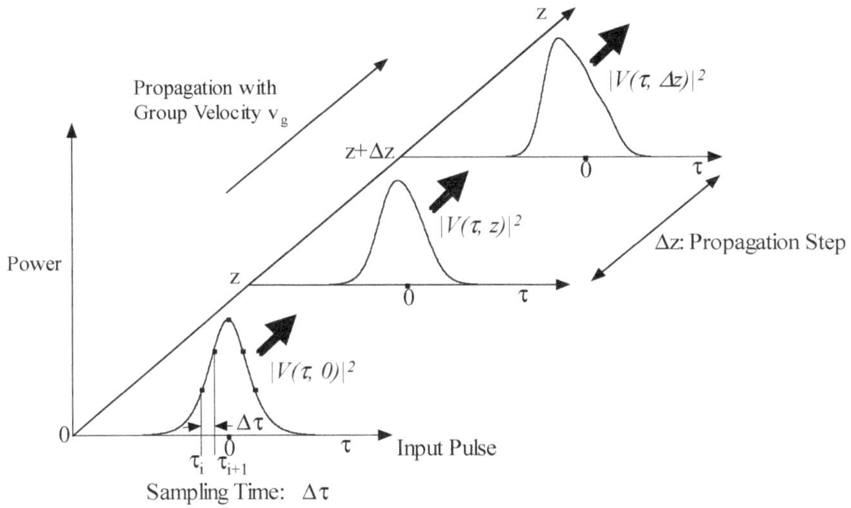

Figure 3. A simple schematic diagram of FD-BPM in the time domain, where, $\tau\,(=t-z/v_g)$ is the local time, which propagates with the group velocity v_g at the center frequency of an optical pulse and $\Delta\tau$ is the sampling time, and z is the propagation direction and Δz is the propagation step.

Figure 3 shows a simple schematic diagram of the FD-BPM in time domain. Here, $\tau\,(=t-z/v_g)$ is the local time, which propagates with the group velocity v_g at the center frequency of an optical pulse and $\Delta\tau$ is the sampling time. z is the propagation direction and Δz is the propagation step. With this procedure, we used up to 3-rd time iteration for more accuracy of the simulations.

The FD-BPM (Conte & Boor, 1980; Chung & Dagli, 1990; Das *et al.*, 2000; Razagi *et al.*, 2009a & 2009b) is used for the simulation of several important characteristics, namely, (1) single pulse propagation in SOAs (Das *et al.*, 2008; Razaghi *et al.*, 2009a & 2009b), (2) two input pulses propagating in SOAs (Das *et al.*, 2000; Connelly *et al.*, 2008), (3) Optical DEMUX characteristics of multi-probe or pump input pulses based on FWM in SOAs (Das *et al.*, 2001), (4) Optical phase-conjugation characteristics of picosecond FWM signal in SOAs (Das *et al.*, 2001), and (5) FWM conversion efficiency with optimum time-delays between the input pump and probe pulses (Das *et al.*, 2007).

2.4. Nonlinear pulse propagation model in SOAs

Nonlinear optical pulse propagation in SOAs has drawn considerable attention due to its potential applications in optical communication systems, such as a wavelength converter based on FWM and switching. The advantages of using SOAs include the amplification of small (weak) optical pulses and the realization of high efficient FWM characteristics.

For the analysis of optical pulse propagation in SOAs using the FD-BPM in conjunction with the MNLSE, where several parameters are taken into account, namely, the group velocity dispersion, SPM, and TPA, as well as the dependencies on the CD, CH, SHB and their dispersions, including the recovery times in an SOA (Hong *et al.*, 1996). We also considered the gain spectrum (as shown in Fig. 2). The gain in an SOA was dynamically changed depending on values used for the carrier density and carrier temperature in the propagation equation (i.e., MNLSE).

Initially, (Hong *et al.*, 1996) used the MNLSE for the simulation of optical pulse propagation in an SOA by FFT-BPM (Okamoto, 1992; Brigham, 1988) but the dynamic gain terms were changing with time. The FD-BPM is capable to simulate the optical pulse propagation taking into consideration the dynamic gain terms in SOAs (Das *et al.*, 2000 & 2007; Razaghi *et al.*, 2009a & 2009b; Aghajanpour *et al.*, 2009). We used the modified MNLSE for nonlinear optical pulse propagation in SOAs by the FD-BPM (Chung & Dagli, 1990; Conte & Boor, 1980). We used the FD-BPM for the simulation of optical phase-conjugation characteristics of picosecond FWM signal pulses in SOAs.

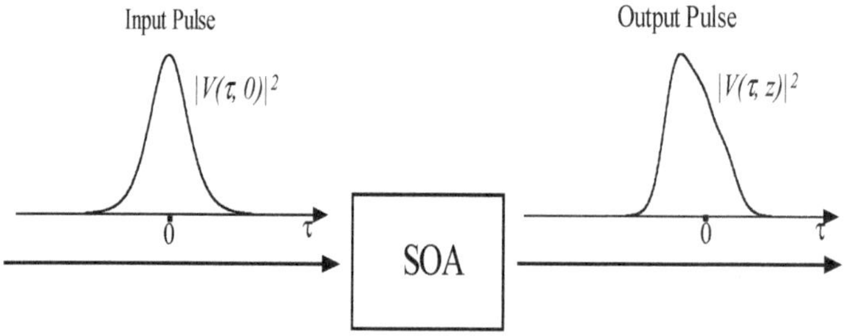

Figure 4. A simple schematic diagram for the simulation of nonlinear single pulse propagation in SOA. Here, $|V(\tau,0)|^2$ is the input (z = 0) pulse intensity and $|V(\tau,z)|^2$ is the output pulse intensity (after propagating a distance z) of SOA.

Figure 4 illustrates a simple model for the simulation of nonlinear optical pulse propagation in an SOA. An optical pulse is injected into the input side of the SOA (z = 0). Here, τ is the local time, $|V(\tau,0)|^2$ is the intensity (power) of input pulse (z = 0) and $|V(\tau,z)|^2$ is the intensity (power) of the output pulse after propagating a distance **z** at the output side of

SOA. We used this model to simulate the FWM and optical phase-conjugation characteristics in SOAs.

3. FWM characteristics between optical pulses in SOAs

In this section, we will discuss the FWM characteristics between optical pulses in SOAs. When two optical pulses with different central frequencies f_p (pump) and f_q (probe) are injected simultaneously into the SOA, an FWM signal is generated at the output of the SOA at a frequency of $2f_p$ - f_q (as shown in Fig. 5). For the analysis (simulation) of FWM characteristics, the total input pump and probe pulse, $V_{in}(\tau)$, is given by the following equation

$$V_{in}(\tau) = V_p(\tau) + V_q(\tau)\exp(-i2\pi\Delta f\tau) \tag{16}$$

where, $V_p(\tau)$ and $V_q(\tau)$ are the complex envelope functions of the input pump and probe pulses respectively, $\tau\,(=t - z/v_g)$ is the local time that propagates with group velocity v_g at the center frequency of an optical pulse, Δf is the detuning frequency between the input pump and probe pulses and expressed as $\Delta f = f_p - f_q$. Using the complex envelope function of (16), we solved the MNLSE and obtained the combined spectrum of the amplified pump, probe and the generated FWM signal at the output of SOA.

For the simulations, we used the parameters of a bulk SOA (AlGaAs/GaAs, double heterostructure) at a wavelength of 0.86 μm. The parameters are listed in Table 1 (Hong *et al.*, 1996; Das *et al.*, 2000). The length of the SOA was assumed to be 350 μm. All the results were obtained for a propagation step Δz of 5 μm. We confirmed that for any step size less than 5 μm the simulation results were almost identical (i.e., independent of the step size).

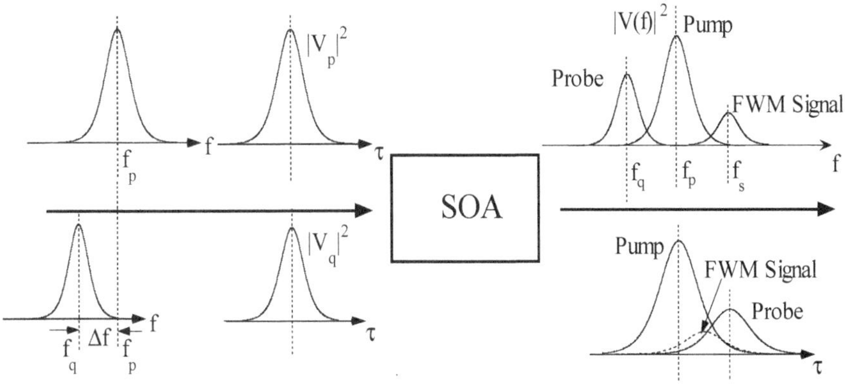

Figure 5. A simple schematic diagram for the simulation of FWM characteristics between pump and probe pulses in SOAs. The input pump and probe pulses with the center frequency of f_p and f_q are injected into the SOA. The pump and probe pulse detuning is Δf. The FWM signal is generated at the output of the SOA.

Name of the Parameters	Symbols	Values	Units
Length of SOA	L	350	μm
Effective area	A	5	μm^2
Center frequency of the pulse	f_0	349	THz
Linear gain	g_0	92	cm^{-1}
Group velocity dispersion	β_2	0.05	$ps^2\ cm^{-1}$
Saturation energy	W_s	80	pJ
Linewidth enhancement factor due to the CD	α_N	3.1	
Linewidth enhancement factor due to the CH	α_T	2.0	
The contribution of stimulated emission and FCA to the CH gain reduction	h_1	0.13	$cm^{-1}pJ^{-1}$
The contribution of TPA	h_2	126	$fs\ cm^{-1}pJ^{-2}$
Carrier lifetime	τ_s	200	ps
CH relaxation time	τ_{ch}	700	fs
SHB relaxation time	τ_{shb}	60	fs
SHB saturation power	P_{shb}	28.3	W
Linear loss	γ	11.5	cm^{-1}
Instantaneous nonlinear Kerr effect	n_2	-0.70	$cm^2\ TW^{-1}$
TPA coefficient	γ_{2p}	1.1	$cm^{-1}\ W^{-1}$
Parameters describing second-order Taylor expansion of the dynamically gain spectrum	A_1 B_1 A_2 B_2	0.15 -80 -60 0	$fs\ \mu m^{-1}$ fs $fs^2\ \mu m^{-1}$ fs^2

Table 1. Simulation parameters of a bulk SOA (AlGaAs/GaAs, double heterostructure) (Hong *et al.*, 1996; Das *et al.*, 2000).

For the simulation of optical phase-conjugate characteristics in SOAs, we used the above model (Fig. 5). Fig. 5 shows a simple schematic diagram illustrating the simulation of the FWM characteristics in an SOA between short optical pulses. In SOA, the FWM signal is generated by mixing between the input pump and probe pulses, whose frequency appears at the symmetry position of the probe pulse with respect to the pump. For our simulation (as shown in Fig. 7), we have selected the detuning frequency between the input pump and probe pulses to +3 THz. The generated FWM signal is filtered using an optical narrow bandpass filter from the optical output spectrum containing the pump and probe signal. Here, the pass-band of the filter is set to be from +2 THz to +4 THz, i.e., a bandwidth of 2 THz is used. The shape of the pass-band was assumed to be rectangular.

4. Optical phase-conjugation of picosecond FWM signals in SOAs

4.1. Chirp of FWM signal pulses for Fourier transform-limited input pulses

In this sub-section, first, we obtain the frequency shift characteristics of FWM signal pulses for Fourier transform-limited input pulses. Fig. 6(a) shows the normalized output power of the pump, probe and FWM signal pulses. Fig. 6(b) shows the center frequency shifts of these pulses. Here, the input pump and probe pulses are Fourier transform limited (non-chirped) Gaussian pulses with a pulsewidth of 2 ps (full width at half maximum: FWHM). The pum-probe detuning is +3 THz. The input pump and probe pulse energy levels are 1 pJ and 0.1 pJ, respectively. The solid, dotted, and dotted-broken curves represent the pulse waveforms of the pump, probe, and generated FWM signal pulses, respectively. We have obtained these output waveforms by the method that have reported by Das et al., (Das et al., 2000). The output waveforms of the pump and probe pulses are close to the input waveforms. However, the peak position of the pump pulse is slightly shifted toward the leading edge due to the gain saturation of teh SOA (Shtaif & Eisenstein, 1995; Das et al., 2000). This is because the pulses have a larger gain at the leading edge than at the trailing edge. The pulsewidth of the FWM signal becomes narrower than that of the pump and probe pulses, because the FWM signal intensity is proportional to $I_p^2 I_q$, were I_p is the pump pulse intensity and I_q is the probe pulse intensity (Das et al., 2000).

The frequency shift characteristics of the pump, probe, and FWM signal pulses are shown in Fig. 6(b). The frequency shift of the FWM signal pulse is plotted for output power, which is greater than 1% of the output peak power. The vertical axis indicates the frequency shifts of the output pulses from the center frequencies of each input pulse. Two components of frequency shifts can be observed; negative frequency shift in the vicinity of the pulse peaks, and a frequency shift that is almost constant with time. In the vicinity of the pulse peaks, all the mixing pulses have negative pulse chirp, indicating that the pulse spectra are mainly red-shifted. A similar frequency shift was reported by Tang & Shore (Tang & Shore, 1999) for SOAs operating at a wavelength of 1.55 μm and attributed to the self- and cross- phase modulation under gain saturation. The origin of another frequency shift that is almost constant with time may be the effect of the gain spectrum of the SOA. The pump pulse frequency was set to near the gain peak. Therefore, both the higher and lower frequency components of the pump pulse have about the same optical gain. Thus the frequency of the pulse does not shift during the propagation in the SOA except for the frequency shift caused by the self- and cross- phase modulation as mentioned above. However, the probe pulse frequency was set at –3 THz from the center frequency of the pump pulse. Therefore, the higher frequency component of the probe pulse was enhanced more than the lower frequency component due to the gain spectrum, and the frequency of the output probe pulse is shifted toward the higher frequency side. This gain spectrum effect caused a +25 GHz shift in the probe pulses. The FWM signal is obtained at about +3 THz from the center frequency of the pump pulse. The lower frequency component of the FWM signal pulse was enhanced more than the higher frequency component. In addition, the probe pulse was shifted toward the higher frequency side. Therefore, the center frequency of the FWM signal pulse is shifted by about –35 GHz.

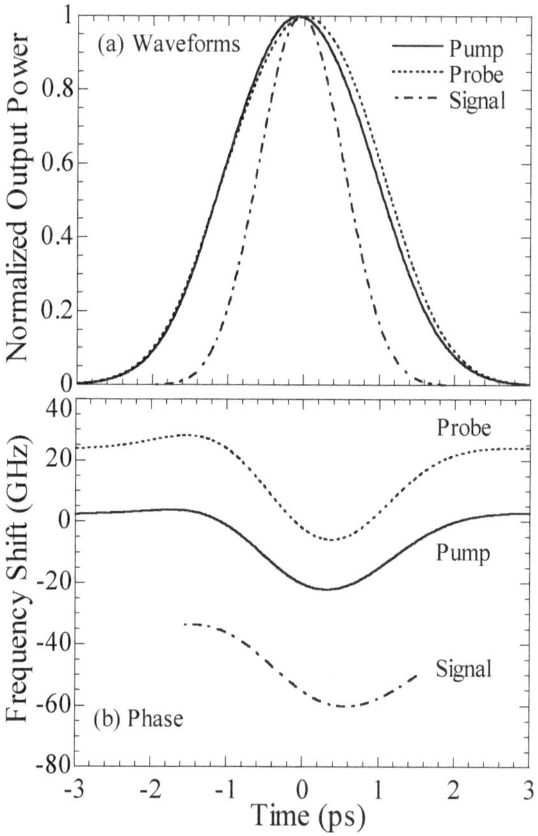

Figure 6. Normalized output power (a) and frequency shift (b) of the pump, probe, and generated FWM signal pulses. Input pump and probe pulses are Fourier transform-limited Gaussian shape with a pulsewidth of 2 ps (FWHM), detuning of the input pump and probe pulse is +3 THz. The input pump and probe energy levels are 1 pJ and 0.1 pJ, respectively.

4.2. Optical phase-conjugate characteristics for linearly-chirped input probe pulse

In this sub-section, we have analyzed the optical phase-conjugate characteristics of the FWM signal pulses. The outline of our simulation model, which is similar to the experimental setup used by Kikuchi & Matsumura (Kikuchi & Matsumura, 1998), is the following one, as shown in Fig. 7, the Fourier transform-limited optical pulse is linearly-chirped by transmission through a fiber (Fiber I) and then injected into the SOA as the probe pulse together with a pump pulse. The FWM signal is generated by the mixing of the probe pulse and the pump pulse, and selected by an optical narrow band-pass filter. The FWM signal is then transmitted through another fiber (Fiber II) that has the same GVD as Fiber I and is an appropriate length.

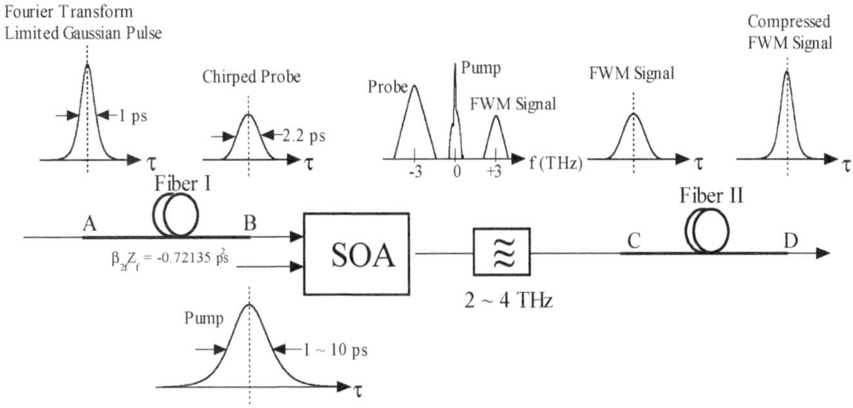

Figure 7. A simple schematic diagram for the simulation of optical phase-conjugate characteristics of picosecond FWM signal pulses in SOAs. Fourier transform-limited Gaussian optical pulse is linearly-chirped by transmitting through a fiber (Fiber I) and injected into the SOA as the probe pulse together with a pump pulse. The FWM signal is generated by the mixing of the chirped-probe pulse and the pump pulse, and it is selected by an optical narrow band-pass filter. Then, the FWM signal is transmitted through the fiber (Fiber II) with the same GVD value as Fiber I and the appropriate fiber length.

We have assumed that the input probe pulse is the Fourier transform-limited Gaussian pulse with a pulsewidth of 1 ps (FWHM). Therefore, the incident field is given by Agrawal (Agrawal, 1995; Das *et al.*, 2001)

$$E(0,T) = \exp\left[-\frac{T^2}{2T_0^2}\right] \tag{17}$$

where, T_0 is the half-width at the $1/e$-intensity (power) point and 0.60056 ps in this case. The input probe pulse energy is chosen to be 0.1 pJ. This unchirped Gaussian pulse has propagated through the fiber (length Z_f) from position A to position B in Fig. 7 and is chirped by the second-order GVD of the fiber, and the duration of the pulse is broadened. The spectrum of the broadened pulse at position B is calculated using the following equation (Agrawal, 1995; Das *et al.*, 2001)

$$\tilde{E}(Z_f,\omega) = \tilde{E}(0,\omega)\exp\left[\frac{i}{2}\beta_{2f}\omega^2 Z_f\right] \tag{18}$$

where, $\tilde{E}(0,\omega)$ is the Fourier transform of the incident field given by equation (17) at $Z_f = 0$ (at position A of Fig. 7), $\omega = 2\pi(f - f_q)$, f_q is the center frequency of the probe pulse, and β_{2f} is the second-order GVD of the fiber. Equation (18) shows that GVD changes the phase of each spectral component of the pulse by an amount that depends on the frequency and the propagation distance. Even though such phase changes do not affect the pulse spectrum, they can modify the pulse shape. Taking the inverse Fourier transform of the Eq. (18), the broadened pulse can be calculated at position B of Fig. 7 using the following equation (Agrawal, 1995; Das *et al.*, 2001):

$$E(Z_f,T) = \frac{1}{2\pi}\int_{-\infty}^{\infty} \tilde{E}(0,\omega)\exp\left[\frac{i}{2}\beta_{2f}\omega^2 Z_f - i\omega T\right]d\omega$$

$$= \frac{T_0}{\left(T_0^2 - i\beta_{2f}Z_f\right)^{1/2}}\exp\left[-\frac{T^2}{\left(T_0^2 - i\beta_{2f}Z_f\right)}\right] \tag{19}$$

A Gaussian pulse preserves its shape on propagation, yet its width increases and becomes (Agrawal, 1995; Das *et al.*, 2001):

$$T_b = T_0\left[1 + \left(Z_f/L_D\right)^2\right]^{1/2} \tag{20}$$

where the dispersion length $L_D = T_0^2/\left|\beta_{2f}\right|$. The input probe pulse is chirped by the Fiber I with $\beta_{2f}Z_f = 2\beta_{2f}L_D = -0.72135$ ps^2 and is broadened to approximately 2.2 ps.

The FWM signal generated by the chirped probe pulse is expressed by Eq. (18) and the Fourier transform-limited Gaussian pump pulses, is obtained from the output facet of a 350 µm length SOA. The detuning between the input pump and the probe pulses is set at 3 THz. The FWM signal is obtained by taking the spectral component between +2 THz and +4 THz (i.e., the bandwidth of the optical filter is 2 THz).

Figure 8 shows the waveforms and the frequency shift of the FWM signal at the output end of the SOA, shown as position C in Fig. 7 together with those of the input probe pulse. The input probe pulse is a chirped Gaussian pulse with a pulsewidth of 2.2 ps (FWHM), as described above, and with an energy of 0.1 pJ. The input pump pulses are the Fourier transform-limited Gaussian pulses with pulsewidths of 1 ps, 2 ps, 3 ps, and 10 ps. The pulsewidth of the FWM signal is increased in step with the increase in the pump pulsewidth. The peak positions of the FWM signals are slightly shifted toward the leading edge due to the gain spectrum of the SOA. As shown in Fig. 8(b) later, the frequency of the probe pulse is linearly chirped from higher frequencies to lower frequencies. As the probe pulse frequency is set to the low frequency side of the SOA gain spectrum, the probe pulse has a larger gain at its leading edge. The center frequency shift of the FWM signal pulses at the output end of the SOA for different input pump pulsewidths is shown in Fig. 8(b), demonstrating the frequency shift of the chirped probe pulse at the input side of the SOA.

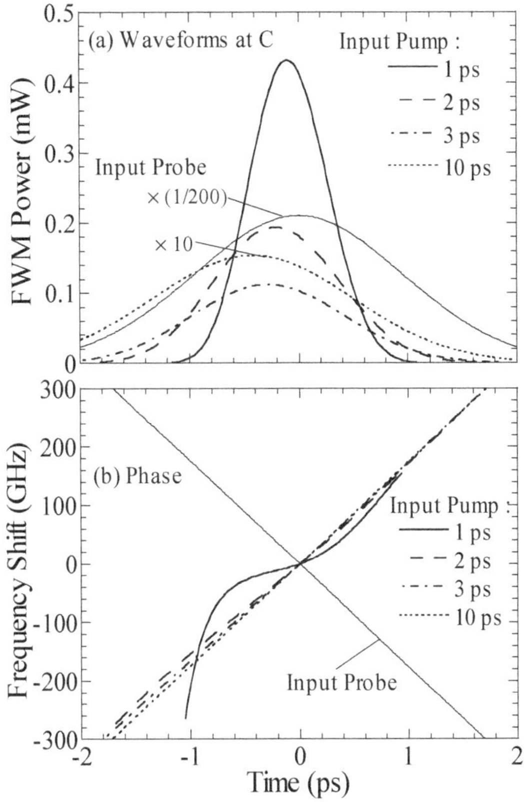

Figure 8. Waveforms and frequency shift of the generated FWM signal at the output end of the SOA shown as the position C in Fig. 7 together with those of the input probe pulse.

We have defined zero frequency shift to be that at 0 ps. If the SOA acts as an ideal optical phase conjugator, the frequency shift of the FWM signal should be symmetrical with that of the input probe pulse. The frequency shift of the FWM signal is plotted for output power, which is greater than 1% of the FWM peak power. For the input pump pulsewidth of 10 ps, the frequency shift of the FWM signal is very similar to the symmetrical shape of the input probe. With a decrease in the input pump pulsewidth, the symmetry breaks due to a change in the refractive index of the SOA, as caused by the pump pulse. For a 1 ps pump pulsewidth, the symmetry is strongly degraded. In the model, we have taken into account carrier depletion, CH, SHB, and the instantaneous nonlinear Kerr effect, as the origins of nonlinear refractive index changes. For the case of a 1 ps pump pulse, the anomalous frequency shift at the leading edge of the pump pulse primarily originates from the Kerr effect. By contrast, all mechanisms contribute to the frequency shift at the trailing edge. As a result of this simulation, we found that the phase-conjugate characteristics are almost entirely preserved, even for a 2 ps input pump pulsewidth.

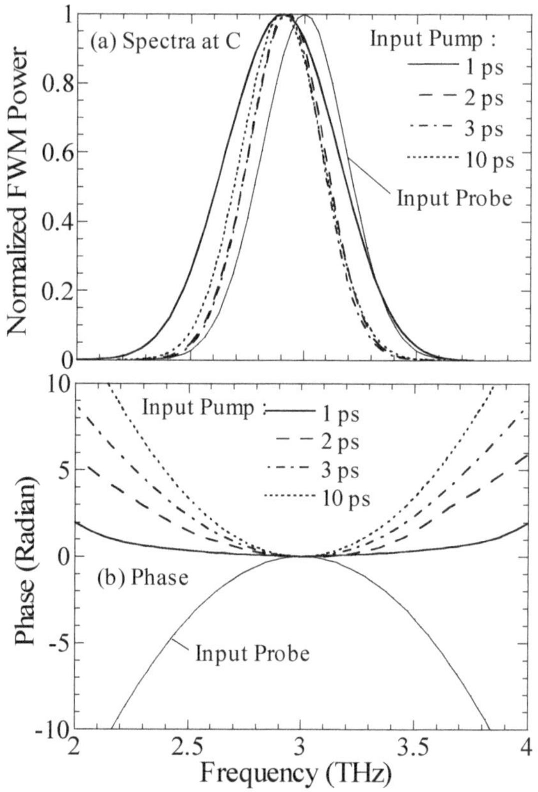

Figure 9. Spectra and phase in the frequency domain of the FWM signal at the output end of the SOA shown as the position C in Fig. 7 together with those of the probe input pulse.

The spectra and phase of the generated FWM signal in the frequency domain at the output end of the SOA shown as the position C in Fig. 7, together with the spectra and phase of the input probe pulse, as shown in Fig. 9. Although the center frequency of the probe pulse is at –3 THz, the spectrum and the phase of the input probe pulse are plotted at +3 THz to aid for comparison with the FWM signal pulse. We assumed a Fourier transform-limited Gaussian pulse as the input of the fiber, and so the power spectrum does not change during the propagation in the fiber. Therefore, the input probe spectrum is the same as the input pulse to the fiber. The phase of the probe pulse in the frequency domain should vary according to $\left(f-f_q\right)^2$ against the frequency f. Here, f_q is the center frequency of the probe pulse. The results shown in Fig. 9 are in complete agreement with the above considerations. The peak frequencies of the FWM spectra are shifted to the lower frequency side of the frequency spectra. These shifts are mainly due to the SPM caused by the gain saturation effect (Das et al., 2000; Das et al., 2001). The phase of the

FWM spectra at the output of the SOA is shown in Fig. 9(b). If the SOA acts as an ideal optical phase conjugator, the phase of the FWM spectra should be symmetrical with that of the input probe pulse spectrum. From the figure, the phase of the FWM signal for a pump pulsewidth of 10 ps is almost symmetrical to the phase of the input probe pulse. With a decrease in the input pump pulsewidth, this symmetry decreases. This tendency is the same as that found for time domain (see Fig. 9(a)).

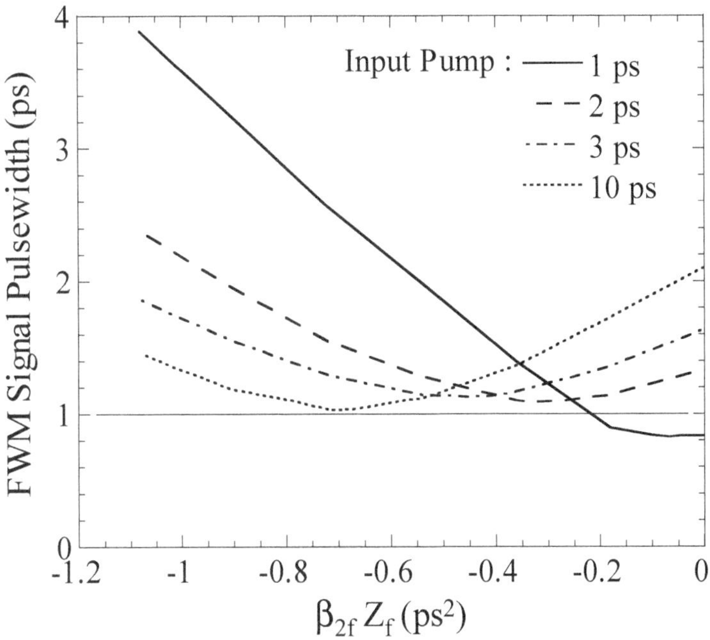

Figure 10. FWM signal pulsewidth versus the $\beta_{2f}Z_f$ value characteristics at position D in Fig. 7.

The chirped pulse can be easily compressed using the phase-conjugate characteristics of the FWM signal pulse and the second-order GVD of a fiber. Figure 10 shows the pulsewidth of the FWM signal versus the $\beta_{2f}Z_f$ value characteristics at position D in Fig. 7. For a 10 ps pump pulse, the SOA acts as a nearly ideal phase conjugator within the confines of reversing the chirp of optical pulses. The pulsewidth of the output pulse becomes the shortest, 1.03 ps, at $\beta_{2f}Z_f$ = –0.70 ps^2 (Dijaili et al., 1990; Kikuchi & Matsumura, 1998), which is slightly smaller (~3%) than the assumed $\beta_{2f}Z_f$ of the input fiber we assumed. The shortest pulsewidth compresses to 1 ps, which is equal to the input pulsewidth of the Fourier transform-limited probe pulse. This result confirms that the SOA acts as a nearly ideal phase conjugator (i.e., reverse chirp) when the input pump pulsewidth is relatively long (10 ps). When the input pump pulsewidth becomes shorter, the $\beta_{2f}Z_f$ value for obtaining the shortest pulsewidth becomes smaller, and the shortest pulsewidth becomes wider than the input pulsewidth of 1 ps. For example, for a 3 ps pump pulse, the shortest pulsewidth is ~1.1 ps which is obtained at $\beta_{2f}Z_f$ = –0.45 ps^2. This result can be understood as follows: The FWM process acts as both a temporal and a spectral window depending on the pulsewidth and the spectral width of the pulses. By the FWM characteristics described in III-A, the pulsewidth of the FWM signal becomes shorter than the chirped probe pulse. Therefore, only a part of the phase information is copied to the FWM signal. For a 1 ps pump pulse, the temporal window effect is enhanced. In addition, the pump pulse loses phase information due to the optical nonlinear effect that is induced by their strong pulse peak intensity, as shown in Fig. 7. Therefore, the FWM signal pulsewidth becomes less than 1 ps at = 0 and the shortest FWM signal pulsewidth was obtained for $\beta_{2f}Z_f$ = –0.07 ps^2.

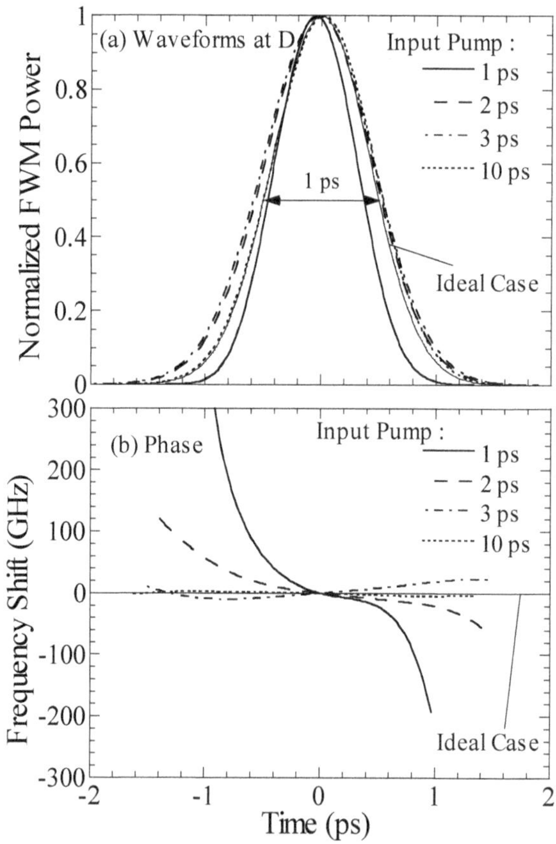

Figure 11. Normalized FWM signal waveforms having minimum pulsewidth (a) and the frequency shift (b) at position D in Fig. 7 after the pulse compression by the fiber. This figure also shows those of the input pulse at position A in Fig. 7, which correspond to the case that the SOA acts as the ideal phase conjugation.

Figure 11 shows the normalized FWM signal waveforms with minimum pulsewidth (a), and the frequency shift (b) at the position D in Fig. 7 after the pulse compression by the same fiber. The normalized waveform of the input pulse at position A in Fig. 7 is also shown, corresponds to the case where the SOA acts as an ideal phase conjugator. The normalized waveforms of the FWM signal pulse for various pump pulsewidths are shown in the figure. Although all the FWM signal pulses are compressed to ~ 1 ps when the values of $\beta_{2f}Z_f$ is optimized, the FWM signal waveform is close to that of the ideal phase conjugator for the longer pump pulse of 10 ps. The frequency shift of the FWM signals after fiber dispersion compensation shows that for the longer pump pulse of 10 ps, the frequency shift becomes nearly zero, which indicates that chirp becomes nearly zero. This suggests that the compressed FWM signal almost becomes a transform-limited pulse. For the pump pulsewidth of 1 ps, the phase of the FWM signal was significantly distorted.

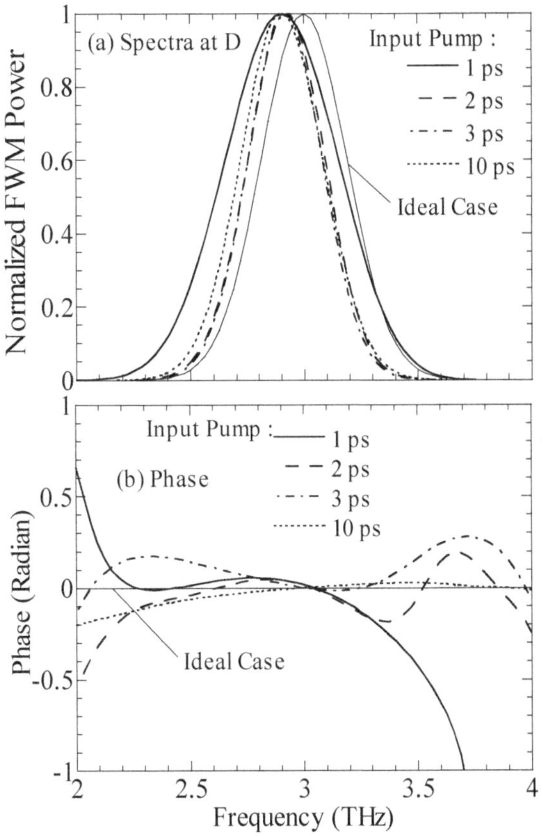

Figure 12. Spectra and phase in the frequency domain of the FWM signal pulse at the output end of the dispersion compensation fiber shown as the position D in Fig. 7, together with those for the ideal case.

The spectra and phase in the frequency domain of the FWM signal pulse at the output end of the dispersion compensation fiber, shown as the position D in Fig. 7, are shown in Fig. 12 with the spectra and phase for the ideal case. For the 10 ps pump pulsewidth, the spectrum is almost identical to the ideal case except for the center frequency. The red shift of the center frequency originates from the gain spectrum of the SOAs. For the shorter pump pulse of 1 ps, the spectral width increased because the pulsewidth of the FWM signal becomes short, as shown in Fig. 11(a). From the figure, the output signal phase becomes nearly constant when the input pump pulsewidth is 10 ps. As a result of this simulation (modeling results), we can conclude that pump pulses longer than 10 ps are needed in order to obtain nearly ideal optical phase-conjugate characteristics for the ~2.2 ps chirped pulse.

5. Conclusion

We have presented a detail numerical analysis of optical phase-conjugate characteristics of the FWM signal pulses generated in SOAs using the FD-BPM. We have shown that the input pump pulsewidth is an important factor in determining the optical phase-conjugate characteristics of the FWM signal pulse. If we use relatively long input pump pulses, nearly ideal phase-conjugate characteristics, within the confines of reversing the chirp of optical pulses, can be obtained even for very short optical probe pulses. From the simulated example, it has been clarified that the nearly ideal phase-conjugate characteristics are obtained for ~2.2 ps chirped probe pulses using a 10 ps pump pulse. When the pulsewidth of pump pulse decreases, the minimum compressed pulsewidth is obtained using the fiber with a smaller $\beta_{2f}Z_f$ than that of the input fiber. For much shorter pump pulses such as 1 ps, short FWM signals can be obtained via the gating characteristics of the FWM. However, only a part of the phase information is copied to the FWM signal, and the phase information is lost due to the nonlinear effect. Thus, the pulsewidth is not compressed by propagation through a dispersive medium.

Author details

Narottam Das
Department of Electrical and Computer Engineering, Curtin University, Australia
School of Engineering, Edith Cowan University, Australia

Hitoshi Kawaguchi
Graduate School of Materials Science, Nara Institute of Science and Technology, Japan

Mohammad Razaghi
Department of Electrical and Computer Engineering, University of Kurdistan, Iran

Acknowledgement

The authors would like to thank Dr. T. Kawazoe and Mr. Y. Yamayoshi for their helpful contribution to this work.

6. References

Agrawal, G. P. (1995). *Nonlinear Fiber Optics*. Academic Press, New York. ISBN 0-12-045142-5.

Agrawal, G. P. & Olsson, N. A. (1989). Self-phase modulation and spectral broadening of optical pulses in semiconductor laser and amplifiers. *IEEE J. Quantum Electron.*, vol. 25, pp. 2297-2306, ISSN 0018–9197.

Aghajanpour, H.; Ahmadi, V. & Razaghi, M. (2009). Ultra-short optical pulse shaping using semiconductor optical amplifier. *Optics & Laser Technology,* vol. 41, pp. 654-658, ISSN 0030-3992.

Brigham, E. Oran (1988). *The Fast Fourier Transform and Its Applications*. Englewood Cliffs, N.J.: Prentice-Hall Inc. ISBN 0-13-307505-2.

Buxens, A.; Poulsen, H. N., Clausen, A. T., & Jeppesen, P. (2000). All-optical OTDM-to-WDM signal-format translation and OTDM add-drop functionality using bidirectional four wave mixing in semiconductor optical amplifier. *Electron. Lett.,* vol. 36, pp. 156-158, ISSN 0013-5194.

Chung, Y. & Dagli, N. (1990). An Assessment of finite difference beam propagation method. *IEEE J. Quantum Electron.*, vol. 26, pp. 1335-39, ISSN 0018–9197.

Connelly, M. J.; Barry, L. P., Kennedy, B. F. & Ried, D. A. (2008). Numerical analysis of four-wave mixing between picosecond mode-locked laser pulses in a tensile-strained bulk SOA. *Optical and Quantum Electronics,* vol. 40, pp. 411-418, ISSN 1572-817X.

Conte, S. D. & Boor, Carl de (1980). *Elementary Numerical Analysis: An Algorithmic Approach,* Third Edition, McGraw-Hill Book Company Co., Singapore. ISBN 0070124477.

Corchia, A.; Antonini, C., D'Ottavi, A., Mecozzi, A., Martelli, F., Spano, P., Guekos, G., & Dall'Ara, R. (1999). Mid-span spectral inversion without frequency shift for fiber dispersion compensation: A system demonstration. *IEEE Photon. Technol. Lett.,* vol. 11, pp. 275-277, ISSN 1041-1135b. 1999.

Das, N. K. (2000). *Numerical simulations of four-wave mixing among short optical pulses in semiconductor optical amplifiers by the beam propagation method.* PhD dissertation, Yamagata University, Japan.

Das, N. K.; Yamayoshi, Y. & Kawaguchi, H. (2000). Analysis of basic four-wave mixing characteristics in a semiconductor optical amplifier by beam propagation method. *IEEE J. Quantum Electron.*, vol. 36, pp. 1184-1192, ISSN 0018–9197.

Das, N. K. & Karmakar, N. C. (2008). Nonlinear propagation and wave mixing characteristics of pulses in semiconductor optical amplifiers. *Microwave and Optical Technology Letters,* vol. 50, pp. 1223-1227, ISSN 0895-2477.

Das, N. K.; Kawaguchi, H. & Alameh, K. (2011). *Advances in Optical Amplifiers*, Paul Urquhart (Ed.), "Ch. 6: Impact of Pump-Probe Time Delay on the Four-Wave Mixing Conversion Efficiency in SOAs", InTech, Austria. ISBN 978-953-307-186-2.

Das, N. K.; Karmakar, N. C., Yamayoshi, Y. & Kawaguchi, H. (2007). Four-wave mixing characteristics in SOAs with optimum time-delays between pump and probe pulses," *Microwave and Optical Technology Letters,* vol. 49, pp. 1182-1185, ISSN 0895-2477.

Das, N. K.; Yamayoshi, Y., Kawazoe, T. & Kawaguchi, H. (2001). Analysis of optical DEMUX characteristics based on four-wave mixing in semiconductor optical amplifiers. *IEEE /OSA J. Lightwave Technol.,* vol. 19, pp. 237-246, ISSN 0733-8724.

Das, N. K.; Kawazoe, T., Yamayoshi, Y. & Kawaguchi, H. (2001). Analysis of optical phase-conjugate characteristics of picosecond four-wave mixing signals in semiconductor optical amplifiers. *IEEE J. Quantum Electron.,* vol. 37, pp. 55-62, ISSN 0018–9197.

Das, N. K.; Karmakar, N. C., Yamayoshi, Y. & Kawaguchi, H. (2005). Four-wave mixing characteristics among short optical pulses in semiconductor optical amplifiers with optimum time-delays, *Proceedings of the 18th Annual Meeting of the IEEE Lasers and Electro-Optics Society 2005 (IEEE-LEOS2005),* pp. 127-128, ISBN 0-7803-9217-5, *Sydney, NSW, Australia*, October 2005, IEEE Press (USA).

Dienes, A.; Heritage, J. P., Jasti, C. & Hong, M. Y. (1996). Femtosecond optical pulse amplification in saturated media. *J. Opt. Soc. Am. B,* vol. 13, pp. 725-734, ISSN 0740-3224.

Diez, S.; Schmidt, C., Ludwig, R., Weber, H. G., Obermann, K., Kindt, S., Koltchanov, I. & Petermann, K. (1997). Four-wave mixing in semiconductor optical amplifiers for frequency conversion and fast optical switching. *IEEE J. Sel. Top. Quantum Electron.,* vol. 3, pp. 1131-1145, ISSN 1939-1404.

Diez, S.; Mecozzi, A. & Mϕrk, J. (1999). Bit rate and pulse width dependence of four-wave mixing of short pulses in semiconductor optical amplifiers. *Opt. Lett.,* vol. 24, pp. 1675-1677, ISSN: 0146-9592.

Dijaili, S. P.; Dienes, A., & Smith, J. S. (1990). ABCD matrices for dispersive pulse propagation. *IEEE J. Quantum. Electron.,* vol. 26, pp. 1158-1164, ISSN 0018–9197.

Ducellier, T. & Bibey, M. B. (1996). Study of optical phase conjugation in bulk travelling wave semiconductor optical amplifier. *IEEE Photon. Technol. Lett.,* vol. 8, pp. 530-32, ISSN 1041-1135.

Eiselt, M. (1995). Optimum pump pulse selection for demultiplexer application of four-wave mixing in semiconductor laser amplifiers. *IEEE Photon. Technol. Lett.,* vol. 7, pp. 1312-1314, ISSN 1041-1135.

Gingrich, H. S.; Chumney, D. R., Sun, S.-Z., Hersee, S. D., Lester, L. F. & Brueck, S. R. (1997). Broadly tunable external cavity laser diodes with staggered thickness multiple quantum wells. *IEEE Photon. Technol. Lett.,* vol. 9, pp. 155-157, ISSN 1041-1135.

Hong, M. Y.; Chang, Y. H., Dienes, A., Heritage, J. P., Delfyett, P. J., Dijaili, Sol & Patterson, F. G. (1996). Femtosecond self- and cross-phase modulation in semiconductor laser amplifiers. *IEEE J. Sel. Top Quantum Electron.,* vol. 2, pp. 523-539, ISSN 1939-1404.

Inoue, J. & Kawaguchi, H. (1998). Highly nondegenerate four-wave mixing among sub-picosecond optical pulses in a semiconductor optical amplifier," *IEEE Photon. Technol. Lett.*, vol. 10, pp. 349–351, ISSN 1041-1135.

Kawaguchi, H. & Inoue, J. (1996a). Four-wave mixing among sub-picosecond optical pulses in semiconductor optical amplifier. *Proc. of the 20th International Quantum Electronics Conference, IQEC '96,* Tul 105, Sydney, Australia, ISSN.

Kawaguchi, H. & Inoue, J. (1996b). Wavelength Conversion of sub-picosecond optical pulses by four-wave mixing in a semiconductor optical amplifier. Conference on Laser and Electro-Optics Europe, CLEO/ Europe '96, CBF5, Hamburg, Germany, ISSN: 0-7803-3169-9.

Kawaguchi, H. & Inoue, J. (1998). Four-wave mixing among sub-picosecond optical pulses in a semiconductor optical amplifier and its applications to optical sampling. *Proc. of SPIE, Physics and Simulation of Optoelectronic Devices VI,* vol. 3283, pp. 447-454, Jan. 26-30, ISSN-13: 978-0819427229.

Kawanishi, S.; Okamoto, K., Ishii, M., Kamatani, T., Takara, H., & Uchiyama, K. (1997). All-optical time-division-multiplexing on four-wave mixing in a travelling-wave semiconductor laser amplifier. *Electron. Lett.,* vol. 33, pp. 976-977, ISSN 0013-5194.

Kawanishi, S.; Morioka, T., Kamatani, O., Takara, H., Jacob, J.M., & Saruwatari, M. (1994). 100 Gbit/s all-optical demultiplexing using four-wave mixing in a travelling wave laser diode amplifier. *Electron. Lett.,* vol. 30, pp. 981-982, ISSN 0013-5194.

Kikuchi, K. & Matsumura, K. (1998). Transmission of 2-ps optical pulses at 1550 nm over 40-km standard fiber using midspan optical phase conjugation in semiconductor optical amplifiers. *IEEE Photon. Technol. Lett.,* vol. 10, pp. 1410-1412, ISSN 1041-1135.

Kirita, H.; Hashimoto, Y., & Yokoyama, H. (1998). All-optical signal processing at over 100 Gbit/s with nonlinear effects in semiconductor lasers. *Tech. Dig. Int'l. Trop. Workshop on Contemporary Technologies (CPT '98),* Paper Pc-14, Tokyo, Japan.

Koltchanov, I.; Kindt, S., Petermann, K., Diez, S., Ludwig, R., Schnabel, R., & Weber, H. G. (1996). Gain dispersion and saturation effects in four-wave mixing in semiconductor laser amplifiers. *IEEE J. Quantum Electron.,* vol. 32, pp. 712-720, ISSN 0018–9197.

Leuthold, J.; Mayer, M., Eckner, J., Guekos, G., Melchior, H. and Zellweger, Ch. (2000). Material gain of bulk 1.55 μm InGaAsP/InP semiconductor optical amplifiers approximated by a polynomial model. *J. Appl. Phys.,* vol. 87, pp. 618-620, ISSN.

Marcenac, D. D.; Nesset, D., Kelly, A. E., & Gavriovic, D. (1998). 40 Gbit/s transmission over 103 km of NDSF using polarization independent mid-span spectral inversion by four-wave mixing in a semiconductor optical amplifier. *Electron. Lett.,* vol. 34, pp. 100-101, ISSN 0013-5194.

Mecozzi, A. & Mϕrk, J. (1997). Saturation effects in nondegenerate four-wave mixing between short optical pulses in semiconductor laser amplifiers. *IEEE J. Sel. Top Quantum Electron.,* vol. 3, pp. 1190-1207, ISSN 1939-1404.

Mecozzi, A.; D'Ottavi, A., Iannone, E., & Spano, P. (1995). Four-wave mixing in travelling-wave semiconductor amplifiers. *IEEE J. Quantum Electron.*, vol. 31, pp. 689-699, ISSN 0018–9197.

Mikami, O.; Noguchi, Y., Yasaka, H., Magari, K. & Kondo, S. (1991). Emission spectral width broadening for InGaAsP/InP superluminescent diodes. *IEE Proc. J Optoelectron.*, vol. 138, pp. 133-137, ISSN: 0267-3932.

Mørk, J. & Mecozzi, A. (1997). Theory of nondegenerate four-wave mixing between pulses in a semiconductor waveguide. *IEEE J. Quantum Electron.*, vol. 33, pp. 545-555, ISSN 0018–9197.

Morioka, T.; Takara, H., Kawanishi, S., Uchiyama, K., & Saruwatari, M. (1996). Polarisation-independent demultiplexing up to 200 Gb/s using four-wave mixing in a semiconductor laser amplifier. *Electron. Lett.*, vol. 32, pp. 840-841, ISSN 0013-5194.

Nesset, D.; Marcenac, D. D., Mason, P. L., Kelly, A. E., Bouchoule, S., & Lach, E. (1998). Simultaneous wavelength conversion of two 40 Gbit/s channels using four-wave mixing in a semiconductor optical amplifier. *Electron. Lett.*, vol. 34, pp. 107-108, ISSN 0013-5194.

Okamoto, K. (1992). *Theory of Optical Waveguides*, Corona Publishing Co., Tokyo; Ch. 7 (in Japanese). ISBN 4-339-00602-5.

Razaghi, M.; Ahmadi, A., Connelly, M. J. & Madanifar, K. A. (2009a). Numerical modelling of sub-picosecond counter propagating pulses in semiconductor optical amplifiers. *Proceedings of the 9th International Conference on Numerical Simulation of Optoelectronic Devices 2009 (NUSOD' 09)*, pp. 59-60, ISBN 978-1-4244-4180-8, GIST, Gwangju, South Korea, September 2009, IEEE Press (USA).

Razaghi, M.; Ahmadi, A., & Connelly, M. J. (2009b). Comprehensive finite-difference time-dependent beam propagation model of counter propagating picosecond pulses in a semiconductor optical amplifier. *IEEE/OSA J. Lightwave Technol.*, vol. 27, pp. 3162-3174, ISSN 0733-8724.

Saleh, A. A. A. & Habbab, I. M. I. (1990). Effects of semiconductor-optical-amplifier nonlinearity on the performance of high-speed intensity-modulation lightwave systems. *IEEE Trans. Commun.*, vol. 38, pp. 839-846, ISSN 0090-6778.

Sauter, E. G. (1996). *Nonlinear Optics.* John Wiley & Sons, Inc. New York. ISBN 0-471-14860-1.

Seki, K.; Kamiya, T. and Yanai, H. (1981). Effect of waveguiding properties on the axial mode competition in stripe-geometry semiconductor lasers. *IEEE J. Quantum Electron.*, vol. 17, pp. 706-713, ISSN 0018–9197.

Shtaif, M. & Eisenstein, G. (1995). Analytical solution of wave mixing between short optical pulses in semiconductor optical amplifier. *Appl Phys. Lett.* 66, pp. 1458-1460, ISSN 0003-6951.

Shtaif, M.; Nagar, R. & Eisenstein, G. (1995). Four-wave mixing among short optical pulses in semiconductor optical amplifiers. *IEEE Photon Technol. Lett.* 7, pp. 1001-1003. ISSN 1041-1135.

Shtaif, M. & Eisenstein, G. (1996). Calculation of bit error rates in all-optical signal processing applications exploiting nondegenerate four-wave mixing in semiconductor optical amplifiers. *IEEE/OSA J. Lightwave. Technol.,* vol. 14, pp. 2069-2077, ISSN 0733-8724.

Summerfield, M. A. & Tucker, S. R. (1995). Noise figure and conversion efficiency of four-wave mixing in semiconductor optical amplifiers. *Electron. Lett.,* vol. 31, pp. 1159-1160, ISSN 0013-5194.

Set, S. Y.; Yamashita, S., Ibsen, M., Laming, R. I., Nesset, D., Kelly A. E. & Gilbertas, C. (1998). Ultra-high bit rate of optical phase conjugation/ wavelength conversion in NDSF and SOA with novel configuration incorporating inline fibre DFB lasers. *Electron. Lett.,* vol. 34, pp. 1681-83, ISSN 0013-5194.

Tang, J. M. & Shore, K. A. (1998). Influence of probe depletion and cross-gain modulation on four-wave mixing of picosecond optical pulses in semiconductor optical amplifiers," *IEEE Photon. Technol. Lett.,* vol. 10, pp. 1563-1565, ISSN 1041-1135.

Tang, J. M. & Shore, K. A. (1999). Active picosecond optical phase compression in semiconductor optical amplifiers. *IEEE J. Quantum. Electron.,* vol. 35, pp. 93-100, ISSN 0018–9197.

Tang, J. M. & Shore, K. A. (1999). Characteristics of optical phase conjugation of picosecond pulses in semiconductor optical amplifiers. *IEEE J. Quantum. Electron.,* vol. 35, pp. 1032-1040, ISSN 0018–9197.

Tang, J. M.; Spencer, P. S. & Shore, K. A. (1998). The influence of gain compression on picosecond optical pulses in semiconductor optical amplifiers. *J. Mod. Opt.,* vol. 45, pp. 1211-1218, ISSN: 0950-0340.

Tatham, M. C.; Sherlock, G., & Westbrook, L. D. (1993). Compensation fibre chromatic dispersion by optical phase conjugation in a semiconductor laser amplifier. *Electron. Lett.,* vol. 29, pp. 1851-52, ISSN 0013-5194Oct. 1993.

Tomkos, I.; Zacharopoulos, I., Syvridis, D., Calvani, R., Cisternino, F., & Riccardi, E. (1999). All-optical demultiplexing/shifting of 40-Gb/s OTDM optical signal using dual-pump wave mixing in bulk semiconductor optical amplifier. *IEEE Photon. Technol. Lett.,* vol. 11, pp. 1464-1466, ISSN 1041-1135.

Uchiyama, K.; Kawanishi, S., & Saruwatari, M. (1998). 100-Gb/s multiple-channel output optical OTDM demultiplexing using multichannel four-wave mixing in a semiconductor optical amplifier. *IEEE Photon. Technol. Lett.,* vol. 10, pp. 890-892, ISSN 1041-1135.

Vahala, K. J.; Zhou, J., Geraghty, D., Lee, R., Newkirk, M., & Miller, B. (1996). Four-wave mixing in semiconductor travelling-wave amplifiers for wavelength conversion in all-optical networks, in T. P. Lee ed. *Current Trends in Optical Amplifiers and Their Applications.* World Scientific, Singapore.

Xie, C.; Ye, P. & Lin, J. (1999). Four-wave mixing between short optical pulses in semiconductor optical amplifiers with the consideration of fast gain saturation. *IEEE Photon. Technol. Lett.,* vol. 11, pp. 560-562, ISSN 1041-1135.

Yariv, A. (1991). *Optical Electronics*, 4th Edition, Saunders College Publishing, San Diego. ISBN 0-03-053239-6.

Chapter 6

Photonic Components for Analog Fiber Links

Alejandro García Juárez, Ignacio Enrique Zaldívar Huerta,
Jorge Rodríguez Asomoza and María del Rocío Gómez Colín

Additional information is available at the end of the chapter

1. Introduction

Currently, several companies are involved in the transport of radiofrequency (RF) or microwave signals radio over fiber links (RoF). RoF links are considered a promising technique in providing broadband wireless access services in the emerging optical-wireless networks. The modulation scheme which decides the way to generate the optical millimeter-wave signal is a key technique in this optical-wireless access network architecture [1–3]. The simplest RoF system consists of a Central Site (CS) and a Remote Site (RS) connected by an optical fiber link or network. The frequencies of the radio signals distributed by RoF systems span a wide range (usually in the GHz region) and the applications for such systems range from in-building distribution of wireless signals (for example in shopping malls and tunnels), wireline interconnections between base stations and microcellular antennas, antenna remoting for various commercial (wing-tip antennas in aircraft) and military radar systems and broadcasting of cable television signals in both hybrid fiber coax (HFC) and triplexer based fiber-to-the-home (FTTH) systems [4]. On the other hand a traditional RF photonic link operating at 1550 nm consists of a directly modulated DFB laser diode, a length of fiber optic as transmission medium, and a photo-detector (PD) [5]. The fiber chromatic dispersion intrinsically occurring in standard 1550 nm single-mode fibers (SMFs) is one of the main drawbacks that limit the transmission distance and operation bandwidth of such RF fiber-optic systems. However if low dispersion fiber is used in conjunction with an appropriate choice of the laser operating parameters the system becomes quasi linear [6]. DFB lasers are the proper light sources for these applications because of their single longitudinal mode, with side-mode suppression ratio (SMSR) in the order of 40 dB, and stable operation with a narrow spectral width. However, the characteristic relaxation oscillations peak in the laser intensity which produce sidebands on both sides of the main laser line, impose a limiting factor on the proximity of two channels in a communication system. Besides these characteristics, limit the high frequency response of semiconductor lasers and the sensitivity in low frequency applications. In order for an optical analog transmission of GHz-range signal to be successful in RoF systems, it is necessary to have a good DFB laser without their inherently serious problem of the modulation distortion caused by several factors: spatial-hole burning (SHB), leakage current, nonlinear interaction of carriers and photons, gain compression, nonlinear I-V characteristics in a p-n heterojunction, and power dependent absorption and including relaxation oscillations frequencies. However, recently the authors have demonstrated in reference [7] that when a DFB laser biased in the low laser threshold current region, relaxation oscillation frequencies in the laser intensity appear. These frequencies were seen as sidebands on both sides of the main laser line when the optical spectrum was analyzed with a spectrum analyzer. This result allowed the generation of microwave signals on C band using direct detection method. In addition with the proposed experimental setup in [7], the authors also demonstrate an analog NTSC TV signal transmitted and satisfactorily received in a short distance wireline communication system by using relaxation oscillation frequency as an information carrier. The proposed microwave photonic link architecture used to transmit simultaneously a TV signal at 4 and 5 GHz and over 30 km of standard optical fiber is shown in Figure 1.

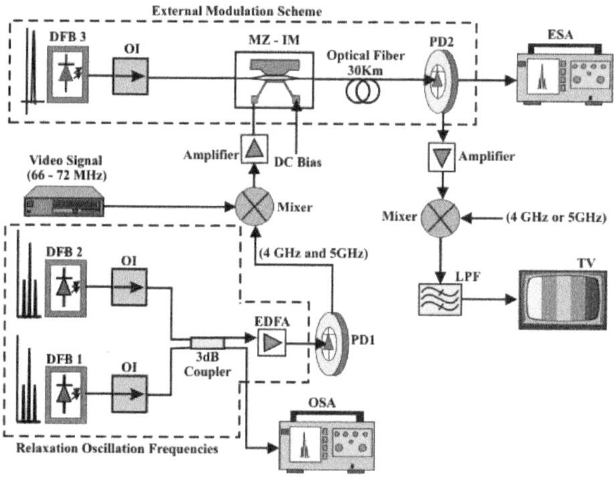

Figure 1. Experimental setup for transmitting a TV signal by using relaxation oscillation frequency as information carrier.

The procedure proposed here starts with the use of two DFB laser diodes emitting at 1550 nm and operating in the low laser threshold current region. When the generated emissions in both lasers are detected by a fast photo-detector using direct detection method, two relaxation oscillation frequencies are obtained. These frequencies acting as information carriers are mixed with TV signals. The signal obtained in the output of the mixer is printed on an optical carrier as an intensity modulation using a MZ-IM. The intensity-modulated optical carrier is then transmitted over 30 km of optical fiber. At the end of the link, a photodiode is used to convert the intensity modulation back to the relaxation oscillation frequencies mixed with TV signals. After that, TV signals are recovered satisfactorily.

In this chapter we start with a brief introduction of microwave photonics. In this first section we describe some schemes for photonic generation of microwave signals with applications in radio-over-fiber systems. After that, in the second section, a detailed theoretical description of the relaxation oscillation frequency in DFB lasers is presented. Moreover, the term photodetection is commonly used here, and we take advantage to introduce the operation principle of direct detection method. In this same section we describe a microwave photonic link by using external modulation. Next, we briefly discuss some passive components suitable for microwave photonics, the main one being optical fiber, but also many of the microwave photonic systems reported in the literature using standard single-mode fiber and associated components such as optical isolators, polarization rotator, polarization control, and optical couplers. In order to show a potential application in the field of communication systems of relaxation oscillation frequencies, it is necessary to report in the last section of this chapter the simultaneous multiplexed transmission of an analog TV signal of 67.25 MHz (TV channel 4) over a photonic link of 30 Km of optical fiber by using two microwave signals at 4 and 5 GHz of our microwave photonic link architecture, giving a signal to noise ratio (SNR) performance of an analog photonic link. Finally our conclusion to this chapter is provided.

2. Microwave photonics

Microwave photonics is an interdisciplinary area that studies the interaction between microwave and optical signals, for applications such as broadband wireless access networks, sensor networks, radar, satellite communications, instrumentation, and warfare systems. In the past few years, there has been an increase effort in researching new microwave photonics techniques for different applications. The major functions of microwave photonic systems include photonic generation, processing, control and distribution of microwave and millimeter-wave (mm-wave) signals. Many research findings have been reported in the last few years. In general, the topics covered by microwave photonics include photonic generation of microwave and mm-wave signals, photonic processing of microwave and mm-wave signals, optically controlled phased array antennas, radio-over-fiber systems, and photonic analog-to-digital conversion. In this section we will emphasize the study of schemes for photonic generation of microwave signals with applications to radio-over-fiber systems.

2.1. Photonic generation of microwave signals

Research into the generation of microwave signals using photonic techniques for various applications such as fiber-radio wireless access networks [8] and phase array antennas [9] has been increasing recently. Techniques which have been proposed and demonstrated include harmonic generation using a MZ-IM [10], heterodyne mixing of two continuous-wave (CW) lasers [11] and resonant enhancement in a monolithic mode-locked laser [12]. Among all these previously reported methods, a favored technique is to use optical heterodyning between two lasing wavelengths whose frequencies are separated by the desired microwave frequency. Optical heterodyning also has the advantage over other generation techniques of generating high RF power, though optical sources with narrow linewidth or feedback loops are required during operation [13]. The following section includes descriptions of the most common techniques to generate microwave signals using the optical heterodyning method, including optical injection locking (OIL) of laser diodes [14], external modulation technique [15], dual-mode semiconductor laser sources [16] as well as using the optical phase locked loop (OPLL) configuration [17].

2.1.1. Optical heterodyning

The basic principle for generating microwave carriers based on the optical heterodyne technique represents a physical process called optical beating or frequency beating. A PD acting as an optical mixer allow us to obtain a photo-current, which results from the difference between two optical sources emitting at frequencies ω_1 and ω_2, where $|\omega_1 - \omega_2| << \omega_1, \omega_2$. This process is possible when both optical fields are overlapped with common polarization, illuminating a PD of responsivity R; the resulting photo-current is given by [18].

$$i = R\left[P_1 + P_2 + 2\sqrt{P_1 P_2}\cos\left((\omega_1 - \omega_2)t + (\phi_1 - \phi_2)\right) \right] \tag{1}$$

where P_1 and P_2 are the powers and correspond to DC terms, ϕ_1 and ϕ_2 are the phases of the two sources incident on the PD. The term $2\sqrt{P_1 P_2}\cos\left((\omega_1 - \omega_2)t + (\phi_1 - \phi_2)\right)$, usually called intermediate frequency (IF), represents the microwave carrier generated. Figure 2 shows the experimental setup used for generating microwave signals.

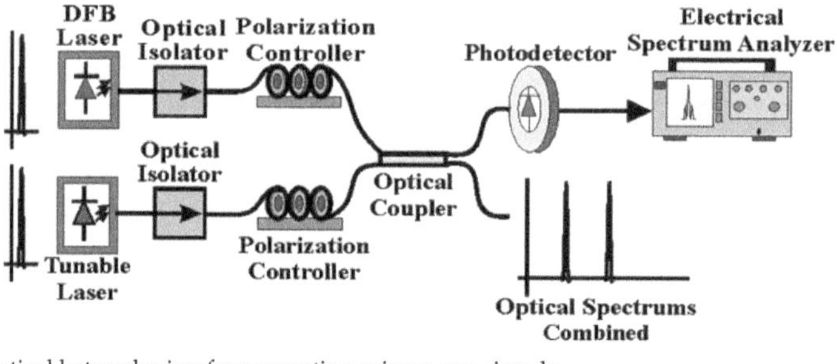

Figure 2. Optical heterodyning for generating microwave signals.

In this experiment two laser diodes emitting at different wavelengths are used. One of them is tunable and can be tuned over the C band with a channel spacing of 25 GHz, and the other is a fiber coupled DFB laser source with a central wavelength at 1550 nm. For the generation of the microwave signals, the outputs of both lasers are coupled to optical isolators to avoid a feedback into the lasers and consequential instabilities to the system. A pair of polarization controllers is used to minimize the angle between the polarization directions of both optical sources. Thus, the polarization of the light issued from each optical source is matched and therefore, there is no degradation of the power levels in the microwave signals generated in the PD. The output of each controller is launched to a 3 dB coupler to combine both optical spectrums. After that, an optical output signal is received by a fast PD. The resulting photo-current from the PD corresponds to the microwave beat signal which is analyzed with an Electrical Spectrum Analyzer (ESA). The other optical output resulting from an optical coupler is applied to an Optical Spectrum Analyzer (OSA) for monitoring the wavelength of the two beams. DFB laser sources provide the ability to control not only the output optical power of the fiber coupled laser diode, but also the precise the temperature at which the laser is operating. Both controls can be used to tune the fiber coupled laser diode to an optimum operating point, providing a stable output. In this way, we can observe that the wavelength of the DFB laser is shifting by varying its temperature with a scale of 1°C. Consequently, the beat signal frequency is continuously tuned in the bandwidth of the fast PD. Figure 3 illustrates the spectrums of three microwave signals generated with optical heterodyne technique. The generated signals are located at 2.4, 4.9 and 9.5 GHz when the temperature of the DFB laser was tuned at different values of temperatures. The frequency difference from both lasers can be expressed by [19]

$$\Delta f = \frac{c}{\lambda_1} - \frac{c}{\lambda_2} = \frac{c(\lambda_2 - \lambda_1)}{\lambda_1 \lambda_2} \approx \frac{c}{\lambda^2}|\Delta\lambda|^2 \tag{2}$$

where λ_1 and λ_2 are the wavelengths of the two beams, respectively, and $\Delta\lambda$ is the difference between the two wavelengths.

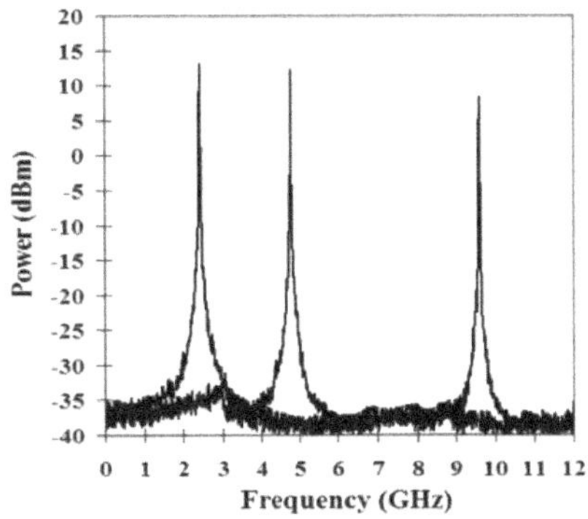

Figure 3. Microwave spectrums continuously tuned from 0 to 10 GHz.

Advantages

Using optical heterodyning, very high frequencies can be generated, limited only by the bandwidth of the photo-detector [20]. Furthermore, heterodyning yields high-detected power (higher link gain) and higher carrier-to-noise ration (CNR). This is so because all the optical powers of the two optical fields contribute to the power of the generated microwave signal.

Disadvantages

Beating two optical waves from two free-running laser diodes would lead to a microwave or mm-wave signal with high phase noise since the phases of the two optical waves are not correlated. They would also be transferred to the generated microwave or mm-wave signal as shown in Eq. (1). Therefore, the generated RF signal is very sensitive to phase noise. Since semiconductor lasers are prone to phase noise, extra measures to minimize the noise have to be taken. Techniques used to reduce phase noise sensitivity, OIL and OPLL, are discussed below.

2.1.2. Optical Injection Locking (OIL)

To generate a high-quality microwave or mm-wave signal, the phase terms of the two optical waves used for heterodyning must be highly correlated. The phase coherence of two laser diodes can be realized by using optical injection locking [21].

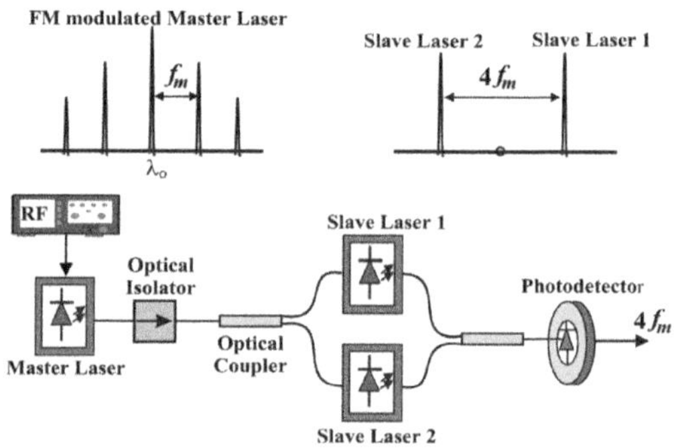

Figure 4. Optical Injection Locking.

Figure 4 shows an optical injection locking system that consists of one master laser and two slave lasers. As can be seen, RF reference is applied to the master laser. Due to frequency modulation (FM) at the master laser, an optical carrier and different orders of optical sidebands are generated at the output of the master laser. The signal at the output of the master laser is then injected into the two slave lasers. The two slave lasers are selected such that their free-running wavelengths are close to two sidebands, that is to say, the +2nd-order and -2nd-order sidebands in Figure 4. Therefore, the wavelengths of the two slave lasers are locked to the +2nd-order and -2nd-order sidebands, optical injection locking is thus achieved [21]. Since the two wavelengths from the two slave lasers are phase correlated, the beating of the two wavelengths at a PD would generate a beat note with low phase noise. In addition, depending on the design, the frequency of the beat note is equal to an integer multiple of the frequency of the RF reference applied to the master laser.

Advantages

One advantage of using OIL is that cheaper broad-linewidth lasers can still be used to generate stable narrow electrical linewidth signals. Secondly, OIL exhibits good phase noise suppression. Thirdly, because the slave laser locks on to a sub-harmonic of its resonance frequency, low frequency reference signals are used. OIL is simpler and easier to achieve.

Disadvantages

The major disadvantage of OIL is that it has a small frequency detuning range. Optimum phase noise suppression occurs only at one point of slave laser detuning relative to the free-running frequency.

2.1.3. Optical Phase Locked-Loops (OPLL)

Another approach to achieving phase coherence between two optical waves is to use an OPLL, in which the phase of one laser is actively locked to that of a second laser by an OPLL, as shown in Figure 5. This technique has been explored extensively in the past few years [22-24]. To achieve effective phase locking, the two lasers should be selected to have narrow linewidths and therefore have phase fluctuations only at low frequencies, which significantly eases the requirement for a very short feedback loop.

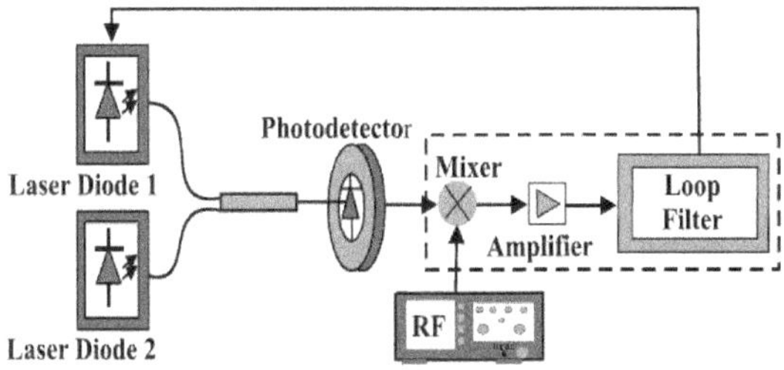

Figure 5. Schematic of an Optical Phase-Locked Loop.

As shown in Figure 5, a beat note is generated at the output of the photo-detector. The phase of the beat note is compared with that of an RF reference from a microwave generator at a mixer followed by a low-pass loop filter. The module in the dotted box is an electrical phase detector, with the output voltage being proportional to the phase difference between the beat note and the RF reference, which is an error voltage that is fed back to control the phase of one of the laser sources by changing the laser cavity length or the injection current. With a proper feedback loop gain and response time, the relative phase fluctuations between the two lasers are significantly reduced and the phase of the beat note is locked to the RF reference.

Advantages

Because OPLL techniques track small phase variations they are capable of producing high quality RF signals with narrow linewidth. OPLLs also have good temperature tracking capabilities. In addition, OPLLs exhibit a wide locking range.

Disadvantages

The major drawback of OPLLs is that they require far more complex laser structures such as 3-contact DFBs. This requirement is a direct consequence of the fact that to track the frequency perturbations of the master laser, the tuning rate of the slave laser must be sufficiently high. This also implies that sufficient feedback bandwidth must be available. The required feedback bandwidth is determined by the summed laser linewidth, the requirements for loop stability, and phase noise requirements placed on the optical microwave signal by the system in which the OPLL is to be applied. A wide feedback bandwidth is necessary if semiconductor lasers are used, because they have a large amount of phase noise. In order to achieve the wide feedback bandwidth, the loop-propagation delay must be small.

2.1.4. Microwave generation based on external modulation

In addition to the techniques using optical injection locking and OPLL, high-quality microwave signals can also be generated based on external modulation [25–27]. A method to generate a mm-wave signal using an external optical modulation technique was first proposed in 1992 [25]. A frequency-doubled electrical signal was optically generated by biasing the MZ-IM to suppress the even-order optical sidebands ($2f$ method). A 36 GHz mm-wave signal was generated when the MZ-IM was driven by an 18 GHz microwave signal. Such a system was employed for a remote delivery of video services [26]. In 1994, another method was proposed to generate a frequency-quadrupled electrical signal ($4f$ method). Instead of biasing the MZ-IM to suppress the even-order optical sidebands, the method in [10] was based on the quadratic response of an optical intensity modulator. The optical carrier and the first and third-order optical sidebands were suppressed by adjusting the drive signal level. A 60 GHz millimeter-wave signal was generated when a 15 GHz drive signal was applied to the MZ-IM.

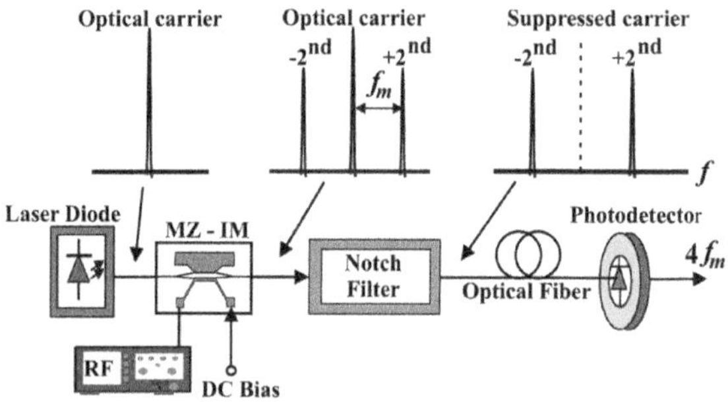

Figure 6. Microwave signal generation based on external modulation.

Figure 6 shows a system to generate a continuously tunable mm-wave signal based on external modulation using a MZ-IM and a wavelength-fixed optical filter [28]. The significance of the technique is that no tunable optical filter is required, which significantly simplifies the system implementation. As can be seen from Figure 6, the system consists of a MZ-IM that is biased at the maximum transmission point of the transfer function to suppress the odd-order optical sidebands. A fiber Bragg grating (FBG) serving as a wavelength-fixed notch filter is then used to filter out the optical carrier. A stable, low-phase noise mm-wave signal that has four times the frequency of the RF drive signal is generated at the output of the PD.

Advantages of the 2f and 4f methods

Both the $2f$ and $4f$ methods rely on optical heterodyning. Therefore, they are capable of generating high frequency mm-waves. Since the same laser generates both optical fields, the phase noise is highly correlated resulting in very narrow linewidth mm-waves. In fact, the performance of these methods in terms of phase noise is comparable to the OPLL system. A key advantage of these approaches in [26,27] is that an optical modulator with a maximum operating frequency of 15 GHz can generate a millimeter-wave signal up to 60 GHz.

Disadvantage of the 2f and 4f methods

The main disadvantage of these techniques is that both approaches use optical filters to select the two optical sidebands. Another disadvantage is that in order to generate tunable mm-wave signals, a tunable optical filter must be used. These two disadvantages significantly increase the complexity and the cost of the system.

2.1.5. Microwave generation using a dual-wavelength laser

Microwave signals can also be generated using a dual wavelength laser source with the two wavelengths separated at a desired frequency [29]. It is different from the techniques of optical injection locking and the OPLL because, the two wavelengths from a dual wavelength laser source are not locked in phase. However, due to the fact that the two wavelengths are generated inside same cavity, the phase correlation between the two wavelengths is better than that using two free-running laser sources. To ensure that the two wavelengths are in single-longitudinal mode, a dual-band filter with ultra-narrow passbands must be used, thus, limiting the number of longitudinal modes to one in each passband. In the experimental demonstration [29], the ultra-narrow dual-band filter was a dual-wavelength ultra-narrow transmission band FBG with two ultra-narrow transmission bands, which was designed and fabricated based on the equivalent phase-shift (EPS) technique [30].

Advantages

The advantage of using a dual-wavelength laser source to generate a microwave or mm-wave signal is that the system is simpler with no need for a microwave reference source, which can significantly reduce the system cost. Due to the fact that the two wavelengths are generated from the same cavity, the phase correlation between the two wavelengths is better than that using two free-running laser sources. The spectral width of the generated microwave signals is as small as 80 kHz with frequency stability better than 1 MHz in the free-running mode at room temperature.

Disadvantages

Due to the fact that three dual-wavelength ultranarrow transmission-bands can be incorporated into the laser cavity by using this technique, then microwave signals continuously tuned can not be generated.

3. Relaxation oscillation frequency

The relaxation oscillation frequency is the frequency of the quasi-sinusoidal oscillations around the steady state output power amplitude that occur during the disturbance of a continuously operating laser. Such oscillations are characterized by exponential damping. The phenomenon of relaxation oscillations has an effect on the performance of an unmodulated laser as well, by producing sidebands in the field fluctuation spectrum. They may be found directly from a small-signal measurement. Figure 7 illustrates the typical standard setup which is used for characterization of laser diodes. We assume that the laser diode is excited by an AC current source with amplitude I_1 and electrical frequency ω, biased at the constant current I_0 by means of a bias-tee. The modulated laser output is coupled into an optical fiber and then into a semiconductor PIN diode, biased through a simple bias network. The detected signal is fed into a spectrum analyzer.

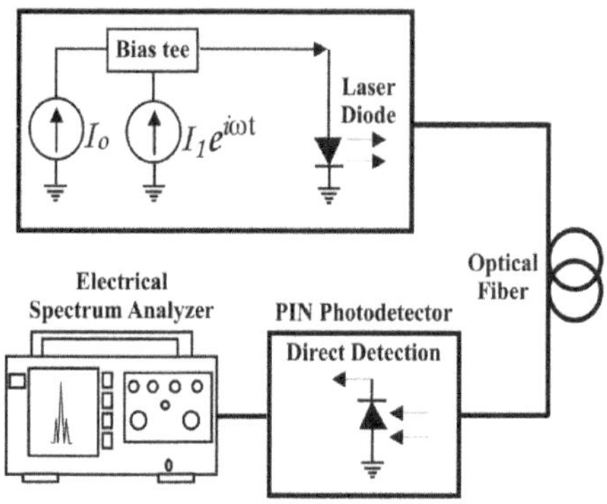

Figure 7. Typical experimental setup for measurement of relaxation oscillation frequencies of laser diodes.

3.1. Analysis of relaxation oscillation frequency

The employed physical model for analyzing the harmonic power can be explained by considering the small signal modulation analysis where the total electric current injected into the laser diode is given by:

$$I = I_0 + I_1 e^{j\omega t} \tag{3}$$

If photon and electron densities inside the active region of a semiconductor laser are denoted, respectively, by N_P and N_E, then expanding N_P and N_E as a Fourier series we find that these densities can be written as:

$$N_E = \sum_{n=0}^{\infty} N_{En} e^{jn\omega t} \tag{4}$$

$$N_P = \sum_{n=0}^{\infty} N_{Pn} e^{jn\omega t} \tag{5}$$

where N_{Pn} and N_{En} are the corresponding expansion coefficients of the photon and electron densities, respectively. On the other hand, the well-known rate equations for a single mode laser diode are [31]

$$\frac{dN_E}{dt} = \eta_i \frac{I}{qV} - \frac{N_E}{\tau_e} - V_g a\left(N_E - N_{tr}\right) N_P \tag{6}$$

$$\frac{dN_P}{dt} = \Gamma V_g a\left(N_E - N_{tr}\right) N_P - \frac{N_P}{\tau_p} \tag{7}$$

here η_i is the efficiency coefficient; τ_e is the carrier lifetime; a is the differential gain; N_{tr} is the threshold electron density; Γ is the cavity confinement factor; V_g is the cavity volume; and τ_p is the photon lifetime. In order to obtain information related with harmonic contents in optical output power of a laser diode, it is necessary to start our approach by substituting Eqs. (3) to (5) in Eq. (6) which results in

$$\begin{aligned} j\omega \sum_{n=1}^{\infty} N_{En} n e^{jn\omega t} &= \eta_i \frac{I_0 + I_1 e^{j\omega t}}{qV} - \frac{1}{\tau_e} \sum_{n=0}^{\infty} N_{En} e^{jn\omega t} \\ &-V_g a\left(\sum_{n=0}^{\infty} N_{En} e^{jn\omega t} - N_{tr} \right) \sum_{n=0}^{\infty} N_{Pn} e^{jn\omega t} \end{aligned} \tag{8}$$

In a similar way, if we substitute Eqs. (4) and (5) in Eq. (7) we obtain

$$\begin{aligned} j\omega \sum_{n=1}^{\infty} N_{Pn} n e^{jn\omega t} &= \Gamma V_g a\left(\sum_{n=0}^{\infty} N_{En} e^{jn\omega t} - N_{tr} \right) \sum_{n=0}^{\infty} N_{Pn} e^{jn\omega t} \\ &- \frac{1}{\tau_p} \sum_{n=0}^{\infty} N_{Pn} e^{jn\omega t} \end{aligned} \tag{9}$$

From DC Analysis (n = 0), however, we have:

$$N_{E_0} = \frac{\eta_i \frac{I_0}{qV} + V_g a N_{tr} N_{P_0}}{\frac{1}{\tau_e} + V_g a N_{P_0}} \tag{10}$$

$$\frac{1}{\Gamma \tau_p} = V_g a \left(N_{E_0} - N_{tr} \right) \tag{11}$$

The first harmonic of the photon density is generated directly by the first harmonic of the current applied to the laser diode, and we easily obtain [32]

$$N_{E_1} = \frac{\eta_i \frac{I_1}{qV}}{j\omega + \frac{1}{\tau_e} + V_g a N_{P_0} \left(1 + \frac{1}{j\omega \tau_p} \right)} \tag{12}$$

$$N_{P_1} = \frac{\Gamma V_g a N_{P_0}}{j\omega} N_{E_1} \tag{13}$$

Higher harmonics are found from recursive equations which in the general case for an arbitrary integer, $k > 1$, may be written as:

$$jk\omega N_{Ek} = -\frac{N_{Ek}}{\tau_e} - V_g a \left[\left(N_{E_0} - N_{tr} \right) N_{P_k} + N_{Ek} N_{P_0} + \sum_{n=1}^{k-1} N_{En} N_{Pk-n} \right] \tag{14}$$

$$jk\omega N_{Pk} = \Gamma V_g a \left(N_{Ek} N_{P_0} + \sum_{n=1}^{k-1} N_{En} N_{Pk-n} \right) \tag{15}$$

Hence, the kth harmonic of the photon density may be calculated once all the lower harmonics of photon and electron densities are known. If we define M_k as the parameter which relates the kth harmonic to the lower ones [33]

$$M_k = \sum_{n=1}^{k-1} N_{En} N_{Pk-n} \tag{16}$$

Therefore, we can find the kth harmonics of photon and electron densities as

$$N_{E_k} = \frac{-V_g a \left(1 + \frac{1}{jk\omega \tau_p} \right)}{jk\omega + \frac{1}{\tau_e} + V_g a N_{P_0} \left(1 + \frac{1}{jk\omega \tau_p} \right)} M_k \tag{17}$$

$$N_{P_k} = \frac{\Gamma V_g a}{jk\omega}\left(N_{P_0} N_{E_k} + M_k\right) \tag{18}$$

The frequency, ω_{RO}, at which the amplitude of the power harmonic content reaches its maximum is called the Relaxation Resonant Frequency. From Eqs. (12) and (13), the primary harmonic of the photon density is obtained as [32]

$$N_{P_1} = \frac{\Gamma V_g a N_{P_0}}{j\omega} \frac{\eta_i \dfrac{I_1}{qV}}{j\omega + \dfrac{1}{\tau_e} + V_g a N_{P_0}\left(1 + \dfrac{1}{j\omega\tau_p}\right)} \tag{19}$$

On the other hand, if the optical output power for the kth harmonic is given by $P_k = P_{outk} = \frac{h\upsilon V_p}{\tau_m} N_{pk}$, where τ_m is the mirror loss time, V_p represents the cavity volume and $h\upsilon$ is the energy per photon, then by using this last equation we can obtain the optical output power for the first harmonic, written as

$$P_1(j\omega) = \frac{h\upsilon V_p}{\tau_m}\left(\frac{\Gamma V_g a N_{P_0} \eta_i \dfrac{I_1}{qV}}{-\omega^2 + \dfrac{j\omega}{\tau_e} + \dfrac{V_g a N_{P_0}}{\tau_p}\left(j\omega\tau_p + 1\right)}\right) \tag{20}$$

The cavity confinement factor, Γ, is equal to the ratio, V/V_p; therefore, the above can be written as:

$$\frac{P_1(j\omega)}{I_1(j\omega)} = \frac{h\upsilon V_g a N_{P_0} \eta_i \dfrac{1}{\tau_m q}}{-\omega^2 + \dfrac{j\omega}{\tau_e} + \dfrac{V_g a N_{P_0}}{\tau_p}\left(j\omega\tau_p + 1\right)} \tag{21}$$

By separating the real and imaginary parts of the denominator of Eq. (21), we finally get [33]

$$\frac{P_1(j\omega)}{I_1(j\omega)} = \frac{h\upsilon V_g a N_{P_0} \eta_i \dfrac{1}{\tau_m q}}{\dfrac{V_g a N_{P_0}}{\tau_p} - \omega^2 + j\omega\left(\dfrac{1}{\tau_e} + V_g a N_{P_0}\right)} \tag{22}$$

This expression clearly has two poles at:

$$\omega = \pm\left[\frac{1}{\tau_p}V_g aN_{P_0} - \frac{1}{4}\left(\frac{1}{\tau_e} + V_g aN_{P_0}\right)^2\right]^{\frac{1}{2}} + j\frac{1}{2}\left(\frac{1}{\tau_e} + V_g aN_{P_0}\right) \tag{23}$$

which are located in the first and second quadrants of the complex ω plane. We now notice that the real part of the denominator of Eq. (22) becomes zero at the frequency:

$$\omega = \omega_{RO}^2 = \frac{V_g aN_{P_0}}{\tau_p} \tag{24}$$

where $\omega_{RO} = 2\pi f_{RO}$. Since usually $\frac{1}{\tau_p} + V_g aN_{P_0} << \omega_{RO}$ holds, we conclude that the magnitude of Eq. (22) also reaches a maximum when $\omega \approx \omega_{RO}$. Hence, the denominator becomes purely imaginary, and the primary power harmonic content reaches its resonance mode. The optimum modulation frequency is the resonant frequency, which is determined from the bias current. Therefore, the optimal operation point for modulation where the power of the primary harmonic reaches its maximum must lie on the curve obtained from the following [34].

$$\omega_{RO} = \sqrt{\frac{V_g a}{qV_p}\eta_i\left(I_0 - I_{th}\right)} \tag{25}$$

From the laser rate equations it is clear that the relaxation oscillation frequency should be proportional to the square root of the difference of the injection current and the threshold current. Thus, from Eq. (25) we can see that there is a linear relationship between relaxation oscillation frequency and the square root of the difference between the injection current and the threshold current. On the other hand, when the laser current density is lightly increased above its threshold value, the laser output power may be written as [35].

$$P_o = \eta_d(I_{in} - I_{th})\frac{1.24}{\lambda_o} \tag{26}$$

where $\eta_d \frac{1.24}{\lambda_o}$ is called the differential responsivity of the laser (W/A).

4. Operation principle of direct detection

When a laser is biased in the threshold current region as shown in Figure 8, the laser intensity presents oscillations know as relaxation frequencies and can be seen as sidebands on both sides of the main laser line. In this case the electric field emitted by the laser diode can be represented as a directly modulated optical signal. After passing through an optical fiber, the optical field incident on the PD can be expressed as [36, 37]

$$E(t) = E_o \cos(2\pi f_o t) + E_{RO} \cos 2\pi(f_o + f_{RO})t + E_{RO} \cos 2\pi(f_o - f_{RO})t \tag{27}$$

Where f_o is the optical frequency of the main laser line, and both $f_o + f_{RO}$ and $f_o - f_{RO}$ represent optical sidebands. This method of detecting optical fields as shown in Figure 8 is called direct detection.

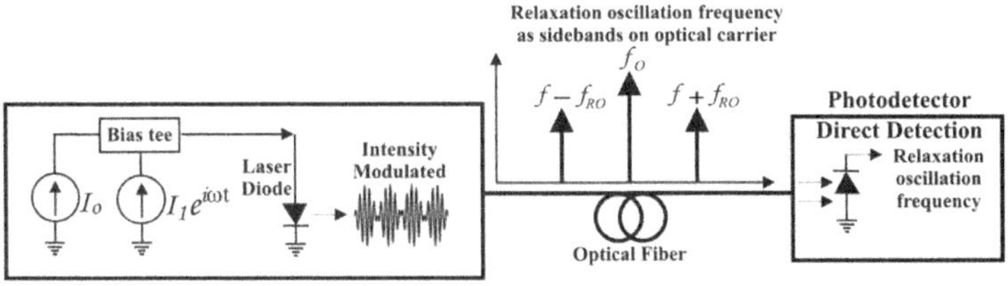

Figure 8. Direct detection to generate relaxation oscillation frequency when laser is operated in low threshold region.

The photo-detector responds to the power in the total field that it collects. The photo-current generated in the detector is proportional to the squared magnitude of the field and it is given by [38]

$$i(t) = \Re |E(t)|^2 = \Re P(t) \tag{28}$$

where $P(t)$ is the optical power and $\Re$ is the detector responsivity given by [38]

$$\Re = \frac{\eta q}{h\upsilon}\left[A/W\right] \tag{29}$$

Where η $(0 < \eta \leq 1)$ is the detector quantum efficiency, the parameters q and $h\upsilon$ are electronic charge (1.6021 x 10^{-19} C) and photon energy (h = 6.6256 x 10^{-34} J, $\upsilon = c/\lambda$) respectively. Substituting Eq. (27) into Eq. (28), and by using trigonometric identities, we obtain

$$\begin{aligned} i(t) &= \Re\left|E_o \cdot \cos(2\pi f_o t) + E_{RO}\left[\cos 2\pi(f_o + f_{RO})t + \cos 2\pi(f_o - f_{RO})t\right]\right|^2 \\ &= \frac{E_o^2}{2} + E_{RO}^2 + \frac{E_o^2}{2}\cos^2 2\pi(2f_o t) + 2E_o E_{RO}\cos 2\pi(f_{RO})t \\ &+ E_o E_{RO}\left[\cos 2\pi(2f_o + f_{RO})t + \cos 2\pi(2f_o - f_{RO})t\right] + \frac{E_{RO}^2}{2}\cos 2\pi(2f_o - 2f_{RO})t \\ &+ \frac{E_{RO}^2}{2}\cos 2\pi(2f_o + 2f_{RO})t + E_{RO}^2\left[\cos 2\pi(2f_o)t + \cos 2\pi(2f_{RO})t\right] \end{aligned} \tag{30}$$

Notice that there are many terms in the detected signal expression where the PD cannot follow currents that are varying at $2f_o$, $2f_o + f_{RO}$, $2f_o - f_{RO}$, $2f_o + 2f_{RO}$ and $2f_o - 2f_{RO}$ so these currents average to zero. The average current from the PD will consist of four terms.

$$i(t) = \Re\left[\frac{E_o^2}{2} + E_{RO}^2 + 2E_oE_{RO}\cos 2\pi(f_{RO})t + E_{RO}^2\cos 2\pi(2f_{RO})t\right] \tag{31}$$

The photo-current generated at the PD is found in a similar way as

$$i(t) = \Re\left[\frac{P_o}{2} + P_{RO} + 2\sqrt{P_oP_{RO}}\cos 2\pi(f_{RO})t + P_{RO}\cos 2\pi(2f_{RO})t\right] \tag{32}$$

The first two terms correspond to DC photo-currents and contain no information, but they do contribute to the shot noise of the signal detected. The next two terms correspond to the signal information which contains the relaxation oscillation frequency f_{RO} and its harmonic located at $2f_{RO}$.

5. Microwave photonic link by using external modulation

The photonic link described in this section is shown in Figure 9. The optical source is a continuous-wave DFB laser, followed by an EDFA. Amplified light is launched into the MZ-IM. Microwave modulation is imposed on the optical carrier through the MZ-IM.

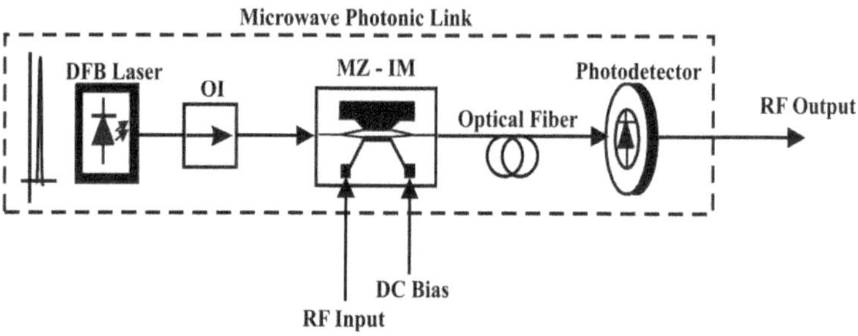

Figure 9. Microwave photonic link.

The modulated signal passes through optical fiber and is then incident upon the photodetector. Photodetector output can be connected to a microwave spectrum analyzer to observe gain, noise figure, and distortion of the microwave signal obtained at the end of the link.

For a modulator with half-wave voltage of V_π, signal voltage V , input power Pin, and insertion loss α, the modulated output power can be written [39, 40]

$$P_{out} = \frac{\alpha P_{in}}{2}\left[1 + \cos\left(\frac{\pi V}{V_\pi}\right)\right] \tag{33}$$

This expression can be examined in more detail by separating the signal voltage into a DC bias, V_b, and a modulation voltage, V_m.

$$P_{out} = \frac{\alpha P_{in}}{2}\left[1 + \cos\left(\frac{\pi V_b}{V_\pi} + \frac{\pi V_m}{V_\pi}\right)\right] \tag{34}$$

Using a trigonometric identity for the sum of cosines, the output power can be rewritten as:

$$P_{out} = \frac{\alpha P_{in}}{2}\left[1 + \cos\left(\frac{\pi V_b}{V_\pi}\right)\cos\left(\frac{\pi V_m}{V_\pi}\right) - \sin\left(\frac{\pi V_b}{V_\pi}\right)\sin\left(\frac{\pi V_m}{V_\pi}\right)\right] \tag{35}$$

Typically, the DC bias point of the MZ-IM is chosen to bring the modulator into quadrature. This is accomplished by operating at $V_b = V_\pi /2$. At this bias voltage, the transfer curve is linear and even order harmonics are suppressed.

Deviation from the quadrature bias point can be described by an angle ϕ.

$$P_{out} = \frac{\alpha P_{in}}{2}\left[1 + \cos\left(\frac{\pi}{2} + \phi\right)\cos\left(\frac{\pi V_m}{V_\pi}\right) - \sin\left(\frac{\pi}{2} + \phi\right)\sin\left(\frac{\pi V_m}{V_\pi}\right)\right] \tag{36}$$

$$P_{out} = \frac{\alpha P_{in}}{2}\left[1 - \sin\phi\cos\left(\frac{\pi V_m}{V_\pi}\right) - \cos\phi\sin\left(\frac{\pi V_m}{V_\pi}\right)\right] \tag{37}$$

At $\phi = 0$, $V_b = V_\pi/2$ and the modulator is in quadrature. This corresponds to the typical linear transmission regime for an electro-optic modulator. The quadrature output power from the modulator is denoted by P_Q.

$$P_Q = \frac{\alpha P_{in}}{2}\left[1 - \sin\left(\frac{\pi V_m}{V_\pi}\right)\right] \tag{38}$$

For single applications, the DC bias point of the MZ-IM can be shifted away from the quadrature point. Even-order distortion is increased, but these terms lie outside the frequency range of interest. As the bias point is shifted, shot noise falls linearly with transmission and RIN is reduced quadratically. However, the link gain is only reduced at the rate of $\sin^2 \phi$. For links dominated by RIN, shifting the bias point toward the transmission null reduces the link noise figure. This improvement continues until the shot noise limit is reached. Consider a MZ-IM biased near $\phi = 90^\circ$. This corresponds to an operating regime with reduced output power. For simplicity, let $\phi = 90^\circ - \delta$ so that $\phi = 0^\circ$ corresponds to the quadrature operating point and $\delta = 0^\circ$ corresponds to the transmission null. These operating regimes and angles are summarized graphically in Figure 10.

By substituting δ for $\pi/2 - \phi$, the modulator output power can be written as [40]

$$P_{out} = \frac{\alpha P_{in}}{2}\left[1 - \cos\delta\cos\left(\frac{\pi V_m}{V_\pi}\right) - \sin\delta\sin\left(\frac{\pi V_m}{V_\pi}\right)\right] \tag{39}$$

The first bracketed term represents continuous wave output power. The second term contains even-order harmonic components. The third term contains the linear signal of interest as well as odd-order harmonic components. Note that for $\delta = 90^{\circ}$, Pout = P_Q as expected. For $\delta << 1$, the output power can be approximated by using a Taylor series expansion to replace sin δ with δ and cos δ with $1 - \delta^2/2$.

$$P_{out} \approx \frac{\alpha P_{in}}{2}\left[1-\left(1-\frac{\delta^2}{2}\right)\cos\left(\frac{\pi V_m}{V_\pi}\right)-\delta\sin\left(\frac{\pi V_m}{V_\pi}\right)\right] \tag{40}$$

The harmonic components can be identified by assuming small signal operation with $\pi V_m/V_\pi << 1$. For the cosine function, the Taylor series expansion can be truncated after the first term. This truncation is justified by operating the modulator over a single octave so that even-order distortion can be neglected.

$$P_{out} \approx \frac{\alpha P_{in}}{2}\left[1-\left(1-\frac{\delta^2}{2}\right)-\delta\sin\left(\frac{\pi V_m}{V_\pi}\right)\right] \tag{41}$$

The sine term can be expressed using a third-order Taylor series expansion.

$$P_{out} \approx \frac{\alpha P_{in}}{2}\left\{\frac{\delta^2}{2}-\delta\left[\frac{\pi V_m}{V_\pi}-\frac{1}{6}\left(\frac{\pi V_m}{V_\pi}\right)^3\cdots\right]\right\} \tag{42}$$

From this expression, it can be seen that the DC term decreases more rapidly than the modulation term for a given increase in δ. By shifting the DC bias point away from quadrature, carrier suppression can be achieved. This can create a double sideband suppressed carrier (DSB-SC) modulation format.

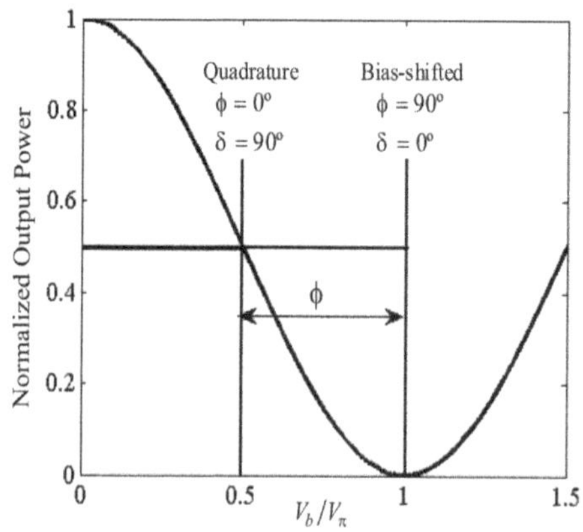

Figure 10. Quadrature and bias-shifted operating regimes.

6. Passive components for microwave photonics

This section presents an overview of the passive components for microwave photonics such as optical fiber, optical isolator, polarization control and optical fiber couplers.

6.1. Optical fiber

The optical fibers used in analog photonic links have a very simple structure. They consist of two sections: the glass core and the cladding layer as shown in Figure 11. The core is a cylindrical structure, and the cladding is a cylinder without a core. Core and cladding have different refractive indices, with the core having a refractive index, n_1, which is slightly higher than that of the cladding, n_2. This difference in refractive indices enables the fiber to guide the light. Because of this guiding property, the fiber is also referred to as an "optical waveguide." Besides these two sections there is also a further layer known as the secondary cladding that does not participate in the propagation but gives the fiber a minimum level of protection.

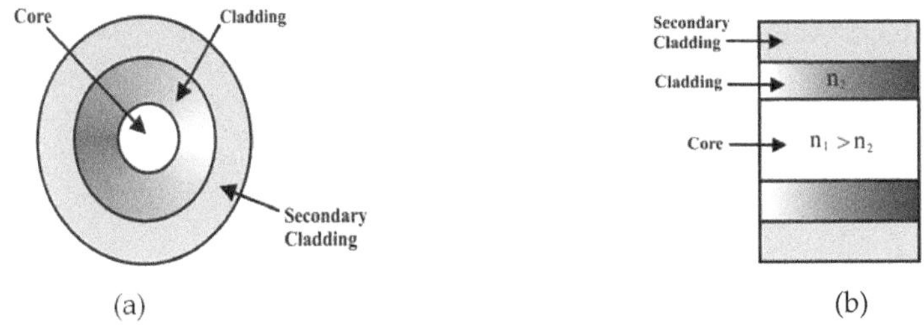

Figure 11. Typical optical fiber (a) cross section, (b) longitudinal cross section.

The basics of light propagation can be discussed with the use of geometric optics. The basic law of light guidance is Snell's law as shown in Figure 12(a). Consider two dielectric media with different refractive indices and with $n_1 > n_2$ and that are in perfect contact. At the interface between the two dielectrics, the incident and refracted rays satisfy Snell's law of refraction and can be written by

$$n_1 \sin\phi_1 = n_2 \sin\phi_2 \tag{43}$$

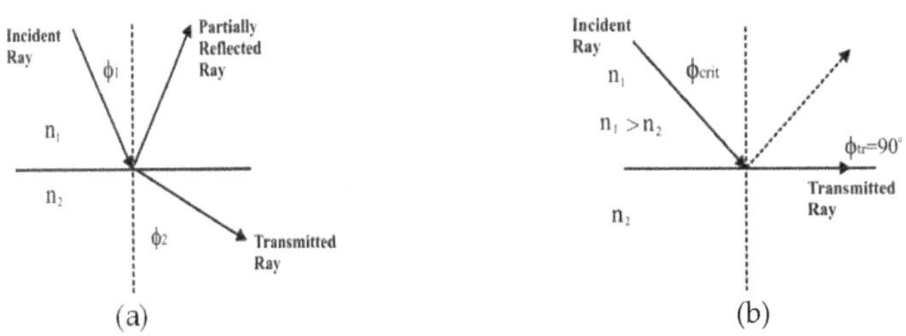

Figure 12. Snell's law, (a) showing transmitted ray, (b) showing total reflection when critical angle is achieved.

In addition to the refracted ray there is a small amount of reflected light in the medium with refractive index n_1. Because $n_2 < n_1$ then always $\phi_2 > \phi_1$. As the angle of the incident ray increases there is an angle at which the refracted ray emerges parallel to the interface between the two dielectrics as shown in Figure 12(b). This angle is known as the critical angle, ϕ_{crit}, and from Snell's law is given by

$$\phi_{crit} = \sin^{-1}\left(\frac{n_2}{n_1}\right) \tag{44}$$

For a ray to be launched into the fiber and propagated, it must arrive at the interface between the two media (with different refractive indices) at an angle that is at minimum equal to ϕ_{crit}. Figure 13 illustrates the geometry for the derivation of the acceptance angle. To satisfy the condition for total internal reflection, the ray arriving at the interface between the fiber and outside medium, like air, must have an angle of incidence less than θ_{acc}, otherwise the internal angle will not satisfy the condition for total reflection, and the energy of the ray will be lost in the cladding. From Snell's law at the interface we obtain

$$n_0 \sin\theta_1 = n_1 \sin\theta_2 \tag{45}$$

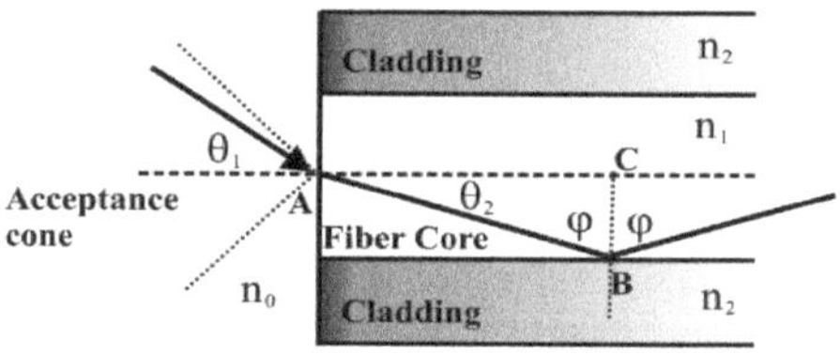

Figure 13. Geometry for the derivation of the acceptance angle.

From the right triangle ABC in Figure 13, the angle $\varphi = \pi/2 - \theta_2$. Substituting θ_2 into Eq. (45), we obtain

$$n_o \sin\theta_1 = n_1 \cos\varphi \tag{46}$$

In the limit as the incident angle, θ_1 approaches θ_{acc}, the internal angle approaches the critical angle for total reflection, ϕ_{crit}. Then, by introducing the trigonometric relation $\sin^2\varphi + \cos^2\varphi = 1$ into Eq. (46), we obtain

$$n_o \sin\theta_1 = n_1 \cos\varphi = n_1\left(1 - \sin^2\varphi\right)^{1/2} = n_1\left(1 - \left(\frac{n_2}{n_1}\right)^2\right)^{1/2} = \left(n_1^2 - n_2^2\right)^{1/2} \tag{47}$$

This equation defines the angle within which the fiber can accept and propagate light and is known as the "Numerical Aperture" (NA).

$$NA = n_o \sin\theta_{acc} = \left(n_1^2 - n_2^2\right)^{1/2} \tag{48}$$

This equation states that for all angles of incident where the inequality $0 \le \theta_1 \le \theta_{acc}$ is satisfied, the incident ray will propagate within the fiber. The parameter *NA* expresses the propensity of the fiber to accept and propagate light within the solid cone defined by an angle, $2\theta_{acc}$. The equation for the NA can be also expressed in terms of the fractional change in the index of refraction of core and cladding Δ. This refractive index difference is given by

$$\Delta = \frac{n_1^2 - n_2^2}{2n_1^2} \approx \frac{n_1 - n_2}{n_1} \tag{49}$$

With these simplifications the NA can now be written as

$$NA = n_1 \left(2\Delta\right)^{1/2} \tag{50}$$

6.1.1. *Attenuation in fiber*

Attenuation in optical fiber leads to a reduction of the signal power as the signal propagates over some distance. When determining the maximum distance that a signal can propagate for a given transmitter power and receiver sensitivity, one must consider attenuation. Let $P(L)$ be the power of the optical pulse at distance L km from the transmitter and α_{fiber} be the attenuation constant of the fiber (in dB/km). Attenuation is characterized by [41]

$$P_R(L) = 10^{-\alpha_{fiber} L/10} P_T(0) \tag{51}$$

Where $P_T(0)$ is the optical power at the transmitter and the received optical power for a link length of L km, is $P_R(L)$. The maximum distance between the transmitter and the receiver depends more heavily on the constant α_{fiber} than on the optical power launched by the transmitter. This distance is given by

$$L_{\max} = \frac{10}{\alpha_{fiber}} \log_{10} \frac{P_T(0)}{P_R} \tag{52}$$

6.2. Optical isolator

An isolator inside an optical fiber link allows the propagation of the light in only one direction (from left to right as is indicated in Figure 14), with minimum loss, avoiding reflected light in sense inverse. Depending on the design, the reflected optical signal has an attenuation value of 40 to 70 dB [42].

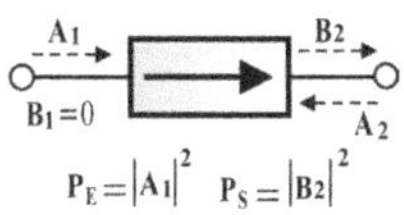

Figure 14. Scheme and flux of signals across an optical isolator.

The optical isolator is a network of 2 ports, and as such, it can be described by a scattering matrix 2x2

$$S = \begin{pmatrix} S_{11} & S_{12} \\ S_{21} & S_{22} \end{pmatrix} \tag{53}$$

This device is characterized by two parameters: the insertion losses (L), and the isolation relationship (I), both expressed in dB. These parameters are related by the optical power at the input and the output of the device, the optical fields in their two ports, and their scattering parameters by means of the following relationships:

$$L = 10Log\left(\frac{P_E}{P_S}\right) = 10Log\left(\frac{1-|S_{11}|^2}{|S_{21}|^2}\right) \tag{54}$$

$$I = 10Log\left(\frac{|A_2|^2}{|B_1|^2}\right) = 10Log\left(\frac{1}{|S_{12}|^2}\right) \tag{55}$$

In an ideal isolator, insertion losses are L = 0 dB, and the isolation relationship is $I = \infty$. Generally, optical isolators are placed at the output of the optical sources in order to avoid reflections of light from returning back to the source avoiding in this way instabilities. Furthermore, isolators are used by the Doped Fiber Amplifiers with the aim of avoiding an oscillating behavior. Figure 15 illustrates a basic diagram of an optical isolator and is used to explain the basic operation of this device [43]. Light propagating from left to right is linearly polarized by means of a vertical polarizer "A" placed at the input port of the isolator. Then the polarization plane of the light is rotated 45° by means of a polarization rotator. Finally, the light is transmitted through a linear polarizer "B" whose axis forms an angle of 45° in relation to the axis of the polarizer "A". The reflecting signal (light propagating from right to left), is linearly polarized into the polarizer "B" 45° with respect to the axis of the polarizer "A". Subsequently, a rotation of the plane of linear polarization at 45° is made, and by consequence at the input of the polarizer "A," its polarization plane forms an angle of 90° (orthogonal) with its axis of polarization. By consequence, the light is blocked.

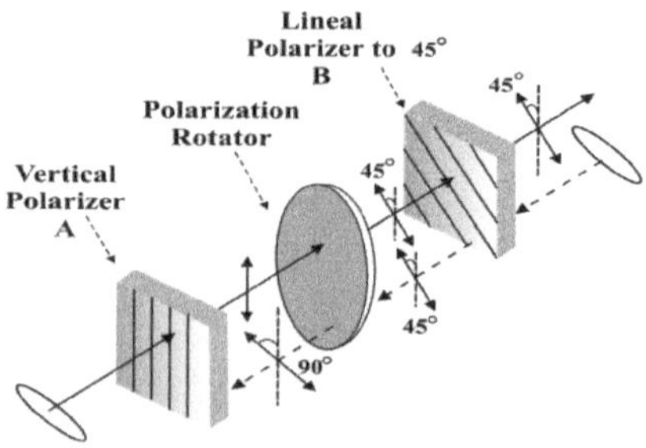

Figure 15. Principle of operation of an optical isolator.

6.3. Polarization rotator

A polarization rotator rotates the polarization plane of an optical field linearly polarized at its input at an angle θ [43]. If the input signal is an optical field linearly polarized forming an angle θ_1 with the axis x, then the Jones Matrix can be expressed as

$$J = \begin{bmatrix} \cos\theta_1 \\ \sin\theta_1 \end{bmatrix} \tag{56}$$

The output signal is an optical field linearly polarized forming an angle θ_2, and by consequence the Jones matrix corresponding to the rotator is given as

$$M = \begin{bmatrix} \cos\theta & -\sin\theta \\ \sin\theta & \cos\theta \end{bmatrix} \tag{57}$$

where: $\theta = \theta_2 - \theta_1$.

6.4. Polarization control

A polarization control changes the polarization state of the light. Currently in the systems, the polarization controls can be inserted in the arms of an optical signal or in the arms of a signal of an optical local oscillator in order to maintain or to match the polarization state of both signals. Figure 16 shows a scheme corresponding to a polarization control. Basically, this scheme is composed of a monomode optical fiber crossing three plates. The optical fiber (from left to right in the figure) is rolled in a circular way one time in the first plate, 2 times in the second plate, and finally 1 time in the third plate.

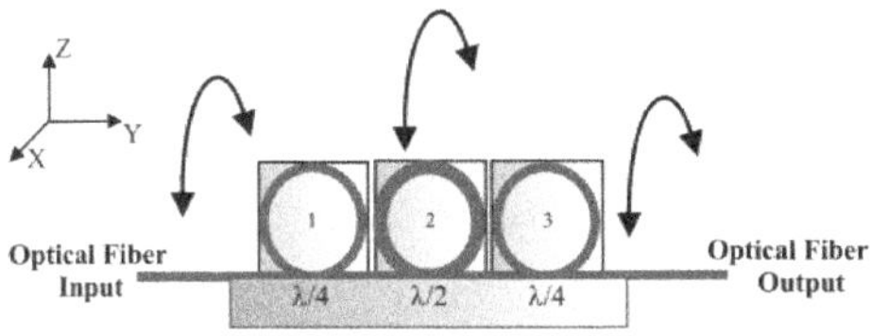

Figure 16. Polarization Control.

For a particular wavelength and an appropriate diameter of the plates, phase changes are introduced in the incident optical field due to the stress applied to the optical fiber, and by consequence, a birefringence is induced. The birefringence degree is function of the material of which the optical fiber, the plate diameter, and the cladding diameter of the optical fiber is composed, and it is defined as [44]

$$\Delta n = a(r / R)^2 \tag{58}$$

where α is a parameter related to the fiber material, r is the cladding diameter, and R is the plate diameter defined

$$R = 8\pi a r^2 / \lambda \tag{59}$$

When an optical field is propagated by the fiber control, this field will suffer a phase change of 90° at the output of the first plate, and by consequence this plate has a similar behavior as a retarder plate of λ/4 [44]. The second plate acts as a retarder plate of λ/2 because at the output the optical field has a phase change of 180° This phase change can be expressed as [44].

$$\Delta\phi = (2\pi / \lambda)\Delta n 2\pi R \tag{60}$$

Taking into account this effect and assuming that the incident optical field has an elliptical polarization state, the first retarder plate λ/4 changes this polarization state to a lineal polarization state. The second retarder plate λ/2 rotates the polarization state toward a specific direction. Finally, the third plate λ/4 changes the linear polarization into a desired polarization state.

6.5. Optical fiber couplers

The optical system communications requires devices capable of coupling light from one fiber or various fibers to others. Typically, this coupling of light has been achieved by using beam splitters fabricated on bulk devices. However, for the case of guided signal transmission this task can be achieved by using couplers [45]. A coupler is a passive device capable of distributing energy from M input guides to N output guides [45]. In this sense, two technologies have been developed to fabricate couplers. One of these is based on optical fibers whereas the other is based on bulk optics, however the principles of operation are similar in both cases. Currently, the coupler most commonly used in optical system communications is the directional coupler 2x2. Figure 17 illustrates a scheme of this type of coupler.

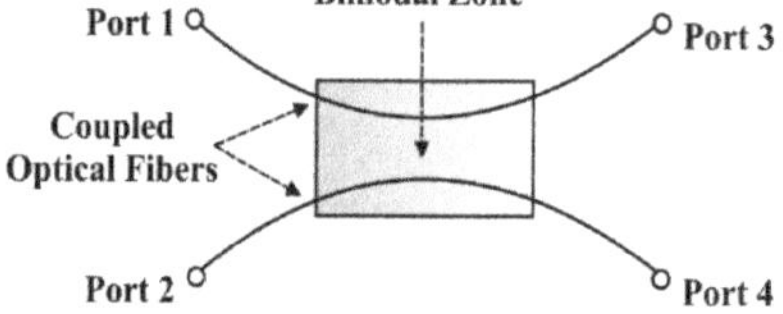

Figure 17. Scheme of an optical coupler 2x2.

From this figure, it is clearly appreciable that this is a 4-port device. In technology of optical fibers, this device can be of two types: (a) Evanescent field [46, 47], and (b) Modal interference. In the first case, the evanescent mode corresponding to the fundamental mode of a guide is used to excite the apparition and propagation of the fundamental mode in another guide producing a controllable energy. In the second case, a common bimodal zone to both guides is generated obtaining in this way the transfer of energy by means of a controlled interference between the modes that are being propagated in this zone. The coupler of evanescent field is generated by placing two guides very near between them as is illustrated in Figure 18(a). The evanescent field of the fundamental mode in one of the two guides is introduced and excites the apparition of the fundamental mode in the other guide whose evanescent field excites and causes a transfer of energy to the first guide. This process is produced in a continuous way generating a periodic signal exchange between the two guides [45].

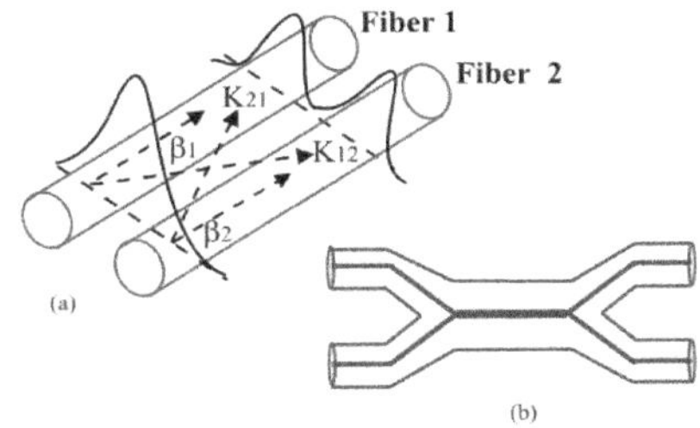

Figure 18. Transfer of energy in an evanescent field coupler (a). Diagram of a coupler based on a bimodal zone (b).

For the coupler based on modal interference, see Figure 18(b), the coupling zone is constituted by a common region of dimensions larger than the input and output of the guides allowing in this way the propagation of two modes (bimodal guide). The coupling energy between both produces the propagation of energy from one guide to another [45].The analysis for these types of devices uses the coupling equations. There are two coupled modes that correspond to the fundamental modes of each guide. If $a_1(z)$ and $a_2(z)$ are the amplitudes corresponding to the fields of the fundamental modes of Guides 1 and 2, and β_1 and β_2 are the propagation constants (not necessarily equal), then the coupling equations between modes are for this case

$$\begin{aligned} \frac{da_1}{dz} &= -j\beta_1 a_1 + K_{12} a_2 \\ \frac{da_2}{dz} &= -j\beta_2 a_2 + K_{21} a_1 \end{aligned} \tag{61}$$

For the case of guides without losses, and taking into account the conservation principle of energy, $K_{12} = K_{21}^*$, where $K_{21} = jc$. The constant c is the coupling coefficient given as

$$c = \frac{K_{12}}{j} = \frac{\omega\varepsilon_0}{4P} \int_{-\infty}^{\infty}\int_{-\infty}^{\infty} (n^2(x,y) - n_2^2) E_1^{t^*}(x,y) E_2^t(x,y)\, dxdy \tag{62}$$

where ω is the angular frequency (rad/sec), P is the power, ε_0 is the permittivity in the free space, n_2 is the index refraction corresponding to the cladding of both fibers, $n(x,y)$ is the transversal variation of the index refraction corresponding to the zone where the guides are very near between them, and $E_1^t(x,y)$ and $E_2^t(x,y)$ are the transversal profiles corresponding to the fundamental modes for the Guides 1 and 2, respectively. To solve Eq. (61), it is necessary to use the parameter $\delta = (\beta_1 - \beta_2)/2$. This parameter indicates the difference between the guides that composes the coupler. The solution of (61) provides the values of the amplitudes of the electric field in any point z for both guides $a_1(z)$ and $a_2(z)$ in function of the values of the amplitudes of electric fields to its input $a_1(0)$ and $a_2(0)$. The solution can be expressed as a matrix [45]

$$\begin{pmatrix} a_1(z) \\ a_2(z) \end{pmatrix} = e^{-j\left(\frac{\beta_1+\beta_2}{2}\right)} \begin{pmatrix} A & jB \\ JB^* & A^* \end{pmatrix} \begin{pmatrix} a_1(0) \\ a_2(0) \end{pmatrix} \tag{63}$$

Where

$$\begin{aligned} A &= \cos\left(z\sqrt{c^2+\delta^2}\right) + j\delta\frac{sen\left(z\sqrt{c^2+\delta^2}\right)}{\sqrt{c^2+\delta^2}} \\ B &= c\frac{sen\left(z\sqrt{c^2+\delta^2}\right)}{\sqrt{c^2+\delta^2}} \end{aligned} \tag{64}$$

The phase factor $\exp -j\left(\frac{\beta_1+\beta_2}{2}\right)$ is a common factor, and so it is not necessary to be considered in subsequent calculations. In practice, optical fiber couplers are made using similar fibers, for this reason $\beta_1 = \beta_2 = \beta$ and $\delta = 0$. Consequently, the matrix relationship (63) is

$$\begin{pmatrix} a_1(z) \\ a_2(z) \end{pmatrix} = \begin{pmatrix} \cos(cz) & jsen(cz) \\ jsen(cz) & \cos(cz) \end{pmatrix} \begin{pmatrix} a_1(0) \\ a_2(0) \end{pmatrix} \tag{65}$$

Defining the coupling constant as $k = sen^2(cz)$, then (65) takes the form

$$\begin{pmatrix} a_1(z) \\ a_2(z) \end{pmatrix} = \begin{pmatrix} \sqrt{1-k} & j\sqrt{k} \\ j\sqrt{k} & \sqrt{1-k} \end{pmatrix} \begin{pmatrix} a_1(0) \\ a_2(0) \end{pmatrix} \tag{66}$$

Since the coupling constant k is generally variable with wavelength, the characteristics of the coupler may vary with the operating frequency. However, this effect can be utilized to create optical filters, multiplexers and wavelength demultiplexers [45]. The energy transfer process can easily be evaluated assuming that at the input of the coupler there is only signal in Guide 1, this is, $a_2(0) = 0$. From equation (65) it is possible to obtain the optical power in each guide for each value of z:

$$\begin{aligned} P_1(z) &= |a_1(z)|^2 = P_1(0)\cos^2(cz) = P_1(0)(1-k) \\ P_2(z) &= |a_2(z)|^2 = P_1(0)sen^2(cz) = P_1(0)k \end{aligned} \tag{67}$$

A special case is when $k = 1/2$. In this case, the optical power in the input guide is divided equally between the two output guides of the coupler. The device is then called a 3 dB coupler.

Until now, the description of the coupler has been made on the assumption that it behaves ideally, this is, without considering losses. In practice this is not real, and from experimental as well as commercial points of view, it is necessary to define a set of parameters that allows the behavior of the coupler to be determined.

Coupling constant

$$k = \frac{P_4}{P_3 + P_4} \tag{68}$$

This constant indicates the optical power division between the output ports.

Excess Loss

$$\Gamma(dB) = 10\log(\gamma) = 10Log\left[\frac{P_1}{P_3 + P_4}\right] \tag{69}$$

This parameter provides the relationship between the total power at the input and the output power of the device.

Directivity coupler

$$D(dB) = 10Log\left[\frac{P_2}{P_1}\right] \tag{70}$$

This parameter indicates the fraction of input power that the coupler returns in the form of signal retro-propagated to Guide 2.

Insertion Losses

$$L_i(dB) = 10Log\left[\frac{P_1}{P_3}\right] \tag{71}$$

This parameter indicates the level of losses introduced by the coupler considering that the signal propagation is through a given configuration of the input port and the output port.

7. Experimental results

According to the theoretical descriptions explained in this chapter, we have assembled in the laboratory the proposed experimental setup shown in Figure 1. For the stage where relaxation oscillation frequencies were measured, we used two fiber-coupled DFB laser sources (Thorlabs, model S3FC1550) with a central wavelength of 1550 nm, a threshold current of ≈8 mA and a differential responsivity of 0.25 mW/mA. An optical isolator OI was connected to each DFB laser to avoid instabilities from feedback into the lasers, and an optical coupler was used to split the optical beam. In order to increase the power level of the microwave signal, it was necessary to connect an EDFA in one of the ports of the optical coupler. After that, the amplified optical signal was launched to fast photodetector PD1 (MITEQ model DR-125G-A) which has a typical optical-to-electrical transfer gain (V/W) of 1900, and –3 dB bandwidth of 12.5 GHz. In order to measure the optical spectrum of the DFB laser, it was necessary to connect the other port of optical coupler to an OSA Anritsu model MS9710C. On the other hand, the obtained photo-current by photo-detector was measured by using an ESA, Agilent model E4407B.

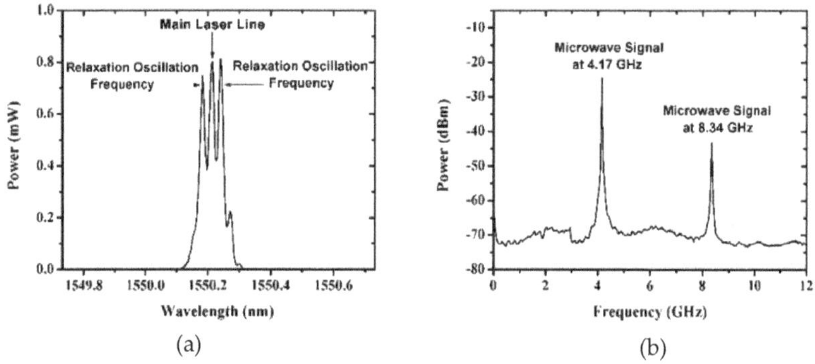

Figure 19. Measured spectrums: (a) showing relaxation oscillation frequencies, (b) microwave signal by using direct detection.

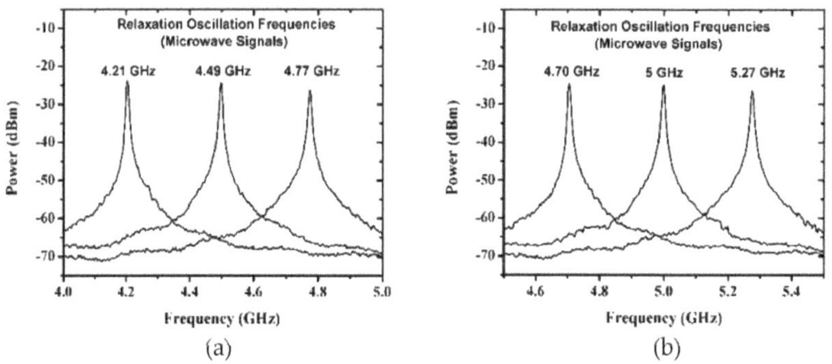

Figure 20. Microwave signal continuously tuned. (a) between 4 and 5 GHz, (b) between 4.5 and 5.5 GHz.

As can be seen from Figure 19(a), the optical spectrum of one of the lasers (DFB 1) exhibits sidebands on both sides of the main laser line as a result of it being biased in the threshold current region. In this case, its optical power was emitted at 1.15 mW at a temperature of 23.5° C. According to Eq. (26), the injection current was approximately at 12.6 mA. With these operation data, a microwave signal was obtained on C band by using direct detection, as shown in Figure 19(b). This spectrum shows a microwave signal located at 4.17 GHz and its harmonic located at 8.34 GHz. This result is in good agreement with Eq. (32). Similar results were found with the DFB 2 laser when it was operated in the low laser threshold current region. Because of its linear relationship between relaxation oscillation frequency and the square root of the difference between the injection current and the threshold current as defined in Eq. (25), and since the optical power of each DFB laser can be tuned by using a power controller system which modifies the injection current through a laser diode correspondingly, we obtained several microwave signals located between 4 and 5 GHz for DFB 1 when it was tuned between 0.7 mW and 0.8 mW at a temperature of 25° C, as can be seen in Figure 20(a). Microwave signal were obtained between 4.5 and 5.5 GHz, when the optical power of DFB 2 was tuned between 0.9 mW and 1 mW with a temperature of 26° C as can be seen in figure 20(b).

To show a potential application in the field of communication systems of relaxation oscillation frequencies, we report herein the simultaneous multiplexed transmission of an analog TV signal of 67.25 MHz (TV channel 4) over a photonic link of 30 Km of optical fiber by using two microwave signals at 4 and 5 GHz. From the experimental setup illustrated in Figure 1, it can be seen that the obtained relaxation oscillation frequencies from the PD1 were mixed with a TV signal by using an electrical frequency mixer (Mini-Circuits, model ZX05-U72MH-s+). Then the obtained modulated signal as shown in Figure 21(a), was amplified and transmitted through a photonic link of 30 Km of optical fiber. Such microwave sub-carriers were applied to an electro-optic modulator which imprints the electric signal on the laser emission. The intensity-modulated optical signal of DFB 3 laser was then transmitted through the optical channel. After that, in the receiver, the transmitted information was photo-detected. The spectrum of the multiplexed microwave signal after photodetection is shown in Figure 21(b). The output electrical spectrums in Figure 21 clearly show the multiplexed microwave sub-carriers. The spectral separation is approximately 1 GHz. According to the experimental setup shown in Figure 1, the photo-detected signal is amplified, filtered and synchronously down-converted to recuperate the analog TV signal of 67.25 MHz. The resulting power spectral density was displayed in an electrical spectrum analyzer, where it was analyzed to measure the power level of recovered information.

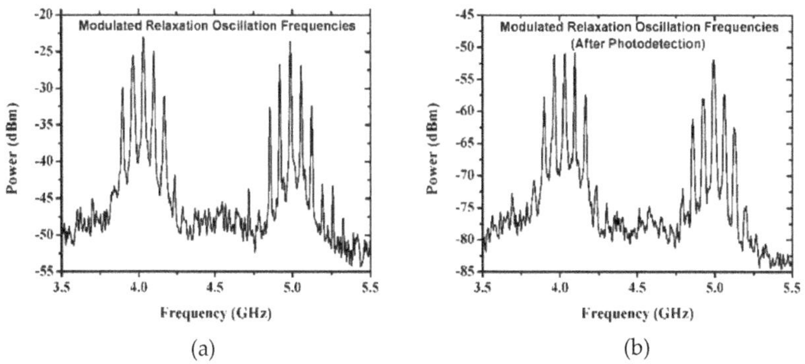

Figure 21. Modulated relaxation oscillation frequencies with a TV signal. (a) applied to electro-optic modulator, (b) after photodetection.

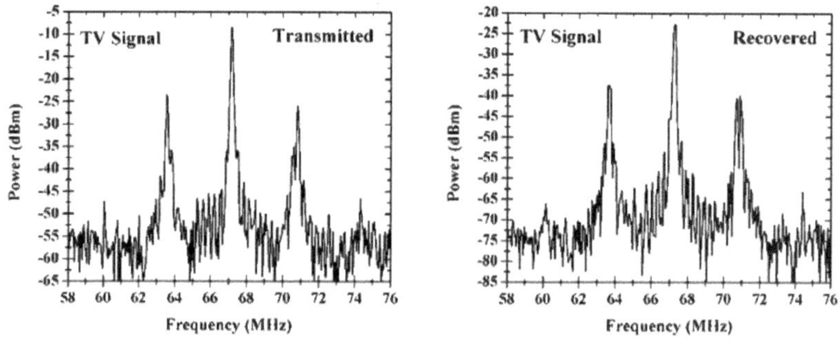

(a) (b)

Figure 22. TV signals at 67.25 MHz, (a) transmitted (b) recovered.

Figure 22(a) shows the frequency spectrum of an analog NTSC TV signal at the input of the transmitter located at 67.25 MHz (before being applied to a frequency mixer). In Figure 22(b), we can see the obtained analog NTSC TV signal at the output of the receiver when the local oscillator was synchronized at 4 GHz. The analog information was successfully transmitted from the transmitter to the receiver, and the received signal was satisfactorily reproduced at the receiver without noticeable degradation compared the original NTSC TV signal. Before experimentally quantifying the parameter of SNR for both cases (TV transmitted and recovered in the proposed photonic link), it is essential to give an analysis of noise in terms of SNR.

7.1. SNR performance of an analog photonic link

To reconstruct a signal at the receiver faithfully, the noise power must be small compared to the signal level. In other words, the SNR is a figure of merit in characterizing a receiver circuit. There are three dominant noise sources on the optical receiver side of an analog photonic link: thermal noise, shot noise, and received relative intensity noise (RIN). The RIN are the fluctuations of the laser intensity caused by random spontaneous light emissions. In a PD, noise can be attributed primarily to two factors. The shot noise, also known as quantum noise, originates from the statistical nature of photon-to-electron conversion. Thermal noise, on the other hand, is an intrinsic property of any electrical circuit that is connected to the PD.

7.1.1. Shot noise

Photon-to-electron conversion is fundamentally a quantum mechanical process. When a photon is absorbed, a pair of electron-holes is generated. Therefore, the photo-generated current is not truly continuous but has a discrete nature. It fluctuates around some average value as a result of the discrete charge of the carriers that contribute to it. Because of the random nature of the current fluctuations, the noise current must be characterized in a statistical manner. It is common to describe the noise current by its mean square value. For a PIN detector, the mean square value of the shot noise is [48].

$$\sigma_s^2 = 2qI_P B \tag{72}$$

where I_P is the photo-current, q is the electron charge, and B is the bandwidth within which the noise is being measured. Eq. (72) implies that shot noise has a constant spectral density, an assumption that holds for all frequencies of interest. Normally, B is set by the bandwidth of the receiver. This shows that one way to minimize the effects of shot noise is to keep the bandwidth of circuit as narrow as possible. The current flowing through a PIN diode is not just photo-generated. Any reverse bias junction has a leakage current. For photo-detectors, the leakage current is called dark current I_D, because it exists even when there is no optical power. As a result, the mean square value of the total shot noise is given by

$$\sigma_s^2 = 2q\left(I_P + I_D\right)B \tag{73}$$

7.1.2. Thermal noise

Shot noise is a consequence of the quantum nature of light detection. Therefore, it is a fundamental property of the photo-detector and sets a maximum limit on the value of SNR. In such a case, the SNR is said to be quantumly limited. In reality, however, there are other sources of noise almost always present. Chief among these is thermal noise, also known as Johnson noise, associated with the electric circuits connected to the detector. The source of this noise is the thermal motion of electrons in the load resistor R_L. The mean squared of the thermal noise in the load resistor is given by

$$\sigma_T^2 = \frac{4kTB}{R_L} \tag{74}$$

where k is Boltzmann's constant, T is the absolute temperature, and B is bandwidth. Like shot noise, thermal noise has a constant spectral density. This is another reason to keep the bandwidth of a receiver as low as possible, i.e., just sufficient to pass the signals of interest.

7.1.3. Relative intensity noise

Any real laser has a certain amount of noise due to spontaneous emission. As a result both the amplitude and the phase are randomly varying in any real laser, then the wave front a DFB laser source may be represented by its electric field as

$$\vec{E}_1(t) = E_o\left(1 + V(t)\right)\exp j\left(2\pi\nu_o t + \varphi(t)\right) \tag{75}$$

where $V(t)$ and $\varphi(t)$ represent the amplitude and phase of the noise, respectively. If the field expressed in Eq. (75) is detected by a photo-detector with responsivity $\Re$, the resulting current is given by [49]

$$\mathrm{I} = \Re \mathrm{E}_0^2[1 + \mathrm{V(t)}]^2 \tag{76}$$

Typically, the amplitude noise, $V(t)$, is very small so that Eq (76) can be rewritten as

$$I \approx \Re E_o^2 [1 + 2V(t)] \tag{77}$$

The noise term $2V(t)$ represents the RIN; it describes the laser power fluctuations. The power spectral density of the RIN is denoted by ζ ; it has the units of 1/Hz. The quantity 10 log ζ is known as the RIN.

$$RIN \equiv 10 \log \zeta, \text{ dB/Hz} \tag{78}$$

The typical values of RIN for a DFB laser are better than -155 dB/Hz. The power spectrum of the RIN is not flat, hence it is not a white noise source. In an analog photonic link, we use a photodiode to detect the optical power from the M-Z modulator. Thus, RIN can be expressed as [50].

$$\text{RIN} = \frac{\left(\frac{i_{RIND}}{\Re}\right)^2}{B\left(\frac{I_D}{\Re}\right)^2} = \frac{i_{RIND}^2}{BI_D^2} \tag{79}$$

From Eq. (79), we obtain the received mean square current RIN noise of the receiver as:

$$i_{RIND}^2 = \text{RIN} \cdot I_D^2 B \tag{80}$$

Comparatively, i_{RIND}^2 is proportional to I_D^2, whereas the shot noise is linearly proportional to I_D. Therefore, the RIN noise will tend to be the dominant noise, when the laser average power is increasing.

7.1.4. Signal-to-Noise Ratio (SNR)

Once we have characterized the noise level at the input of a receiver, it is possible to analyze the SNR. The SNR is an important parameter because it determines the performance of a receiver. In analog receivers, SNR is the main figure of merit and characterizes the quality of the analog link. Assuming that all of the noise sources described previously are uncorrelated, the signal-noise ratio of the link at the single photo-detector can be expressed by:

$$\text{SNR} = \frac{\Re^2 P_R^2}{\sigma_s^2 + \sigma_T^2 + i_{RIND}^2} = \frac{\Re^2 P_R^2}{2q(I_P + I_D)B + \frac{4kTB}{R_L} + \text{RIN} \cdot I_D^2 B} \tag{81}$$

From Eq. (81), P_R is the received optical power at the end of the optical link. On the other hand, according to the Figure 9 and Eq. (42), and considering that the modulated optical power at the output of MZ-IM is propagated through an optical fiber of length L, the P_R can be written by $P_R = 10^{-\alpha_{fiber} L/10} P_{out}$. Here α_{fiber} is the optical fiber loss, and L is the transmission distance of the optical fiber. In that case if we consider that $\pi V_m/V_\pi \ll 1$ then Eq (81), can be rewritten as

$$SNR = \frac{\Re^2 \left(10^{-\alpha_{fiber} L/10} P_{out}\right)^2}{2q(I_P + I_D)B + \frac{4kTB}{R_L} + RIN \cdot I_D^2 B} = \frac{\Re^2 \left(\left(10^{-\alpha_{fiber} L/10}\right) \frac{\alpha P_{in}}{2} \left\{ \frac{\delta^2}{2} - \delta \left[\frac{\pi V_m}{V_\pi} \right] \right\} \right)^2}{2q(I_P + I_D)B + \frac{4kTB}{R_L} + RIN \cdot I_D^2 B} \tag{82}$$

This equation provides several insights into the SNR behavior of a receiver. Notice that as expected, increasing the optical power Pin increases the SNR. On the other hand, increasing the bandwidth of the receiver, B, reduces the SNR. The denominator of Eq. (82) shows the contribution of thermal noise, shot noise and RIN to SNR. An interesting point is that as R_L increases, the effects of thermal noise decrease. In a practical circuit, however, R_L cannot be increased too much because it will reduce the bias headroom of the photo-detector.

7.1.5. Experimental (SNR)

In order to measure the quality of the received signal, it was necessary to experimentally quantify the parameter of signal-to-noise ratio (SNR) for both cases (TV transmitted and recovered in the photonic link).

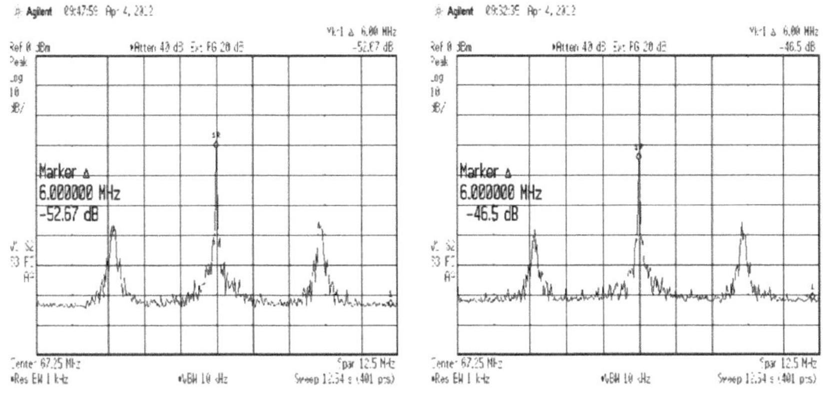

(a) (b)

Figure 23. TV signals at 67.25 MHz in order to measure experimentally the SNR parameter, (a) transmitted (b) recovered.

As can be seen from figure 23(a), the SNR measured in a bandwidth of 6 MHz was of 52.67 dB. On the other hand the SNR measured in the receiver was of 46.5 dB as shown in figure 23(b). Notice that the SNR measured in the receiver was degraded by 6.17 dB due to the optical fiber loss and it could be improved substantially with the addition of another stage of electrical amplification at the end of the link. However, values of SNR > 45 dB ensure a good quality of images [51]. The differential gain and differential phase were not measured in this chapter. Nevertheless we have experimentally demonstrated that the generated microwave signal (relaxation oscillation frequency) by using direct detection can be used as information carriers in a multiplexed transmission system based in an analog photonic link and we have used a TV signal of test to verify it.

8. Conclusions

In this chapter we have described optical components used in the proposed experimental setup, as well as, we have experimentally demonstrated that two DFB lasers biased in the low laser threshold current region showed relaxation oscillation frequencies in the laser intensity. These frequencies were seen as sidebands on both sides of the main laser line when the optical spectrum was analyzed with a spectrum analyzer. This result allowed the generation of microwave signals on C band by using the operation principle of direct detection. Along with the proposed experimental setup in this work, an analog NTSC TV signal was simultaneously transmitted in an analog photonic link by using relaxation oscillation frequency as information carriers located at 4 and 5 GHz. The TV signal was received satisfactorily when the local oscillators were synchronized in the receiver. For our proposed experimental setup to be recovered the transmitted information successfully, it was necessary to have additional amplifiers which adapt the power levels, improving the quality of the signals. The results obtained in this work ensure that as an interesting alternative, several modulation schemes can be used for transmitting not only analog information but also digital information by using relaxation oscillation frequencies as information carriers. The proposed experimental setup described here can generate continually tuned microwaves. In addition, we can use this feature to transmit several TV signals using frequency division multiplexing schemes (FDM) and wavelength division multiplexing (WDM) techniques, not only point to point but also by using bidirectional systems, exploiting the use of mature microwave signal processing techniques, where multiple signals are multiplexed in the microwave domain and transmitted by a single optical carrier when subcarrier multiplexing (SCM) schemes are proposed. To the best of our knowledge, this is the first published work on the employ of relaxation oscillation frequencies as information carriers for transmitting analog TV signals on a long distance photonic link. In addition, as an alternative of analog photonic links, our proposal shows outstanding performance.

Author details

Alejandro García Juárez
University of Sonora, Department of Physics Research, Mexico

Ignacio Enrique Zaldívar Huerta
National Institute of Astrophysics, Optics and Electronics. Department of Electronics, Mexico

Jorge Rodríguez Asomoza
University of the Americas-Puebla, Department of Electronics, Mexico

María del Rocío Gómez Colín
University of Sonora, Department of Physics, Mexico

Acknowledgement

This work was supported by The Mexican Consejo Nacional de Ciencia y Tecnología (CONACyT), (grant No 102046).

9. References

[1] Kai Wang, Xiaoping Zheng, Hanyi Zhang, and Yili Guo. A Radio-Over-Fiber Downstream Link Employing Carrier-Suppressed Modulation Scheme to Regenerate and Transmit Vector Signals. IEEE Photonics Technology Letters 2007; vol. 19, no. 18, pp. 1365-1367.
[2] D. Novak, A. Nirmalathas, C. Lim, C. Marra, and R. B. Waterhouse. Fiber-radio-challenges and possible solutions. Topical Meeting Microwave Photonics 2003, Sep. 10–12, (2003).
[3] X. Zhang, B. Liu, J. Yao, K. Wu, and R. Kashyap. A novel millimeter-wave-band radio-over-fiber system with dense wavelength-division multiplexing bus architecture. IEEE Transactions on Microwave Theory and Techniques 2006; vol. 54, no. 2, pp. 929–937.
[4] Stavros Iezekiel. Microwave Photonics Devices and Applications. John Wiley; 2009.
[5] Xue Jun Meng. A Dispersion Tolerant Analog Optical Receiver for Directly Modulated RF Photonic Links. IEEE Photonics Technology Letters 2007; vol. 19, no. 6, march 15
[6] Spiros Mikroulis, Sotiris Karabetsos, Evaggelos Pikasis, Aristidis Chipouras and Athanase Nassiopoulos. Evaluation on the performance of Distributed Feedback Lasers (DFB) for Radio-Over-Fiber (RoF) Applications. International Conference on Telecommunications & multimedia, Greece 2006.
[7] A. Baylón-Fuentes, P. Hernández-Nava, A.García-Juárez, I. E. Zaldívar-Huerta, J. Rodríguez-Asomoza, G. Aguayo-Rodríguez and R. Gómez-Colín. Modulation of Relaxation Oscillation Frequency of a DFB Laser by Using Direct Detection. Broadband Access Communications Technologies V, Proc. of SPIE Vol. 7958, 2011.
[8] A.J. Seeds. Broadband wireless access using millimetre-wave over fibre systems. IEEE MTTS Int. Microwave Symp.1997; vol.1, pp. 23-25.
[9] W.H. Loh, C.L. Tang. Polarization modulation for optical-microwave applications in phased array antennas. IEEE Photonics Technology Letters 1991; vol.3 no. 5 pp. 478-480.
[10] J.J. O'Reilly, P.M. Lane. Fibre-supported optical generation and delivery of 60 GHz signals. Electronics Letters 1994; vol. 30, no. 16, pp. 1329-1330.
[11] D.C. Scott, D.V. Plant, H.R. Fetterman. 60 GHz sources using optically driven heterojunction bipolar transistors. Applied Physics Letters 1992; vol. 61, no. 1, pp. 1-3.
[12] J.B. Georges, M. -H. Kiang, K. Heppell, M. Sayed, K.Y. Lau. Optical Transmission of Narrow-Band Millimeter Wave Signals by Resonant Modulation of Monolithic Semiconductor Lasers. IEEE Photonics Technology Letters 1994; vol. 6 no. 4 pp 568-570.

[13] C.R. Lima, D. Wake, P.A. Davis. Compact optical millimetre-wave source using a dual-mode semiconductor laser. Electronics Letters 1995; vol. 31 no. 5, pp. 364-366.

[14] L. Goldberg, A.M. Yurek, H.F. Taylor, J.F. Weller. 35 GHz microwave signal generation with an injection-locked laser diode. Electronics Letters 1985; vol. 21, no. 18, pp. 814-815.

[15] P. Shen, N.J. Gomes, P.A. Davis, W.P. Shillue, P.G. Huggard, B.N. Ellison, in: Proc. Int. Top. Microw. Photon. Metting, 2003.

[16] D. Wake, C.R. Lima, P.A. Davies. Optical generation of millimeter-wave signals for fiber-radio systems using a dual-mode DFB semiconductor laser. IEEE Transaction on Microwave Theory and Techniques 1995; vol. 43, no. 9, pp. 2270 – 2276.

[17] R.T. Ramos, A.J. Seeds. Fast heterodyne optical phase-lock loop using double quantum well laser diodes. Electronics Letters 1992; vol. 28 no. 1, pp 82-83.

[18] A. García-Juárez, I. E. Zaldívar-Huerta, G. Aguayo-Rodríguez, J. Rodríguez-Asomoza, R. Gómez-Colín, A. Vera-Marquina, M. C. Acosta-Enriquez, A. Rojas-Hernández. Optical heterodyne technique for generating and distributing microwave signals. Broadband Access Communications Technologies IV, Proc. of SPIE Vol. 7620, 2010.

[19] A. García-Juárez, I. E. Zaldívar-Huerta, G. Aguayo-Rodríguez, J. Rodríguez-Asomoza, R.Gómez-Colín, A. Rojas-Hernández. Coherent demodulation of microwave signals by using optical heterodyne technique with applications to point to point indoor wireless communications systems. 2011 J. Phys.: Conf. Ser. 274 012008, doi:10.1088/1742-6596/274/1/012008. (http://iopscience.iop.org/1742-6596/274/1/012008)

[20] Jianping Yao. Microwave Photonics. Journal of Lightwave Technology 2009; vol. 27, no. 3, pp. 314-335.

[21] L. Goldberg, H. F. Taylor, J. F. Weller, and D. M. Bloom. Microwave signal generation with injection locked laser diodes. Electron. Letters 1983; vol. 19, no. 13, pp. 491-493.

[22] U. Gliese, T. N. Nielsen, M. Bruun, E. L. Christensen, K. E. Stubkjaer, S. Lindgren, and B. Broberg. A wideband heterodyne optical phaselocked loop for generation of 3–18 GHz microwave carriers. IEEE Photon. Technol. Letters 1992; vol. 4, no. 8, pp. 936–938.

[23] A. C. Bordonalli, C.Walton, and A. J. Seeds. High-Performance phase locking of wide linewidth semiconductor lasers by combined use of optical injection locking and optical phase-lock loop. J. Lightwave Technology 1999; vol. 17, no. 2, pp. 328–342.

[24] Z. F. Fan and M. Dagenais. Optical generation of a mHz-linewidth microwave signal using semiconductor lasers and a discriminator-aided phase-locked loop. IEEE Trans. Microw. Theory Tech., 1997; vol. 45, no. 8, pp. 1296–1300.

[25] J. J. O'Reilly, P. M. Lane, R. Heidemann, and R. Hofstetter. Optical generation of very narrow linewidth millimeter wave signals. Electron. Letters 1992; vol. 28, no. 25, pp. 2309–2311.

[26] J. J. O'Reilly and P. M. Lane. Remote delivery of video services using mm-wave and optics. J. Lightw. Technol. 1994; vol. 12, no. 2, pp. 369–375.

[27] P. Shen, N. J. Gomes, P. A. Davies, W. P. Shillue, P. G. Huggard, and B. N. Ellison. High-purity millimeter-wave photonic local oscillator generation and delivery. in Proc. Int. Microw. Photonics Topical Meeting, Sep. 10–12, 2003, pp. 189–192.

[28] G. Qi, J. P. Yao, J. Seregelyi, C. Bélisle, and S. Paquet. Generation and distribution of a wide-band continuously tunable mm-wave signal with an optical external modulation technique. IEEE Trans. Microw. Theory Tech. 2005; vol. 53, no. 10, pp. 3090–3097.
[29] X. Chen, Z. Deng, and J. P. Yao. Photonic generation of microwave signal using a dual-wavelength single-longitudinal-mode fiber ring laser. IEEE Trans. Microw. Theory Tech. 2006; vol. 54, no. 2, pp. 804–809.
[30] X. Chen, J. P. Yao, and Z. Deng. Ultranarrow dual-transmission-band fiber Bragg grating filter and its application in a dual-wavelength single-longitudinal-mode fiber ring laser. Opt. Lett. 2005; vol. 30, no. 16, pp. 2068–2070.
[31] H. Zandi, M. Bavafa, M. Chamanzar, and S. Khorasani. Analysis of power harmonic content and relaxation resonant frequency of a diode laser. Proc. SPIE, Vol. 6468, No. 64680J, 2007.
[32] H. Zandi, M. Bavafa, M.R. Chamanzar and S. Khorasani. Harmonic Content and Relaxation Resonant Frequency of a Modulated Laser Diode. Transactions D: Computer Science & Engineering and Electrical Engineering 2009; Vol. 16, No. 2, pp. 145-156.
[33] M. Bavafa, H. Zandi, M. Chamanzar, and S. Khorasani. Optimal Operation Point for the Primary Harmonic Content of a Modulated Laser Diode. Proceedings of the 6th WSEAS International Conference on Microelectronics, Nanoelectronics, Optoelectronics, Istanbul, Turkey, May 27-29, 2007.
[34] L.A. Coldren, and S.W Corzine. Diode Lasers and Photonic Integrated Circuits. John Wiley & Sons; 1995.
[35] B. E. A. Saleh. Fundamentals of Photonics. John Wiley & Sons, Inc; 1991. pp. 624-626.
[36] A. Baylón-Fuentes, P. Hernández-Nava, A.García-Juárez, I. E. Zaldívar-Huerta, J. Rodríguez-Asomoza, G. Aguayo-Rodríguez and R. Gómez-Colín. Modulation of Relaxation Oscillation Frequency of a DFB Laser by Using Direct Detection. Broadband Access Communications Technologies V, Proc. of SPIE Vol. 7958, 2011.
[37] A.García-Juárez, I. E. Zaldívar-Huerta, J. Rodríguez-Asomoza3, R. Gómez-Colín, A.G Rojas-Hernández, D. Berman-Mendoza, R. Gómez-Fuentes and A. Vera Marquina. Analog Photonic Link by Using DFB lasers Operated in the Low Laser Threshold Current Region and External Modulation. Broadband Access Communications Technologies VI, Proc. of SPIE Vol. 8282, 2012.
[38] D. Derickson. Fiber Optic Test and Measurement. Prentice Hall PTR; 1998. pp. 175-179.
[39] C. H. Cox. Analog Optical Links: Theory and Practice. Cambridge: Cambridge University Press; 2004, pp. 37–38.
[40] X. Jun Meng, A. Karim. Microwave Photonic Link with Carrier Suppression for Increased Dynamic Range. Fiber and Integrated Optics, Taylor & Francis Group; 2006. pp. 161- 174.
[41] P. S. Henry. Lightwave primer," IEEE J. Quantum Electron. 1985; vol. QE-21, pp. 1862–1879.
[42] Jha, A.R. State of the Art Fiber Optic Components and Photonic Devices for Optical Communication and High Speed Transmission. 31st European Microwave Conference, 2001.
[43] B. E. A. Saleh. Fundamentals of Photonics. John Wiley & Sons, Inc; 1991. pp. 193-237

[44] Donald K. Wilson. Polarization control aids fiber component testing. Lasers Focus World 1997; pp. 129 –133.

[45] José Capmany, F. Javier Fraile-Peláez y Javier Martí. Dispositivos de Comunicaciones Ópticas. Editorial Síntesis.

[46] R. A. Bergh, G. Kotler y H. J Shaw. Electron. Lett. , 1980; p 260.

[47] S. K. Sheem y T. Giallorentzi. Opt. Lett. 1979; p 29.

[48] E. Garmire. Sources, modulation, and detectors for fiber-optic communication systems, in Fiber Optics, Handbook. Edited by M. Bass, McGraw-Hill, New York; 2002 pp. 4.1–4.78,

[49] Leonid Kazovsky, Sergio Benedetto and Alan Willner. Optical Communications Systems, Artech House; 1996.

[50] Abhay Joshi, Xinde Wang, Dan Mohr, Don Becker, and Christoph Wree. Balanced photoreceivers for analog and digital fiber optic communications. Proc. of SPIE Vol. 5814, 2005.

[51] M. Kowalczyk and J. Siuzdak. Multi-channel am video transmission beyond the baseband of multimode fiber. Microwave and optical technology letters 2010; vol. 52, no. 2, pp. 435-438.

Chapter 7

Chaotic Dynamics of Semiconductor Lasers for Secure Optical Communication

Muzahim I. Azawe

Additional information is available at the end of the chapter

1. Introduction

Strong optical injection locking of semiconductor lasers can give an output of chaotic behavior which makes them an attractive transmitter for high speed secure optical communication. Data secure and high transmission rates are the most demanded issues in communication network. Optical injection, chaos synchronization of semiconductor lasers had been studied theoretically and experimentally for a decade [1,2], due their durability and very vast progressing technology. The effects of optical injection locking mainly have two aspects: one is to improve the characteristics of the slave and the other is to synchronize the master and the slave. In the former, the locking is able to improve the properties of the slave laser such as a single wavelength emission, high side mode suppression ratio (SMSR), and narrow linewidth. While in the latter, the synchronization of the master and slave in a wavelength, phase and chaos state had led the injection locking to broad applications in the coherent communications. Other improvements of the semiconductor laser by optical injection locking had been reported, these improvements include, increasing modulation bandwidth and reducing chirp [3,4], a high gain of 20-dB with small signal modulation below resonance frequency [5]. Bistability had been reported when a two color Fabry-Perot laser was subjected to optical injection in both modes and could be the basis for an all optical memory element with switching times below 500 ps [6]. The generalized synchronization of chaos based on phenomenon of injection locking characteristics of semiconductor laser and signal amplification in nonlinear systems is an application for secure data transmissions and communications [6,8].

In order to determine in which conditions (optical power, wavelength of the injected signal) the Fabry-Perot laser diode (FP-LD) is locked, it is essential to map the operating regimes on a chart defined by the two parameters, injected power and detuning which corresponds to the difference between the wavelengths of the injected signal and the one of a specific mode of FP-LD that is submitted to optical injection, it is so-called injection map, which is well known for a single-mode laser [9,10].

The maximum available modulation frequency of the laser is in the vicinity of the relaxation oscillation frequency. Optical injection can enhance the relaxation oscillation frequency of the slave laser, and hence the bandwidth. So we would expect higher-speed transmitter for optical communication. On the other hand, a laser with controlled chaos could be obtained. The bandwidth-enhancement of the semiconductor laser by optical injection as well as a chaotic transmitter is the major objective.

This chapter will focus on the improvements of semiconductor lasers by the optical injection locking regimes and its applications for secure optical communication networks. The injection locked semiconductor laser, utilizing such applications, noise properties are of vital importance especially, the relative intensity noise (RIN). The aspects of noise influence on the dynamical operation of the laser with injection locking will be emphasized. The deployment of such a high bandwidth and chaotic carrier transmitter will be feasible without extra protection afforded by other means.

2. Basic concepts of injection locking

Rate equations of the injection locking are given below. Optical injection eventually introduces an extra degree of freedom, and due to the low facet reflectivity of the semiconductor, and perturbation from outside will alter the gain of the laser and may induce nonlinear dynamics. As a point of view, this nonlinear dynamics may be a candidate to chaotic system. The illustration of optical injection is shown in figure 1. A master laser, optically isolated from the slave laser, will inject its single-frequency output into the active region of the slave laser. This optical injection has a variety of effects on the operating characteristics of the slave laser that will be discussed here.

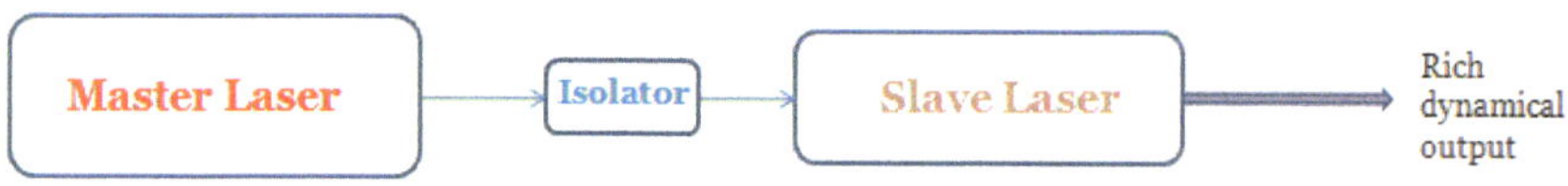

Figure 1. Schematic of optical injection.

The optical injection when it has to be in the locking range, the master and slave lasers have to be precisely identical and have the same oscillation frequencies. The frequency detuning must be within several GHz [11]. Once the appropriate conditions of frequency detuning and the optical injection strength have been achieved, a synchronization state of the two lasers is reached. Optical injection locking will be reached under these conditions, i.e., when the slave laser is forced to oscillate at the injected signal frequency from the master laser and is locked to its phase.

The operating principles of semiconductor laser are similar to those of any other laser system, except the linewidth enhancement factor, which arises because the real refractive index in the active laser medium varies with changing carrier density. The injected carrier-induced refractive index change is associated with the change in the gain, the differential gain. This factor plays an important role in the optical injection regimes of semiconductor lasers. Strong optical injection is usually used in the rate equations so that the impact of the noise and the spontaneous emission rate coupled to the lasing mode are negligible [12].

The rate equations of strong optical injection locking are [13]:

$$\frac{dP(t)}{dt} = \left\{ G(N) - \frac{1}{\tau_p} \right\} P(t) + 2\tau_{in}\sqrt{P(t)P_{inj}}\,\cos\left(\phi_i(t) - \phi(t)\right) \tag{1}$$

$$\frac{d\phi(t)}{dt} = \omega_o(N) - \omega_{inj} - \frac{\alpha}{2}\left(G(N) - \frac{1}{\tau_p} \right) + \tau_{in}\sqrt{\frac{P_{inj}}{P(t)}}\,\sin\left(\phi_i(t) - \phi(t)\right) \tag{2}$$

$$\frac{dN(t)}{dt} = J - \frac{N(t)}{\tau_{sp}} - G(N)P(t) \tag{3}$$

where $\omega_O(N)$ is the angular optical frequency (carrier dependent) of the slave laser, ω_{inj} is of the injected signal, P_{inj} is the injected signal power, $G(N)$ is the carrier dependent gain, τ_{sp} is the carrier lifetime, α is the linewidth enhancement factor, J is the injection current, and $\left(\phi_i(t) - \phi(t)\right)$ is the phase difference between the injected and the free-running laser fields.

The frequency detuning between the master and slave lasers is defined as $2\pi.\Delta f = \left(\omega_{inj} - \omega_o\right)$

The injection ratio is defined as the ratio of optical powers of the master and free-running slave laser inside the slave laser cavity $\frac{P_{inj}}{P}$.

The static state of the slave laser can be found from solving the steady solutions in the above equations by setting the time derivatives to zero and this will lead to the state of locking frequency within the master-slave frequency detuning. This static locking range is:

$$-\frac{1}{2\pi.\tau_{in}}\sqrt{\frac{P_{inj}}{P}}\sqrt{1+\alpha^2} < \Delta f < \frac{1}{2\pi.\tau_{in}}\sqrt{\frac{P_{inj}}{P}} \tag{4}$$

Equation (4) gives the locking range of the slave laser and is increased with the injection ratio and is also inversely proportional to the slave laser cavity round-trip time τ_{in}. The locking frequencies depend strongly on the linewidth enhancement factor. Figure 2 shows the locking range of the slave laser. Asymmetric locking range is obvious from the graph. This can be interpreted as; the locking properties depend on the gain profile of the salve laser supporting many longitudinal modes, and when injection occurs at frequency close to the side mode instead of the free-running dominant mode of the slave laser. Now, gain is carrier density dependent, and refractive index also depends on carrier density the in the active region which results in this asymmetry in the locking curve.

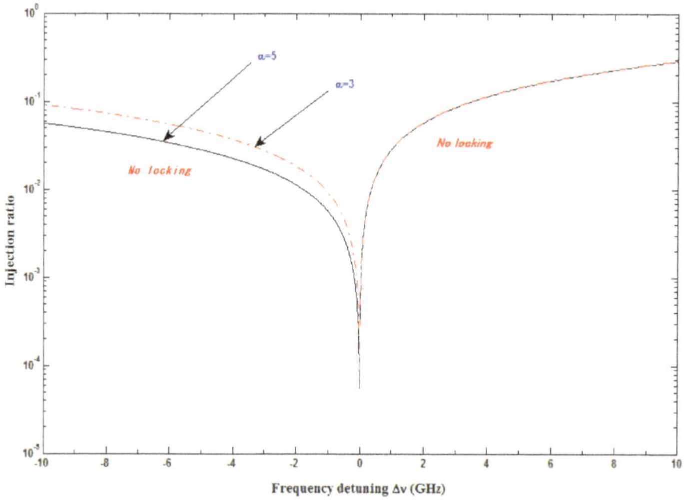

Figure 2. Locking characteristics of the slave laser showing the stable (locking) and unlocking regions for two values of linewidth enhancement factor.

3. Bandwidth enhancement by injection locking

The bandwidth of semiconductor laser is limited by the relaxation oscillation frequency. When a small-signal transient response is applied to rate equations of semiconductor laser as a step increment in the injection current and solved analytically, we will observe that the photon and electron numbers approach asymptotically to a new steady-state after an evanescent oscillation frequency (called relaxation) and with high damping factor. The relaxation oscillation frequency depends on the laser structure and the operation conditions. Any expression of the relaxation oscillation will show that it depends on the differential gain and the volume density of the injection current, as well as the photon lifetime. We will show the power dependence of the relaxation oscillations and the dependence on differential gain, while neglecting the gain saturation (intensity dependence) for simplicity in this monograph.

The frequency response of the intensity modulation of the laser describes the transfer of the modulation from current to optical power output. Figure 3 shows the frequency response of the semiconductor laser with and without strong optical injection. The response is flat in the low-frequency region and with a peak in the vicinity of the relaxation oscillation frequency. The response shows a rapid decrease for frequencies above the relaxation frequency. This means that the maximum frequency response of the laser when directly modulated is limited by the relaxation oscillation frequency.

Strong optical injection, as shown in figure 3, will shift the relaxation oscillation (RO) frequency (cutoff frequency) of the semiconductor to higher values depending on the operating conditions of the laser. This shift due to optical injection is effective for enhancing the modulation characteristics of semiconductor lasers. The shift is more obvious when plotted as a function of injection strength, as in figure 4. The laser was biased above threshold at 1.6 J_{th} and frequency detuning was of -7.962 GHz. The cutoff frequency can be said to be proportional to the injection strength, and hence modulation response of the laser can be enhanced due to strong optical injection [16].

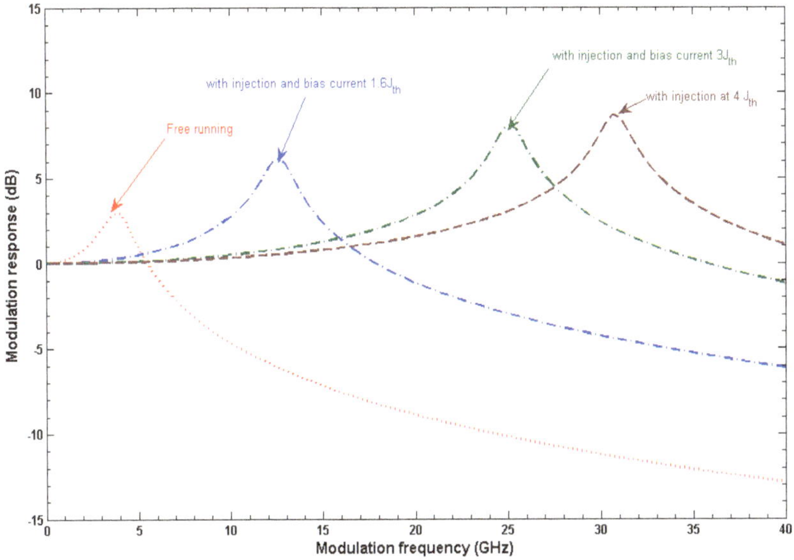

Figure 3. Modulation response of free-running semiconductor laser and with injection-locked laser at different bias currents . Frequency detuning was -7.962 GHz.

Modulation response of the semiconductor laser as being dependent on both the operating conditions and optical injection is illustrated in figure 5. As can be seen from the graph, the shift of RO to a higher frequencies with increasing the injection strength. The strength of RO resonance plays a crucial role in harmonic distortions. It follows that the lasers of having a high RO would result in large harmonic distortions. The resonance peaks can be explained as a result of frequency domain manifestation of the time domain of the optical field of the laser. The fall-off in the modulation response is due to a combination of the intrinsic laser response and the effects due to parasitic elements in the device.

The theory of modulation characteristics of strongly injected semiconductor laser will be stated. Modulation of semiconductor laser had been studied since the invention of this laser and the approach was using the rate equation with noise free and with periodic injection current. Steady-state solutions of the rate equations can be found for constant injection rate. Small fluctuations (first-order perturbations) terms were then added to steady-state as given below. Modulation response when evaluated can be a treated as a direct measure of the rate at which information can be transmitted (primarily baseband). Modulation bandwidth at any biasing current or optical power is an important for optical communication systems.

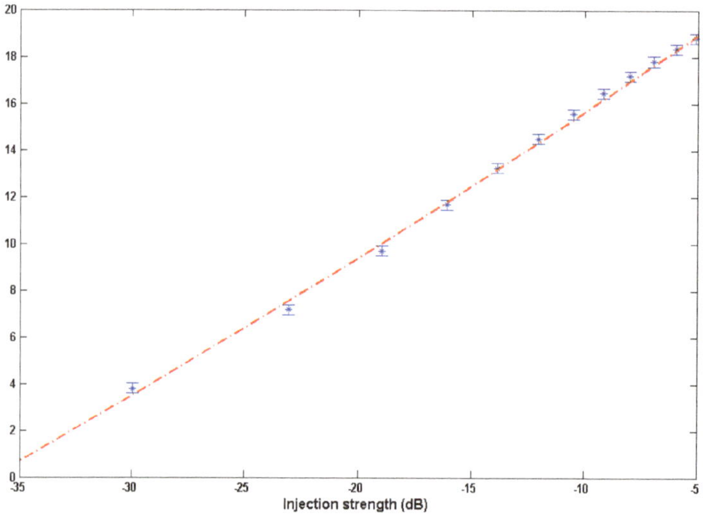

Figure 4. Cutoff frequency dependence on the injection strength of the injection locked semiconductor laser. The solid line is the best fit of the numerical results.

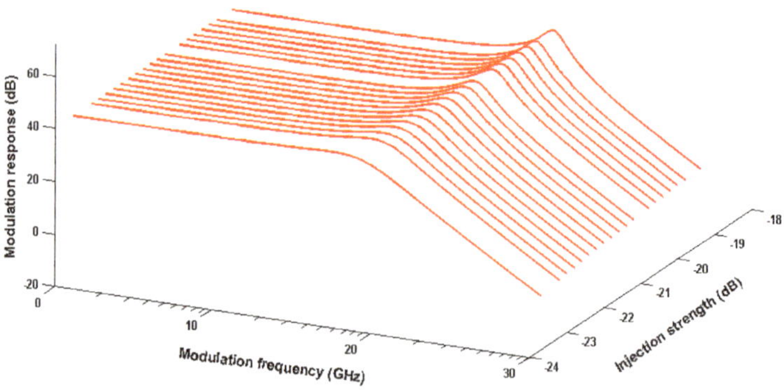

Figure 5. Modulation response of injection locked semiconductor laser as a function of injection strength and modulation frequency. The slave laser was detuned at -7.962 GHz

Single mode rate equations of free-running laser can be regarded as either as a model for purely single-mode operation or as an approximate model for the dynamics of the total photon number. The carrier number N and photon number P can be written as:

$$\frac{dN(t)}{dt} = injected\ \ carrier - spont.emission - stimulated\ \ emission$$

$$\frac{dP(t)}{dt} = (Gain - loss) + spont.term\ \ coupled\ to\ \ lasing\ mode$$

Now, the two above equations will read as:

$$\frac{dN(t)}{dt} = J(t) - \frac{N}{\tau_{sp}} - G(N,P)P \tag{5}$$

$$\frac{dP(t)}{dt} = \left(G(N,P) - \frac{1}{\tau_p}\right)P + \beta\frac{N}{\tau_{sp}} \tag{6}$$

where the gain *G (N, P)* can be expressed, assuming linear gain dependence on the carriers, as:

$$G(N,P) = \frac{G_o + G_N(N - N_o)}{1 + \varepsilon P} \tag{7}$$

where G_o is the linear gain, $G_N = \left(\frac{\partial G}{\partial N}\right)_{N=N_o}$ is the differential gain at N_o , and the factor $1+\varepsilon P$ accounts for nonlinear gain saturation. The gain saturation becomes important at high photon numbers. The factor ε is called the gain suppression coefficient.

β is the fraction of spontaneous emission coupled to the lasing mode. The steady-state equations can be found by setting $\frac{d}{dt}$ equal to zero.

First-order perturbation is written as:

$$N = \bar{N} + \delta N \quad , |\delta N| \ll \bar{N} \tag{8}$$

$$P = \bar{P} + \delta P \quad , |\delta P| \ll \bar{P} \tag{9}$$

Assuming the gain is a function of the carriers N and photons P, and can be approximated by Taylor expansion around the bias as:

$$G(N,P) = G_o + G_N\delta N + G_P\delta P \tag{10}$$

where G_N is the differential gain, and G_P is the saturation gain.

Substituting Eqs.(8), (9), and (10) into the rate equations (5) & (6) and omitting the small quantities of the second and higher order terms, a differential equations result as:

$$\frac{d}{dt}\delta N = -\left(G_0 + G_p\bar{P}\right)\delta P - \left(G_N\bar{P} + \frac{1}{\tau_p}\right)\delta N \tag{11}$$

$$\frac{d}{dt}\delta P = -\left(\frac{\beta\bar{N}}{\tau_p\bar{P}} - G_p\bar{P}\right)\delta P + \left(G_N\bar{P} + \frac{\beta}{\tau_p}\right)\delta N \tag{12}$$

Eliminating δN from the above equations we get an equation for the modulation in the photon number as:

$$\frac{d^2}{dt^2}\delta P + 2\Gamma\frac{d}{dt}\delta P + \omega_R^2\delta P = G_N\bar{P}J \tag{13}$$

where the Γ is the damping factor and ω_R is the resonance frequency given as:

$$\omega_R^2 \simeq \frac{G_N}{\tau_P}P \tag{14}$$

This resonance frequency is for free-running semiconductor laser, i.e., without optical injection. The modulation response can be found from Eqs. (7) and (8) by assuming an exponential solution for both δP and δN as $(\exp(i\omega t))$, with ω is the modulation frequency. The response is written as:

$$|H(\omega)| = \frac{\omega_R^2}{\left[\left(\omega_R^2 - \omega^2\right)^2 + 4\Gamma^2\omega^2\right]^{1/2}} \tag{15}$$

Now, the resonance frequency and frequency response of semiconductor laser with strong optical injection will be presented.

To estimate the modulation response of optically injected semiconductor laser, the rate equations have to be solved again with strong optical injection, a resonance frequency with optical injection formula can be derived as [16]:

$$\omega_{R_{inj}}^2 \approx \omega_R^2 + \left(\frac{1}{\tau_{in}}\right)^2\left(\frac{P_{inj}}{P}\right)\sin^2\theta \tag{16}$$

where $\omega_{R_{inj}}$ is the resonance frequency with optical injection, and θ is the is phase difference between the slave laser and the master laser.

With given theoretical analysis above, the transfer function for optically injected laser was obtained and plotted in the above figures. The frequency response of the slave laser with the highest enhanced resonance frequency can be achieved by optical injection and this found to be more than five times increased as compared to experimental results [17]. The bandwidth accordingly will be increased due to this increase in the frequency response of the slave laser. This enhancement in the bandwidth is dependent on the optical injection parameters and when the slave laser is operated in the stable locking region. Modulation enhancements characteristics of strongly injected slave laser are mainly due to the shift in the resonance frequency have to be understudied thoroughly. The theory of semiconductor laser, as a matter of fact, always simplify the arguments and make assumptions in order to make the an analytical solutions available. Numerical simulations for the laser nowadays can be found in easy used packages in the market. We have analyzed the response of semiconductor laser based on the rate equations and solved numerically. The change in the photon number due to optical injection is not the only source of increasing bandwidth but also the competition between the frequency of dominant mode of the slave laser and the frequency shift induced by strong optical injection. This bandwidth enhanced of slave laser is ultimately a candidate source for optical communications, not due to its enhanced bandwidth but also to its chaotic output as secure transmitter.

4. Chaos by injection locking

The chaotic behavior of the slave laser under strong optical injection will be given in terms of the important parameters of locking regimes which are the frequency detuning and injection ratio. The optical chaos have been observed, as we will see later, and developed through period-doubling, i.e., route-to-chaos. The analysis will show that the chaotic behavior is dependent on the injection strength and frequency detuning. The bandwidth of semiconductor laser will be verified and enhanced by optical injection. The bandwidth of such chaotic laser transmitter is enhanced roughly three times by optical injection compared with the bandwidth when there is no optical injection [14]. Chaotic dynamics and period doubling were observed experimentally in the VCSELs lately [15], in addition to edge emitting lasers.

The differential rate equations (1)-(3) were subjected to numerical solutions using fourth-order Runge-Kutta algorithm with strong optical injection and the chaotic output of the slave laser is shown in figure 6. This output was originated from period-doubling route-to-chaos. This chaotic behavior is optical injection locking parameters dependent, i.e., injection strength, and frequency detuning. This is illustrated in the following diagram.

Figure 7 shows the chaotic dynamics of the slave laser under strong optical injection, with the same parameters as in the previous figure. The chaos is a phenomenon of the generic properties of the extra degree of freedom introduced by the optical injection in rate equations. The scenario of period doubling route-to-chaos was initiated by excitability of the chaotic attractor. This attractor triggers the system in deterministic sequence reaching the chaos. This observation of bifurcation in the laser power output (the response system) related to the stimulus (the drive) when coupled, synchronization is established. The ability of such synchronizing system offers the opportunity as a chaotic transmitter in optical communications. Any further increase in the injection strength will eventually be an extreme output of chaotic dynamics.

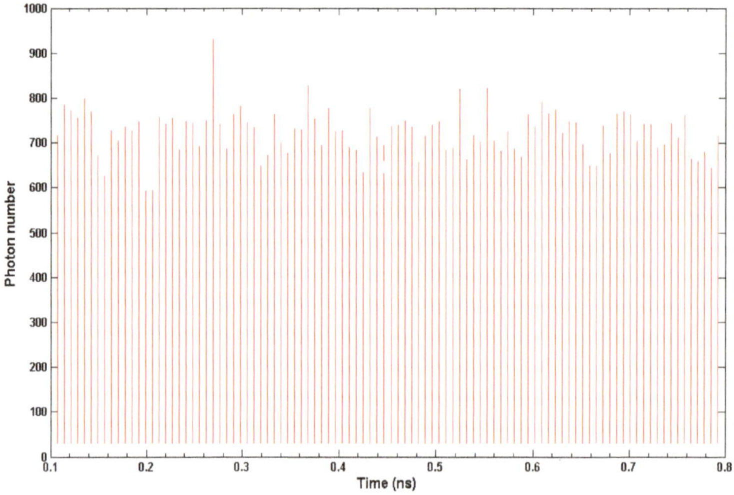

Figure 6. Time series of chaotic photon number of the slave laser at detuning (--7.962 GHz), linewidth enhancement factor (α =3), and injection strength (-22 dB).

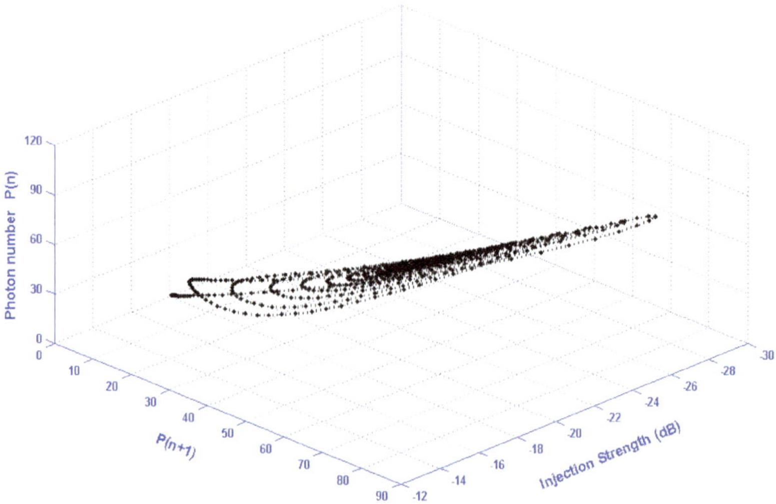

Figure 7. Phase portraits of chaotic output of the slave laser as a function of injection strength.

Injection strength plays a great role in the chaotic behavior of the slave laser through period-doubling route-to-chaos. This chaotic dynamics, when explained by nonlinear bifurcation theory, is established as the relaxation oscillations (RO) become undamped via Hopf bifurcations. This bifurcations originating from the undamped RO for Fabry-Perot lasers, while the enhanced RO damping of quantum dot lasers results in the removal of these chaotic regions [18]. Period doubling in the above figure has predicted the dependence on the injection parameters. Recently, the origin of such periodicity was explained as the beating between two wavelengths, namely, the injected wavelength and the cavity resonance wavelength [19].

The above theoretical results and predictions of chaotic dynamics agree well with the experiments on a qualitative level [20]. Many theoretical and experimental studies revealed the necessity in viewing the dynamics from a broad perspective [21-25].

5. Frequency chirping

Frequency chirping in semiconductor lasers can be suppressed by strong optical injection and hopefully this laser could have better modulation characteristics than free-running laser. As the carrier density increases, resulting from the current injection into the active region will change the refractive index of the region and generate the frequency chirping. This phenomenon can have a considerable limitations on the modulation of semiconductor laser at high bit rate. An important figure-of-merit for chirp is the chirp-to -power ratio (CPR), which is defined as the ratio of lasing frequency deviation to power deviation [26], and is defined as [27]:

$$CPR = \frac{1}{2\pi H(\omega)}\left|\frac{d\theta}{dt}\right| \cong \frac{\omega}{2\pi}\alpha\sqrt{\frac{\omega^2 + (U - V/\alpha)^2}{\omega^2 + (U + V\alpha)^2}} \tag{17}$$

Where

$$U = \frac{1}{\tau_{in}}\left|\frac{P_{inj}}{P}\right|\cos\phi_L \tag{18}$$

$$V = \frac{1}{\tau_{in}}\left|\frac{P_{inj}}{P}\right|\sin\phi_L \tag{19}$$

with ϕ_L is the phase of the intracavity laser field relative to the injection field. This equation provides with the fact that CPR is dependent entirely on the modulation response and the phase. The stated above equations of the injection-locked semiconductor explains the chirp suppression or reduction and its dependence on injection parameters, with this arguments it can optimize the performance of the laser.

Chirp-to-power ratio of injection locked semiconductor at locking range of stable operation and as a function of modulated frequency is shown in figure 8, three different injection strength were taken and for two values of linewidth enhancement factor (α). This factor has influenced the frequency chirping characteristics of the laser under direct current modulation, since this factor plays great role in the refractive change with injected carriers. This influence can be realized from the relation stated in Eq. (17).

The frequency chirping when considered in the output power of the slave laser, the detuning was assumed to be constant, but it is dependent on the optical input power, hence the CPR has to take into account this dependent. Also, the laser cavity frequency $\omega_o(N)$, is carrier dependent. So the effects of variations in frequency detuning, and laser cavity frequency have to be included in the CPR.

Frequency chirping had been decreased with increasing injection strength and with lowered value of linewidth enhancement factor. The simulation reveals the dramatic influence of injection locking on the frequency chirping characteristics. Dynamical properties have been found, experimentally lately [28], to depend on the injection strength through the evolution of the optical and electrical spectral distribution. Substantial reduction in frequency chirping was observed in the direct modulation injection-locked laser, and this reduction was much more pronounced at low modulation frequency in experiment [29].

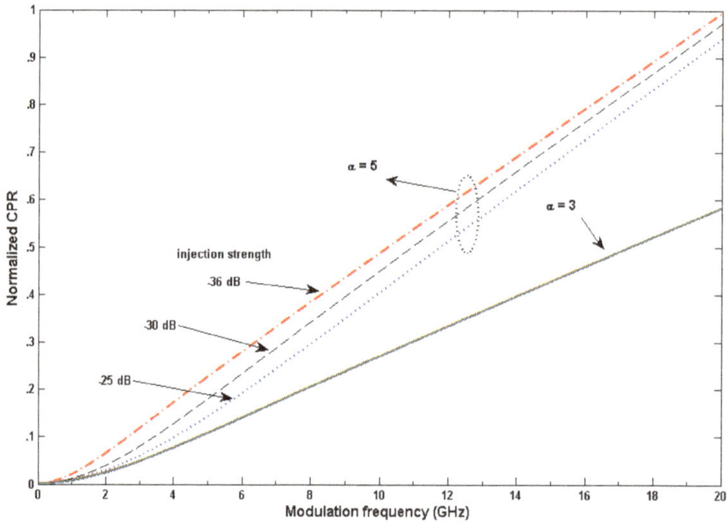

Figure 8. Frequency chirping characteristics of injection locked semiconductor showing its dependence on injection strength and linewidth enhancement factor.

6. Relative intensity noise

The relative intensity noise (RIN), intrinsic noise inherited in the device due to spontaneous emission noise, is of major importance for optical communication systems, whereas low RIN is needed to achieve high signal-to-noise ratio (SNR). It is observed that RIN shows a peak near the relaxation oscillation frequency and this would be related to the bandwidth of the laser [30]. If the injection locked slave laser could show a suppression in the spectrum of the

RIN near the resonance and more damping due to the increase of injection strength, more enhancement in bandwidth will be realized. Experimentally has verified the enhancement bandwidth of sampled grating distributed Bragg reflector laser (SG-DBR), lower of intensity noise and higher resonance frequency due to strong optical injection [31]. The dependence of RIN on both bias current and detuning have been studied [32].

The fluctuations in the power spectrum $\Theta_P(\omega)$ can be defined as:

$$\Theta_P(\omega) = < \left|\Delta\tilde{P}(\omega)\right|^2 > \tag{20}$$

This, by normalization with $(\bar{P})^2$, would give the relative intensity noise spectrum:

$$RIN = \frac{\Theta_P(\omega)}{\left(\bar{P}\right)^2} \tag{21}$$

The relative intensity noise spectrum, RIN, for free-running laser and with strong optical injection is shown in figure 9. With injection locking, the noise peak was shifted to a higher frequency.

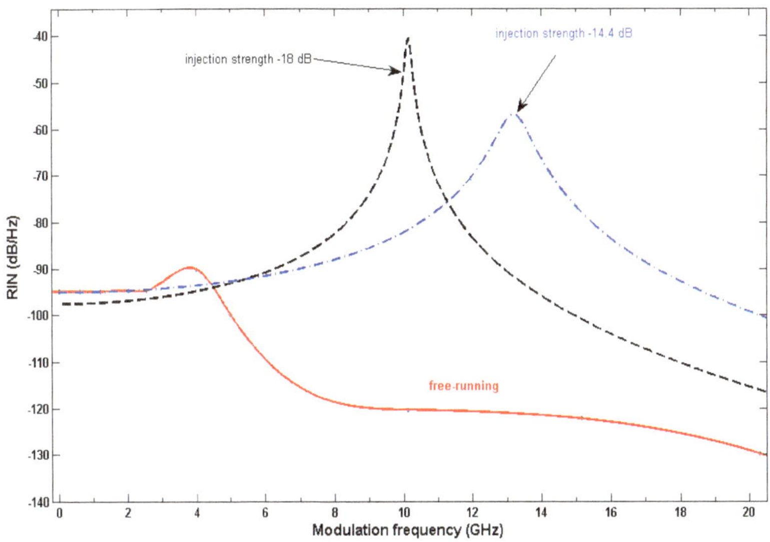

Figure 9. Relative intensity noise spectrum of free-running semiconductor laser and injection-locked for two different injection strengths.

When injection locking was increased from -18 dB to -14.4 dB, the noise peak was shifted form 10 GHz to at least 13 GHz. Hence, injection locking can enhance the bandwidth of optical communication systems. Also, a reduction 14 dB/Hz in RIN had occurred when injection strength was increased at the mentioned values above.

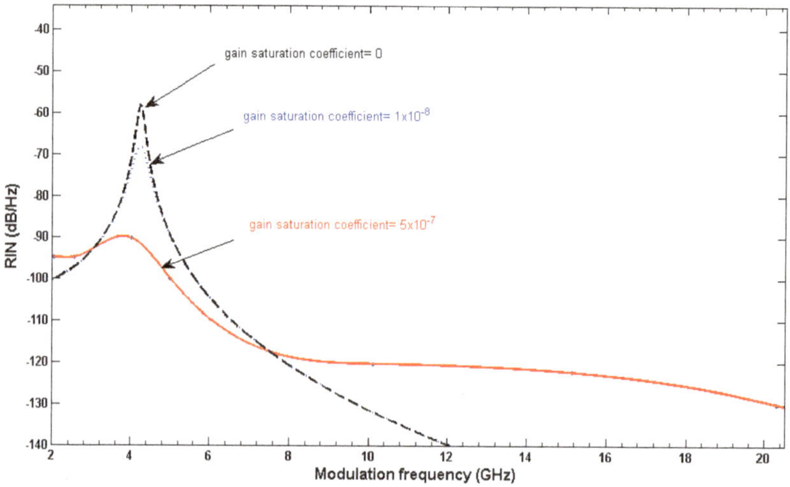

Figure 10. The influence of gain saturation coefficient on the RIN spectrum of semiconductor laser.

To illustrate how gain saturation coefficient can alter the RIN spectrum of the slave laser, figure 10 shows this effect. When the gain saturation coefficient was increased, it is noticed that the RIN spectrum at the resonance peak was suppressed and this can be interpreted as damping due to the gain saturation coefficient. We would expect that the strong injection locking phenomenon may give extra damping of RIN peak in addition to the gain saturation coefficient. This has to be verified experimentally. So, both the methods will exhibit large significant RIN suppression, and meet the demands for higher bit rates and longer optical communication network.

7. Conclusions

This chapter, as I intend to, gave a brief description and theory of strong optical injection and its influence on the characteristics of the slave laser. Secure optical communications can be verified by such operation of the laser. Modulation bandwidth enhancement has been found by numerical simulation of the rate equations. The enhancement was found to be dependent on the injection parameters, frequency detuning and injection strength. Period doubling and relaxation oscillation frequency are the physical mechanisms of bandwidth enhancement by injection locking. Substantial reduction in frequency chirping, and a suppression of 14 dB/Hz in the RIN were observed in the direct modulation injection-locked laser. These characteristics of such a laser are the demands for high bit rate and longer optical communication network. Chaotic dynamics of the slave laser is the challenging candidate for secure optical systems.

Author details

Muzahim I. Azawe
Mosul University, Mosul, Iraq

8. References

[1] Gavrielides A, Kovanis V, Nizette M, Erneux T, and Simpson T B (2002) Period Three Limit-Cycles in Injected Semiconductor Lasers. J. Opt. B: Quant. Semiclass. Opt. 4:pp. 20–26.

[2] Hong Y, Shore K A, Lawrence J S, and Kane D M (2000) Wavelength Switching by Positively Detuned Side-Mode Injection in Semiconductor Lasers. Appl. Phys. Lett. 76: pp. 3170-3172.

[3] Chan S-C (2010) Analysis of an Optically Injected Semiconductor Laser for Microwave Generation. IEEE J. Quant. Elect. QE-46: pp. 421-428.

[4] Chang C-H, Chrostowski L, Chang-Hasain C, and Chow W (2002) Study of Long-Wavelength VCSEL–VCSEL Injection Locking for 2.5-Gb/s Transmission. IEEE Photon. Techn. Lett. 14: pp. 1635-1637.

[5] Chrostowski L, Zhao X, Chang-Hasain C, Shau R, Ortsiefer M, and Amann M-C (2006) 50GHz Optically Injection-Locked 1.55 μm VCSELs. IEEE Photon. Techn. Lett. 16: pp. 367-369.

[6] Heinricht P, Wetzel B, O'Brien S, Amann A, and Osborne S (2011) Bistability in an Injection Locked Two Color Laser With Dual Injection. Appl. Phys. Lett. 99: pp. 011104-011106.

[7] Kusumoto K, and Ohtsubo J (2003) Anticipating Synchronization Based on Optical Injection-Locking in Chaotic Semiconductor Lasers. IEEE J. Quant. Elect. QE-39: pp. 1531-1539.

[8] Bavard X, Locquet A, Larger L, and Goedgebuer J-P (2007) Influence of Digitization on Master–Slave Synchronization in Chaos-Encrypted Data Transmission. IET Optoelect. 1:pp. 3-8.

[9] Blina S,Guignarda C, Besnarda P, Gabetb R, Stephana G, and Bondiouc M (2003) Phase and Spectral Properties of Optically Injected Semiconductor Lasers. Comptes Rendus Physique 4: pp. 687-699.

[10] Chlouverakis K E, and Adams M (2003) Stability Maps of Injection-Locked Laser Diodes Using the Largest Lyapunov Exponent. Opt. Commun. 216:pp. 405-412.

[11] Chan S-C, and Liu J-M (2006) Frequency Modulation on Single Sideband Using Controlled Dynamics of an Optically Injected Semiconductor Laser. IEEE J. Quant. Elect. QE-42: pp. 699-705.

[12] Lau E K, Sung H K, and Wu M C (2008) Frequency Response Enhancement of Optical Injection-Locked Lasers. IEEE J. Quant. Elect. QE-44: pp. 90-99.

[13] Kobayashi S (1991) Injection-Locked Semiconductor Laser Amplifiers in *Coherence Amplification and Quantum Effects in Semiconductors Lasers*, Yamamoto Y (Ed), John Wiley and Sons, Inc. New York.

[14] Wang A, Wanh Y, and He H (2008) Enhancing the Bandwidth of the Optical Chaotic Signal Generated by a Semiconductor Laser With Optical Feedback. IEEE Photon. Tech. Lett. 20:pp. 1633-1635.

[15] Hurtado A, Quirce A, Valle A, Pesquera L, and Adams M J (2010) Nonlinear Dynamics Induced by Parallel and Orthogonal Optical Injection in 1550 nm Vertical- Cavity Surface-Emitting Lasers (VCSELs). Opt. Exp. 18:pp. 923-928.

[16] Murakami A, Kawashima K, and Atsuki K, (2003) Cavity Resonance Shift and Bandwidth Enhancement in Semiconductor Lasers with Strong Light Injection. IEEE J. Quant. Elect. QE-39:pp. 1190-1204.

[17] Sung H-K, Lau E K, and Wu M C, (2007) Optical Properties and Modulation Characteristics of Ultra-Strong Injection-Locked Distributed Feedback Lasers. IEEE J. Select. Top. Quant. Elect. 13, pp. 1215-1221.

[18] Kelleher B, Bonatto C, Huyet G, and Hegarty S P, (2011) Excitability in Optically Injected Semiconductor Lasers: Contrasting Quantum-well and Quantum-Dot-Based Devices. Phys. Rev. E 83, pp. 026207-026213.

[19] Chan S-C, (2010) Analysis of an Optically Injected Semiconductor Laser for Microwave Generation. IEEE J. Quant. Elect. QE- 46, pp. 421-428.

[20] Thèvenin J, Romanelli M, Vallet M, and Brunel M, (2011) Resonance Assisted Synchronization of Coupled Oscillators: Frequency Locking Without Phase Locking. Phys. Rev. Lett. 107, pp. 104101-104106.

[21] Christoph S, Labukhin D, Nickolay Z, and Adams M, (2010) Dynamics of Optically-Injected Semiconductor Lasers Using the Travelling-Wave Approach. IEEE J. Quant. Elect. QE-46,pp. 220-227.

[22] Hwang S k, Gao J B, and Liu J M (2000) Noise-Induced Chaos in an Optically Injected Semiconductor Laser Model. Phys. Rev. E 61, pp. 5162-5170.

[23] Kéfélian F, and Gallion P, (2008) Locking and Noise Properties of Multisection Semiconductor Lasers With Optical Injection. Application to Fabry–Pérot and DFB Cavities. IEEE J. Quant. Elect. QE-44, pp. 547-560.

[24] Martínez-Zérega B E, Jaimes-Reategui R, Pisarchik A N, and Liu J M (2005) Experimental Study of Self-Oscillation Frequency in a Semiconductor Laser With Optical Injection. J. Phys. 23, pp. 62-67.

[25] Kovanis V, Gavrielides A, and Gallas J A C, (2010) Labyrinth Bifurcations in Optically Injected Diode Lasers. Eur. Phys. J. D 58, pp. 181-186.

[26] Wang X, and Chrostowski L, (2011) High-speed Q-modulation of Injection-Locked Semiconductor Lasers. IEEE Photon. J. 3, pp. 936-945.

[27] Chen H F, Liu J M, and Simpson T B, (2000) Response Characteristics of Direct Current Modulation on a Bandwidth-Enhanced Semiconductor Laser Under Strong Injection Locking. Opt. Commun. 173, pp. 349-355.

[28] Schires K, Hurtado A, Henning I D, and Adams M J, (2011), Comprehensive Experimental Analysis of Nonlinear Dynamics in an Optically-Injected Semiconductor Laser. AIP Advance. 1, pp. 032131-12.

[29] Okajima Y, Hwang S K, and Liu J M, (2003), Experimental Observation of Chirp Reduction in Bandwidth-Enhanced Semiconductor Lasers Subject to Strong Optical Injection. Opt. Commun. 219, pp. 357-364.

[30] Jin X, Tarng B Y, and Chuang S L, (2007), Relative Intensity Noise Study in the Injection-Locked Integrated Electroabsorption Modulator-Lasers, ISDRS, December 12-14, College Park, MD, USA.

[31] Bhardwaj A, Ristic S, Johansson L A, Althouse C, and Coldren L A, (2010), Direct Modulation Bandwidth Enhancement of Strongly Injection-Locked SG-DBR Laser. Elect. Lett. 46, pp. 362 – 363.

[32] Park K Y, and Lee C H, (2008), Noise Characteristics of a Wavelength-Locked Fabry–Perot Laser Diode. IEEE J. Quant. Elect. 44, pp. 995-1002.

Section 3

Optical Communication: Optical Systems

Chapter 8

Terrestrial Coherent Free-Space Optical Communication Systems

Mingbo Niu, Julian Cheng and Jonathan F. Holzman

Additional information is available at the end of the chapter

1. Introduction

In the past three decades, the demand for high-speed communications has increased dramatically, while fiber optical communications has been applied in the majority of data transmission networks. Optical fiber has advantages over existing copper wire in long distance and high demand applications. The ever increasing need for higher bandwidth and higher speed optical data and communications transmission is driving the development of 100 gigabit per second (Gbit/s) communication links. However, infrastructure development within cities is relatively difficult and time-consuming, and fiber-optic systems are complex and expensive. Due to these difficulties, fiber-optic communication systems have primarily been installed in long-distance applications, where they can be used to their full transmission capacity, offsetting the increased cost.

The original free-space optical (FSO) communications white paper by Dr. Erhard Kube, "Information transmission by light beams through the atmosphere," was published in German in *Nachrichtentechnik*, June 1968. New advances in FSO technologies have led to a recent rebirth of optical broadband access as an attractive alternative for ultra high-speed networking. This can push forward the seamless development of the promising all-optical networks. In dense urban areas or places where optical fiber infrastructure does not exist, FSO communication systems have been shown to be a viable alternative [1]. Lower costs, larger license-free bandwidths, better information security, greater link flexibility, and a reduced time-to-market are all significant benefits of FSO communication systems [1]-[3]. FSO communications is a promising candidate to satisfy the new communication requirements due to its ability to transmit information at extremely high data rates using compact, low-mass terminals, while avoiding interference problems.

FSO communications, also known as the wireless optical communications, transmits optical signals through free-space. It requires line-of-sight (LOS) transmission, which means the transmitter and receiver at both networking locations must see each other. Whereas existing optical fiber is a predictable medium, FSO communications can suffer from cloud coverage and harsh weather conditions leading to atmospheric effects which degrade the designed system availability and performance. Rain, snow, sleet, fog, etc. are limiting factors which can affect the transmission of laser beams through the atmospheric channels.

There are three primary atmospheric factors that can affect optical beam propagation: absorbtion, scattering, and refractive-index fluctuations (i.e., optical turbulence). Absorption and scattering are often grouped together under the topic of extinction, defined as the reduction or attenuation in the amount of radiation transmitting through the atmosphere. They are both deterministic effects that are fairly well-known and can be predicted by software packages such as FASCODE [4] or MODTRAN [5]. Nonetheless, optical turbulence is generally considered as the most serious optical effect on a propagating beam through atmospheric channels. In this chapter, we will focus on terrestrial coherent FSO systems in the presence of atmospheric turbulence.

1.1. Preliminaries on coherent free-space optical communications

Coherent fiber optical communications attracted considerable attention in the late 1980s for its ability to approach the theoretical receiver sensitivity limit. Similarly, FSO systems have great potential on improving channel usage when implemented with coherent detection [6]. One scheme of coherent FSO detection is called homodyne detection, where the receiver demodulates the optical signal directly to the baseband because the local oscillator laser frequency is synchronized to the optical signal carrier frequency. However, it can be unstable to perform optical synchronization in practice. As a result, heterodyne detection was introduced to simplify the receiver design and make coherent FSO systems more applicable. In heterodyne detection, the optical signal is first converted to an electrical signal with an intermediate frequency. Then a phase noise compensation scheme is used to track the phase noise of the IF signal. The received signals in coherent FSO systems can be made to be limited only by the shot noise of the incident optical power (given a sufficiently large local oscillator beam power). Furthermore, the extraction of phase information with these processes allows for a greater variety of modulation formats in comparison to irradiance-dependent detection schemes such as irradiance modulation with direct detection (IM/DD). The advantages of coherent FSO systems with phase noise compensation over IM/DD systems are excellent background noise rejection [7], higher sensitivity, and improved spectral efficiency (at the cost of higher system complexity). Note that the implementation and tradeoff of coherent FSO systems are beyond the scope of this chapter. More details can be found in [8]–[10]. Unlike coherent radio frequency (RF) communication links, coherent FSO systems utilize local oscillator and signal optical fields together with optical/electrical synchronization for the signal recovery.

An optical wave propagating through the atmosphere will experience irradiance fluctuations, also referred to as scintillation or turbulence-induced fading. Scintillation is caused by random fluctuations of refractive index due to temperature, pressure, and wind variations along the optical propagation path of the channel in the atmosphere. Theoretical and experimental studies generally center around the scintillation index, which is defined as the normalized variance of irradiance fluctuations

$$\sigma_{si}^2 = \frac{E[I^2] - (E[I])^2}{(E[I])^2} = \frac{E[I^2]}{(E[I])^2} - 1 \tag{1}$$

where the quantity I is the instantaneous optical irradiance and $E\bullet\cdot\bullet$ denotes the expectation operation.

In weak turbulence regimes (when the scintillation index is less than unity), the scintillation index is found to be proportional to the Rytov variance [12], which is defined as

$$\sigma_R^2 \bullet\ 1.23 C_n^2 k^{\frac{7}{6}} L_t^{\frac{11}{6}} \tag{2}$$

where C_n^2 stands for the index of refraction structure parameter in $m^{-2/3}$, $k \bullet\ 2\pi/\lambda_w$ is the optical wave number (λ_w denotes the wavelength), and L_t is the transmission path length between the transmitter and receiver. In (2), C_n^2 is an altitude-dependent variable, and the most commonly used Hufnagle-Valley model for C_n^2 is given by [12]

$$C_n^2 \bullet\ 0.00594 \left(\frac{v}{27}\right)^2 \bullet h \times 10^{-5\bullet 10} e^{\frac{h}{1000}} \bullet\ 2.7 e^{\frac{-h}{1500}} \times 10^{-6} \bullet\ A_c e^{\frac{-h}{1000}} \tag{3}$$

where v is the root-mean-square wind speed in meters per second, h is the altitude in meters, and A_c is a nominal value of C_n^2 at the ground. Typically, the value of C_n^2 varies from approximately $10^{-17} m^{-2/3}$ for weak turbulence conditions to $10^{-13} m^{-2/3}$ for strong turbulence conditions (with $10^{-15} m^{-2/3}$ as a typical average value) [13].

When the optical turbulence strength extends to moderate-to-strong irradiance fluctuation regimes (when the scintillation index is greater than unity) by increasing C_n^2 and/or path length L_t, the scintillation index for a plane wave and that for a spherical wave are, respectively, related to the Rytov variance by [11]

$$\bullet\sigma_{si}^2\bullet_{plane} \bullet\ \exp\left[\frac{0.54\sigma_R^2}{\bullet 1\bullet\ 1.22\sigma_R^{\frac{12}{5}}\bullet^{\frac{7}{6}}} \bullet\ \frac{0.509\sigma_R^2}{\bullet 1\bullet\ 0.69\sigma_R^{\frac{12}{5}}\bullet^{\frac{5}{6}}}\right] - 1 \tag{4}$$

and

$$\bullet\sigma_{si}^2\bullet_{sphere} \bullet\ \exp\left[\frac{0.17\sigma_R^2}{\bullet 1\bullet\ 0.167\sigma_R^{\frac{12}{5}}\bullet^{\frac{7}{6}}} \bullet\ \frac{0.225\sigma_R^2}{\bullet 1\bullet\ 0.259\sigma_R^{\frac{12}{5}}\bullet^{\frac{5}{6}}}\right] - 1. \tag{5}$$

The performance of FSO communication systems can be significantly degraded by turbulence-induced scintillation. To be more specific, scintillation can lead to power loss at the photodetector and random fluctuations of the received signal below a predetermined detection threshold. The reliability of such FSO systems can be predicted by introducing mathematical models for the probability density function (PDF) of the instantaneous fading irradiance of the optical signal.

For the past several decades, the scientific community has investigated statistical models of turbulence-induced scintillation in FSO communications through the atmosphere. Of the turbulence-induced scintillation models introduced so far, the most commonly accepted models are the log-normal turbulence model (typically describing irradiance fluctuations in weak turbulence conditions), K-distributed turbulence model (typically describing irradiance fluctuations in strong turbulence conditions), and Gamma-Gamma turbulence model (providing a description of much wider irradiance fluctuation ranges across the weak to strong turbulence regimes). In the rest of this chapter, turbulence-induced scintillation is referred as (atmospheric) turbulence for simplicity unless stated otherwise.

This chapter is structured as follows. We first introduce the concept of coherent FSO communication and conduct a detailed literature review in the rest of Section 1. Section 2 will describe the system model of a coherent FSO link over atmospheric turbulence channels. We will analyze the signal-to-noise ratio (SNR) at the coherent FSO receiver and derive its expression in Section 2.1. A generalized atmospheric turbulence model and its characteristics will be presented in Section 2.2. In Section 3.1, we will present exact error rates for binary phase-shift keying (BPSK) using a moment generating function approach and a characteristic function approach. Furthermore, we will present asymptotic error rate analyses in Section 3.2 to offer further insights into system performance behavior in large SNR regimes. We will analyze outage probability in Section 3.3. Diversity techniques such as maximum ratio combining (MRC) and equal gain combining (EGC) will be employed to mitigate the turbulence effects in Section 4.1 and Section 4.2, respectively, and a performance comparison using numerical case studies will be presented in Section 4.3. We will also study the error rate performance degradation caused by phase noise compensation error in Section 5. When the standard deviation of the phase noise compensation error is large, we propose differential PSK (DPSK) as an effective alternative to BPSK in Section 5.3, as DPSK does not require the estimation of optical carrier phase. In Section 6, we summarize and conclude our important findings for coherent FSO systems and suggest further research topics on coherent FSO communications.

1.2. Literature review

FSO communications can be a key building block for future wide-area wireless data networks [14], [15] and can have great potential for applications in fourth-generation (4G) wireless systems [16]. Such systems are currently being deployed and will encompass a number of complementary access technologies with higher channel capacities, multiple antennas, and Gbit/s data rates [16]. Since the wireless optical transmission links are over the atmosphere, the laser beam propagating through turbulent channels is affected by scintillation and phase aberration, which can impair the system performance significantly.

In order to evaluate the system performance, an accurate model of turbulence is needed in error rate studies. In early studies of FSO communications, the log-normal distribution was used as the turbulence model [6], [17]–[22]. Although the log-normal distribution is one of the most widely used turbulence models, this PDF is only applicable for weak turbulence conditions. It was shown in [23]–[26] that the K-distributed turbulence model provides good agreement with experimental data in a variety of FSO experiments involving radiation scattered by strong turbulence. In a recent series of papers on scintillation theory [11], [27], Andrews *et al.* introduced the modified Rytov theory and the Gamma-Gamma PDF was proposed as a tractable mathematical model for a wide range of atmospheric turbulence levels. Other statistical models in the FSO literature to describe atmospheric turbulence are the log-normal Rician, Rayleigh and $I-K$ models [28], [29], [44].

The performance of IM/DD FSO systems for different turbulence models has been well studied in the literature. Zhu and Kahn studied the employment of maximum likelihood sequence detection (MLSD) for IM/DD FSO links [21]. They further studied the pairwise error probability of coded FSO links assuming the turbulence to be log-normal distributed [22]. In [30], Uysal *et al.* studied the pairwise error probability of on-off keying (OOK) with temporally correlated K-distributed turbulence. Since FSO communications requires LOS links, pointing errors can affect the FSO system performance if the detector aperture size is finite (non-negligible compared to the beam spot size). In [31], Farid and Hranilovic presented an FSO channel model which models the fading due to log-normal/Gamma-Gamma atmospheric turbulence and pointing errors by considering beam width, pointing error variance and detector size. A closed-form expression of fading PDF, including the combined effects of K-distributed turbulence and pointing errors as well as the bit-error rate (BER) expression for OOK were obtained by Sandalidis *et al.* in terms of the Meijer's G-function [32]. Sandalidis *et al.* further studied the BER performance of the same links but considered the misalignment (pointing errors) effects [32]. Later, Uysal extended their discussion of pairwise error probability for coded OOK FSO links to the cases with independent Gamma-Gamma turbulence [33]. Riediger *et al.* investigated a multiple symbol detection decision metric for OOK in both log-normal and Gamma-Gamma turbulence [34]. The results in these papers demonstrate that the performance of a single branch FSO link severely suffers from atmospheric turbulence and is far from satisfying the typical BER requirements for communication applications with practical SNRs. This necessitates the deployment of powerful fading-mitigation techniques.

In the existing literature on FSO communications, two techniques have been proposed to mitigate the degrading effects of atmospheric turbulence: error control coding in conjunction with interleaving [30], [33], and maximum likelihood sequence detection with the knowledge of joint temporal statistics of the turbulence [21]. However, both approaches come with some practical limitations. The first one requires large-size interleavers whereas the latter suffers from high computational complexity [20].

FSO systems using diversity reception can achieve significant performance improvements by mitigating the atmospheric channel turbulence. The use of spatial diversity was first proposed for FSO systems by Ibrahim and Ibrahim [43]. EGC and optimal combining have been shown to enhance the link outage performance with independent log-normal turbulence by Lee and Chan [17]. BERs of multiple-input multiple-output (MIMO) FSO systems with both independent and correlated log-normal turbulence was studied by Navidpour *et al.* [20]. In [21], a symbol-by-symbol maximum likelihood detector with spatial diversity in correlated log-normal turbulence was studied by Zhu and Kahn. In [37] and [38], Wilson *et al.* investigated MIMO FSO links employing pulse-position modulation (PPM) and Q-ary PPM with both Rayleigh and log-normal turbulence-induced fading. In a recent work, Tsiftsis *et al.* studied the K turbulence FSO link performance for an IM/DD system with OOK using optimal combining, EGC, and selection combining [40]. BER solutions that require multi-dimensional integrations were presented, and approximate BER expressions were also given using the Gaussian quadrature rule and an error function approximation based on the trapezoidal rule. Bayaki *et al.* studied MIMO IM/DD FSO links over the Gamma-Gamma turbulence and demonstrated a significant performance improvement by exploiting both transmitter and receiver diversity [41]. In a recent work, Abou-Rjeily and Slim studied the system performance for Q-ary PPM FSO systems with cooperative diversity (parallel multi-hop) over both Rayleigh and log-normal turbulence channels [42].

It is known that the best signal modulation format for IM/DD FSO systems is OOK, and the existing literature on IM/DD FSO mostly considers OOK modulation. However, to minimize the error rate, the OOK receiver requires the knowledge of the turbulence state in order to choose an optimum threshold. This implies the need for an adaptive decision threshold, which may be difficult to implement in a practical FSO system and is subject to channel estimation errors. As a result, practical FSO systems typically adopt a fixed threshold to reduce the complexity. Such systems will utilize large transmit powers to overcome the impact of atmospheric turbulence. This leads to costly operation. In [48], Li *et al.* theoretically showed that a fixed threshold for IM/DD systems with OOK modulation will lead to suboptimal system performance. It is pointed out that the BER of OOK modulation is determined by both the turbulence level and the fixed threshold, and, therefore, can not be made arbitrarily small in the presence of atmospheric turbulence even when the SNR is asymptotically large. PPM modulation has been proposed as an error-floor-free alternative to the OOK modulation [37]–[39]. However, PPM modulation needs a complex transceiver design because of the tight synchronization requirements, and it also suffers from a poor bandwidth efficiency.

Coherent FSO communication is an attractive alternative to FSO communication systems using direct detection. It offers an improved frequency/spatial selectivity, higher spectral efficiency, better background noise rejection and increased detector sensitivity (compared to direct detection) while eliminating the need of the adaptive threshold in the IM/DD OOK systems. The main feature of coherent FSO systems is that the receiver of a coherent FSO system is limited only by local oscillator induced quantum noise when the power of the local oscillator (as will be discussed in Section 2.1) is sufficiently high. This is a significant difference from the intensity modulated FSO systems, for which background and thermal noise are the dominant factors affecting the error rate performance. Some comprehensive references to early work in this coherent FSO area can be found in [6] and [36]. Recently, a comparison study was carried out by Lee and Chan [18] and showed performance improvement of coherent detection over IM/DD detection in a log-normal environment. They compared the IM/DD and coherent FSO systems with their corresponding best modulation schemes and demonstrated theoretically that coherent FSO systems can lead to lower error rates. It was also found that coherent detection can provide additional outage probability improvement over direct detection [19].

With the benefits of coherent FSO in mind, exact BER expressions have been developed for DPSK over K-distributed turbulence [45]. As an extension to the work done in [45], Tsiftsis evaluated the BER performance of coherent FSO with DPSK in Gamma-Gamma distributed turbulence [46]. In both theses works, however, a detailed system model and receiver SNR analysis for coherent FSO communication links were not presented. More recently, a heterodyne FSO system with pointing errors was studied by Sandalidis *et al.* for Gamma-Gamma turbulence channels [47]. In [47], closed-form fading statistics expressions that take into account both the turbulence and pointing error effects were derived in terms of the Meijer's G-function, and the BER expressions of DPSK for such cases were also developed. A statistical model was developed in [49] considering spatial phase noise with log-normal turbulence and analyzed for capacity evaluation [50], [51]. In [52] and [53], the error rate performance of coherent FSO systems with MRC, EGC, and selection diversity reception in strong turbulence regions was studied. Spatial diversity techniques have been shown to be effective in mitigating the atmospheric turbulence-induced fading. Recently, Belmonte and Kahn studied the performance of a coherent FSO link with a large effective aperture achieved by signal combining from multiple aperture arrays where a Gamma distribution is used to model irradiance fluctuations [54]. Aghajanzadeh and Uysal adopted the receiver model in [50] and studied the diversity-multiplexing tradeoff and the finite-SNR diversity gain for a single-input multiple-output coherent WOC system [55]. Later, they extended the work in [55] to a decode-and-forward multi-hop FSO relay link [56], which was shown to have an impressive power saving over the multi-hop IM/DD system in log-normal turbulence via numerical studies. A recent experiment carried out by Lange *et al.* has demonstrated a 142 km terrestrial coherent FSO link using a homodyne BPSK scheme successfully with a data rate of 5.625 Gbit/s [57].

2. Coherent free-space optical system model

In this section, we present some background knowledge concerning wireless optical channels and the coherent FSO system model. We first address the fading characteristics of wireless propagation environments. We then introduce some basic concepts and the composition of a coherent FSO system, as shown in Fig. 1. Finally, we point out some technical challenges associated with coherent FSO communication systems.

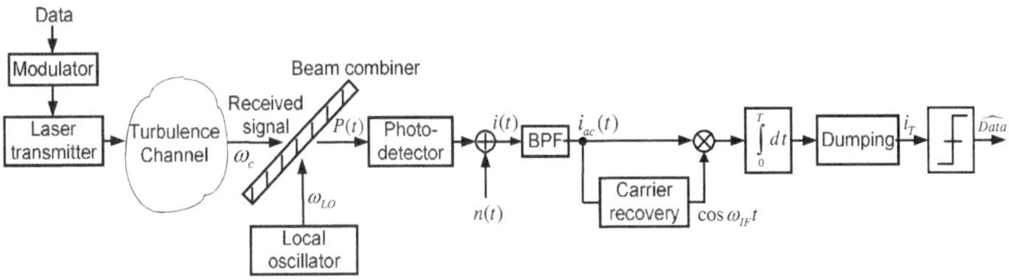

Figure 1. Block diagram of a typical coherent FSO system through an atmospheric turbulence channel.

2.1. Coherent free-space optical receiver

The main idea behind coherent optical systems lies in the mixing of the optical signal coherently with a continuous-wave local oscillator beam before it strikes the photodetector. By employing an optical local oscillator as well as a beamsplitter, we introduce a coherent FSO system for BPSK modulation in turbulence channels. The system model developed here gives a general idea on how coherent FSO systems are operated in turbulence channels.

In general optical communication systems, the current $i_e(t)$ at the output of photodetector is comprised of multiple components [35], [58]

$$i_e(t) = i_p(t) + i_b + i_d + n_e(t) = \frac{\eta q}{h\nu}(P_{in}(t) + P_b) + i_d + n_e(t) \quad (6)$$

where the photocurrent $i_p(t)$ represents the electrical photocurrent converted from the photodetector, i_b is the undesired background noise, i_d is the dark current (which is independent of the signal irradiance), $n_e(t)$ is the total noise in a receiver circuit, $P_{in}(t)$ is the received optical power, P_b is the background noise power and $R = \eta q/(h\nu)$ is the responsivity defined in [59]. The total noise variance can be expressed as $\sigma_{tot}^2 = \sigma^2 + \sigma_T^2$, where σ^2 and σ_T^2 denote the variance of shot noise and thermal noise, respectively. Since σ^2 and σ_T^2 are independent of each other, we consider them separately. The variance σ^2 of the shot noise is given by [59]

$$\sigma^2 = 2q(i_p(t) + i_b + i_d)\Delta f = 2q(RP_{in}(t) + i_b + i_d)\Delta f \quad (7)$$

where q is the electronic charge, and Δf is the noise equivalent bandwidth (NEB) of the photodetector. The thermal noise variance σ_T^2 is given by

$$\sigma_T^2 = \frac{4kT\Delta f}{R_L} \tag{8}$$

where k is the Boltzmann constant, T is the absolute temperature in Kelvin, and R_L is the load resistance. In thermal noise dominant FSO systems, it can be readily seen that the total noise variance $\sigma_{tot}^2 \doteq \sigma_T^2$ is independent of the incident optical power. In shot noise dominant FSO systems, the total noise variance $\sigma_{tot}^2 \doteq \sigma^2$ is dependent on the incident optical power on the photodetector.

Since the ambient noise power is much stronger than the signal power in the free-space channel, shot noise due to ambient light and/or thermal noise is dominant in many IM/DD or subcarrier intensity modulated FSO systems. With the aid of a sufficiently large local oscillator power in a coherent FSO system, however, (local oscillator induced) shot noise can be made to be dominant, and the noise variance depends on the incident mixed optical irradiance. Thus, in a coherent FSO communication system, using the fact that the photocurrent from the receiver is the product of the responsivity R and incident optical power, we write the detected photocurrent from Fig. 1 as [53]

$$i(t) = i_{dc} + i_{ac}(t) + n(t) \tag{9}$$

where

$$i_{dc} = R(P_{in}(t) + P_{LO}) \doteq RP_{LO} \tag{10}$$

and

$$i_{ac}(t) = 2R\sqrt{P_{in}(t)P_{LO}}\cos(\omega_{IF}t + \phi_m) \tag{11}$$

represent the DC and AC terms at the receiver, respectively, and $n(t)$ is a zero-mean AWGN process due to shot noise. In (11), ϕ_m denotes the encoded phase information. In practice, an FSO system can be employed with $P_{LO} \gg P_{in}(t)$, and the DC term in (10) can be approximated by the dominant term RP_{LO}. For the same reason, photocurrent due to thermal noise and the dark current are negligible compared to RP_{LO}. The variance of the shot noise process $n(t)$ is therefore found as [35]

$$\sigma_{n(t)}^2 = 2qRP_{LO}\Delta f. \tag{12}$$

The SNR of an optical receiver is ultimately defined as the ratio of the time-averaged AC photocurrent power to the total noise variance [59], and it can be expressed as

$$\gamma = \frac{\langle i_{ac}^2(t)\rangle}{\sigma_{tot}^2} \tag{13}$$

where $\langle\cdot\rangle$ denotes the time average.

For coherent synchronous detection with M-ary constant amplitude modulation, we have $P_{in}(t) = P_s$ and obtain that the instantaneous SNR as

$$\gamma = \frac{4R^2 P_s P_{LO}}{2\sigma_{n(t)}^2} = \frac{RP_s}{q\Delta f}. \tag{14}$$

Let us define E_s to be the symbol energy of the M-ary constellation. Using the relationship that the optical power is the product of optical signal irradiance and photodetector area A, we can also write the SNR per symbol as

$$\gamma \bullet \frac{RE_sA}{q\Delta f}I \bullet \frac{\eta_e E_s A}{h\nu\Delta f}I \bullet E_sCI \tag{15}$$

where η_e denotes the quantum efficiency of the photodetector, h is Planck's constant, ν denotes the frequency of the received optical signal, and $C \bullet \eta_e A/\bullet h\nu\Delta f\bullet$ is a multiplicative constant for a given FSO system. Note that the instantaneous SNR is independent of the local oscillator power P_{LO}.

2.2. Terrestrial free-space optical channels

The random variation in signal irradiance resulting from atmospheric turbulence is a major source of system performance degradation in FSO communication systems. To address and mitigate such link performance degradation caused by turbulence channels, researchers have studied wireless optical channels extensively and proposed different channel models [14]-[22]. The turbulence effects, which are mainly due to the fluctuation of the refractive index, can be categorized by the range of scintillation index indicating the strength of turbulence-induced fading. In this section, we will briefly describe log-normal, K, Gamma-Gamma and negative exponential turbulence models.

2.2.1. Log-normal turbulence

When the optical channel is considered as a clear-sky atmospheric turbulence channel with several hundred meters propagation distance, the optical turbulence can be modeled as log-normal distribution. The corresponding weak turbulence range of scintillation index for log-normal is less than unity. Thus, although a log-normal model can be valid for longer propagation distances, the condition $\sigma_{si}^2 < 1$ limits the log-normal model with longer propagation distances to be only used for a small index of refraction structure constant C_n^2 [45]. This can be clearly observed from (2), which describes the relationship between scintillation index (i.e., Rytov variance) and propagation distance under weak turbulence conditions.

For weak turbulence conditions, Parry [24], Phillips and Andrews [25] independently suggested a log-normal PDF to model the irradiance, which is the power density of the optical beam. With a log-scale parameter λ, the log-normal PDF of the irradiance I can be expressed as [12]

$$f_{LN}\bullet I\bullet \bullet \frac{1}{I\sqrt{2\pi\sigma_{si}^2}}\exp\left\{-\frac{\bullet \ln I - \lambda\bullet^2}{2\sigma_{si}^2}\right\}, \quad I > 0. \tag{16}$$

If we let $\lambda \bullet -\frac{1}{2}\sigma_{si}^2$, the mean irradiance can be normalized to be $E\bullet I\bullet\bullet \exp\bullet\lambda \bullet \sigma_{si}^2/2\bullet \bullet e^0 \bullet 1$. The nth moment of the log-normal PDF is

$$E\bullet I^n\bullet\bullet \exp\bullet n\lambda \bullet n^2\sigma_{si}^2/2\bullet. \tag{17}$$

When turbulence levels increase further, the negative exponential distribution can be applied as a limiting distribution for the irradiance fluctuation. This limiting distribution can only provide sufficient accuracy when the system goes into the deep saturation regime [28].

2.2.2. *K-distributed turbulence*

One of the widely accepted models under the strong turbulence regime is the *K*-distributed turbulence model. In the 1970s, Jakeman *et al.* introduced this turbulence model for a non-Rayleigh sea echo. They have shown that the *K* distribution arises from the limiting form when the average number of multi-path fluctuations becomes large in the random sinusoid model [60]. Then Phillips and Andrews proved the validity of this *K* distribution by experiments in the strong turbulence regime [24], [25]. The *K*-distribution is an accurate model of turbulence if moderate propagation distances are encountered (about 1km) or the scintillation index [11] is confined to the range (2, 3) [45].

The *K*-distributed FSO turbulence is modeled as follows [45]:

$$f_K(I) = \frac{2}{\Gamma(\alpha)\eta^{\alpha+1}}\alpha^{\frac{\alpha+1}{2}} I^{\frac{\alpha-1}{2}} K_{\alpha-1}\left(\frac{2}{\eta}\sqrt{\alpha I}\right), \quad I \geq 0 \tag{18}$$

where $\Gamma(\cdot)$ denotes the gamma function, $K_x(\cdot)$ is the modified Bessel function of the second kind of order x, η^2 is the mean irradiance of the optical signal, α is a channel parameter related to the effective number of discrete time scatterers and/or scintillation index. The nth moment of the *K* distribution can be shown to be

$$E(I^n) = \frac{\Gamma(\alpha+n)\eta^{2n} n!}{\alpha^n \Gamma(\alpha)}. \tag{19}$$

2.2.3. *Gamma-gamma turbulence*

In [11], Andrews *et al.* proposed the modified Rytov theory which defines the optical field as a function of perturbations due to large-scale and small-scale atmospheric effects. This leads to the Gamma-Gamma turbulence model. The PDF of the Gamma-Gamma distribution is

$$f_G(I) = \frac{2(\alpha\beta)^{(\alpha+\beta)/2}}{\Gamma(\alpha)\Gamma(\beta)I_0}\left(\frac{I}{I_0}\right)^{\frac{\alpha+\beta}{2}-1} K_{\alpha-\beta}\left(2\sqrt{\alpha\beta\frac{I}{I_0}}\right), \quad \alpha > 0,\ \beta > 0,\ I \geq 0 \tag{20}$$

where I_0 denotes the mean irradiance, α and β represent, respectively, the effective number of large-scale and small-scale cells of the scattering process. Note that by setting the shape parameter $\beta = 1$, the Gamma-Gamma distribution will degenerate to the *K*-distribution. The Gamma-Gamma turbulence model is desirable for both weak and strong turbulence scenarios, because this model can provide a good fit to such experimental measurements of irradiance [27]. Therefore, the key advantage of using the Gamma-Gamma turbulence model is that it covers a wide-range of turbulence conditions.

The nth moment of the Gamma-Gamma PDF is found as

$$E(I^n) = \frac{\Gamma(\alpha+n)\Gamma(\beta+n)}{\Gamma(\alpha)\Gamma(\beta)}\left(\frac{I_0}{\alpha\beta}\right)^n. \tag{21}$$

It can be shown that the channel parameters α and β are determined through the Rytov variance σ_R^2 defined in (2) [12], so that they are not arbitrarily chosen. The relationship $\alpha > \beta$ holds in most FSO scenarios, as shown in Fig. 2 and Fig. 3, though the reverse relationship

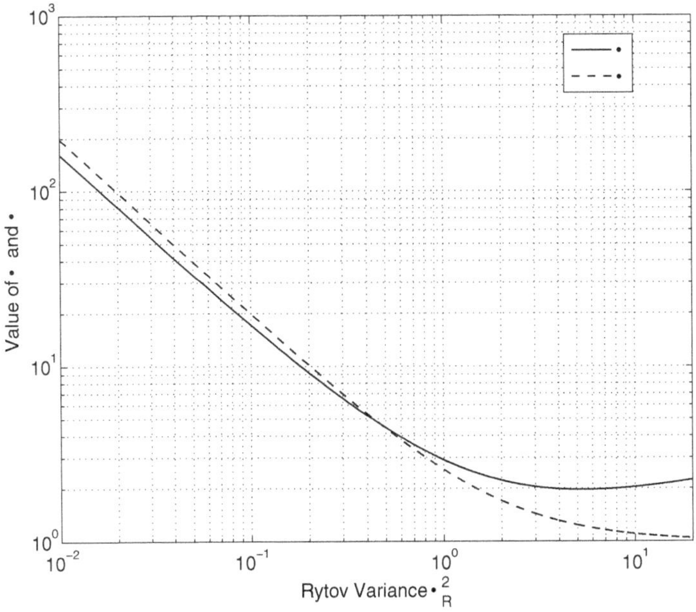

Figure 2. The relationship of the Gamma-Gamma turbulence channel parameters α and β with the Rytov variance σ_R^2 for a finite inner scale $l_0 = R_F/2$ where R_F is the radius of the first fresnel zone.

$\beta > \alpha$ may appear in weak turbulence regimes when the inner-scale l_0 is non-negligible. When considering spherical wave propagation, α and β can be directly linked to the physical parameters through [12]

$$\alpha = \left\{ \exp\left[\frac{0.49\chi^2}{(1 + 0.18d^2 + 0.56\chi^{\frac{12}{5}})^{\frac{7}{6}}} \right] - 1 \right\}^{-1} \tag{22}$$

and

$$\beta = \left\{ \exp\left[\frac{0.51\chi^2(1 + 0.69\chi^{\frac{12}{5}})^{-\frac{5}{6}}}{(1 + 0.9d^2 + 0.62d^2\chi^{\frac{12}{5}})^{\frac{5}{6}}} \right] - 1 \right\}^{-1} \tag{23}$$

where $\chi^2 = 0.5C_n^2 k^{7/6} L_t^{11/6}$ and $d = (kD_a^2/4L_t)^{1/2}$, and where D_a denotes the diameter of the receiver collecting lens aperture.

2.2.4. *Negative exponential turbulence*

The negative exponential distribution is considered to be a limiting turbulence model case for describing irradiance fluctuations in the deep saturation regime. In negative exponential turbulence environments, the irradiance is modeled as a negative exponential RV having a PDF [44]

$$f_{\mathrm{NE}}(I) = \lambda \exp(-\lambda I), \quad I \geq 0 \tag{24}$$

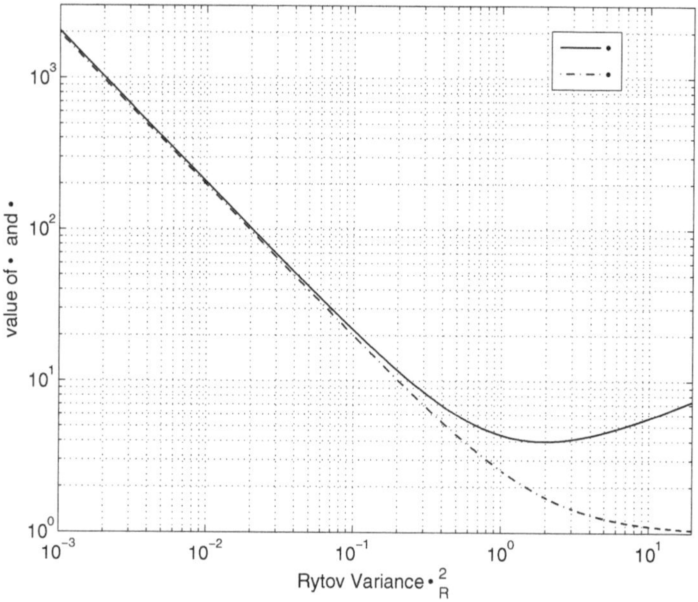

Figure 3. The relationship of Gamma-Gamma turbulence channel parameters α and β as a function of the Rytov variance σ_R^2 with negligible inner scale, i.e., $l_0 \longrightarrow 0$.

where $1/\lambda > 0$ is the mean irradiance. The nth moment of the negative exponential distribution is

$$E[I^n] = \frac{n!}{\lambda^n}. \tag{25}$$

The negative exponential model can be considered to be a special case of the K-distributed model. The K-distribution approaches the negative exponential distribution when the channel parameter α approaches ∞, i.e., irradiance fluctuations approach the deep saturation regime.

3. System performance analysis over a single free-space optical link

In this section, we present error rate results on coherent FSO communications in general Gamma-Gamma turbulence channels (weak-to-strong regimes). The average error rate of FSO systems over a turbulence channel can be expressed as

$$P_e = \int_0^\infty P_e(I) f(I) dI \tag{26}$$

where $P_e(I)$ denotes the conditional symbol error probability and $f(I)$ is the PDF of the received optical irradiance. To facilitate the system analyses, we study the statistics of the Gamma-Gamma distribution related to the instantaneous SNR.

The moment generating function (MGF) of irradiance I is defined as

$$M_I(s) = \int_0^\infty e^{sI} f(I) dI = E[e^{sI}]. \tag{27}$$

With [63, Eq. 6.643(3), Eq. 9.220(2), Eq. 9.220(3), Eq. 9.220(4)], we obtain the MGF of I as

$$M_I(s) = \frac{(\alpha\beta)^{\frac{\alpha+\beta}{2}} \exp\left(-\frac{\alpha\beta}{s}\right)}{\sqrt{\alpha\beta}} (-s)^{-\frac{\alpha+\beta-1}{2}} \left[\frac{\Gamma(\beta-\alpha)}{\Gamma(\beta)} \left(-\frac{\alpha\beta}{s}\right)^{\frac{\alpha-\beta+1}{2}} {}_1F_1\left(\alpha, \alpha-\beta+1; -\frac{\alpha\beta}{s}\right) \right.$$
$$\left. + \frac{\Gamma(\alpha-\beta)}{\Gamma(\alpha)} \left(-\frac{\alpha\beta}{s}\right)^{\frac{\beta-\alpha+1}{2}} {}_1F_1\left(\beta, \beta-\alpha+1; -\frac{\alpha\beta}{s}\right) \right] \tag{28}$$

where ${}_1F_1(\cdot,\cdot;\cdot)$ denotes the confluent hypergeometric function. We comment that by replacing s with $j\omega$ in (28) one can readily obtain the characteristic function (CHF) of the irradiance I.

3.1. Error rate analysis

Using the alternative form of the Gaussian Q-function,

$$Q(x) = \frac{1}{\pi} \int_0^{\frac{\pi}{2}} \exp\left(-\frac{x^2}{2\sin^2\theta}\right) d\theta, \ x > 0 \tag{29}$$

we can express the average symbol-error rate (SER) of M-ary phase-shift keying (MPSK) as [61]

$$P_{e,\,MPSK} = \frac{1}{\pi} \int_0^{\frac{(M-1)\pi}{M}} M_I\left(-\frac{\sin(\pi/M)^2\overline{\gamma}}{2\sin^2\theta}\right) d\theta \tag{30}$$

where $\overline{\gamma} = CE_s$ is the average SNR. When $M = 2$, we can express the average BER with BPSK as

$$P_{e,\,BPSK} = \frac{1}{\pi} \int_0^{\frac{\pi}{2}} M_I\left(-\frac{\overline{\gamma}}{2\sin^2\theta}\right) d\theta. \tag{31}$$

Closed-form SER and BER expressions can then be obtained from (30) and (31) via a series expansion approach. Substituting (20) and a series expansion of the modified Bessel function of the second kind [41, Eq. (6)], [62, Eq. (03.04.06.0002.01)],

$$K_v(x) = \frac{\pi}{2\sin(\pi v)} \sum_{p=0}^{\infty} \left(\frac{(x/2)^{2p-v}}{\Gamma(p-v+1)p!} - \frac{(x/2)^{2p+v}}{\Gamma(p+v+1)p!} \right), \ v \notin \mathbf{Z}, \ |x| < \infty \tag{32}$$

into (26) and using integral identities [63, Eq. 3.621(1), Eq. 8.384(4)], we can express the SER for MPSK modulated coherent FSO systems as

$$P_{e,\,MPSK} = \frac{B(\alpha-\beta, 1-\alpha+\beta)}{\pi} \sum_{p=0}^{\infty} \left[\frac{a_p(\alpha,\beta)\Gamma(p+\beta)\varphi(p+\beta, M)}{\sin^{2p+2\beta}(\frac{\pi}{M})} \left(\frac{\overline{\gamma}}{2}\right)^{-(p+\beta)} \right.$$
$$\left. - \frac{a_p(\beta,\alpha)\Gamma(p+\alpha)\varphi(p+\alpha, M)}{\sin^{2p+2\alpha}(\frac{\pi}{M})} \left(\frac{\overline{\gamma}}{2}\right)^{-(p+\alpha)} \right] \tag{33}$$

where we define

$$a_p(\alpha,\beta) \triangleq \frac{(\alpha\beta)^{p+\beta}\Gamma(\alpha-\beta)\Gamma(\beta-\alpha+1)}{\Gamma(\alpha)\Gamma(\beta)\Gamma(p-\alpha+\beta+1)p!}. \tag{34}$$

Here, $\varphi(x,M)$ denotes an integral identity from Mathematica® defined as

$$\begin{aligned}\varphi(x,M) &= \int_0^{\frac{M-1}{M}\pi} \sin^{2x}\theta d\theta \\ &= \frac{\pi^{\frac{3}{2}}\sec(\pi x)}{2\Gamma(x+1)\Gamma\left(\frac{1}{2}-x\right)} - \cos\left(\frac{M-1}{M}\pi\right) {}_2F_1\left(\frac{1}{2},\frac{1}{2}-x;\frac{3}{2};\cos^2\frac{(M-1)\pi}{M}\right)\end{aligned} \tag{35}$$

where ${}_2F_1(\cdot,\cdot;\cdot;\cdot)$ denotes the Gaussian hypergeometric function [63, Eq. 9.100]. Again, substituting (32) into (26) and using [63, Eq. 3.478(1), Eq. 3.621(1), Eq. 8.384(1), Eq. 8.384(4)], one can similarly obtain the closed-form average BER for BPSK coherent FSO systems as

$$\begin{aligned}P_{e,\,BPSK} =& \frac{B(\alpha-\beta,1-\alpha+\beta)}{2\pi}\sum_{p=0}^{\infty}\left[a_p(\alpha,\beta)\Gamma(p+\beta)B\left(\frac{1}{2},p+\beta+\frac{1}{2}\right)\left(\frac{\overline{\gamma}}{2}\right)^{-(p+\beta)}\right. \\ &\left. -a_p(\beta,\alpha)\Gamma(p+\alpha)B\left(\frac{1}{2},p+\alpha+\frac{1}{2}\right)\left(\frac{\overline{\gamma}}{2}\right)^{-(p+\alpha)}\right]\end{aligned} \tag{36}$$

where $B(x,y) = \Gamma(x)\Gamma(y)/\Gamma(x+y)$ denotes the Beta function [63, Eq. 8.384(1)] which is defined as

$$B(x,y) \triangleq \int_0^1 t^{x-1}(1-t)^{y-1}dt, \quad \Re\{x\}>0, \Re\{y\}>0. \tag{37}$$

Note that the condition $(\alpha-\beta)\notin \mathbf{Z}$ holds for most values of σ_R^2, i.e., most terrestrial FSO senecios, as shown in Fig. 2 and Fig. 3. If needed, we can change the difference between α and β via varying σ_R^2 by a small constant ϵ to approximate the scenario when $(\alpha-\beta)\in\mathbf{Z}$ (corresponding to a specific σ_R^2 value).

The error rates of coherent MPSK systems are shown for both weak ($\alpha=3.0$, $\beta=2.7$) and strong ($\alpha=2.3$, $\beta=1.1$) turbulence channels in Fig. 4. As the modulation index M increases, it is clearly seen that the error rate degrades due to the symbol power constraints.

3.2. Asymptotic performance analysis

The asymptotic approach is a powerful method since the asymptotic solutions can often be used to reveal important insights of the target system behavior in large SNR regimes. For the Gamma-Gamma turbulence model, we have $\alpha>\beta$ in most scenarios. Therefore, without losing of generality, we assume $\alpha>\beta$ in asymptotic studies. We observe from (33) and (36) that the term $(0.5\overline{\gamma})^{-\alpha}$ decreases faster than the term $(0.5\overline{\gamma})^{-\beta}$ as $\overline{\gamma}$ increases for given p values. Consequently, when $\overline{\gamma}$ increases, the SER of MPSK modulation in large SNR regimes for coherent FSO systems over the Gamma-Gamma channel can be approximated by

$$P_{e,\,asym}^{\mathrm{MPSK}} = \frac{a_0(\alpha,\beta)B(\alpha-\beta,1-\alpha+\beta)\Gamma(\beta)\varphi(\beta,M)}{\pi\sin^{2\beta}(\frac{\pi}{M})}\left(\frac{\overline{\gamma}}{2}\right)^{-\beta}. \tag{38}$$

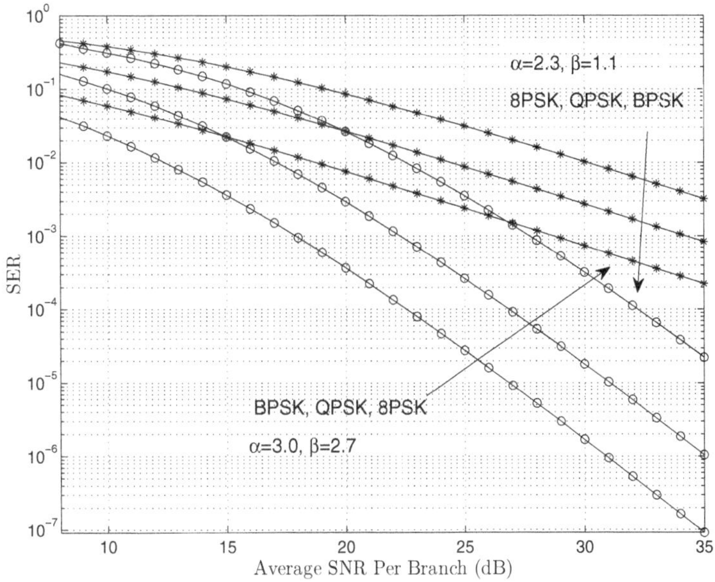

Figure 4. SER comparison of coherent MPSK optical communication over weakly ($\alpha = 3.0, \beta = 2.7$) and strongly ($\alpha = 2.3, \beta = 1.1$) turbulent Gamma-Gamma channels.

The BER of BPSK coherent FSO systems in large SNR regimes can be found as

$$P_{e,\, asym}^{\mathrm{BPSK}} = \frac{\Gamma(\alpha-\beta)B\left(\frac{1}{2},\beta+\frac{1}{2}\right)}{2\pi\Gamma(\alpha)}\left(\frac{\overline{\gamma}}{2\alpha\beta}\right)^{-\beta}. \tag{39}$$

From (38) and (39), we conclude that the diversity order for MPSK coherent FSO systems is equal to the smaller channel parameter, i.e., $\beta = \min\{\alpha,\beta\}$, in the Gamma-Gamma turbulence.

3.3. Outage probability analysis

Outage probability is an important criterion for digital wireless communication networks. The outage probability for a given FSO link is defined as

$$P_{outage}(\Lambda) = \Pr(\gamma < \Lambda) = \int_0^{\Lambda} f_\gamma(\gamma)d\gamma \tag{40}$$

where $\Pr(\cdot)$ denotes the probability of a event, Λ is a predefined outage threshold, and $f_\gamma(\gamma)$ is the PDF of the instantaneous SNR. Thus, substituting the relationship $\gamma = \overline{\gamma}I$ and (32) into (40), one obtains the outage probability for coherent FSO systems as

$$P_{outage}(\Lambda) = B(\alpha-\beta,1-\alpha+\beta)\sum_{p=0}^{\infty}\left[\frac{a_p(\alpha,\beta)}{(p+\beta)\overline{\gamma}^{p+\beta}}\Lambda^{p+\beta} - \frac{a_p(\beta,\alpha)}{(p+\alpha)\overline{\gamma}^{p+\alpha}}\Lambda^{p+\alpha}\right]. \tag{41}$$

4. System performance analysis of multichannel coherent free-space optical links

In this section, we present unified error rate studies for MRC and EGC for coherent FSO systems using an MGF approach. In a multichannel coherent FSO link, the average error rate can be expressed as [52]

$$P_e = \frac{1}{\pi}\int_0^{\frac{\pi}{2}} M_{\gamma_{mc}}\left(-\frac{\overline{\gamma}}{2\sin^2\theta}\right)d\theta \tag{42}$$

where $M_{\gamma_{mc}}(\cdot)$ denotes the MGF of the multichannel combiner output instantaneous SNR. We will use (42) in the error performance study of multichannel coherent FSO systems. Here we will only mention the BER expressions for the most commonly used BPSK modulation.

4.1. Analysis with maximum ratio combining

The instantaneous SNR at the output of the MRC combiner can be found as

$$\gamma_{mc,M} = \frac{RA\left(\sum\limits_{l=1}^{L} I_l\right)}{q\Delta f} = \overline{\gamma}\left(\sum_{l=1}^{L} I_l\right) \tag{43}$$

where I_l denotes the optical signal irradiance at the lth branch. Substituting (32) into (27), we can write the MGF of I_l as power series by [41]

$$M_I(s) = \left(\sum_{p=0}^{\infty} a_p(\alpha,\beta)\Gamma(p+\beta)(-s)^{-(p+\beta)} + \sum_{p=0}^{\infty} a_p(\beta,\alpha)\Gamma(p+\alpha)(-s)^{-(p+\alpha)}\right). \tag{44}$$

With the help of a binomial expansion, the MGF of the summed RV $\sum_{l=1}^{L} I_l$, i.e., $M_{\sum_{l=1}^{L} I_l}(s) = (M_I(s))^L$, in terms of power series can be expressed as

$$M_{\sum_{l=1}^{L} I_l}(s) = \sum_{q=0}^{L}\binom{L}{q}\sum_{p=0}^{\infty}(\Gamma(p+\beta)a_p(\alpha,\beta))^{(L-q)} * (\Gamma(p+\alpha)a_p(\beta,\alpha))^{(q)}(-s)^{-p-L\beta-q(\alpha-\beta)} \tag{45}$$

where $*$ denotes the convolution operator and $(a_p(\alpha,\beta)\Gamma(p+\beta))^{(n)}$ denotes $a_p(\alpha,\beta)\Gamma(p+\beta)$ convolved $n-1$ times with itself. When n={0,1}, we have $(a_p(\alpha,\beta)\Gamma(p+\beta))^{(1)} = a_p(\alpha,\beta)\Gamma(p+\beta)$ and $(a_p(\alpha,\beta)\Gamma(p+\beta))^{(0)} = 1$. For such an L-branch MRC coherent FSO system, making use of (32), (42) and Laplace transforms, we can obtain the closed-form expression of the average BER as

$$P_{e,MRC} = \frac{1}{2\pi}\sum_{q=0}^{L}\binom{L}{q}\sum_{p=0}^{\infty}(\Gamma(p+\beta)a_p(\alpha,\beta))^{(L-q)} * (\Gamma(p+\alpha)a_p(\beta,\alpha))^{(q)} \times B\left(\frac{1}{2}, p+L\beta+q(\alpha-\beta)+\frac{1}{2}\right)\left(\frac{\overline{\gamma}}{2}\right)^{-p-L\beta-q(\alpha-\beta)}. \tag{46}$$

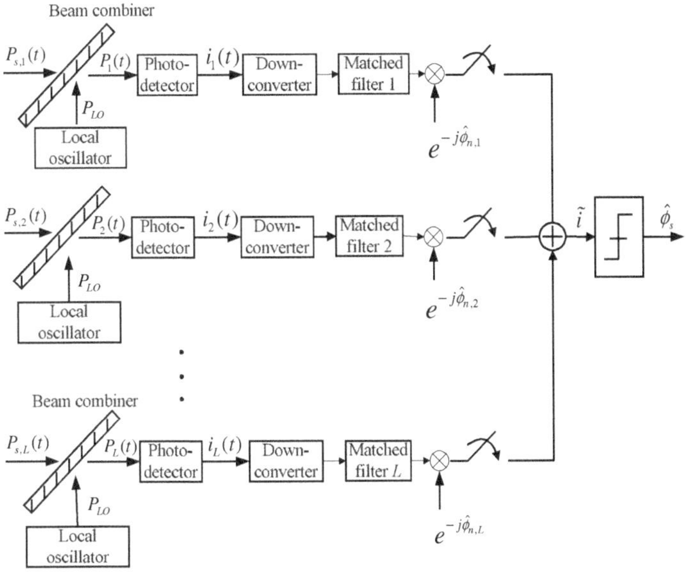

Figure 5. Block diagram of a coherent EGC FSO system through an atmospheric turbulence channel.

4.2. Analysis with equal gain combining

For coherent FSO systems using EGC, based on the block diagram given in Fig. 5, the instantaneous SNR at the output of the combiner is found as [65]

$$\gamma_{mc,E} = \frac{RA\left(\sum_{l=1}^{L}\sqrt{I_l}\right)^2}{Lq\Delta f} = \frac{\overline{\gamma}}{L}\left(\sum_{l=1}^{L}\sqrt{I_l}\right)^2. \tag{47}$$

Based on the coherent EGC combiner SNR expression in (47), we derive the MGF of this summed RV $X = \sum_{l=1}^{L}\sqrt{I_l}$. With the PDF of $\sqrt{I}$ and power series expansion of the modified Bessel function of the second kind, we obtain the MGF of $\sqrt{I}$ in terms of a series expansion as

$$M_{\sqrt{I}}(s) = 2\sum_{p=0}^{\infty}\left[a_p(\alpha,\beta)\Gamma(2p+2\beta)(-s)^{-2(p+\beta)} + a_p(\beta,\alpha)\Gamma(2p+2\alpha)(-s)^{-2(p+\alpha)}\right]. \tag{48}$$

Making use of a binomial expansion, we can express the MGF of X as

$$\begin{aligned} M_X(s) =& 2^L\sum_{q=0}^{L}\binom{L}{q}\left(\sum_{p=0}^{\infty}a_p(\alpha,\beta)\Gamma(2p+2\beta)(-s)^{-2(p+\beta)}\right)^{L-q} \\ &\times\left(\sum_{p=0}^{\infty}a_p(\beta,\alpha)\Gamma(2p+2\alpha)(-s)^{-2(p+\alpha)}\right)^{q}. \end{aligned} \tag{49}$$

By taking the inverse Laplace transform of (49), we derive the PDF of X as

$$f_X(x) = 2^L \sum_{q=0}^{L} \binom{L}{q} \sum_{p=0}^{\infty} \frac{b_p^{*(L-q)}(\alpha,\beta) * b_p^{*q}(\beta,\alpha)}{\Gamma(2p+2(L-q)\beta+2q\alpha)} x^{2p+2(L-q)\beta+2q\alpha} \quad (50)$$

where $b_p(\alpha,\beta) \triangleq a_p(\alpha,\beta)\Gamma(2p+2\beta)$. Again, with $Y = X^2$, the PDF of Y is found to be

$$f_Y(y) = 2^{L-1} \sum_{q=0}^{L} \binom{L}{q} \sum_{p=0}^{\infty} \frac{b_p^{*(L-q)}(\alpha,\beta) * b_p^{*q}(\beta,\alpha)}{\Gamma(2(p+(L-q)\beta+q\alpha))} y^{p+(L-q)\beta+q\alpha-1}. \quad (51)$$

The MGF of Y is finally obtained as

$$M_Y(s) = 2^{L-1} \sum_{q=0}^{L} \binom{L}{q} \sum_{p=0}^{\infty} \frac{b_p^{*(L-q)}(\alpha,\beta) * b_p^{*q}(\beta,\alpha)\Gamma((p+(L-q)\beta+q\alpha))}{\Gamma(2(p+(L-q)\beta+q\alpha))} (-s)^{-(p+(L-q)\beta+q\alpha)}. \quad (52)$$

Substituting (52) into (42), we can obtain the BER for coherent FSO links with EGC as

$$\begin{aligned} P_{e,EGC} = \frac{2^{L-2}}{\pi} \sum_{q=0}^{L} \binom{L}{q} \sum_{p=0}^{\infty} \frac{b_p^{*(L-q)}(\alpha,\beta) * b_p^{*q}(\beta,\alpha)\Gamma((p+(L-q)\beta+q\alpha))}{\Gamma(2(p+(L-q)\beta+q\alpha))} \\ \times B\left(\frac{1}{2}, p+(L-q)\beta+q\alpha+\frac{1}{2}\right)\left(\frac{\overline{\gamma}}{2L}\right)^{-(p+(L-q)\beta+q\alpha)}. \end{aligned} \quad (53)$$

Figure 6 plots and compares the error rates of coherent MRC and EGC systems with L receiver branches. As shown in Fig. 6, coherent MRC systems outperform EGC systems for $L > 1$. As the average SNR increases, the benefits from diversity reception of coherent systems are increasingly clear. It is notable that EGC has a close error performance compared to the optimal MRC systems. Therefore, EGC can be a preferable choice in designing a coherent FSO system as it provides a comparable performance to MRC and can be implemented with reduced cost, as EGC does not require the estimation of instantaneous irradiance fluctuations.

4.3. Truncation error analysis with a coherent diversity receiver

Up to this point, the presented error rate results are given in terms of power series. For practical calculation, we need to eliminate the infinite terms in the obtained series solutions. In this subsection, we present detailed truncation error studies for coherent FSO systems with diversity. Due to space limitations, we will study the error rate solution truncation error only in the MRC case for illustration purposes. Truncation error studies for other cases can be readily carried out.

We define the truncation error caused by eliminating the infinite terms after the first $J+1$ terms in the MRC series solution in (46). The error is

$$\begin{aligned} \varepsilon_{e,MRC}(J) = \frac{1}{2\pi} \sum_{q=0}^{L} \binom{L}{q} \sum_{p=J+1}^{\infty} (\Gamma(p+\beta)a_p(\alpha,\beta))^{*(L-q)} * (\Gamma(p+\alpha)a_p(\beta,\alpha))^{*q} \\ \times B\left(\frac{1}{2}, p+L\beta+q(\alpha-\beta)+\frac{1}{2}\right)\left(\frac{\overline{\gamma}}{2}\right)^{-p-L\beta-q(\alpha-\beta)}. \end{aligned} \quad (54)$$

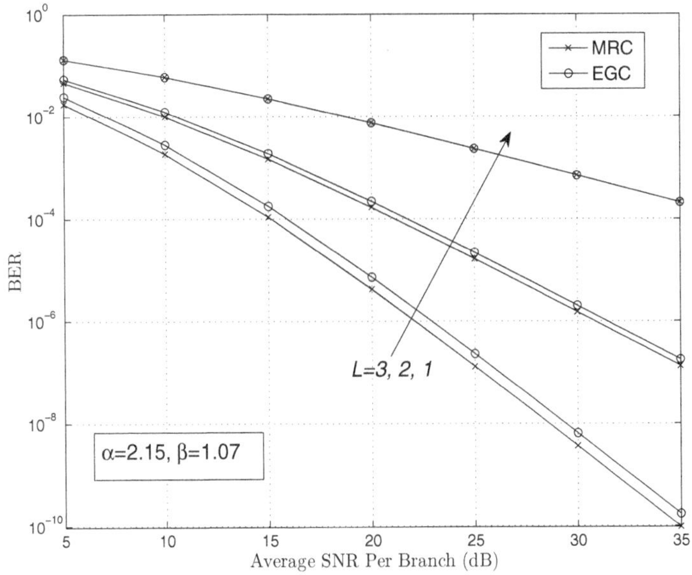

Figure 6. BER comparison of coherent BPSK optical communication with MRC and EGC in strongly turbulent Gamma-Gamma channels.

To facilitate the truncation error analysis, we can rewrite (54) as

$$\varepsilon_{e,MRC}\bullet J\bullet\ \bullet\ \frac{1}{2\pi}\sum_{q\bullet\ 0}^{L}\binom{L}{q}\sum_{p\bullet\ J\bullet\ 1}^{\infty}\eta_p\bullet\alpha,\beta,L,q\bullet\left(\frac{\overline{\gamma}}{2}\right)^{-p} \tag{55}$$

where

$$\begin{aligned}\eta_p\bullet\alpha,\beta,L,q\bullet\ \bullet\ 2^{2\bullet p\bullet\ L\beta\bullet\ q\bullet\alpha-\beta\bullet\bullet}&\frac{\Gamma^2\bullet p\bullet\ L\beta\bullet\ q\bullet\alpha-\beta\bullet\bullet\ \frac{1}{2}\bullet}{\Gamma\bullet 2p\bullet\ 2L\beta\bullet\ 2q\bullet\alpha-\beta\bullet\bullet\ 1\bullet}\bullet\Gamma\bullet p\bullet\ \beta\bullet a_p\bullet\alpha,\beta\bullet\bullet^{L-q\bullet}\\ &*\bullet\Gamma\bullet p\bullet\ \alpha\bullet a_p\bullet\beta,\alpha\bullet\bullet^{q\bullet}\left(\frac{\overline{\gamma}}{2}\right)^{-L\beta-q\bullet\alpha-\beta\bullet}.\end{aligned} \tag{56}$$

Making use of a Taylor series expansion by multiplying $x^{J\bullet\ 1}$ to both sides as

$$x^{J\bullet\ 1}\frac{1}{1-x}\ \bullet\ x^{J\bullet\ 1}\bullet 1\bullet\ x\bullet\ x^2\bullet\ \cdots\bullet\ x^n\bullet\ \cdots\bullet,\quad |x|<1 \tag{57}$$

one can obtain an upper bound for the truncation error in (55) as

$$\varepsilon_{e,MRC}\bullet J\bullet \leqq \frac{1}{\pi\bullet\overline{\gamma}-2\bullet}\left(\frac{\overline{\gamma}}{2}\right)^{-J}\sum_{q\bullet\ 0}^{L}\frac{\max\{\eta_p\bullet\alpha,\beta,L,q\bullet\}L!}{\bullet L-q\bullet!q!} \tag{58}$$

where we have used the identity

$$\binom{L}{q} = \frac{L!}{(L-q)!q!}. \tag{59}$$

Let us consider the first term in the infinite summation in (55). We note that the term $\eta_p(\alpha, \beta, L, q)$ in (56) approaches zero when p (and/or $\overline{\gamma}$) approaches infinity. This implies that the value of $\eta_p(\alpha, \beta, L, q)$ deceases as the value of J increases. Therefore, it is apparent that $\eta_p(\alpha, \beta, L, q)$ is bounded as the value of p increases. This validates our upper bound for the truncation error.

More importantly, we observe that the truncation error upper bound decreases approximately on the order of $\overline{\gamma}^{-(J+1)}$. This suggests an increasing accuracy of the error rate solution along with an increasing of average SNR, which is desired in a practical FSO link performance estimation.

5. Coherent free-space optical systems with phase noise impacts

It should be mentioned that phase noise, which can impair the error rate performance of coherent FSO systems, is assumed to be fully compensated in our previous studies in Section 3 and Section 4. The effects of spatial phase noise and atmospheric turbulence on coherent FSO systems performance have been considered together for the first time in [49]. Belmonte and Kahn studied the performance of coherent FSO links using modal phase compensation in log-normal turbulence channels [49]-[51]. In [64], the spectrum and SNR efficiencies of a variety of modulation methods were compared for coherent FSO communication through the log-normal turbulent channels. As shown in [49] and [64], phase distortion becomes an important performance limiting factor for the coherent system besides the turbulence-induced fading.

5.1. A revised coherent free-space optical receiver model

Here we present a revised coherent receiver model to facilitate the analysis with phase noise impacts. We consider FSO applications where a phase-locked loop (PLL) phase noise compensation mechanism is implemented at the receiver for the Gamma-Gamma turbulence channels. The optical power incident on the lth photodetector can be rewritten as [65]

$$P_l(t) = P_{s,l} + P_{LO} + 2\sqrt{P_{s,l}P_{LO}}g(t)\cos(\omega_{IF}t + \phi_s + \phi_{n,l}), \quad l = 1, 2, ..., L \tag{60}$$

where $P_{s,l}$ is the instantaneous incident optical signal power on the beamsplitter at the lth branch, P_{LO} denotes the local oscillator power which is assumed to be the same for all branches, ϕ_s is the encoded phase information, $\phi_{n,l}$ denotes the phase noise for the lth branch, and $\omega_{IF} = \omega_0 - \omega_{LO}$ is the intermediate frequency, where ω_0 and ω_{LO} denote the carrier frequency and local oscillator frequency, respectively. In (60), $g(t)$ represents the signal pulse, which is defined as

$$g(t) = \begin{cases} \sqrt{\frac{1}{T}}, & 0 \le t \le T \\ 0, & \text{elsewhere} \end{cases} \tag{61}$$

where T denotes the symbol duration. In obtaining (60), we have assumed that the received optical beam and the local oscillator beam are mixed in perfect spatial coherence over a sufficiently small photodetector area. Thus, the effect of spatial phase variation is negligible, while the temporal phase variation is considered in the ensuing analysis. The incident optical power results in the photocurrent

$$i_l \bullet t \bullet \bullet \; RP_l \bullet t \bullet \bullet \; i_{dc,l} \bullet \; i_{ac,l} \bullet t \bullet \bullet \; n_l \bullet t \bullet, \quad l \bullet \; 1,2,...,L \tag{62}$$

where we have

$$i_{dc,l} \bullet \; R \bullet P_{s,l} \bullet \; P_{LO} \bullet, \quad l \bullet \; 1,2,...,L \tag{63}$$

and

$$i_{ac,l} \bullet t \bullet \bullet \; 2R\sqrt{P_{s,l}P_{LO}}g \bullet t \bullet \cos \bullet \omega_{IF}t \bullet \; \phi_s \bullet \; \phi_{n,l} \bullet, \quad l \bullet \; 1,2,...,L \tag{64}$$

representing, respectively, the DC and AC terms at the receiver, and $n_l \bullet t \bullet$ is an AWGN process with equal variance σ^2 for all branches. In practice, an FSO system is operated in the regime $P_{LO} \gg P_{s,l}$, and the DC term in (62) can be approximated by the dominant term RP_{LO}. The variance of the shot noise process $n_l \bullet t \bullet$ can then be expressed by [59]

$$\sigma^2 \bullet \; 2qRP_{LO}\Delta f. \tag{65}$$

Note that the DC term can be removed easily using an appropriate bandpass filter.

5.2. Impacts of imperfect phase noise compensation

In this section, we study the impact of phase estimation error on system performance for a coherent FSO link. We first derive the demodulator output decision statistics in the presence of phase noise compensation error. If we let $\xi_l \triangleq 2R\sqrt{P_{s,l}P_{LO}}$, we can express the AC current as

$$i_{ac,l} \bullet t \bullet \bullet \; \xi_l g \bullet t \bullet \left[\cos \phi_s \cos \bullet \omega_{IF}t \bullet \; \phi_{n,l} \bullet - \sin \phi_s \sin \bullet \omega_{IF}t \bullet \; \phi_{n,l} \bullet \right]. \tag{66}$$

Two real filters are then used to implement the complex filtering in the down-conversion process. The real and imaginary parts of the baseband signal are, respectively, obtained as

$$y_{c,l} \bullet t \bullet \bullet \; \sqrt{2}\left\{\xi_l g \bullet t \bullet \bullet \cos \phi_s \cos \bullet \omega_{IF}t \bullet \; \phi_{n,l} \bullet - \sin \phi_s \sin \bullet \omega_{IF}t \bullet \; \phi_{n,l} \bullet \bullet \right\} \cos \bullet \omega_{IF}t \bullet \tag{67}$$

and

$$y_{s,l} \bullet t \bullet \bullet \; -\sqrt{2}\left\{\xi_l g \bullet t \bullet \bullet \cos \phi_s \cos \bullet \omega_{IF}t \bullet \; \phi_{n,l} \bullet - \sin \phi_s \sin \bullet \omega_{IF}t \bullet \; \phi_{n,l} \bullet \bullet \right\} \sin \bullet \omega_{IF}t \bullet. \tag{68}$$

After passing through a lowpass filter, we obtain the equivalent baseband signal $\tilde{i}_{ac,l} \bullet t \bullet$. With the relationship $P_{s,l} \bullet \; AI_l$ where I_l denotes the lth optical signal irradiance incident on the beamsplitter, the equivalent baseband signal of $i_l \bullet t \bullet$ can be found as

$$\tilde{i}_l \bullet t \bullet \bullet \; \tilde{i}_{ac,l} \bullet t \bullet \bullet \; \tilde{n}_l \bullet t \bullet \bullet \; \sqrt{2}R\sqrt{AP_{LO}}g \bullet t \bullet \sqrt{I_l}e^{j\phi_s}e^{j\phi_{n,l}} \bullet \; \tilde{n}_l \bullet t \bullet, \quad l \bullet \; 1,2,...,L \tag{69}$$

where $\tilde{n}_l \bullet t \bullet$ is the complex-envelope of the real white Gaussian noise process with power spectral density (PSD) $4qRP_{LO}u \bullet \omega \bullet \; \omega_{IF} \bullet$ with $u \bullet \cdot \bullet$ denoting the unit step function.

After correlation and sampling, assuming perfect bit synchronization, we obtain

$$\begin{aligned}\tilde{i}_l &= \int_0^T \sqrt{2A}R\sqrt{I_l P_{LO}}g(t)e^{j\phi_{n,l}}e^{j\phi_s}g(t)dt + \int_0^T \tilde{n}_l(t)g(t)dt \\ &= \sqrt{2A}R\sqrt{I_l P_{LO}}e^{j\phi_{n,l}}e^{j\phi_s} + \tilde{n}_l, \quad l = 1,2,...,L \end{aligned} \tag{70}$$

where $\tilde{n}_l$ is a zero mean complex Gaussian RV, and its real and imaginary parts are Gaussian RVs with equal variance σ^2. The receiver removes the random phase noise in the optical links on all diversity branches by multiplying the received signals with the complex conjugate of the phase noise estimates from the respective channels. The output of the combiner can then be found as

$$\begin{aligned}\tilde{i} &= \sum_{l=1}^{L} e^{-j\hat{\phi}_{n,l}}\sqrt{2}R\sqrt{P_{s,l}P_{LO}}e^{j\phi_{n,l}}e^{j\phi_s} + \sum_{l=1}^{L} e^{-j\hat{\phi}_{n,l}}\tilde{n}_l \\ &= \sum_{l=1}^{L} e^{j\Delta\phi_l}\sqrt{2}R\sqrt{I_l A P_{LO}}e^{j\phi_s} + \nu \end{aligned} \tag{71}$$

where $\hat{\phi}_{n,l}$ is the estimation of $\phi_{n,l}$ at the lth branch, $\Delta\phi_l = \phi_{n,l} - \hat{\phi}_{n,l}$ denotes the phase noise compensation error, $\nu = \sum_{l=1}^{L} e^{-j\hat{\phi}_{n,l}}\tilde{n}_l$ is the complex noise term at the output of the combiner. The real and imaginary parts of the noise term ν are Gaussian RVs with equal variance $L\sigma^2$. We assume that the phase noise estimations are derived from an unmodulated carrier using a first-order PLL and only Gaussian noise is present in the PLL circuit. In this case, the PDF of the phase noise compensation error $\Delta\phi$ is given by [66]

$$f_{\Delta\phi}(\Delta\phi_l) = \frac{\exp\left(\frac{\cos(\Delta\phi_l)}{\sigma^2_{\Delta\phi}}\right)}{2\pi I_0\left(\frac{1}{\sigma^2_{\Delta\phi}}\right)}, \quad |\Delta\phi| \le \pi \tag{72}$$

where $\sigma_{\Delta\phi}$ denotes the standard deviation of the phase noise compensation error $\Delta\phi_l$ for $l = 1, \cdots, L$. We note that $\sigma_{\Delta\phi}$ is the standard deviation of $\Delta\phi_l$ as long as the loop SNR is large [66], which is true for practical communication links. We assume that $\Delta\phi_1, \Delta\phi_2, \cdots, \Delta\phi_L$ are independent and identical distributed (i.i.d.) RVs and the irradiance I_l is independent of $\Delta\phi_l$. In a typical FSO link, the turbulence has little change over the duration of hundreds of consecutive information bits, and the phase noise varies slowly compared to the high data rates in FSO systems. Therefore, β_l can be assumed to be a constant over the duration.

Since BPSK is assumed here, one finally obtains the demodulator decision variable by taking the real part of (71) as

$$D = \sum_{l=1}^{L} \sqrt{2}R\sqrt{AP_{LO}}\cos\phi_s\sqrt{I_l}\cos\Delta\phi_l + \Re\{\nu\} = \cos\phi_s\sum_{l=1}^{L} S_l + \nu_R$$

where $S_l = \sqrt{2}R\sqrt{AP_{LO}}\sqrt{I_l}\cos\Delta\phi_l$, and $\nu_R = \Re\{\nu\}$ is a real valued zero-mean Gaussian noise RV with variance $\sigma^2_{\nu_R} = L\sigma^2$. Based on the decision variable at the output of the combiner, we can find the SNR for EGC reception with phase noise compensation error as

$$\tilde{\gamma}_{EGC} = \frac{2R^2AP_{LO}}{L\sigma^2}\left(\sum_{n=1}^{L}\sqrt{I_l}\cos\Delta\phi_l\right)^2 = \frac{\overline{\gamma}}{L}\left(\sum_{n=1}^{L}\sqrt{I_l}\cos\Delta\phi_l\right)^2. \tag{73}$$

Of great importance to the on-going investigation is the fact that the SNR in (73) is related to $\sqrt{I_l}$ and $\cos\Delta\phi_l$, but it is independent of the local oscillator power.

Without loss of generality, we assume $\phi_s = 0$. From the expression of the decision variable in (73), we now derive the average BER for EGC with phase noise compensation error through a CHF approach. We define the cumulative distribution function (CDF) of the decision variable as

$$F_D(\xi|\phi_s = 0) = \Pr\{D < \xi|\phi_s = 0\}. \tag{74}$$

The average BER can, thus, be written as $P_e = F_D(0|\phi_s = 0)$ when (74) is evaluated at $\xi = 0$.

To find $F_D(\cdot|\phi_s = 0)$, we write the conditional CHF of D as

$$\Phi_D(\omega|\phi_s = 0) = \Phi_{\nu_R}(\omega)\prod_{l=1}^{L}\Phi_{S_l}(\omega) = \Phi_{\nu_R}(\omega)[\Phi_{S_1}(\omega)]^L \tag{75}$$

for i.i.d. RVs S_l's ($l = 1, \cdots, L$), where $\Phi_{S_l}(\omega)$ is the CHF of S_l for $l = 1, 2, \cdots, L$, and $\Phi_{\nu_R}(\omega) = \exp(-L\sigma^2\omega^2/2)$ is the CHF of the Gaussian RV ν_R. The CHF of S_1 conditioned on $\Delta\phi_1$ can be found to be

$$\Phi_{S_1|\Delta\phi_1}(\omega) = \Phi_z\left(\omega\sqrt{2AP_{LO}}R\cos\Delta\phi_1\right). \tag{76}$$

Averaging (76) over $\Delta\phi_1$ gives the CHF of S_1 as $\Phi_{S_1}(\omega) = E_{\Delta\phi_1}[\Phi_z(\omega\sqrt{2AP_{LO}}R\cos\Delta\phi_1)]$. Then, the CHF of D can be found as

$$\begin{aligned}\Phi_D(\omega|\phi_s = 0) &= [\Phi_{S_1}(\omega)]^L\Phi_{\nu_R}(\omega)\\ &= (E_{\Delta\phi_1}[\Phi_z(\omega\sqrt{2AP_{LO}}R\cos\Delta\phi_1)])^L\Phi_{\nu_R}(\omega).\end{aligned} \tag{77}$$

The CDF $F_D(\xi|\phi_s = 0)$ can be calculated through the Gil-Pelaez formula

$$F_D(\xi|\phi_s = 0) = \frac{1}{2} - \frac{1}{\pi}\int_0^\infty \frac{\Im\{\Phi_D(\omega|\phi_s = 0)e^{-j\omega\xi}\}}{\omega}d\omega. \tag{78}$$

Finally, we find the BER with phase noise compensation errors to be

$$P_e = \frac{1}{2} - \frac{1}{\pi}\int_0^\infty \frac{\Im\{\Phi_D(\omega|\phi_s = 0)\}}{\omega}d\omega. \tag{79}$$

Substituting (77) into (79) gives the BER expression for EGC reception with phase noise compensation error. Two integrations are required to evaluate the BER performance of an L-branch EGC system using (77)-(79). In Fig. 7 and Fig. 8, the impact of phase noise is presented for $L = 2$ and $L = 3$ EGC FSO systems, respectively. As shown, the system will not suffer a considerable performance loss when a relatively small phase noise compensation error appears in an FSO link.

5.3. Differential phase-shift keying for coherent free-space optical systems

Since phase noise levels from laser sources and from atmospheric turbulence channel are time variant, the phase tracking device may be subject to carrier phase estimation error. This can lead to system performance losses and may reduce the diversity order. In this section, we present an efficient technique using DPSK for coherent FSO communications over

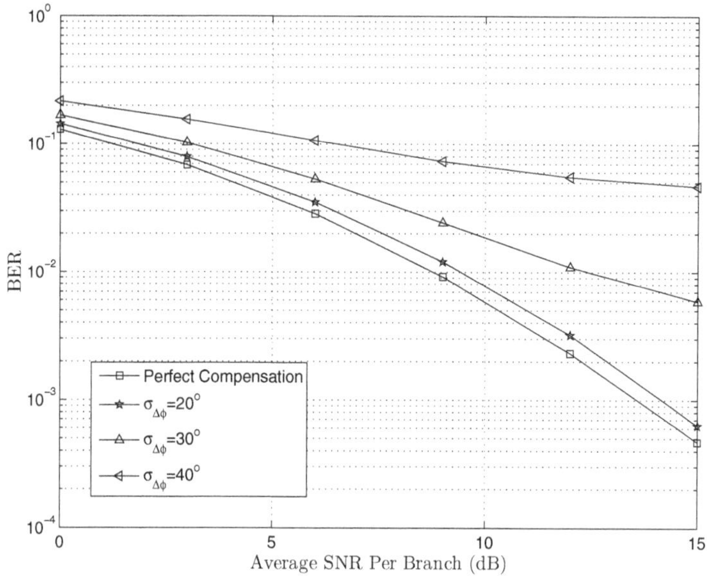

Figure 7. Error performance of a two-branch EGC coherent FSO system through an atmospheric turbulence channel with imperfect phase noise compensation.

atmospheric turbulence channels, which does not require estimation of the phase noise. DPSK modulation is used instead of coherent PSK for diversity reception since there is no need to estimate the phase noise.

Here, we consider a coherent FSO system employing postdetection EGC to mitigate amplitude fading as it is more suitable for differential coherent detection. Its receiver block diagram is shown in Fig. 9. The proposed FSO system is set up with L-branch wireless optical links through Gamma-Gamma turbulence. From (69) the received complex envelope at the lth branch in the kth bit interval can be written as

$$\tilde{i}_{k,l}(t) = \tilde{i}_{ac,k,l}(t) + \tilde{n}_{k,l}(t) = \sqrt{2}R\sqrt{AP_{LO}}g(t)\sqrt{I_l}e^{j\phi_{s,k}}e^{j\phi_{n,l}} + \tilde{n}_{k,l}(t) \tag{80}$$

where $\phi_{s,k} = \phi_{s,k-1} + \Delta\phi_{s,k}$ is the differentially coded phase. Here, $\Delta\phi_{s,k} \in \{0, \pi\}$ denotes the differential carrier phase, and the encoded phase differences are assumed to be equally likely transmitted. Due to the high data rate (on the order of Gbit/s), one can assume a "frozen atmosphere" model [45], where the characteristics of atmospheric turbulence remain constant over at least two successive symbol intervals. At the same branch, the signal in the $(k-1)$th bit interval can, therefore, be obtained as

$$\tilde{i}_{k-1,l}(t) = \tilde{i}_{ac,k-1,l}(t) + \tilde{n}_{k-1,l}(t) = \sqrt{2}R\sqrt{AP_{LO}}g(t)\sqrt{I_l}e^{j\phi_{s,k-1}}e^{j\phi_{n,l}} + \tilde{n}_{k-1,l}(t).$$

The shot noise processes $\tilde{n}_{k,l}(t)$ and $\tilde{n}_{k-1,l}(t)$ are i.i.d. complex Gaussian random processes with power spectral density $4qRP_{LO}u(\omega + \omega_{IF})$. In postdetection EGC reception, we can obtain the outputs of the correlator at the lth branch as

$$V_{k,l} = \int_0^T \tilde{i}_{k,l}(t)g(t)dt = \sqrt{2}R\sqrt{AP_{LO}}\sqrt{I_l}e^{j\phi_{s,k}}e^{j\phi_{n,l}} + \mu_{k,l} \tag{81}$$

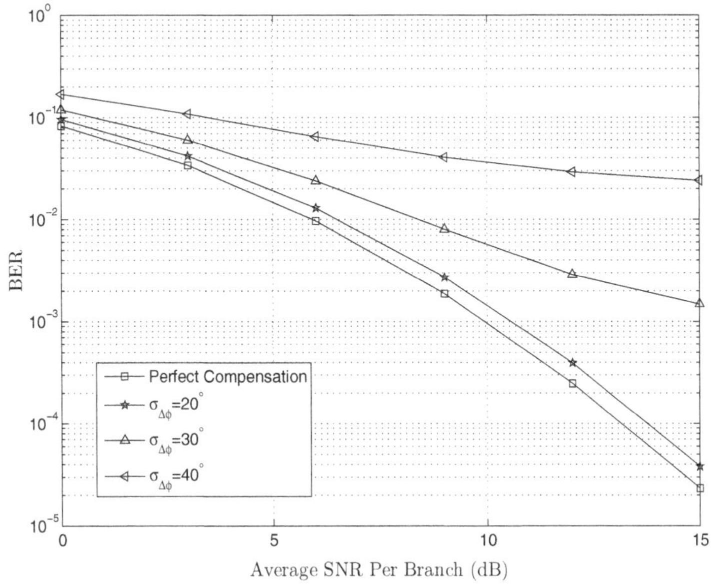

Figure 8. Error performance of a three-branch EGC coherent FSO system through an atmospheric turbulence channel with imperfect phase noise compensation.

and

$$V_{k-1,l} = \int_0^T \tilde{i}_{k-1,l}(t)g(t)dt = \sqrt{2}R\sqrt{AP_{LO}}\sqrt{I_l}e^{j\phi_{s,k-1}}e^{j\phi_{n,l}} + \mu_{k-1,l} \tag{82}$$

respectively. Here, $\mu_{k,l}$ and $\mu_{k-1,l}$ are filtered complex-valued zero mean Gaussian RVs with equal variance $\sigma^2 = 2qRP_{LO}\Delta f$ for both the real and imaginary parts. Hence, for simplicity we drop the subscript l for the noise terms. Without loss of generality, we normalize the variance of μ_k and μ_{k-1} to be unity for convenience of later applications. After normalization, we obtain

$$\tilde{V}_{k,l} = \frac{R\sqrt{A}}{\sqrt{qR\Delta f}}\sqrt{I_l}e^{j\phi_{s,k}}e^{j\phi_{n,l}} + \tilde{\mu}_k = \sqrt{\bar{\gamma}}\sqrt{I_l}e^{j\phi_{s,k}}e^{j\phi_{n,l}} + \tilde{\mu}_k \tag{83}$$

and

$$\tilde{V}_{k-1,l} = \frac{R\sqrt{A}}{\sqrt{qR\Delta f}}\sqrt{I_l}e^{j\phi_{s,k-1}}e^{j\phi_{n,l}} + \tilde{\mu}_{k-1} = \sqrt{\bar{\gamma}}\sqrt{I_l}e^{j\phi_{s,k-1}}e^{j\phi_{n,l}} + \tilde{\mu}_{k-1} \tag{84}$$

where $\tilde{\mu}_k$ and $\tilde{\mu}_{k-1}$ are i.i.d. Gaussian RVs with unity variance. Therefore, the decision variable $\tilde{D}$ at the output of the postdetection combiner is obtained as

$$\tilde{D} = \sum_{l=1}^{L} \tilde{U}_l = \sum_{l=1}^{L} \Re\{\tilde{V}_{k-1,l}^{*}\tilde{V}_{k,l}\} \tag{85}$$

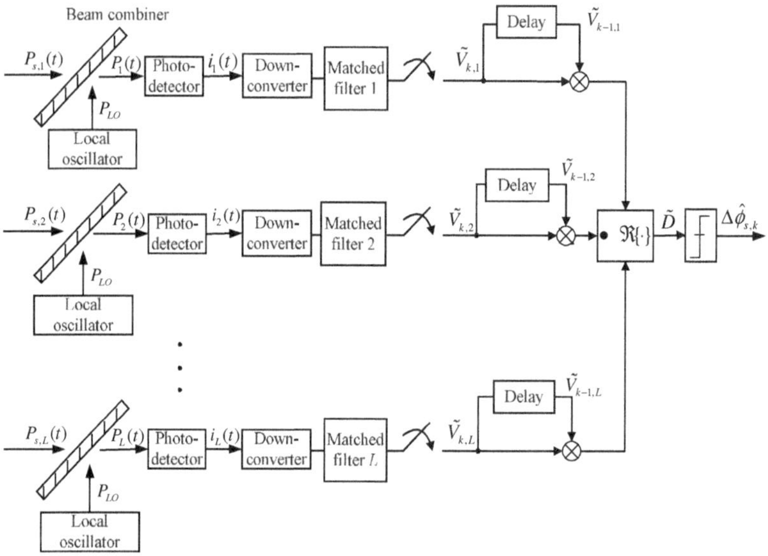

Figure 9. Block diagram of a coherent FSO system adopting postdetection EGC through an atmospheric turbulence channel.

where $\tilde{U}_l = \Re\{\tilde{V}^*_{k-1,l}\tilde{V}_{k,l}\}$. With [66, Eq. (B.5)], the CHF of $\tilde{U}_l$ conditioned on I_l can be found as

$$\Phi_{\tilde{U}|I_l}(\omega|\Delta\phi_{s,k}=0) = \frac{1}{\omega^2+1}\exp\left(-\bar{\gamma}\frac{\omega^2-j\omega}{\omega^2+1}I_l\right). \tag{86}$$

With the help of [63, Eq. 6.643(3)], averaging (86) over I_l gives the CHF of $\tilde{U}_l$

$$\Phi_{\tilde{U}}(\omega|\Delta\phi_{s,k}=0) = \frac{1}{\omega^2+1}M_I\left(-\bar{\gamma}\frac{\omega^2-j\omega}{\omega^2+1}\right) \tag{87}$$

where $M_I(\cdot)$ is given by (28) or (44). For i.i.d. Gamma-Gamma turbulence, we can express the CHF of $\tilde{D}$ as

$$\Phi_{\tilde{D}}(\omega|\Delta\phi_{s,k}=0) = \Phi^L_{\tilde{U}}(\omega|\Delta\phi_{s,k}=0) = \frac{1}{(\omega^2+1)^L}\left[M_I\left(-\bar{\gamma}\frac{\omega^2-j\omega}{\omega^2+1}\right)\right]^L. \tag{88}$$

With the Gil-Pelaez formula, we can obtain the average BER for DPSK with postdetection EGC as

$$P_e = \Pr\{\tilde{D}<0|\Delta\phi_{s,k}=0\} = \frac{1}{2}-\frac{1}{\pi}\int_0^\infty \frac{\Im\left\{\left[M_I\left(-\bar{\gamma}\cdot\frac{\omega^2-j\omega}{\omega^2+1}\right)\right]^L\right\}}{\omega(\omega^2+1)^L}d\omega \tag{89}$$

which can be used in the BER calculation for coherent DPSK systems with (28) or (45). Figure 10 and Fig. 11 show the error performance of postdetection EGC employing DPSK in weak and strong turbulence conditions, respectively.

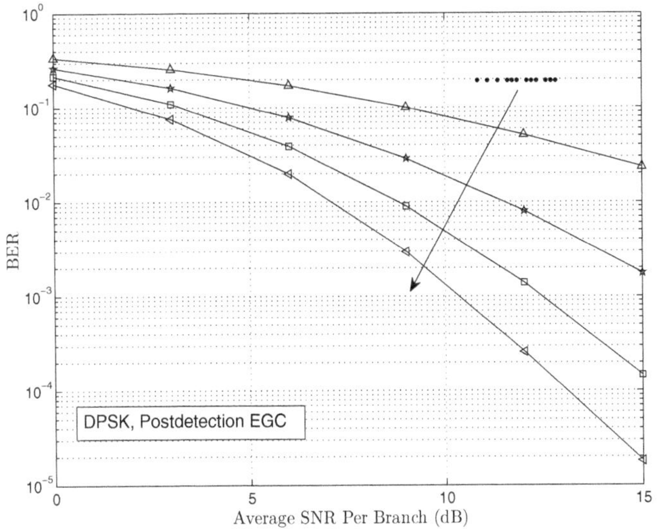

Figure 10. Error performance of a multi-branch postdetection EGC coherent FSO system through a strong turbulence channel with $\alpha = 2.2, \beta = 1.7$.

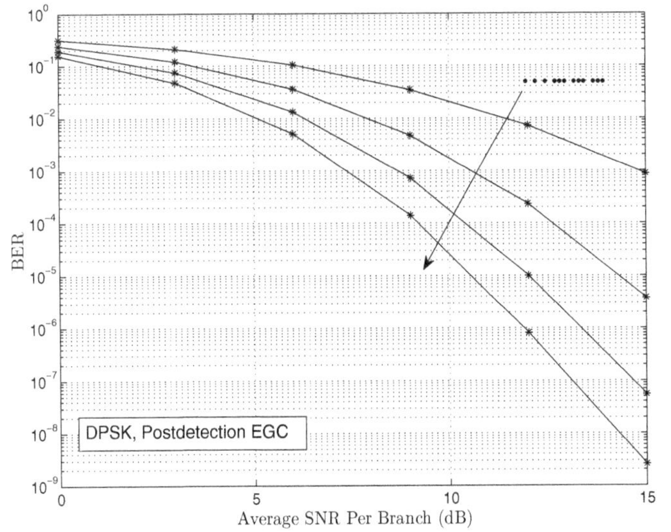

Figure 11. Error performance of a multi-branch postdetection EGC coherent FSO system through a weak turbulence channel with $\alpha = 6.2, \beta = 6.6$.

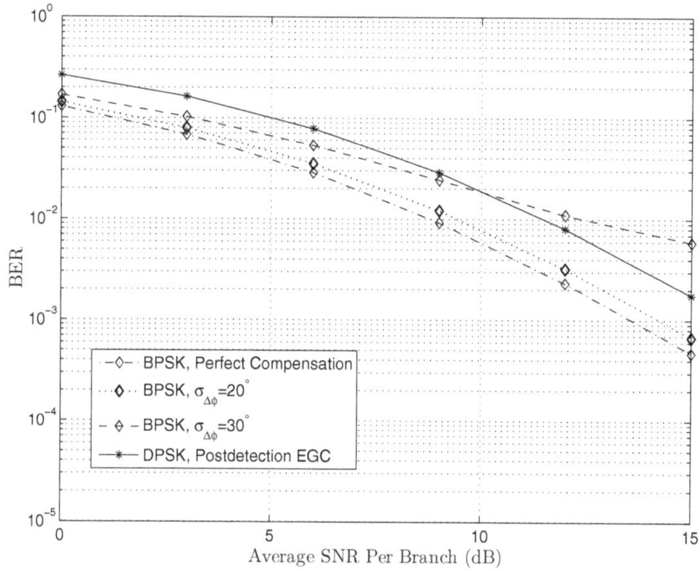

Figure 12. Performance comparison of two-branch EGC BPSK and postdetection DPSK FSO systems through a strong atmospheric turbulence channel with imperfect phase noise compensation.

To further illustrate the usefulness of the proposed postdetection DPSK FSO system in the presence of phase noise, we compare its error performance with that of the coherent EGC system with different standard deviations of phase noise compensation errors. As shown in Fig. 12, the proposed DPSK system is robust to the optical phase noise variations and can outperform the coherent PSK system where large phase noise compensation errors are present.

6. Concluding remarks and future research directions

In this chapter, we performed case studies on coherent systems for terrestrial FSO applications. We demonstrated the effectiveness of spatial diversity techniques in mitigating atmospheric turbulence effects. Important diversity reception techniques, MRC and EGC, were studied, and we showed that the EGC systems can provide comparable error performance to the optimal MRC FSO systems. We compared coherent PSK using EGC to DPSK using postdetection EGC when phase noise compensation error is present. It was found that the coherent PSK based system outperforms the DPSK based system when phase noise compensation error is small. However, it was also demonstrated that the DPSK is an excellent alternative to coherent PSK with EGC in terrestrial coherent diversity FSO communication systems where large phase noise compensation errors exist. These results can be useful in coherent FSO system design and performance evaluation. The rest of this section outlines some directions of future research on coherent FSO systems.

For practical reasons, real-time estimation of the instantaneous SNR may be difficult or expensive. Therefore, a more practical selection based combining can be proposed for coherent FSO systems. Traditional selection combining systems must choose the branch with the largest instantaneous SNR at high operational rates. To gain the benefits of reduced complexity for selection combining, without introducing the practical challenges of maximum branch selection, signal-plus-noise (S+N) selects the largest S+N power from the multiple receiver branches. A theoretical error rate analysis as well as asymptotic analysis for S+N selection combining systems will be worthwhile to investigate to demonstrate benefits from this simplified combining scheme. Furthermore, the use of multiple optical beams to transmit information provides additional degrees of freedom to improve the optical channel liability and/or capacity. However, unlike direct detection FSO systems, the presence of phase noise will prevent multiple beams from ideal electric field mixing at the coherent receiver. To overcome this problem, the desired performance improvements can be obtained by 1. using space-time type codes; 2. implementing polarized beam transmission/reception (by applying different states of polarization to either the optical signal or carrier); 3. using wavelength multiplexing for coherent FSO systems. Specifically, for subcarrier intensity modulation and IM/DD FSO systems, Alamouti type space time coding (STC) does not offer advantages over the repetition coding. A coherent FSO system with Alamouti type STC can be of interest to study for performance comparison between different coding schemes. As an attribute of the coherent FSO technique, coherent FSO data information can also be carried through the polarization state of the electric field. Polarization multiplexing coherent FSO system can therefore be used to increase the data rate per wavelength. In addition, quantifying system performance loss due to phase noise distortion is of interest and of importance for coherent FSO links. One can study the synchronous processing with coherent PSK and differential processing with DPSK respectively for a variety of turbulence channels with a proper assumption of phase noise model (for instance, a zero mean Gaussian distribution).

Acknowledgements

The research work presented in this chapter was supported by the University of British Columbia Internal Research Grants and the Natural Sciences and Engineering Research Council of Canada (NSERC).

Author detail

Mingbo Niu, Julian Cheng and Jonathan F. Holzman
The University of British Columbia (UBC), School of Engineering, Faculty of Applied Science, Kelowna, British Columbia, Canada

7. References

[1] Willebrand H, Ghuman B S (2002) *Free Space Optics: Enabling Optical Connectivity in Today's Networks*. Indianapolis, IN: Sams Publishing.
[2] Pratt W K (1969) *Laser Communication System*. New York: Wiley.
[3] Liu Q, Qiao C, Mitchell G, Stanton S (2005) Optical wireless communication networks for first- and last-mile broadband access, OSA J. Opt. Netw., 4(12): 807-828.
[4] Clough S A, Kneizys F X, Shettle E P, Anderson G P (1986) Atmospheric radiance and transmittance: FASCOD2, Proceedings of the Sixth Conference on Atmospheric Radiation, Williamsburg, VA, May 13th-16th. American Meteorological Society.

[5] Anderson G P, Berk A, Acharya P K, Matthew M W, Bernstein L S, Chetwynd J H, Dothe H, Adler-Golden S M, Ratkowski A J, Felde G W, Gardner J A, Hoke M L, Richtsmeier S C, Pukall B, Mello J, Jeong L S (2000) MODTRAN4: Radiative transfer modeling for remote sensing, Proceedings of SPIE 4049: 176-183.
[6] Karp S, Gagliardi R, Moran S E, Stotts L B (1988) *Optical Channels*. New York: Plenum.
[7] Win M Z, Chen C -C, Scholtz R A (1995) Optical phase-locked loop (OPLL) for an amplitude modulated communications link using solid-state lasers, IEEE J. Select. Areas Commun. 13(3): 569-576.
[8] Jafar M, O'Brien D C, Stevens C J, Edwards D J (2008) Evaluation of coverage area for a wide line-of-sight indoor optical free-space communication system empolying coherent detection, IET Commun. 2(1): 18-26.
[9] Kahn J M (2006) Modulation and detection techniques for optical communication systems, in Optical Amplifiers and Their Applications/Coherent Optical Technologies and Applications, Technical Digest (CD) (Optical Society of America), paper CThC1.
[10] Seimetz M (2005) Multi-format transmitters for coherent optical M-PSK and M-QAM transmission, Proceedings of 7th International Conference on Transparent Optical Networks 2: 225-229, July 3rd-7th, Barcelona, Spain.
[11] Andrews L C, Phillips R L, Hopen C Y, Al-Habash M A (1999) Theory of optical scintillation, J. Opt. Soc. Am. A 16(6): 1417-1429.
[12] Andrews L C, Phillips R L, Hopen C Y (2001) *Laser Beam Scintillation with Applications*. Bellingham, WA: SPIE Press.
[13] Goodman J W (1985) *Statistical Optics*, 1st ed. New York: Wiley-Interscience.
[14] Chan V W S (2006) Free-space optical communications, IEEE/OSA J. Lightwave Technol. 24(12): 4750-4762.
[15] Wireless Communications Association International. Available: http://www.wcai.com. Accessed 2010 March.
[16] O'Brien D C, Katz M (2005) Optical wireless communication within fourth-generation wireless systems, OSA J. Opt. Netw. 4(6): 312-322.
[17] Lee E J, Chan V W S (2004) Part 1: Optical communication over the clear turbulent atmospheric channel using diversity, IEEE J. Select. Areas Commun. 22(9): 1896-1906.
[18] Lee E J, Chan V W S (2007) Diversity coherent receivers for optical communication over the clear turbulent atmosphere, IEEE International Conference on Communications (ICC'07), June 24th-28th, Glasgow. pp. 2485-2492.
[19] Lee E J, Chan V W S (2009) Diversity coherent and incoherent receivers for free-space optical communication in the presence and absence of interference, IEEE/OSA J. Opt. Commun. Netw. 1(5): 463-483.
[20] Navidpour S M, Uysal M, Kavehrad M (2007) BER performance of free-space optical transmission with spatial diversity, IEEE Trans. Wireless Commun. 6(8): 2813-2819.
[21] Zhu X, Kahn J M (2002) Free-space optical communication through atmospheric turbulence channels, IEEE Trans. Commun. 50(8): 1293-1300.
[22] Zhu X, Kahn J M (2003) Performance bounds for coded free-space optical communications through atmospheric turbulence channels, IEEE Trans. Commun. 51(8): 1233-1239.
[23] Jakeman E, Pusey P N (1978) Significance of *K* distributions in scattering experiments, Phys. Rev. Lett. 40(9): 546-550.
[24] Parry G (1981) Measurement of atmospheric turbulence induced intensity fluctuations in a laser beam, Optica Acta 28(5): 715-728.
[25] Phillips R L, Andrews L C (1981) Measured statistics of laser light scattering in atmospheric turbulence, J. Opt. Soc. Am. 71(12): 1440-1445.

[26] Phillips R L, Andrews L C (1982) Universal statistical model for irradiance fluctuations in a turbulent medium, J. Opt. Soc. Am. 72(7): 864-870.

[27] Al-Habash M A, Andrews L C, Phillips R L (2001) Mathematical model for the irradiance probability density function of a laser beam propagating through turbulent media, Opt. Eng. 40(8): 1554-1562.

[28] Churnside J H, Clifford S F (1987) Log-normal Rician probability-density function of optical scintillations in the turbulent atmosphere, J. Opt. Soc. Am. A 4(10): 1923-1930.

[29] Churnside J H, Frehlich R G (1989) Experimental evaluation of lognormally modulated Rician and IK models of optical scintillation in the atmosphere, J. Opt. Soc. Am. A 6(11): 1760-1766.

[30] Uysal M, Navidpour S M, Li J (2004) Error rate performance of coded free-space optical links over strong turbulence channels, IEEE Commun. Lett. 8(10): 635-637.

[31] Farid A A, Hranilovic S (2007) Outage capacity optimization for free-space optical links with pointing errors, IEEE/OSA J. Lightwave Technol. 25(7): 1702-1710.

[32] Sandalidis H G, Tsiftsis T A, Karagiannidis G K, Uysal M (2008) BER performance of FSO links over strong atmospheric turbulence channels with pointing errors, IEEE Commun. Lett. 12(1): 44-46.

[33] Uysal M, Li J, Yu M (2006) Error rate performance analysis of coded free-space optical links over Gamma-Gamma atmospheric turbulence channels, IEEE Trans. Wireless Commun. 5(6): 1229-1233.

[34] Riediger M L B, Schober R, Lampe L (2009) Fast multiple-symbol detection for free-space optical communications, IEEE Trans. Wireless Commun. 57(4): 1119-1128.

[35] Kazovsky L, Benedetto S, Willner A (1996) *Optical Firber Communication Systems*. MA: Artech House.

[36] Fried D L (1967) Optical heterodyne detection of an atmospherically distorted signal wave front, Proceedings of the IEEE 55(1): 57-68.

[37] Wilson S G, Brandt-Pearce M, Cao Q, Baedke M (2005) Free-space optical MIMO transmission with *Q*-ary PPM, IEEE Trans. Commun. 53(8): 1402-1412.

[38] Wilson S G, Brandt-Pearce M, Cao Q, Baedke M (2005) Optical repetition MIMO transmission with multipulse PPM, IEEE J. Select. Areas Commun. 9(9): 1901-1910.

[39] Letzepis N, Holland I, Cowley W (2008) The Gaussian free space optical MIMO channel with *Q*-ary pulse position modulation, IEEE Trans. Wireless Commun. 7(5): 1744-1753.

[40] Tsiftsis T A, Sandalidis H G, Karagiannidis G K, Uysal M (2009) Optical wireless links with spatial diversity over strong atmospheric turbulence channels, IEEE Trans. Wireless Commun. 8(2): 951-957.

[41] Bayaki E, Schober R, Mallik R K (2009) Performance analysis of MIMO free-space optical systems in Gamma-Gamma fading, IEEE Trans. Commun. 57(11): 3415-3424.

[42] Abou-Rjeily C, Slim A (2011) Cooperative diversity for free-space optical communications: transceiver design and performance analysis, IEEE Trans. Commun. 59(3): 658-663.

[43] Ibrahim M M, Ibrahim A M (1996) Performance analysis of optical receivers with space diversity reception, IEE Proc.-Commun. 143(6): 369-372.

[44] Andrews L C, Phillips R L (2005) *Laser Beam Propagation through Random Media*, 2nd ed. Bellingham, WA: SPIE Press.

[45] Kiasaleh K. (2006) Performance of coherent DPSK free-space optical communication systems in *K*-distributed turbulence, IEEE Trans. Commun. 54(4): 604-607.

[46] Tsiftsis T A (2008) Performance of heterodyne wireless optical communication systems over Gamma-Gamma atmospheric turbulence channels, Electron. Lett. 44(5): 373-375.

[47] Sandalidis H G, Tsiftsis T A, Karagiannidis G K (2009) Optical wireless communications with heterodyne detection over turbulence channels with pointing errors, IEEE/OSA J. Lightwave Technol. 27(20): 4440-4445.
[48] Li J, Liu J Q, Taylor D P (2007) Optical communication using subcarrier PSK intensity modulation through atmospheric turbulence channels, IEEE Trans. Commun. 55(8): 1598-1606.
[49] Belmonte A, Kahn J M (2008) Performance of synchronous optical receivers using atmospheric compensation techniques, Opt. Express 16(18): 14151-14162.
[50] Belmonte A, Kahn J M (2009) Capacity of coherent free-space optical links using atmospheric compensation techniques, Opt. Express 17(4): 2763-2773.
[51] Belmonte A, Kahn J M (2009) Capacity of coherent free-space optical links using diversity combining techniques, Opt. Express 17(15): 12601-12611.
[52] Niu M, Cheng J, Holzman J F, McPhail L (2009) Performance analysis of coherent free space optical communication systems with *K*-distributed turbulence, IEEE International Conference on Communications (ICC'09), June 14th-18th, Dresden, Germany. pp. 1-5.
[53] Niu M, Cheng J, Holzman J F (2010) Exact error rate analysis of equal gain and selection diversity for coherent free-space optical systems on strong turbulence channels, Opt. Express 18(13): 13915-13926.
[54] Belmonte A, Kahn J M (2011) Field conjugation adaptive arrays in free-space coherent laser communications, IEEE/OSA J. Opt. Commun. Netw. 3(11): 830-838.
[55] Aghajanzadeh S M, Uysal M (2010) Diversity-multiplexing trade-off in coherent free-space optical systems with multiple receivers, IEEE/OSA J. Opt. Commun. Netw. 2(12): 1087-1094.
[56] Aghajanzadeh S M, Uysal M (2011) Multi-hop coherent free-space optical communications over atmospheric turbulence channels, IEEE Trans. Commun. 59(6): 1657-1663.
[57] Lange R, Smutny B, Wandernoth B, Czichy R, Giggenbach D (2006) 142 km, 5.625 Gbps free-space optical link based on homodyne BPSK modulation, Proceedings of SPIE 6105: 61050A(1-9).
[58] Kopeika N S, Bordogna J (1970) Background noise in optical communication systems, Proceedings of the IEEE 58(10): pp. 1571-1577.
[59] Agrawal G P (2002) *Fiber-Optical Communication Systems*, 3rd ed. New York: Wiley.
[60] Jakeman E, Pusey P N (1976) A model for non-Reyleigh sea echo, IEEE Trans. Antennas Propagat. AP-24(6): 806-814.
[61] Niu M, Cheng J, Holzman J F (2011) Error rate analysis of *M*-ary coherent free-space optical communication systems with *K*-distributed turbulence, IEEE Trans. Commun. 59(3): 664-668.
[62] Wolfram Mathworld (2011). Available: http://functions.wolfram.com/03.04.06.0002.01. Accessed 2011 August.
[63] Gradshteyn I S, Ryzhik I M (2000) *Table of Integrals, Series, and Products*, 6th ed. San Diego: Academic Press.
[64] Belmonte A, Kahn J M (2010) Efficiency of complex modulation methods in coherent free-space optical links, Opt. Express 18(4): 3928-3937.
[65] Niu M, Schlenker J, Cheng J, Holzman J F, Schober R (2011) Coherent wireless optical communications with predetection and postdetection EGC over Gamma-Gamma atmospheric turbulence channels, IEEE/OSA J. Opt. Commun. Netw. 3(11): 860-869.
[66] Proakis J G (2007) *Digital Communications*, 5th ed. New York: McGraw-Hill.

Chapter 9

Realization of Mixed WDM Transmission System

Vjaceslavs Bobrovs and Girts Ivanovs

Additional information is available at the end of the chapter

1. Introduction

Wavelength division multiplexing (WDM) technology for fiber optic transmission systems has been developed and introduced in order to make use of approximately 60 THz bandwidth that is offered by silica optical fibers "in [1,3]". In recent years total information carrying capacity of transmission systems was increased for the account of channels number, channel spacing and per channel bit rates. However in this case must take into the account total amount of optical power coupled into a fiber. Coupled power increase resulting in additional transmission impairments caused by nonlinear optical effects (NOE) and its combination with linear distortion mechanisms "in [4,6]". Generally it leads to distorted transmission in some channels of fiber optic transmission system (FOTS) or even to complete failure of system's channels. It means that informative signals cannot be detected on the other end of the fiber with a required error probability. This reduces system's total carrying capacity and channel's data throughput. Consequently, another system's total transmission capacity increment solution must be found out.

Currently one of the most intensively studied system's total transmission capacity increment solutions is the increasing of system's channel spectral efficiency "in references [7,11]". Actually it is more efficient utilization of available bandwidth. It means that more informative bits are transmitted using one hertz from available frequency band. It ensures that a smaller number of channels must be used to transmit the same amount of informative bits. Channel's spectral efficiency can be increased in three different ways. The first one, the reduction of used system's channel spacing "in [2]". This means that a larger number of transmission channels can be allocated in available frequency band. The second one, the increase of per channel bit rate maintaining previously used channel spacing values for separation of transmission channels "in [10]". And finally the third one is the combination of pervious two ways.

Obviously that it is easier to achieve a larger channel's spectral efficiency if for optical signal modulation and coding some of novel modulation formats are used. This novel (or advanced) modulation formats provide narrower optical signals spectrum or multilevel encoding schemes that ensure more bits per one symbol than it is in traditional modulation formats, for example, on – off keying (OOK) with non – return to zero (NRZ) encoding format (NRZ – OOK) "in reference[10]". Maximal spectral efficiency, which can be obtained with traditional OOK modulation formats, is about 0.4 bit/s/Hz "in [2]". "It has been reported in ref. [7,8]" that using such novel modulation formats as quadrature amplitude modulation (16 – QAM particularly) and orthogonal frequency – division multiplexing(OFDM) together with polarization division multiplexing (PDM) technique it can be achieved SE larger than 6 bit/s/Hz and even reaches 7 bit/s/Hz.

Study object of this chapter is optimal mixed WDM system configuration that provides lowest in system's channels detected signals BER values. This developed mixed WDM system's model is offered for the future design of backbone optical networks and can be considered under the concept of next generation optical network (NGON). Chosen optical signal modulations formats and per channel bit rates, according to authors' thoughts, are the most appropriate and probable at this moment. It was concluded after careful evaluation of current state of optical telecommunication networks, their possible and the most likely development strategy and trends in the future.

2. Necessity of mixed WDM transmission system

With the explosive growth in demand for capacity in national, regional, and metropolitan optical networks, high spectral efficiency wavelength division multiplexing transmission becomes essential. Current WDM optical transport systems are primarily based on 10 Gbit/s channels that are modulated with on – off – keying on a 50 GHz channel grid. Capacity upgrade of these systems calls for 40 Gbit/s wavelength channels to be carried in the same system "in [5,9]". To achieve this, several technical challenges need to be resolved.

2.1. Basic principle of WDM

Wavelength division multiplexing (WDM) is a technology used to combine or retrieve two or more optical signals of different optical center wavelengths in a fiber. It allows fiber capacity to be expanded in the frequency domain from one channel to more than 100 channels. This is accomplished by first converting standard, non – WDM optical signals to signals with unique WDM wavelengths that correspond to the available channel center wavelengths in the WDM multiplexer and demultiplexer. Typically, this is done by replacing non – WDM transceivers with the proper WDM channel transceivers. WDM channels are defined and labeled by their center wavelength or frequency and channel spacing. The WDM channel wavelength assignment is an industry standard defined in ITU – T documentation. Then the different WDM signal wavelengths are combined into one fiber by the WDM multiplexer. In the fiber, the individual signals propagate with little interaction assuming low signal power. For high powers, interchannel interaction can occur.

Once the signals reach the fiber link end, the WDM demultiplexer separates the signals by their wavelengths, back to individual fibers that are connected to their respective equipment receivers. Optical receivers have a broad reception spectrum, which includes all of C (1530-1565 nm) band. Many receivers can also receive signals with wavelengths down to O (1260-1360) band "in [11]".

The process of combining and separating the different wavelengths in the fiber is the basis of WDM technology. As a rough analogy, a WDM demultiplexer can be thought of as a prism separating different colors of light from the incident white light ray. Each color of light can represent a unique WDM signal. The white light is the aggregate of all colors of light and represents all the WDM signals propagating in the fiber. It should be noted that WDM wavelength and WDM channel have different meanings. WDM wavelength refers to the center wavelength of an optical signal or a WDM channel. WDM channel refers to an optical signal communications path that is defined by a center wavelength and a spectral pass band. It can be full duplex, half duplex, or simplex. Most telecommunication systems require full duplex communication channels. All full duplex fiber communication channels require two optical signals, one in each direction, to work properly. For a two fiber full duplex WDM system, the two laser signals (one at both ends of the fiber link on different fibers) are assigned the same wavelength, which is referred to as the channel assignment. However, in cases where only one fiber is used for full duplex communications, both lasers will have different wavelength assignments and therefore communication channel uses two WDM channels. Typically, most WDM systems will use two fibers for a WDM link. Therefore, the WDM wavelength assignment is the same for both lasers for that channel and is referred to as the channel assignment.

The heart of any WDM system is the basic WDM (also referred to as passive WDM). The basic WDM only consist of optical filters without any electronics. Therefore, it is completely passive and highly reliable. The filters are designed to pass a selected light spectral range, referred to as a channel, with low loss and reject or reflect all other spectrum. A combination of these filters results in a multichannel WDM multiplexer or demultiplexer. At the transmit end, the multiplexer combines the unique wavelengths from each channel port into one common fiber and at the receive end the demultiplexer coupler separates the combined signals from the common fiber to their respective channel ports. For some systems, the multiplexer and demultiplexer units are the same and are interchangeable (universal units) "in [11]".

In recent years the dramatic increase of demand for transmission capacity is observed and to secure an appropriate quality of service (QoS) level telecommunications services providers must constantly and continuously develop their transmission systems in use "in references [4,12]". Currently, as a study object of many scientific works have been chosen and focused directly on the fibre total transmission capacity increases, and this can happen in a three different ways. The first one, the existing 10 Gbit/s Dense WDM system upgrade, but in fact it is the substitution of existing system with 40 Gbit/s DWDM system or faster, because the only 10 Gbit/s system components, which can be used in new 40 Gbit/s system, are fibre, boosters and some external modulated lasers, but all transmitter and receiver electrical parts with bandpass filters must be changed to a new one. The second one, channels compaction by location them closer to each other using smaller channel spacing between them, in that way increasing the number of channel in available transmission frequency spectrum "in [2,15]". In this case, the total transmission capacity increment is achieved only because of increasing the number of channels, as the individual transmission rate in each channel remains unchanged. And the third way, total transmission capacity increment, using channel compaction with simultaneous increment of individual channel's transmission bit rate.

It is clear, that none of the proposed fibre's transmission capacity increment solution can be realized immediately, but it requires a certain amount of time and work, as any solution should be implemented gradually in several stages to avoid unnecessary problems.

2.2. Application of optical modulation formats

Since the introduction of the first optical transmission systems, capacities have steadily increased and the cost per transmitted bit has gradually decreased. The core of the global telecommunication network consists today of wavelength division multiplexed optical transmission systems. WDM is for these systems the technology of choice as it allows for a high spectral efficiency, i.e. the transmitted capacity per unit bandwidth. Commercial WDM systems generally use up to 80 wavelength channels with 50 GHz channel spacing and a bit rate of 10 Gbit/s or sometimes 40 Gbit/s per wavelength channel. This translates into a spectral efficiency between 0.2 and 0.8 bit/s/Hz. However, to cope with the forecasted increase in data traffic it will be necessary to develop next – generation transmission systems with even higher capacities. These transmission systems are expected to have a 40 Gbit/s or more bit rate per wavelength channel, with a spectral efficiency of between 0.8 and 2.0 bit/s/Hz. At the same time, such systems should be robust, i.e. provide a tolerance towards transmission impairments similar to currently deployed systems "in [9]".

Traditionally, optical transmission systems have used amplitude modulation. However, for the next generation of transmission systems this is not a suitable choice. They normally require too large channel spacing, have a high optical – signal – to – noise ratio (OSNR) requirement and generate significant nonlinear impairments. Current state-of-the-art transmission systems therefore often use differential phase shift keying (DPSK). Compared to amplitude modulation, DPSK generates less nonlinear impairments and has a lower OSNR requirement. However, due to the high symbol rate (e.g. 40-Gbaud) the robustness against the most significant linear transmission impairments, i.e. chromatic dispersion and polarization – mode dispersion (PMD) is still small. Tunable optical dispersion compensators can be used to improve the chromatic dispersion tolerance and PMD impairments are generally avoided through fiber selection. But optical compensation and fiber selection are generally not suitable for cost-sensitive applications "in references [17,18]". Another potential candidate to reduce the symbol rate is polarization multiplexing (POLMUX). This doubles the number of bits per symbol by transmitting independent information in each of the two orthogonal polarizations of an optical fiber "in [5]".

Our issue in a mixed system solution is offered as a part of common transmission system development, during the transition from traditional use of NRZ – OOK modulation format to alternative modulation formats, such as NRZ – DPSK and 2 – POLSK. Such hybrid solution can be topical in the case of combination or even in the case of different transmission systems merger, which results in the necessity to make a different modulated optical signal transmission over a single optical bus. As well as, such a need may occur in the future, switching traffic from a variety of WDM systems with the help of reconfigurable optical add – drop multiplexers (ROADM) and transmitting it further over common fibre to its destination or to the next ROADM "in [9]". The shift towards alternative optical signal modulation formats is necessary, because one of the major problem need to be overcome, in order to increase the total transmission capacity of core networks and a single fibre, are the reduction of transmission impairments and signal modulation format capability to resist against such impairments.

In high density WDM (HDWDM) systems with a large fibre span length between two optical amplifiers, signal form distortion causes such effects as linear chromatic dispersion, polarization mode dispersion, fibre non – linear effects or thereof combinations. In WDM system channel spacing reduction limiting factor is interchannel crosstalk, which originate due to optical fibre nonlinearities, such as crossphase modulation (XPM), selfphase modulation (SPM) and four – wave mixing (FWM) "in [13]". In order to reduce the impact of those effects, various optical modulation formats are increasingly being studied and offered, which could serve as an alternative to currently used traditional on – off keying. In this way manipulated signals are significantly distorted at high speed and high spectral density transmission conditions "in [14]".

3. Measurement technique and accuracy

This research is based on powerful and accepted mathematical simulation software OptSim. It solves complex differential nonlinear Schrödinger equation (NLSE) using split-step Fourier method (SSFM). This equation describes optical signal propagation over the fiber and can be written as Eq. (1) "in [19]":

$$\frac{\partial}{\partial z}\cdot A+\frac{\alpha^{l}}{2}\cdot A+j\cdot\frac{\beta_2}{2}\cdot\frac{\partial^2}{\partial t^2}\cdot A-\frac{\beta_3}{6}\frac{\partial^3}{\partial t^3}\cdot A=$$

$$=j\cdot\gamma\cdot|A|^2\cdot A \tag{1}$$

where $A(t, z)$ is complex optical field; z is fiber length, [km]; α^l is linear attenuation coefficient of an optical fiber, [km^{-1}]; β_2 is the second order parameter of chromatic dispersion, [ps^2/nm]; β_3 is the third order parameter of chromatic dispersion, [ps^3/nm]; γ is nonlinear coefficient, [W$^{-1}\cdot$km^{-1}]; t is time, [s]. NLSE takes into the account linear and nonlinear affects and they influence to optical signal distortions. The principle of split-step method is better illustrated by (1), which can be written as follows "in [19]":

$$\frac{\partial}{\partial z} \cdot A(t,z) = \left(\widehat{D} + \widehat{N}\right) \cdot A(t,z) \tag{2}$$

$\widehat{D}$ is linear operator responsible for linear effects such as dispersion and attenuation "in [19]":

$$\widehat{D} = -\frac{\alpha^l}{2} - j \cdot \frac{\beta_2}{2} \cdot \frac{\partial^2}{\partial t^2} + \frac{\beta_3}{6} \cdot \frac{\partial^3}{\partial t^3} \tag{3}$$

$\widehat{N}$ is nonlinear operator, which takes into account Kerr and other nonlinear effects (NOEs) "in [19]":

$$\widehat{N} = j \cdot \gamma \cdot |A|^2 \cdot A \tag{4}$$

In general split-step method is based on assumption that linear and nonlinear effects affect optical signals independently. This statement can be considered as true if we assume that all fiber length z is being divided into sufficiently small spans Δz, and only then these linear and nonlinear effects by turns are taken into account for each $\frac{\Delta z}{2}$ segment.

There are two basic algorithms for realization of SSFM: time domain split step (TDSS) and frequency domain split step (FDSS). These two algorithms differ only with an approach that is being used for calculation of linear operator $\widehat{D}$. While in both cases nonlinear operator $\widehat{N}$ is being calculated in time domain "in [19]".

Operator $\widehat{D}$ is being fully characterized by its impulse response *h(t)* and it is mathematically correct to calculate its influence to *A(t, z)* optical field using products of mathematical convolution. In TDSS case it can be written as follows "in [19]":

$$A_L[n] = A[n] * h[n] = \sum_{k=-\infty}^{\infty} A[k] \cdot h[n-k] \tag{5}$$

This algorithm calculates this convolution in time domain and precisely obtains time delay values between signals with different wavelength. In OptSim software this TDSS algorithm is realized using finite impulse reaction (FIR) filters. This sophisticated technique provides complete control of an overall mistake that may occur during all process of calculating. By contrast FDSS calculates $\widehat{D}$ in frequency domain but firstly for this algorithm is necessary to calculate fast Fourier transformation (FFT) from *A[n]* signal samples and from *h(t)* impulse reaction. Then it is necessary to use invers FFT (FFT^{-1}) to convert obtained data array to time sample domain. FDSS algorithm can be mathematically described using following equation "in [19]":

$$A'_L[n] = A[n] \otimes h[n] =$$

$$= FFT^{-1}\{FFT(A[n]) \times FFT(h[n])\} \tag{6}$$

As one can see, then in this case circular convolution is used for obtaining signal sample array $A'_L[n]$. This array may contain fewer samples than it is necessary to obtain actual convolution products—$A_L[n]$ sample array. Hence this algorithm is easier to implement than TDSS and it requires less computation time and resources but serious errors may occur during calculation "in [19]".

Eq. (2) can be solved assuming that $\widehat{D}$ and $\widehat{N}$ operators are independent and fiber span Δz length is small enough (5-100 m depending on the simulation accuracy requirements). Then optical signal after propagation over Δz span can be described in the following manner "in [15]":

$$A(t, z+\Delta z) \cong \exp\left[\frac{\Delta z}{2}\cdot \widehat{D}\right]\cdot \exp\left\{\Delta z\cdot \widehat{N}\left[A\left(t, z+\frac{\Delta z}{2}\right)\right]\right\}\cdot \exp\left(\frac{\Delta z}{2}\cdot \widehat{D}\right)\cdot A(t,z) \quad (7)$$

For the evaluation of system performance will be used such parameter as Q-factor and BER value. Q = 7.03 (16.94 dB) corresponds to the commonly used reference for BER of 10^{-12}.

Q factor uncertainty range (see Fig. 1) and BER confidence interval magnitude depends on the total number of simulated bits N_{total} "in [15]":

$$\text{dev}[Q^*] \equiv \sigma_Q \cong \frac{Q}{\sqrt{2\cdot N_{total}}}, \quad (8)$$

where Q is Q-factor value that can be calculated using following Eq. (9) "in references [2, 15]":

$$Q = \frac{|\mu_1-\mu_0|}{\sigma_1+\sigma_0}, \quad (9)$$

where $\mu_{1,0}$ and $\sigma_{1,0}$ are the mean and the standard deviation of the received signal, when a logical "1" and "0" is transmitted, and $\pi \approx 3.14$ "in references [3, 15]".

$$\text{BER} = \frac{1}{2}\cdot \text{erfc}\left(\frac{Q}{\sqrt{2}}\right) \quad (10)$$

Using Eq. (8) Q-factor uncertainty range can be expressed as "in [15]":

$$\text{range} = 20\cdot \log_{10}\left(\frac{1+\sqrt{\frac{2}{N_{total}}}}{1-\sqrt{\frac{2}{N_{total}}}}\right) \quad (11)$$

As one can see from Fig. 1 Q-factor uncertainty range for 1,024 simulated bits that is used in our schemes is equal to 0.77 dB.

Q-factor and BER value 95% ($\pm 2\sigma_Q$) confidence intervals for 16.94 dB nominal value can be obtained using Eq. (8) and Eq. (10) assuming that we are dealing with Gaussian distribution. For 1024 of simulated bits these intervals are:

$$\text{Q}_{\text{for 16.94 dB}} \in [16.55; 17.31], [\text{dB}] \quad (12)$$

$$\log 10\{\text{BER}_{\text{for }10^{-12}}\} \in [-12.97; -11.04] \quad (13)$$

As an example we will give BER 95 % confidence interval as a function from the number of simulated bits for 10^{-12} nominal. This value will be used as reference for transmission channel with 10 Gbit/s per channel bitrate (see Fig. 1).

As one can see, when simulating 1,024 bits at BER = 10^{-12}, the confidence interval magnitude is less than ±1 order. It points to the conclusion that OptSim simulation software allows obtaining sufficiently accurate preliminary results and there is no point to increase the total number of simulated bits, because obtained results accuracy does not improve sufficiently.

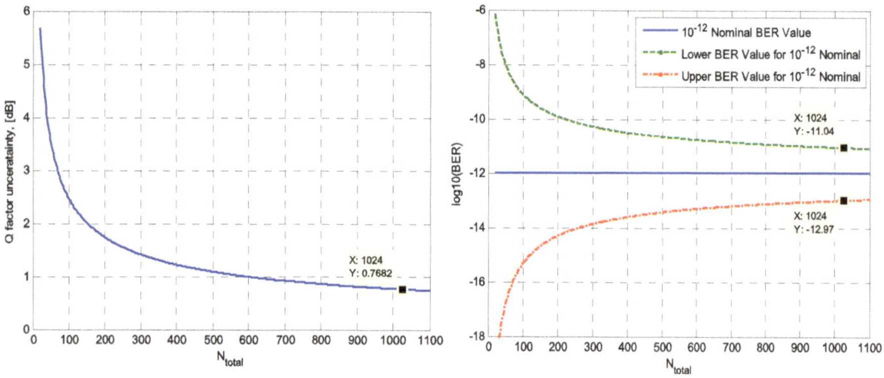

Figure 1. Q-factor uncertainty as a function form total number of simulated bits, and BER value 95% confidence intervals for 10^{-12} nominal.

4. Mixed WDM transmission system models

As a start model for our mixed WDM transmission system, 10 Gbit/s three – channel system was chosen. In this system optical signal modulation formats applied in each system's channel are different. Per channel transmission bit rate is chosen equal to 10 Gbit/s, because our research is focused on ultra – long haul mixed HDWDM transmission system development, which expecting existing 10 Gbit/s system's infrastructure use and further it's development.

4.1. Simplest model of mixed WDM

In the mentioned system's first channel optical signal is transmitted for which modulation differential phase shift keying with non – return – to – zero encoding technique is used (NRZ – DPSK). For the system's second channel on – off keying method and NRZ encoding is used, despite the fact, that NRZ – OOK modulation format is not well situated for high density WDM systems with a large number of transmission channels and high transmission rate and, as consequence of that, a high total transmission capacity. This modulation format can be used as a good foundation and reference point for comparison of different modulation formats, because it's traditionally used modulation format in optical transmission systems, due to its relatively simple realization and historical domination "in [2]". As modulation format for the system's third channel binary polarization shift keying (2 – POLSK) was chosen. It's the newest modulation format and in the same times the most promising "in [16]".

Then all three differently modulated optical signals are combined and transmitted through 50 km long standard single mode fibre (SSMF), without using optical amplifiers. On the other fibre end optical signals are filtered with optical Gaussian filters, converted to electrical signals and then electrically filtered using Bessel electrical filters (see Fig. 2).

SSMF length was chosen equal to 50 km, because it's better permissible length between two EDFA in an ultra – long haul transmission system, if fibre attenuation coefficient is 0.2 dB/km "in [4]". Large amplifier spacing in such system would result in a prohibitive increase in amplified spontaneous emission (ASE) noise and in order to achieve the greater range and information capacity, the amplifiers must be located close together with gain no greater than about 10 dB and preferably less "in [17]". Amplifier spacing further increment will lead to increase of ASE noise influence and as a result BER grow for each system channel. As well as, we must take into account system's accumulated dispersion level, because 10 Gbit/s network, where for optical signal modulation NRZ format is used, operates error free only if residual system dispersion is below 1000 ps/nm. SSMF has 17 ps/nm*km dispersion and this mean that mentioned above dispersion level threshold won't be exceeded if length of the used fibre is below 58 km. In our case, we start studied optical signal transmission over only one span of ultra – long haul transmission system and that's why fibre length was taken from its possible optimal configuration.

In the first previously mentioned system for optical signal modulation in all three channels differential binary phase shift keying (DPSK) was used, in the second – the intensity modulation (IM) and in the third – polarization shift keying were used, while the fourth is a mixed transmission system, where for the first channels optical signal modulation NRZ – DPSK was used, for the second channel – NRZ – OOK and third – 2 – POLSK.

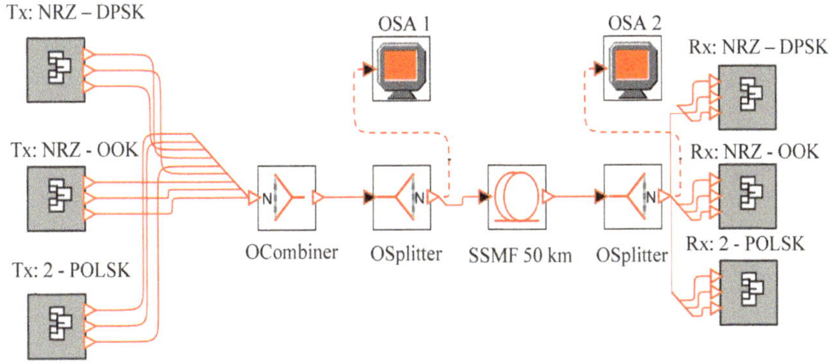

Figure 2. Simulation model of 3 – channel mixed WDM system

10 Gbit/s WDM System		Channel Spacing, GHz				
		25	37,5	50	75	100
		BER				
NRZ - DPSK	1st	$5{,}11 \times 10^{-12}$	$6{,}49 \times 10^{-13}$	$1{,}78 \times 10^{-16}$	$5{,}90 \times 10^{-17}$	$6{,}27 \times 10^{-17}$
	2nd	$3{,}25 \times 10^{-9}$	$6{,}12 \times 10^{-10}$	$1{,}00 \times 10^{-15}$	$1{,}98 \times 10^{-17}$	$4{,}00 \times 10^{-17}$
	3rd	$3{,}18 \times 10^{-12}$	$2{,}64 \times 10^{-13}$	$3{,}07 \times 10^{-16}$	$1{,}23 \times 10^{-16}$	$1{,}30 \times 10^{-16}$
NRZ - OOK	1st	$4{,}19 \times 10^{-29}$	$9{,}00 \times 10^{-18}$	$4{,}56 \times 10^{-22}$	$4{,}54 \times 10^{-26}$	$3{,}03 \times 10^{-26}$
	2nd	$1{,}06 \times 10^{-21}$	$2{,}53 \times 10^{-13}$	$1{,}04 \times 10^{-21}$	$4{,}85 \times 10^{-26}$	$1{,}84 \times 10^{-26}$
	3rd	$6{,}79 \times 10^{-31}$	$2{,}60 \times 10^{-16}$	$1{,}00 \times 10^{-22}$	$1{,}62 \times 10^{-25}$	$6{,}82 \times 10^{-26}$
2POLSK	1st	$1{,}00 \times 10^{-40}$	$1{,}71 \times 10^{-29}$	$1{,}00 \times 10^{-40}$	$1{,}00 \times 10^{-40}$	$1{,}00 \times 10^{-40}$
	2nd		$4{,}57 \times 10^{-18}$			
	3rd		$7{,}12 \times 10^{-30}$			
Mixed	1st (DPSK)	$1{,}88 \times 10^{-25}$	$9{,}95 \times 10^{-17}$	$7{,}70 \times 10^{-17}$	$3{,}49 \times 10^{-17}$	$2{,}55 \times 10^{-17}$
	2nd (NRZ)	$3{,}23 \times 10^{-12}$	$2{,}63 \times 10^{-10}$	$3{,}68 \times 10^{-24}$	$3{,}54 \times 10^{-25}$	$2{,}18 \times 10^{-26}$
	3rd (2POLSK)	$1{,}00 \times 10^{-40}$	$5{,}12 \times 10^{-27}$	$1{,}00 \times 10^{-40}$	$1{,}00 \times 10^{-40}$	$1{,}00 \times 10^{-40}$

Table 1. Simulation results

The configuration type of this system was chosen precisely in order to clarify interchannel crosstalk influence effect to transmission in adjacent channels, if modulation formats applied for each channel are different. Number of channels in systems, where just one modulation format is used for the optical signal modulations, was chosen equal to the number of channels of mixed system under study. It was specially done, in order to provide, that a total amount of input optical power coupled into the fibre would be approximately equal. This condition was specially held, in order to provide, that fibre nonlinearities could become apparent to the same extent and transmission would take place under same conditions, to make a comparison of these four different transmission systems for a range of channel spacing values. Each system simulation was performed for five different channels spacing, whose values were chosen based on the establishment principle of ITU – T Recommendation G.694.1. As the result, systems were simulated at following values of channel intervals: 25, 37.5, 50, 75, 100 GHz. Systems channels were grouped around 193.1 THz central frequency value and were located in C – Band (1530 – 1565 nm). The simulation results are summarized (see Table 1.).

Let's also note, that optical bandpass Gaussian filters with -3 dB bandwidth equal to 0.11 nm were used for signal filtering at 25 GHz channel spacing, rather than in the other cases, where -3 dB bandwidth is equal to 0.3 nm. For electrical signal filtering Bessel filters with number of poles equal to 5 and -3 dB bandwidth equal to 10 GHz were used.

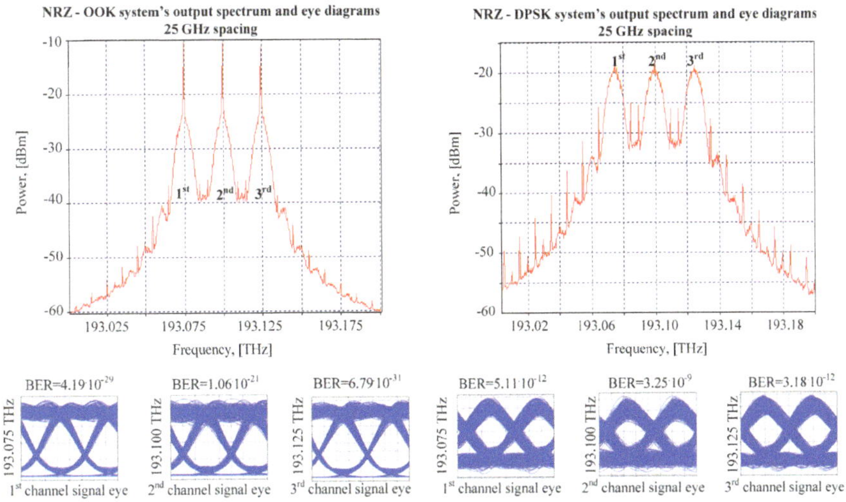

Figure 3. 3 – Channel NRZ – OOK system's output spectrum, signal eye diagrams and BER value in case of 25 GHz channel spacing, and 3 – channel NRZ – DPSK system's output spectrum, signal eye diagrams and BER value in case of 25 GHz channel spacing

At the beginning of the results analysis we will focus on traditionally used NRZ – OOK modulation format. As one can conclude form the simulation results, in the case of small channel spacing values the worst systems bit – error – ratio is for the second system channel. This is explained by the fact, that in this case interchannel crosstalk effects are more quintessential and signal spectrum compaction is maximal affordable (see Fig. 3). As one can see from this figure, further compaction leads to different signal spectrum overlapping and as a consequence imminent grow of BER values. If we increase the value of channel spacing, this difference between BER values of each channel disappears. But if we increase channel spacing up to 37.5 GHz, the worst channel BER is already less than desired 10^{-12} at the same filter characteristics. In the data transmission networks with 10 Gbit/s bitrates and higher, if forward error correction techniques (FEC) are not used, BER value must be $<10^{-12}$.

If we increase channel spacing value form 37.5 GHz up to 50 GHz, the worst channel BER improves till $1.04{\cdot}10^{-21}$. Further increment of spacing form 50 to 75 GHz or even up to 100 GHz at the given modulation and coding format, as well as bit rate, is not needed, because the BER improvement is not significant (see Fig. 4 – 5).

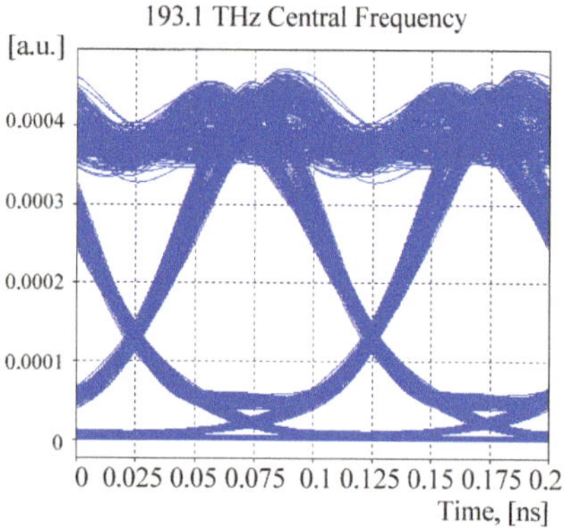

Figure 4. Eye diagram of the worst (2nd) NRZ – OOK system's channel, 50 GHz channel spacing and BER=$1.04 \cdot 10^{-21}$

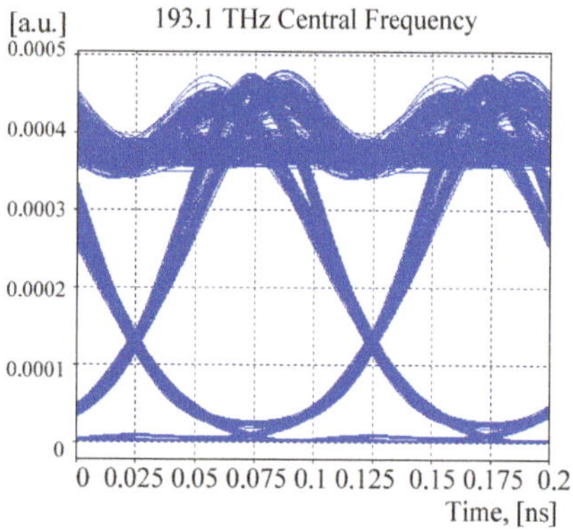

Figure 5. Eye diagram of the best (2nd) NRZ – OOK system's channel, 100 GHz channel spacing and BER=$1.84 \cdot 10^{-26}$

If we use for optical signal modulation NRZ – DPSK format, the resulting BER values for each simulated channel at certain channel spacing values is several orders worse than it is in the NRZ – OOK format cases (see Fig. 6 – 7). Channel spacing reduction form 100 GHz to 50 GHz, leads to reduction of the worst channel bit – error – rate by one order. This lets make a conclusion about NRZ – DPSK modulation formats suitability to high spectral density transmission conditions. It's non-susceptible to channel spacing decreases or increases, if it happens to specified threshold values, above which a sudden channel degradation process is unavoidable.

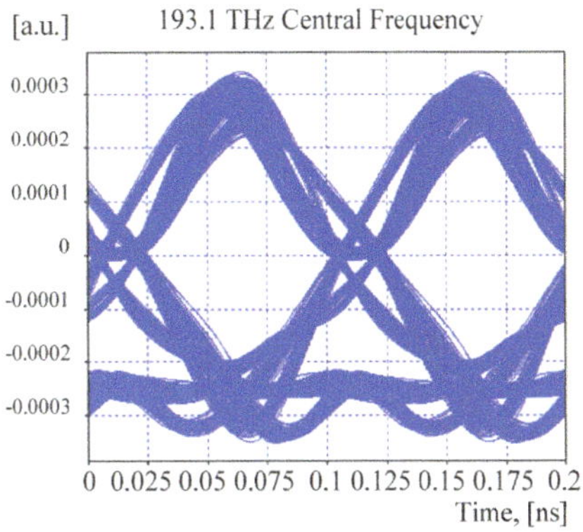

Figure 6. Eye diagram of the best (2nd) NRZ – DPSK systems channel, Δf=100 GHz and BER=$4.00 \cdot 10^{-17}$

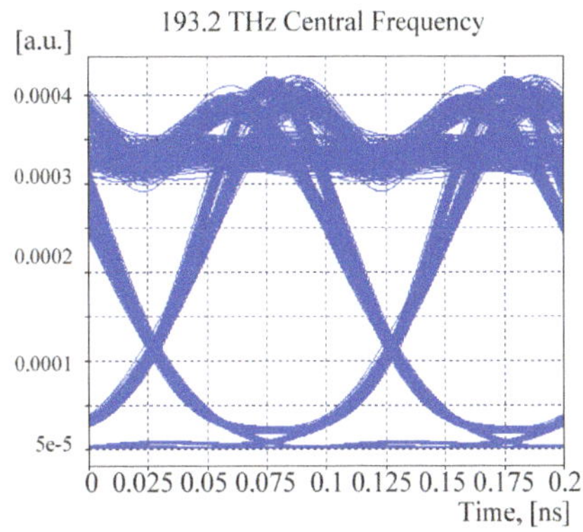

Figure 7. Eye diagram of the worst (3rd) NRZ – OOK systems channel, Δf=100 GHz and BER=$6.82 \cdot 10^{-26}$

However, if the channel spacing is reduced to 25 GHz, then all three channels BER values are greater than required 10^{-12} but middle channels BER is even greater than ITU – T defined 10^{-9}. If the modulation format, applied for optical signal in each WDM system transmission channel, is polarization shift keying (2 – POLSK), it is possible to achieve the best possible of channel BER values, irrespective to the channel spacing values as compared to other modulation formats. This is possible due to 2 – POLSK modulated signal spectrum (see Fig. 8). As can been seen, 2 –POLSK modulated optical signal spectrum is narrower than NRZ – DPSK and NRZ – OOK modulated signal spectrum. This property provides to a data signals greater error protection, when it is spread through the optical fibre transmission systems, and WDM signal spectrum lines at the beginning and at the end differ only by the level, spectrum extension and nonlinear effect influence are minimal.

If in multi-channel communication system for optical signal modulations different modulation formats are used, then obtained BER values for each channel will depend not only on the individual modulation format capability to resist from interchannel distortion, but also form that, which modulation format is used in the channel, which is the source of this disorders. This feature gets stronger on small (<50 GHz) channel spacing values, and the obtained simulation results for mixed system allow us to conclude this.

If in a mixed system 50, 75 or 100 GHz channel spacing are used for channel separation, then the channel BER values corresponding to BER values obtained for systems, where only one modulation format is used for the optical signal modulation. It is approximately 10^{-17} in NRZ – DPSK case, about 10^{-25} in NRZ – OOK and 10^{-40} in 2 – POLSK. Reducing channel spacing to 37.5 GHz, become evident special features of combined transmission and they stand out even more against the background, if 25 GHz interval is used for channel separation. As it can be seen from the obtained results, the first channel, where is used phase modulation, BER level is several orders lower (10^{-25}) than it is for the first channel of 3 – channel NRZ – DPSK system (10^{-12}) (see Fig. 8), the same can be applied to the second channel of the mixed system (10^{-12}) and 2nd channel (10^{-21}) of 3 – channel NRZ – OOK system (see Fig. 8).

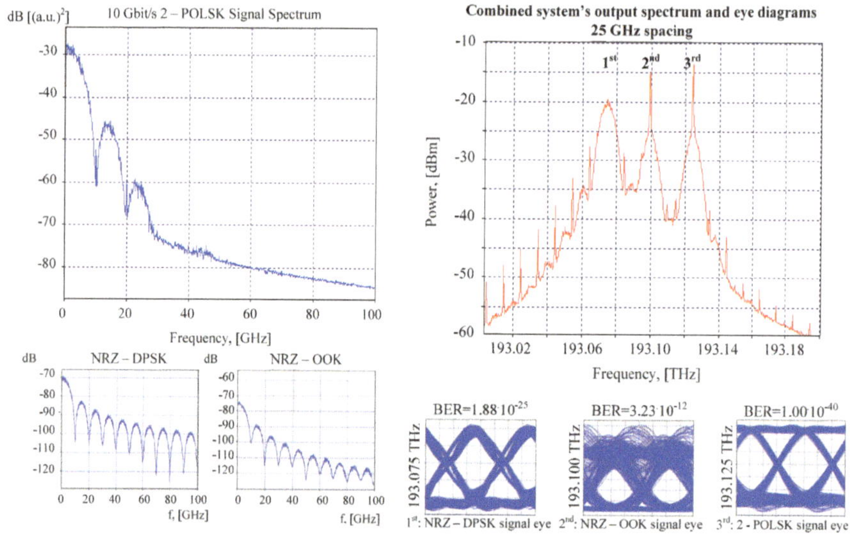

Figure 8. NRZ – DPSK, NRZ – OOK, 2 – POLSK optical signal spectrums at the transmitter end after electrical conversion, and 3 – Channel mixed system's output spectrum, signal eye diagrams and BER value in case of 25 GHz channel spacing

As one can see from these figures, eye opening in both cases are materially different, eye opening for mixed system is narrower than for traditional NRZ – OOK system's 2nd channel signal, if 25 GHz interval is used for channel separation.

The next step for mixed WDM system performance evaluation is to increase the channel quantity, bitrate and also transmission line length. Upgraded mixed WDM scheme contains nine channels, which are grouped in three groups with an identical transmitter and receiver block configuration but with different channel central wavelengths. It was specially done in order to take into account linear and nonlinear crosstalk influence, which experience central's group channels from adjacent channels. The central group consists from channels number 1 to 3, left group consists from channels number 4 to 6 and right group – from 7 to 9. For further system's analysis we will use only channels number 1 – 3, but 4 – 6 and 7 – 9 is used as sources of transmission impairments.

In the first channel as in previous model the NRZ – OOK signal optical modulation format is used. This modulation format can be used as a reference point for comparison of different modulation formats. It is traditionally used modulation format in optical transmission systems due to its relatively simple realization and historical domination "in references [8, 12]". Data transmission rate for this channel is chosen equal to 40 Gbit/s. For the system's second channel 2 – POLSK modulation format was chosen, because it's the newest and in the same time the most promising modulation formats for optical transmission systems "in [2]". In this case per channel bit rate was chosen equal to 40 Gbit/s. And finally, as modulation format for the system's third channel NRZ – DPSK was chosen. Per channel bit rate was chosen equal to 10 Gbit/s.

Modulation format allocations to each system's channels was performed on the basis of the fact that such modulation formats distribution among channels provides the lowest possible average BER value for system channels at 10 Gbit/s per channel bit rate and 25 GHz channel spacing, comparing to other five possible formats distribution variants. As for per channel bit rate assignment, then such variant provides SE = 0.4 bit/s/Hz and an average BER value not greater than 10^{-40}, if for CD compensation ideal FBG is used (see Fig. 9). Comparing with two other mixed data rates mixed system variants, which can provide SE = 0.4 bit/s/Hz, then their can secure BER < 10^{-12} or even < 10^{-14}, but their average channel BER >10^{-40} "in [3]". Thus, as a simulation model of mixed system was chosen a transmission system with a following configuration: [NRZ – OOK (40 Gbit/s, 193.025 THz)] – [2 – POLSK (40 Gbit/s, 193.100 THz)] – [NRZ – DPSK (10 Gbit/s, 193.175 THz)].

As one can see (see Fig. 9), then SE equal to 0.4 bit/s/Hz was obtained in case, if 75 GHz interval is used for channel separation. That channel spacing value was selected based on the establishment principle of ITU – T Recommendation G.694.1, which provides a frequency grid for dense wavelength division multiplexing applications. The frequency grid, anchored to 193.1 THz, supports a variety of channel spacing's ranging from 12.5 GHz to 100 GHz and wider. Afterwards all optical signals from nine channels are combined and transmitted through 50 km SSMF using 4dBm fixed output power optical amplifier, which operates on basis of EDFA. This amplifier is necessary, because we simulate optical signal transmission through one span of ultra-long haul system.

The output power level is chosen equal to 4 dBm due to the fact that this level can ensure the lowest average BER values for transmission channels of that configuration mixed transmission systems but with bit rate of 10 Gbit/s per channel and 25 GHz channel spacing comparing to other fixed output power levels "in [3]". On the other fiber end optical signals are filtered with optical Super Gaussian filters, converted to electrical signals and then electrically filtered using Bessel electrical filters.

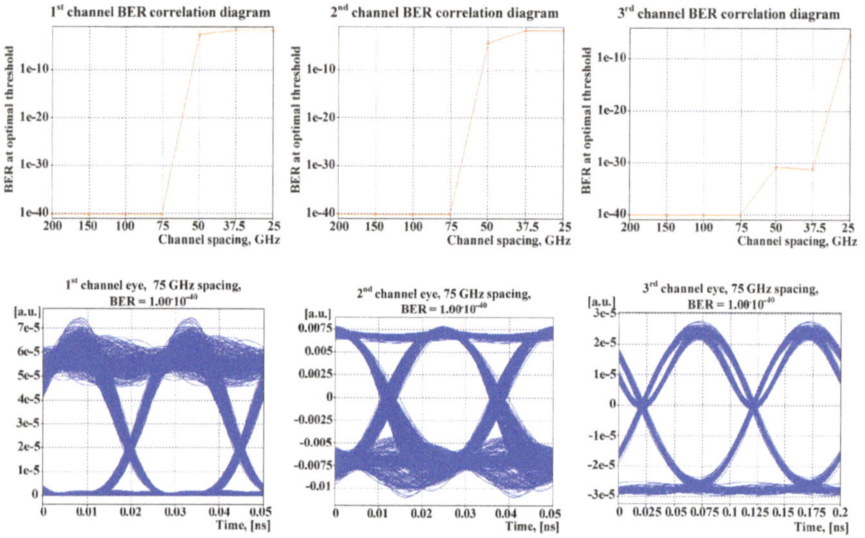

Figure 9. Mixed HDWDM system 1st/2nd/3rd channel's BER correlation diagram, detected signals eye diagrams and BER values at 75 GHz channel spacing

In order to achieve the greater range and information capacity, the amplifiers must be located close together with gain no greater than 10 dB and preferably less "in [2]". Amplifier spacing further increment will lead to increase of ASE noise influence and as a result BER growth for each system channel. As well as, we must take into account system's accumulated chromatic dispersion management strategy, which will be the next sections main goal.

4.2. Chromatic dispersion management strategies

Chromatic dispersion divides into material and waveguide dispersion.Waveguide dispersion is caused by physical structure of optical fiber core and cladding (refractive index profile), and as a result different wavelengths propagate at different velocities in the core and cladding. Material dispersion is dominant part of chromatic dispersion, and it is caused by change of optical fiber core and cladding refractive index with wavelength "in [4]".

Dispersion compensating fiber (DCF), fiber Bragg grating (FBG) and optical phase conjugator (OPC) can be used for chromatic dispersion compensation.

DCF has large negative dispersion (D = -80 ps/(nm*km), that helps to compensate chromatic dispersion. Such an optical fiber with negative dispersion is achieved by developing a complex refractive index profile. The effective core area (A_{eff}) of a DCF is much smaller than standard ITU – T G.652 single mode fiber, thereby dispersion compensating fiber experience much higher optical signal distortions caused by nonlinear optical effects (NOE). Typical dispersion compensating fiber has small effective core area $A_{eff} = 12\ \mu m^2$ whereas standard single mode optical fiber has $A_{eff} = 80\ \mu m^2$, and DCF has attenuation coefficient up to $\alpha = 0.6$ dB/km, whereas standard single mode optical fiber has $\alpha = 0.2$ dB/km. Impact of nonlinear optical effects can be reduced by lowering optical power "in [4]".

Chirped fiber Bragg grating (FBG) is effective technology for chromatic dispersion compensation, because it is more suitable for large transmission capacity WDM systems. It has grating period which is not constant, but changes linearly over the length of the grating with the shorter grating period located at the beginning of the grating. FBG grating period is distance between two adjacent maximum values of the refractive index.

The fiber grating reflects a narrow spectrum of wavelengths, that are centred at reflected wavelength (λ_B) and passes all the other wavelengths. Dispersion affected input pulse with width τ is passing Chirped fiber Bragg grating and at output its width is decreased by $\Delta\tau$ and shape is restored. Chirped fiber Bragg grating has shorter grating periods at beginning, but over the length of the grating these periods linearly increase. Therefore shorter signal wavelengths are reflected sooner and have less propagation delay through the FBG, but longer signal wavelengths travel further into the fiber grating before they are reflected back and have more propagation delay through the FBG. Typically the length of the fiber grating is from 10 to 100 cm "in [4]".

A significant advantage of using fiber Bragg grating over DCF fiber is its relatively small insertion loss. For comparison, commercial DCF specified to compensate accumulated chromatic dispersion of 100 to 120 km standard single mode fiber span have about 10 dB of insertion loss, whereas a FBG based dispersion compensation unit, capable to compensate the same fiber span length, insertion loss is only up to 4 dB. In contrast to DCF fiber Bragg grating can be used at higher optical powers without inducing nonlinear optical effects. These CD compensation methods described above will be practically implemented in our investigated mixed system model.

4.3. Advanced model of mixed WDM

There are three different combined chromatic dispersion compensation methods and one combined method described in this work. These simple CD compensation methods are DCF and FBG, but as a combined CD compensation method for practical realization we offer the compound solution of these two methods (DCF – FBG). In this work we are studied chromatic dispersion compensation solutions that can be implemented in transmitter or receiver side, as close as possible to beginning or end of fiber optical link. In these places it is possible to access directly to the optical fiber and place dispersion compensation modules.

The aim of this research is to find the best chromatic dispersion compensation solution that can be used for implementation and adaptation in already working mixed fiber optical transmission systems to improve the performance of these systems (see Fig. 10). As

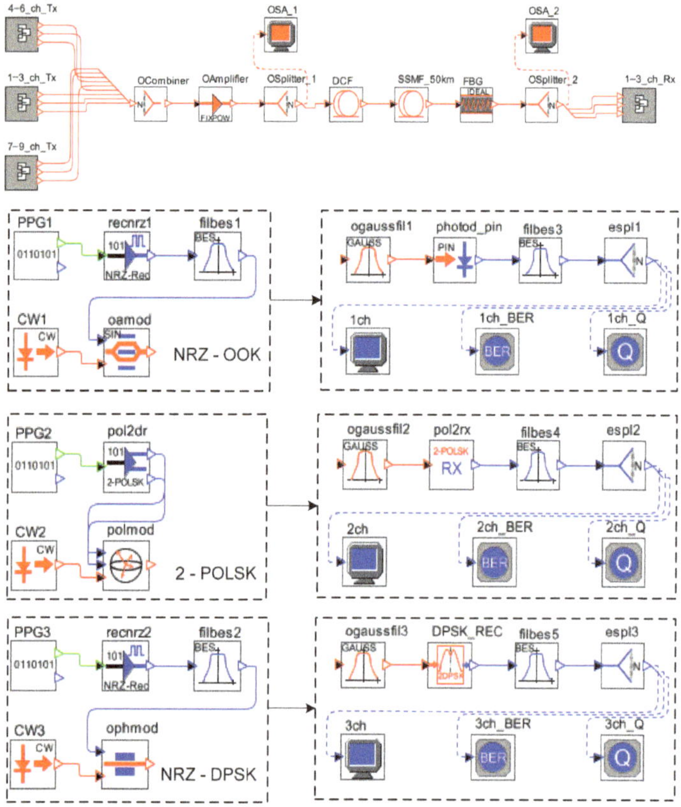

Figure 10. Simulation scheme of 9 – channel mixed HDWDM system and channels' transmitting and receiving parts of NRZ – OOK/ 2 – POLSK/ NRZ – DPSK modulated optical signals

performance improvement we mean the improvement of data transmission rate and / or transmission distance, simultaneously guaranteeing a stable system working condition with the recommended bit error ratio BER<10^{-12}.

If dispersion compensation is not used in developed mixed WDM model, then performance of 1st and 2nd channel is seriously affected by accumulated dispersion. The 3rd channel is affected by accumulated CD at the same level, but this channel's tolerance to CD is much higher, because there is used NRZ – DPSK as a coding format and optical signals are transmitted with 10 Gbit/s per channel bit rate. Without dispersion compensation the BER value of the 1st and 2nd channel are high (there are many bit errors) and we can assume it is because of inter-symbol interference (ISI), which causes pulse overlapping and receiver has difficulties to separate transmitted bit sequence. In this case BER > 10^{-12}, system performance is poor and fiber optical WDM transmission system is not able to qualitatively transmit information over distance of 50 km, until CD compensation will not be realized (see Fig. 11).

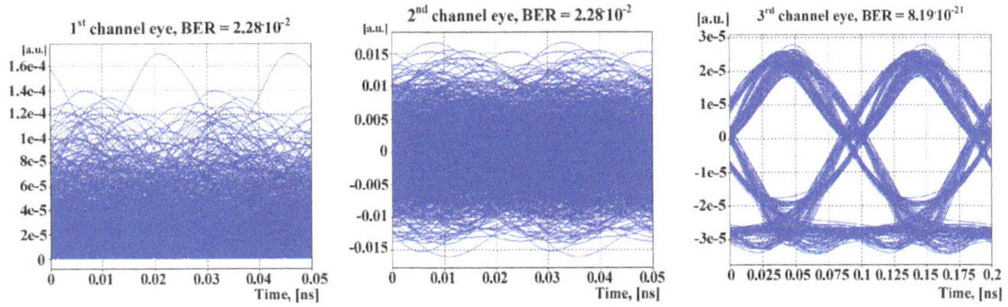

Figure 11. Output eye diagram for all three channels of mixed WDM system without CD compensation

As one can see from obtained detected signal eye diagrams, then 3rd channel's eye is wide open and has almost ideal form. This fact let us to conclude, that in this channel it is possible an error-free transmission even without using some CD compensation schemes. As for 1st and 2nd system's channel, then CD compensation is vital necessary, because eye openings are completely closed for these transmission system channel's eye diagrams. Without CD compensation signal transmission in these channels over 50 km of SSMF with BER<10^{-12} is impossible.

The first realized compensation method includes the implementation of dispersion compensating fiber (DCF). We used pre- and post-compensation schemes for effective chromatic dispersion compensation in our mixed WDM system. The best proportion of L_{DCF1} and L_{DCF2} was studied, where L_{DCF1} is a DCF fiber length at transmitter side (pre-compensation) and L_{DCF2} is DCF fiber length at receiver side (post-compensation), in km. As shown in Fig. 12, we found that the optimal required DCF fiber length proportion is 5/5 km.

That proportion of DCF length was found by analyzing Q – value correlation diagrams. As one can see from 2nd channel's (system's worst channel) Q – value correlation diagram, then the highest - value (20.75 dB) is for fiber length proportion equal to 5/5 km. DCF proportion length was chosen on basis of 2nd channel Q – value due to the fact, that this channel is the most affected by transmission impairments and it has BER values higher than for 1st and 3rd channels. Implementation of two DCF fiber spans, with length equal to 5km each, provides the best CD compensation results for our investigated mixed fiber optical transmission system.

The second realized compensation method includes the implementation of fiber Bragg grating (FBG). At the first stage we changed FBG compensated CD value from -1000 ps/nm to -600 ps/nm with 25 ps/nm step and found that optimal compensation level providing BER<10^{-12} is equal to -750 ps/nm. As displayed in Fig. 13, the best BER results for the 1st and 2nd channel can be achieved, if we compensate all CD level, which is accumulated during optical signal transmission over the 50 km of SSMF (800 ps/nm). The optimal compensation level was chosen on the basis of BER results obtained from the 1st channel of mixed WDM system. If we overcompensate accumulated dispersion (if compensated CD level exceeds -875 ps/nm then BER value grows rapidly and exceeds 10^{-12} value for the systems first channel.

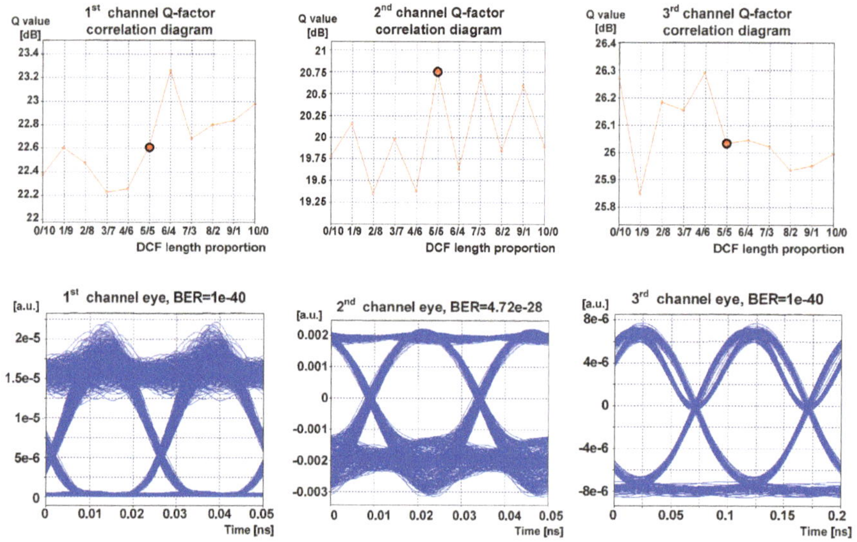

Figure 12. Mixed WDM system 1st/2nd/3rd channel's Q-factor correlation diagrams, detected signals eye diagrams and BER values at 75 GHz channel spacing

At the second stage the optimal dispersion slope compensation value for FBG used in our simulation scheme was investigated. Dispersion slope value was changed from -6 ps/nm^2 to -1 ps/nm^2 with 0.5 ps/nm^2 step. As shown in Fig. 13, the optimal dispersion slope value was chosen equal to -2.5 ps/nm^2. Such a conclusion was obtained basis on the first channel BER value at above mentioned dispersion slope value. This was done due to the fact that BER values for this channel are higher than for the second channel's BER values, but BER values for the third system's channel do not vary depending on compensated CD and dispersion slope levels (see Fig. 13).

In comparison, if we use DCF pre- and post-compensation modules to compensate 750 ps/nm of accumulated CD, then the worst system's channel (1st) BER value is equal to $2.78 \cdot 10^{-14}$ (Q=17.60 dB). This value was obtained in case, if used DCF length proportion is 7:3. Numerically this means that pre-compensation module compensates 525 ps/nm, but post-compensation module compensates 225 ps/nm. As one can see, despite the fact that DCF length proportion 5:5 provides better BER results for the 2nd system's channel, it is not providing the optimal Q values (or BER values) in each system's channel. The system's 1st (worst) channel BER=$2.34 \cdot 10^{-13}$ (Q=17.28 dB), if DCF length proportion, which is used in pre- and post-compensation modules, is equal to 5:5.

The last inspected dispersion compensation method is combined and includes the common implementation of dispersion compensating fiber (DCF) and fiber Bragg grating (FBG) in our combined fiber optical transmission system. This combined CD compensation method will be named as DCF-FBG. In pre-compensation module DCF fiber will be used, but in post-compensation module FBG will be used. For dispersion compensation we changed DCF length L_{DCF} from 0 km to 15 km, with step 1 km to find out optimal DCF length and FBG dispersion compensation level proportion, see Fig. 13. The DCF dispersion D=-80 ps/nm/km and total accumulated dispersion amount in fiber optical link that must be compensated is -750 ps/nm to obtain BER values in each system's channel smaller than 10^{-12}, as we found out in previous section.

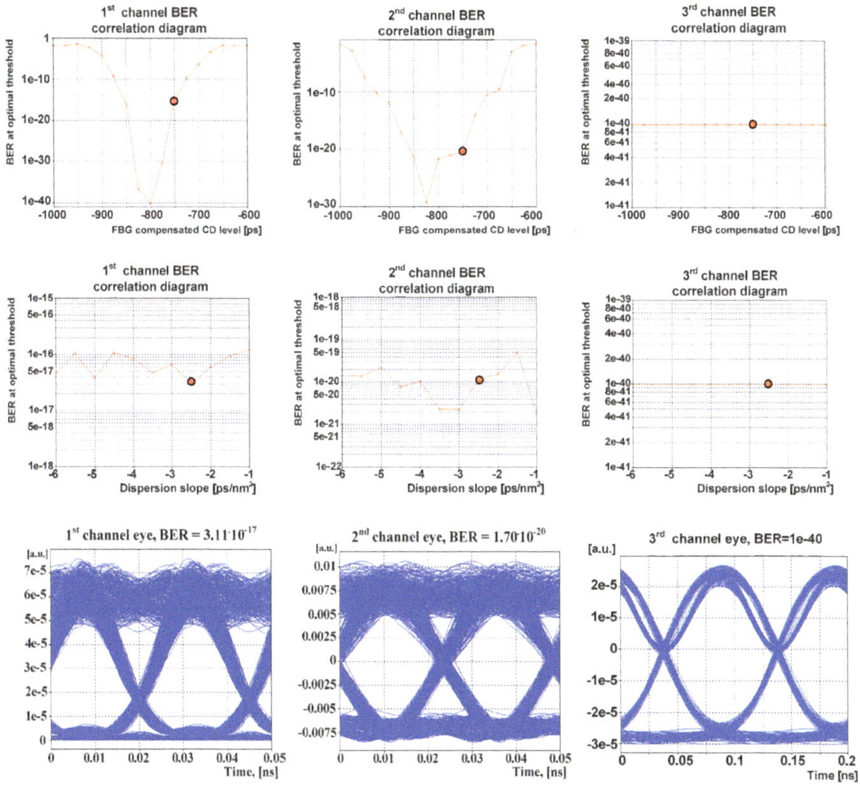

Figure 13. Mixed WDM system 1st/2nd/3rd channel's BER correlation diagrams, which represents channel's BER value as a function from compensated CD level and BER as a function from compensated dispersion slope value at the dispersion compensation level equal to 750 ps/nm; eye diagrams of detected signals and BER values at chosen compensated CD and slope values

Therefore we used a following formula: $D_{FBG} = -750 + 80 \cdot (L_{DCF})$, where D_{FBG} is FBG CD compensation amount provided by FBG. Dispersion amount that can be compensated by DCF fiber can be expressed as $D_{DCF} = -80 \cdot L_{DCF}$, where L_{DCF} is the length of used DCF.

As one can see, in Fig. 14, the optimal proportion $|D_{DCF}/D_{FBG}|$ is 21/79 %. It means that 21% of -750 ps/nm must be compensated by DCF, but remaining 79% must be compensated by FBG. Numerically, in our investigated system, it means that DCF compensate 160 ps/nm (2 km of DCF are used), but FBG compensates remaining 590 ps/nm of accumulated CD amount. This proportion was found basis on the 1st system's channel BER values. This channel is the most affected by chromatic dispersion due to the fact that for optical signal modulation in this channel NRZ – OOK modulation format is used. This modulation format has smaller CD tolerance comparing to 2 – POLSK and NRZ – DPSK formats.

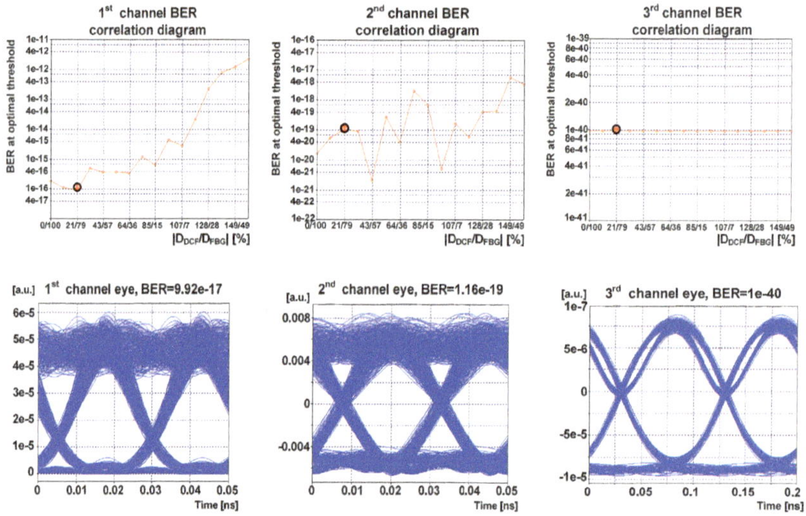

Figure 14. Mixed WDM system 1st/2nd/3rd channel's BER correlation diagrams, eye diagrams of detected signals and BER values at optimal DCF-FBG dispersion compensation proportion level

This is recommended DCF-FBG proportion for optimal accumulated CD compensation in [NRZ – OOK (40 Gbit/s, 193.025 THz)] – [2 – POLSK (40 Gbit/s, 193.100 THz)] – [NRZ – DPSK (10 Gbit/s, 193.175 THz)] mixed fiber optical transmission system.

To identify channel that is a source of larger amount of interchannel crosstalk noise than the rest of system's channels six mixed systems were investigated. These systems differ from each other only with modulation formats distribution among channels. This distribution scheme is as follows: [NRZ-DPSK (1, 2, 3, 3, 2, 1)]-[NRZ – OOK (2, 1, 1, 2, 3, 3)]-[2 – POLSK (3, 3, 2, 1, 2)]. This configuration represents modulation format and channel's number where one on these formats is used. The system's channels central frequencies are anchored to 193.1 THz according to ITU-T Recommendation G.694.1 and the first channel's central frequency is equal to 193.075 THz, the second is193.100 THz and the third is193.125 THz. After this crosstalk source have been detected simulation model were updated in order to find out the optimal modulation format distribution, which provides the lowest in system's channels detected signals' BER values. For this purpose existing transmission systems model were updated to nine-channel WDM system. These channels are grouped by three and these groups have identical transmitter and receiver blocks configuration but with different channels' central wavelengths. It was specially done to take into account linear and nonlinear crosstalk influences to optical signal transmission which are experience central's group channel (1st-3rd) from adjacent groups (4th-6th and 7th-9th). For system's further analysis we will use only channels number 1-3, but 4-6 and 7-9 are used as sources of interchannel crosstalk.

Then NRZ – DPSK, 2 – POLSK and NRZ – OOK modulated optical signals are combined, optically preamplified with fixed output power erbium-doped fiber amplifier (EDFA) and send over 50 km of single mode optical fiber. There are two different types of single mode fiber used in this research: standard single mode fiber or SSMF (according to ITU – T Recommendation G.652 D) and non-zero dispersion shifted fiber or NZ – DSF (according to ITU – T Recommendation G.655). Then optical signals are filtered with Super Gaussian optical filters, converted to electrical signals and then electrically filtered using Bessel electrical filters. Fiber span length was chosen equal to 50 km in order to avoid increase of amplified spontaneous emission (ASE) noise. Larger amplifier spacing would require gain greater than 10 dB, but this in a prohibitive leads to growth of ASE noise "in [14]". The aim of this task was to investigate optimal configuration for mixed WDM systems where differently modulated optical signals are transmitted. To achieve this goal several objectives must be solved.

Firstly it is necessary to identify channel in [1st: NRZ – DPSK, 10 Gbit/s, 193.075 THz]-[2nd: 2 – POLSK, 10 Gbit/s, 193.100 THz]-[3rd: NRZ – OOK, 10 Gbit/s, 193.125 THz] mixed WDM FOTS that is a source of larger amount of interchannel crosstalk noise than the rest of channels.

For this purpose six different systems were studied. These systems differ from each other only with modulation formats that are used in each particular system's channels. These systems have following configurations:

[NRZ-DPSK]-[NRZ-OOK]-[2-POLSK];
[NRZ-OOK]-[NRZ-DPSK]-[2-POLSK];
[NRZ-OOK]-[2-POLSK]-[NRZ-DPSK];
[2-POLSK]-[NRZ-OOK]-[NRZ-DPSK];
[2-POLSK]-[NRZ-DPSK]-[NRZ-OOK];
[NRZ-DPSK]-[2-POLSK]-[NRZ-OOK].

In each systems channel were determined signals BER values that further were used for system's performance analyze. The obtained results are summarized below (see Table 2). Using these results for each configuration system's average detected signals BER values were calculated. As one can see, sufficiently smaller BER value is for the third mixed system configuration then it is for the rest of possible configuration. The third configuration is as follows: [1st: NRZ – OOK\193.075 THz]-[2nd: 2 – POLSK\193.100 THz]-[3rd: NRZ – DPSK\193.125 THz]. After careful analysis of these obtained results it was found that investigated mixed system channel, where for optical signals modulation NRZ-DPSK modulation format is used, is a source of larger amount of interchannel crosstalk than channels, where NRZ – OOK or 2 – POLSK format is used. This was concluded based on obtained NRZ – OOK and 2 – POLSK channels BER results for different system's configurations. This become evident if we analyze obtained BER values for the fourth, fifth and sixth system.

f (THz)	1st system	2nd system	3rd system
193.075	3×10^{-24}	2×10^{-8}	1×10^{-40}
193.100	9×10^{-12}	9×10^{-25}	1×10^{-18}
193.125	1×10^{-40}	1×10^{-13}	3×10^{-27}
Average	3×10^{-12}	7×10^{-9}	5×10^{-19}
f (THz)	4th systems	5th system	6th system
193.075	1×10^{-40}	1×10^{-21}	4×10^{-27}
193.100	2×10^{-8}	6×10^{-25}	8×10^{-14}
193.125	3×10^{-25}	6×10^{-12}	1×10^{-40}
Average	5×10^{-9}	2×10^{-12}	3×10^{-14}

Table 2. BER values for different mixed systems channels

Firstly let's focus to the fifth system's BER values. As one can see from this configuration scheme then in this case NRZ – DPSK modulated optical signals are transmitted in central system's channel. As a result detected signals BER values in adjacent channels are sufficiently higher than they are in cases, when NRZ – OOK or 2 – POLSK modulated optical signals are located further from NRZ – DPSK channel as it is in the sixth system. Comparing BER results obtained for 2 – POLSK modulated signals in the fourth and sixth system (1×10^{-40} and 9×10^{-14} respectively), we can conclude that in mixed system detected signals BER value decreases if channel, where these signals are transmitted, is located further from NRZ – DPSK channel.

To assess NRZ-DPSK channel created crosstalk impact to optical signals transmission in all others mixed system's channels previously studied three-channel mixed systems model was modified and supplemented with 2 × 3 channels that have appropriate system's configuration. As before, in system channels detected signals BER values were obtained for six different mixed system configurations (see Table 3).

As well as using these data two different channels average BER values were calculated: system's average BER that takes into account all system channels (1st-9th); central group channels' average BER that takes into account only channels number one to three.

No.	f (THz)	1st system	2nd system	3rd system
4	193.000	5×10^{-23}	2×10^{-11}	1×10^{-40}
5	193.025	5×10^{-11}	4×10^{-24}	1×10^{-11}
6	193.050	1×10^{-13}	3×10^{-16}	9×10^{-24}
1	193.075	4×10^{-22}	2×10^{-8}	1×10^{-10}
2	193.100	5×10^{-10}	2×10^{-23}	4×10^{-15}
3	193.125	3×10^{-16}	3×10^{-13}	4×10^{-23}
7	193.150	4×10^{-24}	8×10^{-8}	1×10^{-11}
8	193.175	7×10^{-9}	2×10^{-23}	4×10^{-17}
9	193.200	1×10^{-40}	1×10^{-14}	4×10^{-29}
Average (1st-3rd)		2×10^{-10}	7×10^{-9}	4×10^{-11}
Average (1st-9th)		9×10^{-10}	1×10^{-8}	2×10^{-11}
No.	f (THz)	4th systems	5th system	6th system
4	193.000	1×10^{-40}	3×10^{-31}	9×10^{-27}
5	193.025	3×10^{-11}	2×10^{-23}	6×10^{-15}
6	193.050	3×10^{-24}	2×10^{-11}	3×10^{-11}
1	193.075	1×10^{-11}	2×10^{-15}	4×10^{-23}
2	193.100	1×10^{-8}	3×10^{-26}	1×10^{-14}
3	193.125	4×10^{-22}	3×10^{-10}	5×10^{-10}
7	193.150	1×10^{-12}	2×10^{-15}	3×10^{-22}
8	193.175	3×10^{-8}	5×10^{-24}	2×10^{-13}
9	193.200	1×10^{-23}	3×10^{-12}	1×10^{-40}
Average (1st-3rd)		3×10^{-9}	9×10^{-11}	2×10^{-10}
Average (1st-9th)		4×10^{-9}	3×10^{-11}	5×10^{-11}

Table 3. BER values for different 9-channel mixed systems' channels

As one can see from obtained data (see Table 3), then the lowest average BER values for 1st till 3rd and 1st till 9th channel are for the third mixed system configuration and they are equal to $BER_{1st\text{-}3rd} = 4 \times 10^{-11}$and $BER_{1st\text{-}9th} = 2 \times 10^{-11}$ respectively. But the highest BER values are for the second configuration and they are equal to $BER_{1st\text{-}3rd} = 7 \times 10^{-9}$ and $BER_{1st\text{-}9th} = 1 \times 10^{-8}$. So, BER difference between the best and worst case scenario, corresponding to [(NRZ – OOK)\193.075 THz]-[(2 – POLSK)\193.100 THz]-[(NRZ – DPSK)\ 193.125 THz] and [(NRZ – OOK)\193.075 THz]-[(NRZ – DPSK)\193.100 THz]-[(2 – POLSK)\ 193.125 THz] configuration respectively, is approximately three orders.

In these both cases channel with highest detected signal error probability is the first one, where by the way NRZ-OOK modulated optical signals are transmitted. Comparing BER values obtained for NRZ-OOK and 2-POLSK modulated optical signals for these two systems configuration, we have to conclude that these values differ by no more than two orders (1×10^{-10} and 2×10^{-8} in NRZ – OOK case and 4×10^{-15} and 3×10^{-13} for 2 – POLSK channels). As for NRZ – DPSK channel then the resulting BER values differences in both cases are not significant: 4×10^{-23} and 2×10^{-23} (see Fig. 15).

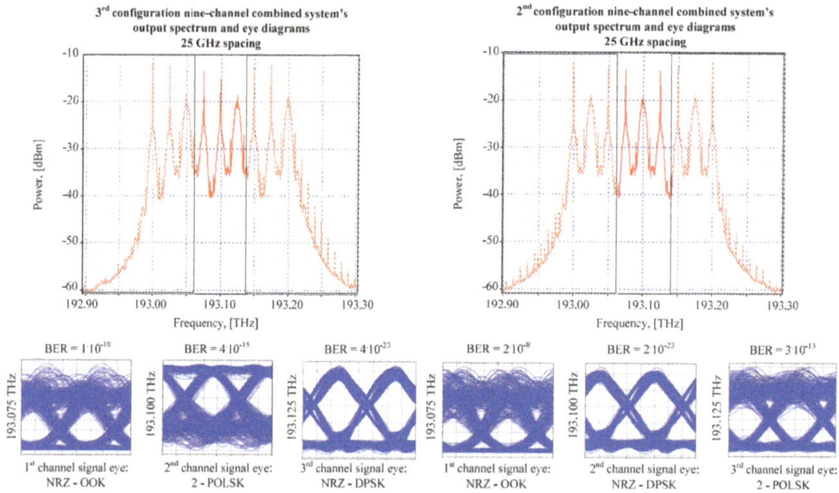

Figure 15. Nine-channel mixed system's with the third configuration output spectrum and eye diagrams in case of 10 Gbit/s per channel bitrates and 25 GHz channel spacing, and nine-channel mixed system's with the second configuration output spectrum and eye diagrams in case of 10 Gbit/s per channel bitrates and 25 GHz channel spacing.

As a result, for further research of optimal mixed system configuration will be used as a starting point nine-channel mixed WDM system with the third configuration.

Previously it has been detected that channel, where NRZ – DPSK modulated optical signals are transmitted, is larger amount of interchannel crosstalk source than NRZ – OOK or 2 – POLSK channels. So, to reduce that type of noise it has been decided to decrease optical power level radiated by distributed feedback lasers (DFB) in continuous wavelength (CW) regime that are used in these channels.

As previously, using these BER results for each system channel average BER value for central channels were calculated. It revealed that in system channels detected signals average BER values are below 10^{-12} if NRZ – DPSK channels' lasers output power level is in the range from 3.5 to 4.5 dBm. The lowest average channels' BER value is reached if these lasers output power is equal to 3.5 dBm. In this case $BER_{1st\text{-}3rd} = 3 \times 10^{-18}$ and the worst channel is the second one (2 – POLSK) and its $BER_{2nd} = 1 \times 10^{-17}$ (see Fig. 16).

Assuming that we are dealing with one sector of ultra-long haul backbone optical network, it was decided to supplement this model of mixed WDM system with additional optical element is fixed output power optical amplifier. It allowed take into an account ASE noise arising from EDFA which is the most widely used optical amplifier. To find out optimal amplifier output power level, that provides minimal channels' BER values, BER correlation diagram for each were obtained. It represents in systems channels detected signals BER values as a function from amplifier fixed output power level (see Fig. 17). Let is note, that in this case NRZ-DPSK channel laser output power level remains unchanged as it was in initial mixed WDM system model "in [2]".

As one can see from Fig. 17, then BER value for the system's first channel varies around 10^{-11} value, for the second channel around 10^{-16} and for the third is10^{-24}. Knowing that the worst mixed system's channel is the second one, where 2 – POLSK modulated optical signals are transmitted, and then was decided to choose amplifier output power level that provides minimal BER value exactly in this channel.

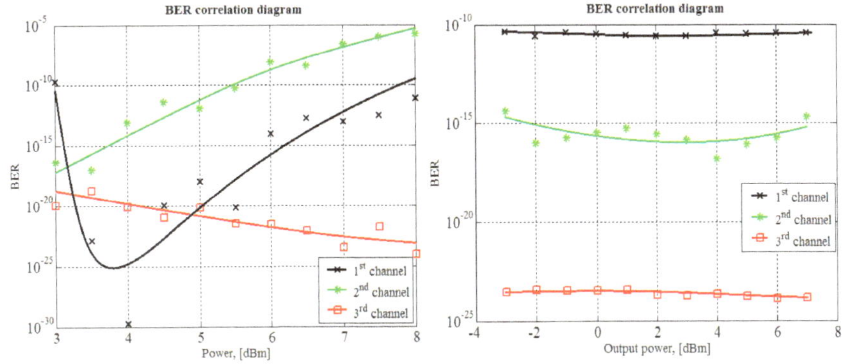

Figure 16. BER as a function from NRZ – DPSK channel laser radiated output power level, and BER as a function form optical amplifier fixed output power level.

Consequently, optical amplifier fixed output power level equal to 4 dBm was chosen. This level provides in the second system's channel detected signals $BER_{2nd} = 1 \times 10^{-17}$.

If both these optimal parameters are used in mixed system model configuration, then in system channels detected signals BER values are well below the maximal acceptable BER threshold 10^{-12}, that is defined for 10 Gbit/s per channel bitrate (see Table 4). Channels BER values for revealed optimal system configuration were obtained for two types of single mode fiber. The first fiber was standard single mode fiber (SSMF) according to ITU-T Recommendation G.652 D and the second was non-zero dispersion shifted fiber (NZ – DSF) according to ITU – T Recommendation G.655.

System description	1st	2nd	3rd
Initial	1×10^{-10}	4×10^{-15}	4×10^{-23}
$P_{NRZ\text{-}DPSK}$ = 3.5 dBm	1×10^{-23}	1×10^{-17}	2×10^{-19}
FIXPOWER = 4 dBm	4×10^{-11}	1×10^{-17}	2×10^{-24}
Optimal and SSMF	5×10^{-26}	7×10^{-19}	6×10^{-22}
Optimal and NZ-DSF	9×10^{-20}	6×10^{-21}	1×10^{-40}

Table 4. BER value for different system configurations

For these two cases BER results as well as detected signals eye diagrams were compared one to another (see Fig. 17).

BER value for the first system channel sufficiently dropped from 1×10^{-10} to 5×10^{-26} if SSMF is used and to 9×10^{-20} if NZ-DSF is used. Exactly for this first channel, where NRZ – OOK modulated optical signals are transmitted, experiencing the most radical BER value improvement comparing to 2 – POLSK and NRZ – DPSK channels. In these channels detected signal BER values do not improve so noticeably. The second channel's BER value decreases from 4×10^{-15} to 7×10^{-19} for SSMF and to 6×10^{-21} for NZ – DSF, but the third channel's BER value variation is not essential from 4×10^{-23} to 6×10^{-22} if SSMF is used. But if in this system instead of SSMF NZ – DSF is used then it is possible to obtain lower BER values for NRZ – DPSK channels. In this channel detected signals BER value decreases to 1×10^{-40}. In addition, coherence between detected signals BER values and channels' central frequency position in C-band (191.6-195.9 THz) was investigated. As well as, for each system channel the worst and the best position in C-band, that provides the highest and the lowest possible signals BER values, respectively, for previously found optimal mixed system's configuration, was revealed (see Table 4). As previously, this research was held for two types of optical fiber: SSMF (G.652 D) and NZ – DSF (G.655).

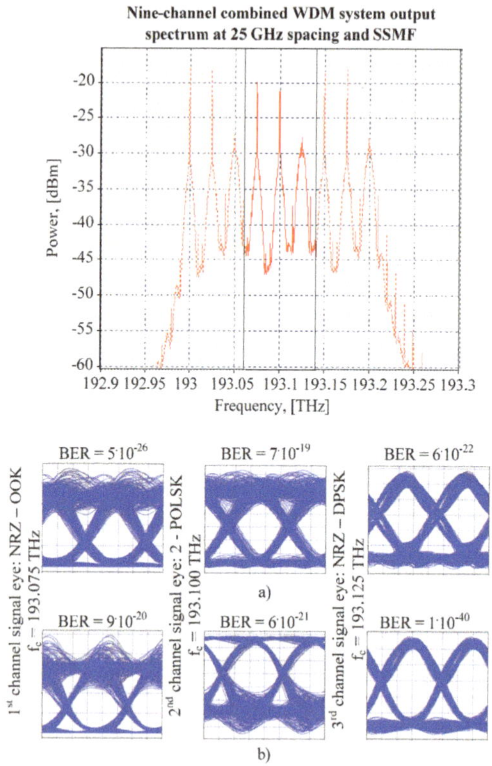

Figure 17. Optimal configuration nine-channel mixed WDM system output optical spectrum and eye diagrams: (a) SSMF; (b) NZ – DSF.

It showed that depending on channel central frequency the first channel's BER value varies around nominal value of 10^{-20} if SSMF is used and around 10^{-30} if NZ – DSF. But BER values obtained for the second channel and NZ – DSF are for several orders worse comparing to transmission over SSMF. They vary around 10^{-20} and 10^{-25}, respectively.

In addition, as one can see from Table 5, then BER value obtained for NZ-DSF and the worst case of 2nd channel central frequency is approximately for six orders larger comparing to the worst case of SSMF.

These let us conclude that 2-POLSK modulated signals are not suitable for transmission over NZ – DSF fiber in [NRZ – OOK]-[2 – POLSK]-[NRZ – DPSK] mixed WDM systems and they are sufficiently distorted at appropriate channel central frequency.

No.	Fiber	B/W	f_c (THz)	BER
1st	SSMF (G.652 D)	best	195.725	8×10^{-26}
		worst	192.350	2×10^{-16}
2nd		best	193.950	3×10^{-31}
		worst	192.150	2×10^{-20}
3rd		best	195.100	8×10^{-25}
		worst	192.400	7×10^{-19}
1st	NZ-DSF (G.655)	best	193.475	1×10^{-39}
		worst	193.925	5×10^{-26}
2nd		best	191.925	2×10^{-23}
		worst	195.750	3×10^{-14}
3rd		best	C-band	1×10^{-40}
		worst		

Table 5. Best and worst channels positions in C-band and their BER values

As result system's average BER value is significantly higher than it might be in case of SSMF (see Table 6). In same time the lowest 1st-3rd channels average BER value is gained with NZ – DSF and it is about two orders lower than in case of SSMF. As for the third channel, then in case of SSMF and different channel central frequencies obtained BER values vary somewhere around 10^{-23}. Whereas in case of NZ – DSF these values remain constant and approximately equal to 10^{-40} in all C-band. It allows to judge about NRZ – DPSK modulated optical signals transmission suitability over NZ – DSF single mode optical fiber.

Let is note, that Corning LEAF non-zero dispersion shifted fiber characteristics and parameters were used in OptSim in order to obtain mathematical model of a NZ-DSF. This fiber is the world's most widely deployed NZ-DSF and is specially optimized for high-speed and high capacity long-haul and metro networks.

Characteristics	BER	2nd channel f_c (THz)	1st-3rd channel
MIN average BER/ SSMF	3×10^{-23}	195.750	8×10^{-26}
			2×10^{-24}
			8×10^{-23}
MAx average BER/ SSMF	6×10^{-17}	192.375	2×10^{-16}
			7×10^{-28}
			7×10^{-19}
MIN average BER/ NZ-DSF	5×10^{-25}	192.825	1×10^{-31}
			1×10^{-24}
			1×10^{-40}
MAx average BER/ NZ-DSF	1×10^{-14}	195.750	1×10^{-27}
			3×10^{-14}
			1×10^{-40}

Table 6. Minimal and maximal in system's 1st-3rd channel detected signals BER values

5. Results and discussions

In summary it can be concluded, that the mixed WDM solution allows combining channels with a variety of modulation formats, which are used for optical signal modulation, in one single transmission system, preserving a previously used channel spacing values. We would like to point out one more time, that such mixed solution of transmission system is being offered as the transition state form traditionally used NRZ – OOK modulation format to the alternative modulation formats, such as NRZ – DPSK or 2 – POLSK, which use provides a number of superior properties due to their abilities of providing greater protection from interchannel crosstalk, less exposed to expression of non-linear effects and better exposed to channel filtration, as well as less exposed to chromatic dispersion effect. By gradually introducing new system channels, can be increased the total transmission capacity of fibre, thus avoiding of core networks bottleneck effect and in the same time minimize growth of non-linear optical effect influences, because the alternative modulation formats are able to provide the same BER levels as traditionally used NRZ – OOK, but only at lower input power levels.

The suitability of different chromatic dispersion compensation methods for the most efficient high – speed mixed HDWDM systems, with a 50 km of SSMF span length and following system's configuration: [1st channel: NRZ – OOK, 40 Gbit/s, 193.025 THz] – [2nd channel: 2 – POLSK, 40 Gbit/s, 193.100 THz] – [3rd channel: NRZ – DPSK, 10 Gbit/s, 193.175 THz], has been evaluated. This configuration represents channel's number, modulation format, per-channel data rate and channels' central frequencies at 75 GHz channel spacing. The following recommendations and conclusions are summarized below for the next generation of WDM transmission systems.

- The use of both CD pre- and post-compensation modules together, which might contain DCF of different length, allow to achieve better system's BER values comparing with pre-compensation or post-compensation schemes used separately. We have found, that the best CD compensation and lowest channels BER results for investigated mixed HDWDM systemwith mixed data rates , mentioned above, can be achieved using 5 km of DCF in pre- and post-compensation modules. The similar results can be achieved using 7 km of DCF in pre-compensation module and 3 km in post-compensation module.

- Minimal CD level that needs to be compensated, using FBG, to achieve investigated mixed HDWDM system channels' BER<10^{-12}, is equal to 750 ps/nm, without dispersion slope compensation. If CD slope compensation of -2.5 ps/nm^2 is realized in addition to dispersion level compensation of 750 ps/nm, then the worst channel's (NRZ – OOK channel) Q-value increases by 0.17 dB, from 18.25 dB to 18.42 dB.
- As combined CD compensation methods we have investigated DCF – FBG solution. DCF is used in pre-compensation module, but FBG is used in post-compensation module. The minimal average system's channel BER value can be achieved, if dispersion amount compensated by DCF and FBG is equal to 21 % and 79 % from minimal accumulated CD level, which must be compensated to obtain BER < 10^{-12} in each system's channel. In our case this accumulated CD level is equal to 750 ps/nm. Similar results could be obtained, if 57/43% dispersion level compensation proportion is used for realized mixed HDWDM system.

Investigating coherence between BER values and channels' central frequencies position in C-band, it have been stated several facts. Firstly, obtained channels central frequency values for each system channel the worst and the best position in C-band, that provide the highest and the lowest possible detected signals BER values. Secondly, 2 – POLSK modulated signals are not suitable for transmission over NZ – DSF fiber in investigated mixed WDM system, because these signals are sufficiently distorted at appropriate channel central frequency. If mixed system's channels are allocated around nominal of 195.750 THz, then obtained average detected signals BER value is about 10^{-14} and it is the worst possible case. Whereas the lowest possible average BER value (5×10^{-25}) can be reached anchored mixed systems channels frequency grid to 192.825 THz. Thirdly, NRZ – DPSK modulated optical signals are well suitable for transmission over NZ – DSF.

Author details

Vjaceslavs Bobrovs and Girts Ivanovs

Riga Technical University, Institute of Telecommunications, Latvia

6. References

[1] Cisco Systems (2010) Cisco Visual Networking Index – Forecast and Methodology 2009–2014, White paper, no.1, pp. 1-17.

[2] Udalcovs, A., Bobrovs, V. and Ivanovs, G. (2011) Investigation of Allowed Channel Spacing for Differently Modulated Optical Signals in Combined HDWDM systems, Electronics and Electrical Engineering 6 (112): 19-24.

[3] Dutta, A.K., Dutta, N.K. and Fujiwara, M. (2004) "WDM Technologies: Optical Networks" – USA, Elsevier Inc. 336 p.
[4] Bobrovs, V., Spolitis, S., Udalcovs, A. and Ivanovs, G. (2011) Investigation of Chromatic Dispersion Compensation Methods for Combined HDWDM Systems, Latvian Journal of Physics and Technical Sciences, no. 5: 13-27
[5] Miyamoto, Y. (2011) Ultra-High Capacity Transmission for Optical Transport Network, OFC/NFOEC Conference proceedings pp. 1-28.
[6] Sano, A., Masuda, H., Kobayashi, T., Fujiwara, M., Horikoshi, K., Yoshida, E., Miyamoto, Y., Matsui, M., Mizoguchi, M., Yamazaki, H., Sakamaki, Y. and Ishii, H. (2011) Ultra-High Capacity WDM Transmission Using Spectrally-Efficient PDM 16-QAM Modulation and C- and Extended L-Band Wideband Optical Amplification, Journal of Lightwave Technology, vol. 29, no. 4: 578-586.
[7] Takahashi, H., Al Amin, A., Jansen, S. L., Morita, I., and Tanaka, H. (2010) Highly Spectrally Efficient DWDM Transmission at 7.0 b/s/Hz Using 8x65.1-Gb/s Coherent PDM-OFDM, Journal of Lightwave Technology, vol. 28: 406-414.
[8] Kikuchi, K. (2008) Coherent transmission systems, in: 34th European Conference on Optical Communication, pp. 1-39.
[9] Bottacchi, S., Beling, A., Matiss A., Nielsen, M. L., Steffan, A.G. and Unterborsch, G. (2010) Advanced Photo receivers for High-Speed Optical Fiber Transmission Systems, IEEE Journal of Selected Topics in Quantum Electronics, 16 (5): 1099-1112.
[10] Udalcovs A., Bobrovs V. and Ivanovs G. (2011) Investigation of Differently Modulated Optical Signals Transmission in HDWDM Systems, Journal of Computer Technology and Application, vol. 2, no. 10: 801-812.
[11] Peucheret, C. (2004) Fibre and component induced limitations in high capacity optical networks. Doctoral thesis, pp. 1-8.
[12] Hamano, H., Ishikawa, G. & Yamashita, K. (1999) High – speed transmission systems. Fujitsu Sci. Tech. J., 35 (1): 91-99.
[13] Keiser G. (2007) Optical Communications Essentials. McGraw-Hill, 372 p.
[14] Clarke, A. (2007) Optical pulse processing towards Tb/s high – speed photonic systems. Doctoral thesis, pp. 1-14.
[15] Bobrovs, V. & Ivanovs, G. (2008) Investigation of Mixed Data Rate and Format Transmission in WDM Networks, Electronics and Electrical Engineering. 4(84): 63-66.
[16] Xu, C., Liu, X. & Wei, X. (2004) Differential Phase-Shift Keying for High Spectral Efficiency Optical Transmissions. IEEE Journal of Selected Topics in Quantum Electronics, 10(2): 281-293.
[17] Bobrovs, V. & Ivanovs, G. (2009) Investigation of Minimal Channel Spacing in HDWDM Systems, Electronics and Electrical Engineering. 4(92): 53-56.
[18] Chomycz, B. (2009) Planning Fiber Optic Networks. McGraw-Hill, 401 p.
[19] Bobrovs, V., Ivanovs, G. & Spolitis S. (2011) Realization of Combined Chromatic Dispersion Compensation Methods in High Speed WDM Optical Transmission Systems, Electronics and Electrical Engineering. 5(111): 33-38.

Chapter 10

Next Generation Optical Wireless Communication Systems Using Fiber Direct Coupled Optical Antennas

Koichi Takahashi

Additional information is available at the end of the chapter

1. Introduction

The broadband communication environment becomes the essential social infrastructure when we perform a social activity. Above all, the optical fiber network was the main artery which supported a communications infrastructure, and various technology developments have been performed to take in traffic to increase rapidly [Cochennec 2002][ITU-T 2001][ITU-T 2011]. On the other hand, wireless communication technology is essential element technology to realize ubiquitous communication environment. And new generation service development by new technology is beginning in the fields of mobile communication and broadcast [Lee 2007]. In addition, the research and development of ultra-high speed photonic network [Ben 2006][Zhang 2005][Kitayama 2005] performing the information transmission from the edge of the network to the edge in an optical domain is pushed forward, but an application of the radio technology to raise the flexibility in photonic network is an important problem, and advancement of optical wireless technology having possibility is expected [Kim 1999][Bouchet 1999]. However, it becomes further ultra-high speed/huge-capacity of the optical wireless and wavelength division multiplexing (WDM) limitation factor of the flexibility corresponding to various transmission methods because the current optical wireless system performs from optical to electrical (O/E) conversion of a transmitting and receiving part respectively. Therefore the realization of the optical wireless method that is not dependent on in ultra-high speed/huge-capacity and the interface (a protocol) of the device is enabled more than before if we can connect optical fiber to optical wireless as optical signal without performing O/E conversion, and it is expected that it can be it with the effective means of flexible photonic network construction. It is anticipated that compact optical antennas with an advanced fine optical tracking system will be used for the ultra-fast free space optical communication system in the next generation. Such laser communication systems will be used not only for the space communications [Kaliski 1999][Borocom 2005][Koyama 2004] but also the terrestrial long-distance photonic network. In this paper, we discuss demand specifications of optical antennas considering phenomena such as the scintillation which occurs by atmospheric turbulence. And then we explain optics design and the design results to satisfy demand specifications. We also mention the fine tracking mechanism using the fine pointing mirror (FPM) and the feedback of the signal's incident angle detection by the quadrant detector (QD) for the antenna. The influences of fluctuation of laser beam angle-of-arrival (AOA) are reduced effectively. We mention the following research studies of the two optical wireless communication systems such as next generation free-space optical communication (NG-FSO) system and radio on FSO (RoFSO) system.

2. Transition of the Free Space Optical Communication systems

The various concepts and architectures of Free Space Optical Communication (FSO) systems are illustrated in Figure 1. Figure 1(a) shows a conventional FSO system [Infrared][Free-Space]. Since it uses as a transmission media the difference wavelength as an optical fiber, it includes optical and electrical conversion between an optical fiber side and the free-space side. Furthermore, it is necessary to prepare a different interface which performs individual modulation and demodulation, coding, etc. for every communications service. Figure 1(b) shows the concept of a NG-FSO system [Takahashi 2008][Matsumoto 2008][Kazaura 2007]. In seamless connection of free-space and fiber systems an optical beam is emitted directly from a single mode fiber (SMF) termination to free-space using a new concept FSO antenna. Loss of the optical signal power caused by space transmission can be compensated using a fiber amplifier using the same wavelength band of 1550 nm as an optical fiber network [Khaleghi 1996][Luo 1998]. In this method the need to convert the optical signal from electrical to optical formats or vice versa for transmitting or receiving through space is eliminated. Figure 1(c) shows the concept of a Radio on Free Space Optical communication (RoFSO) system for realizing application in radio service combining with Radio on Fiber (RoF) technology [Al-Raweshidy 2002][Hai 2006]. Main advantage and goal of RoFSO systems is they can be used to quickly and effectively provide heterogeneous wireless service [Komaki 2003][Tsukamoto 2006] for example WiFi (IEEE 802.11) ,WiMAX (IEEE 802.16), cellular based 3G signals etc, simultaneously.

However, the challenge in NG-FSO system design is making seamless connection of free space propagated beam to the SMF. The optical signal which has been propagated through free-space is reduced by the optical elements like lens etc. in the antenna and focused to the core of SMF. The optical beam transmitted through the free-space is influenced by various weather conditions, such as attenuation by rain, fog, snow etc., and by atmospheric absorption. Furthermore, the beam experiences atmospheric turbulence as it propagates through free-space, as well as vibrations of the device at the installation site and beam distortion occurrence. The consequence of these effects is the fluctuation of the beam AOA which in turn leads to significant variation in the power of the light focused into the SMF.

Although AOA is not a problem for conventional FSO system thanks to a large area of the photo detector, it is a problem for NG-FSO because of little margin between the arrival beam and SMF. It is therefore difficult to maintain a stable link performance.

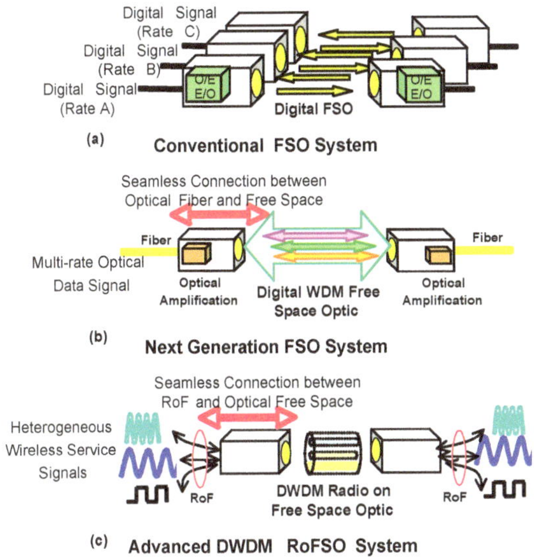

Figure 1. The concept of various FSO systems, (a) conventional, (b) NG-FSO and (c) RoFSO

3. Laser beam propagation in atmosphere

When laser beam propagates in the atmosphere [Tatarskii 1971][Dayton 1992][Hufnagel 1974][Friend 1967][Andrew 1998], its phase is fluctuated by the scatter of cloud, fog and rain, also by the change of local refractive index of the atmosphere. The fluctuation is varied depending on medium, and it becomes the largest when its size is similar to the size of light, also the medium constant changes spatially and temporally at random. When refractive index slowly and continuously changes due to atmospheric turbulence, the forward scattering dominates because the scale of fluctuation is larger than the wave length. In this case, it is possible to ignore the fluctuation of plane of polarization, but the fluctuation of phase/angle of arrival and the amplitude and the scintillation should be taken into consideration. On the other hand, when the medium is distributed discretely such as cloud, fog and rain and so on, the refractive index changes drastically. Therefore, the equivalent damping which occurs due to absorbing/scattering of particles and the influence against polarization property by multiple scattering with multiple particles should be considered. Modified von Karman spectrum density function is used not to overflow an integral value in theoretical analyses [Andrew 1995]:

$$\Phi_n(K) = 0.033C_n^2(K^2 + \kappa_0^2)^{-\frac{11}{6}} \exp\left(-\frac{K^2}{\kappa_i^2}\right) \quad (1)$$

Herein, $\kappa_i = 5.92 / l_0$, $\kappa_0 = 2\pi / L_0$

l_0 is the inner scale, L_0 is the outer scale. Based on the actual measurements of the change of C_n^2 (refractive index structure constant) in high altitude, Hufnagel-Valley model (H-$V_{5/7}$ model) parameterized by the scaling factors of wind speed in upper layer of the atmosphere and the size of the fluctuation of the atmosphere are used:

$$C_n^2(h) = 0.00594\left(\frac{v}{27}\right)^2 \left(10^{-5}h\right)^{10} \exp\left(-\frac{h}{1000}\right) + 2.7\times10^{-16} \exp\left(-\frac{h}{1500}\right) + A\exp\left(-\frac{h}{100}\right) \quad (2)$$

Herein, h represents the height of atmosphere (m), v represents the wind velocity in upper layer of atmosphere (m/s), A is the scaling factor which represents the fluctuation of the earth's surface ($m^{-2/3}$). In H-V day model, the scaling factors are v = 21 m/s and $A = 1.7\times10^{-14}$ $m^{-2/3}$. This is equivalent to looking up the sky from the earth's surface, and the coherence length of the fluctuation of the atmosphere is r_0 = 5 cm. We firstly consider the averaged influence of fluctuation which also means the extensity of beam and degradation of the strength of the central part. We use H-V model to represent the fluctuation of the refractive index of the atmosphere. And we calculate the averaged intensity of light of the receiving point when the Gaussian beam is transmitted from the earth to the sky (uplink). If the atmosphere is not disturbed and it is possible to apply Rytov approximation to the wave equation, the averaged strength of the Gaussian beam after propagating the distance L is :

$$\langle I(r,L)\rangle = \frac{W_0^2}{W_e^2}\exp\left(-\frac{2r^2}{W_e^2}\right) \quad (3)$$

Therefore, the anticipated value of the received beam profile still remains Gaussian distribution, and the variant is given by the effective beam radius as follows:

$$W_e = W\left(1+G_u\right)^{1/2} \quad (4)$$

And,

$$G_u = 4.35\mu_1\Lambda^{5/6}k^{7/6}\left(H-h_0\right)^{5/6}\sec^{11/6}\left(\xi\right) \quad (5)$$

$$\mu_1 = \int_{h_0}^{H} C_n^2(h)\xi_1^{5/3}dh,\ \xi_1 = 1-\frac{h-h_0}{H-h_0} \quad (6)$$

are the parameters which represents the influence of fluctuation, W_0 is the radius of transmitting beam, κ is the wave number, L is the distance between the transmitting system and the receiving system, h_0 is the height between the earth's surface and the transmitting system, ξ is zenith angle and $H = h_0 + L\cos(\xi)$ is the height of the receiving system in the sky. Also, W is the radius without the fluctuation of the atmosphere, and it is possible to describe W using F_0 (the curvature radius of exit point of the transmitting light) as follows:

$$W = W_0\sqrt{\left(1-\frac{L}{F_0}\right)^2 + \left(\frac{2L}{kW_0^2}\right)^2} = W_0\sqrt{\Theta_0^2 + \Lambda_0^2} \tag{7}$$

Herein:

$$\Lambda = \frac{2L}{kW^2} = \frac{\Lambda_0}{\Theta_0^2 + \Lambda_0^2},\ \Lambda_0 = \frac{2L}{kW_0^2} \tag{8}$$

$$\Theta = 1 + \frac{L}{F} = \frac{\Theta_0}{\Theta_0^2 + \Lambda_0^2},\ \Theta_0 = 1 - \frac{L}{F_0} \tag{9}$$

$$W_e = W(1+G_d)^{1/2} \tag{10}$$

On the other hand, regarding the downlink, the following values are substituted into the equation (3).

$$G_d = 4.35\mu_2\Lambda^{5/6}k^{7/6}(H-h_0)^{5/6}\sec^{11/6}(\xi) \tag{11}$$

$$\mu_2 = \int_{h_0}^{H} C_n^2(h)\xi_0^{5/3}dh,\ \xi_0 = \frac{h-h_0}{H-h_0} \tag{12}$$

With the above-mentioned formulas, we will discuss in the next section with the demand of optical specifications of the optical antenna available for NG-FSO system and RoFSO system.

4. Fine tracking technology

The challenge in all-optical connection of FSO and SMF systems is not only to design an effective beam tracking and optical antenna alignment technique, but also an efficient method for focusing the light into the SMF at the receiver. Active tracking is required to maintain alignment of the received optical signal to the SMF. To achieve stable beam spot position control as well as compensate for beam AOA fluctuation caused by atmospheric turbulence we adopted a small size two axis galvanometer type mirror drive mechanism. This mirror device called fine pointing mirror (FPM) is shown in figure 2. The tracking system constitutes the feedback system which considers the direction of the optic axis of a received optical system which changes the arrival directions of beacon light with an input and inclination of FPM as an output. Figure 3 shows a block diagram of a tracking servo system. Although the actual servo system consists of two horizontal independent axes this

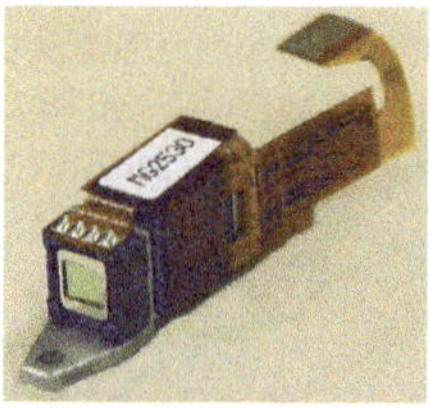

Figure 2. Photograph of FPM

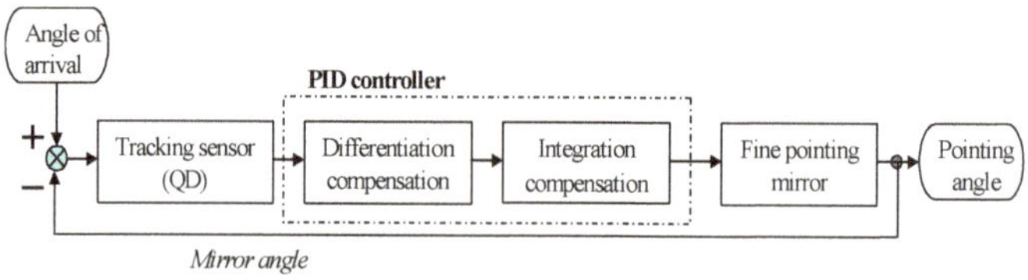

Figure 3. Block diagram of the fine tracking servo system

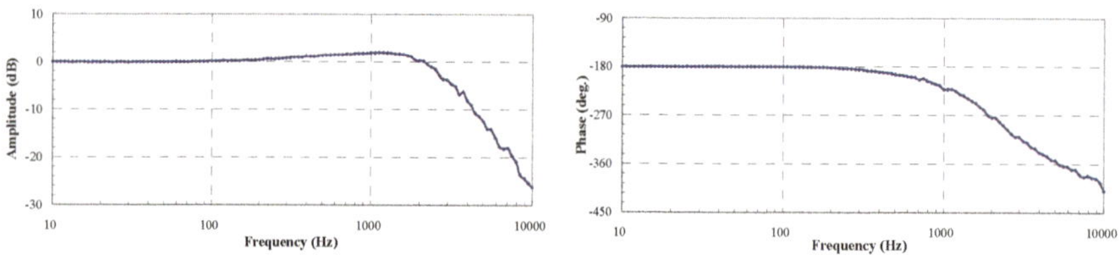

(a) Amplitude (b) Phase

Figure 4. Frequency characteristics of closed loop feedback system. (a) Amplitude (b) Phase

figure shows only one axis. The position of the received beam is determined by calculation the position of the beacon spot on the four elements of the QD. This signal serves as an input of an analog proportional–integral–derivative (PID) controller. The transfer characteristic of FPM is carrying out the response of a typical secondary system (azimuth (Az) axis is 100 Hz and elevation (El) axis of resonance frequency is 80 Hz at the weak oscillating system of dumping), and the PID controller by progress and delay compensation is designed so that a control zone may become the largest in consideration of this characteristic and the nonlinearity of a tracking sensor. As a result, the control zone of the closed loop response has attained the performance of about 2 kHz. In the case of a feedback system, disturbance suppression performance is proportional to a closed loop zone. Therefore, this tracking system shows that tracking control more highly efficient a single figure than the conventional tracking system which was an about 100 Hz control zone was realizable. Disturbance is poured in from the integration input of Figure 4, and the result of having measured the tracking response which appears in a mirror drive output about Az axis is shown in Figure 5. A vertical axis is the amount of suppression of disturbance amplitude. The characteristic is the same also about El axis. On the frequency of 100 Hz or less, it turns out that it has the capability that disturbance can be suppressed or less to 1/100.

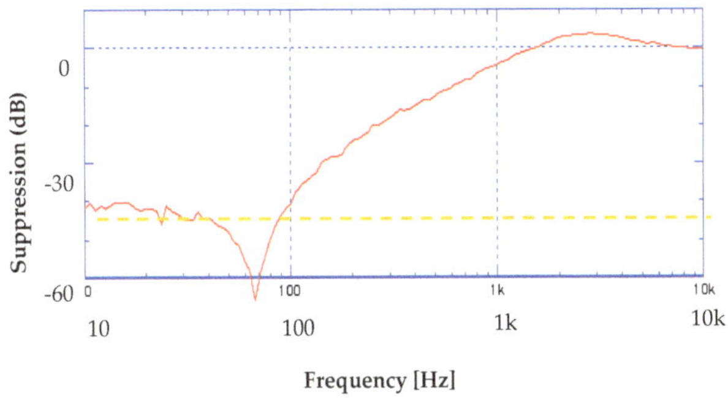

Figure 5. Turbulence suppression characteristics of feedback system

5. Research and development on NG-FSO system

5.1. The basic concept of the NG-FSO

In research and development of the NG-FSO system, we developed a compact optical antenna suitable for optical communication between satellite in space and deployment in stratospheric platform [Arimoto 2003][Katsuo 2005]. Because of the limitation in payload and equipment size in such usage, a lightweight optical antenna was required designed using off-axis free-form optics without the obscuration [Takahashi 2006]. In general, a Galileo type optical system, composed of a convex lens and a concave lens, is used to convert input/output diameter of laser beam in an optical antenna. However, when using such kind of refractive optical system it is difficult to suppress reflection of each surface of lenses in a wide wavelength range, so measurement errors as a result of ghost and/or flare can arise. In addition, since this system is accompanied with a chromatic aberration, adjustment of the focal length based on a wavelength is necessary. Therefore, a reflection based optical system which does not have a chromatic aberration is suitable for an optical system that is incorporated into a highly precise optical antenna. Although there are many cases that a Cassegrain type configuration is used for beam expander of reflecting mirror system, it is not suitable for the optics of the antenna of this purpose because of an obscuration in the beam center caused by secondary mirror. We abandoned the constitution on co-axial optical system and devised off-axial reflection optical system instead. In case of optical wireless communication on ground or between ground and space, the fluctuation of arrival beam angle is caused by the atmospheric turbulence. Therefore, if comatic aberration of the optical antenna is large, beam expander causes peculiar aberrations according to an incident angle of laser beam. Because rotational asymmetric aberration (shown in Figure 6(b)) by the decentering of the optical element occurs as well as radial aberration such as the comatic aberration, in the case of an off-axial optical system, all those aberration becomes able to be compensated by using free-form surface for all mirrors. The free-form surface [Takahashi 2011] is curved surface defined in XY polynomial such as equation 13.

$$Z = \sum_{n=0}^{k} \sum_{m=0}^{n} C_{nm} X^{m} Y^{n-m} \tag{13}$$

In one aspect, free-form surface becomes able to have positive and negative power (Figure 6(a)). Furthermore, the curvature of the tangential direction can be different from azimuth direction.

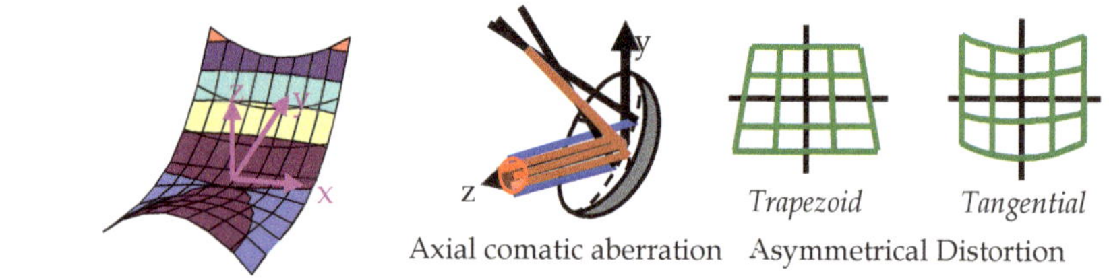

(a) Example of a free-form surface (b) Asymmetrical aberrations

Figure 6. A free-form surface and Asymmetrical aberrations

5.2. Consideration of the optical specifications

The average expansion and fluctuation of light intensity of the laser beam at 1.550 μm wavelength with the atmospheric turbulence are calculated by using the expressions of Section 3 stated above. In this calculation, the transmission beam is Gaussian beam which have beam divergence of diffractive limit, the difference in altitude between the two stations (low altitude stratospheric platform experiment [Katsuo 2005] condition) is 4 km, zenith angle is 60 degree, the transmission distance of the beam is 8 km maximum, and the pointing errors is 0 μradian. When the transmission beam radius is 40 mm, the free space loss is approximately 1 dB. So that if the optical antenna achieves an output transmission of around 100 mW and the receiver sensitivity of -20 dB, we can expect a margin of 40 dB and it is possible to communicate under thin cloud or fog conditions. The relation between the beam radius at the transmitter (W0) and the average beam expansion at receiver point is shown in Figure 7. The mean intensity fluctuation of the received beam under similar conditions is shown in Figure 8. According to these results, the radius of received beam is varied slightly and the expansion rate of the radius of received beam is constant when the radius of transmit beam is more than 60 mm. Moreover, the larger the transmission beam diameter is, the smaller the amount of the decrease of center strength of the received beam, and the free space transmission loss is less than 5 dB for a transmit beam diameter of uplink 70 mm or downlink 50 mm or more. Assuming Cn^2 is 1.7×10^{-14} (H/V model), we calculated the fluctuation of the received light intensity (scintillation index) by varying the transmitted beam radius. The result is shown in figure 9.

The change is small in weak fluctuation case and there is no big received optical power changes for propagation distance of ~4 km with laser beam transmit and receive aperture diameter of 10~100 mm therefore an optical antenna having this range of aperture beam transmission is possible. Because this fluctuation is significantly small and in the region of weak fluctuation and there is not large fluctuation of intensity of the received beam, the optical antenna which has such diameter of aperture transmits the optical beam have the possibility. The calculation result of receive light intensity fluctuation when the transmit beam radius is set at 20 mm and pointing error is 1 μradian are shown in figure 10. On the assumption that the inoperable rate is 10^{-7}, it is necessary to consider the fading margins which are 3.6 dB for the uplink and 5.3 dB for the downlink. From these calculations, it is desirable for the pointing error to be around 1 μradian in consideration of fading when a diffraction limited Gaussian beam is transmitted.

The pointing error of the FPM used for the fine tracking of the NG-FSO system in this research is estimated to be approximately 20 μradian. The angular magnification of the optical antenna is 20, and the tracking accuracy in the system can expect the improvement of the atmosphere fluctuation to become 1.0 μradian in a calculation. Considering the above outlined analytical results, the target specifications of the optical system are as follows:

1. Optical antenna aperture diameter is 40 mm
2. Angular magnification is 20
3. Coupling efficiency with SMF on axis is around -1 dB
4. Resolution is diffraction limit of field angle ±0.2 degrees (less than wave aberration $\lambda/10$)

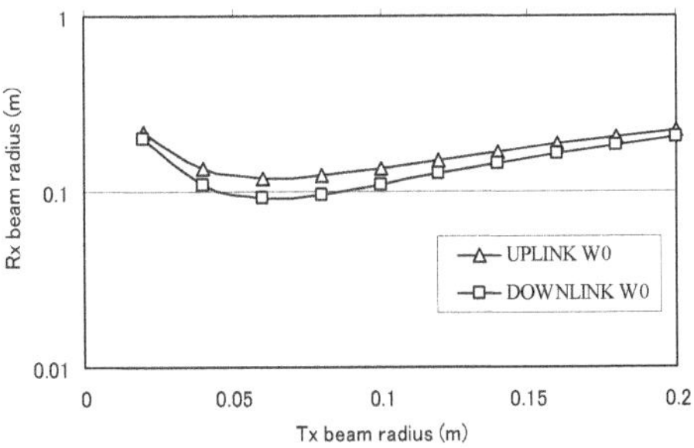

Figure 7. Rx beam radius versus Tx beam radius

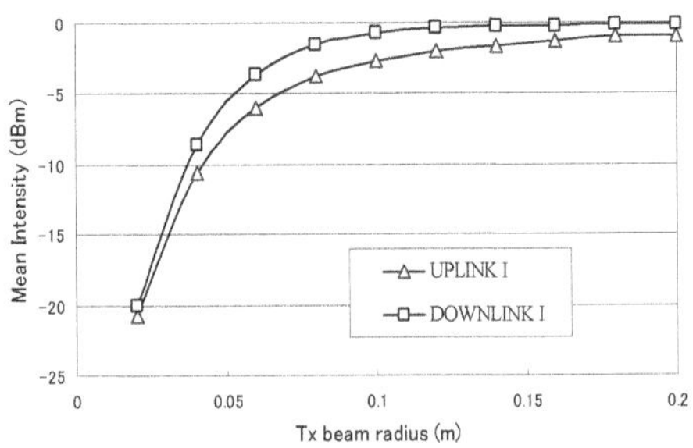

Figure 8. Mean intensity depend on Tx beam radius

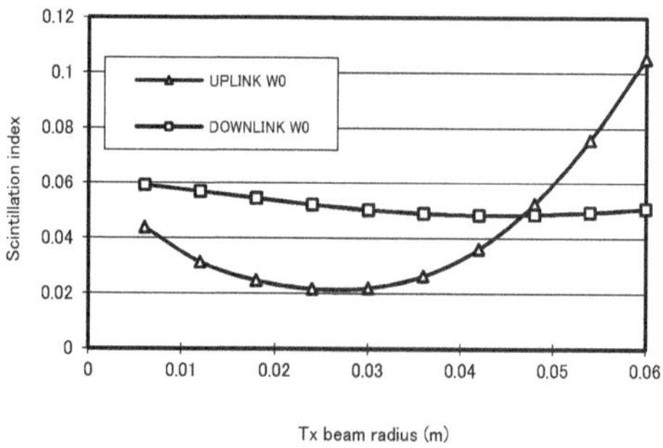

Figure 9. Scintillation index depend on Tx beam radius

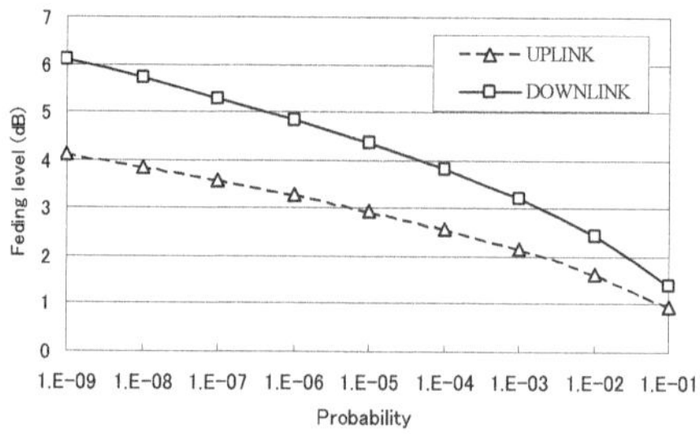

Figure 10. Probability of fading level

5.3. Result of optical design

In order to conduct an optical system without obscuration, having less deterioration of wavefront accuracy when the incident beam angle varies, we designed an optimum optical system comprising off-axial free form surface triple mirror which have enlargement/reduction rate of 20 times. Optical layout of triple mirror is shown in figure 7. Fundamental power placement in an optical design is a primary concave mirror plus a secondary convex mirror. With this optical configuration, a primary image is formed whose spherical and coma aberrations are compensated, and by using a collimator mirror which is arranged after the primary image a beam of parallel rays of light is producing. We set the paraxial focal length of the primary mirror and the secondary mirror at f_1=230 mm, the collimator mirror at f_2=11.8 mm, and the angular magnification of the triple mirror at approximately 20. Moreover, we have decentered aberrations, such as axial comatic aberration and axial astigmatism caused by tilt and decentered of both reflective surfaces, compensated by the three pieces of free-form mirror. A figure of constitution of the prototype of the optical antenna module which had a QD feedback type built-in fine tracking system by a microminiaturized fine pointing mirror (FPM) is shown in figure 11, the photograph is shown in figure 12, and the photograph is shown in figure 13. Based on the design result, the calculations of coupling efficiency when an ideal lens is deployed in the radiation side are shown in figure 14.

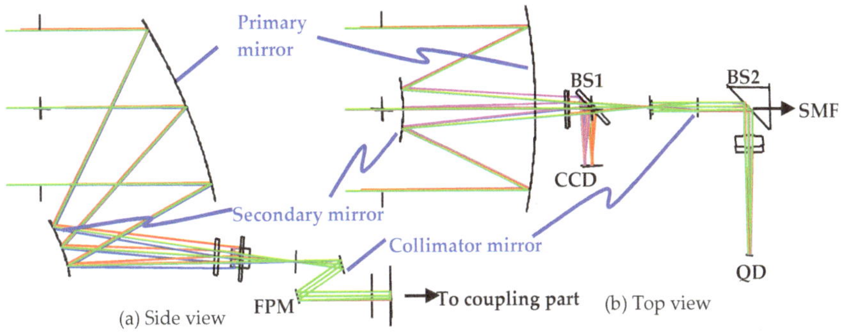

Figure 11. Optical layout of an optical antenna with off-axial free form optics.

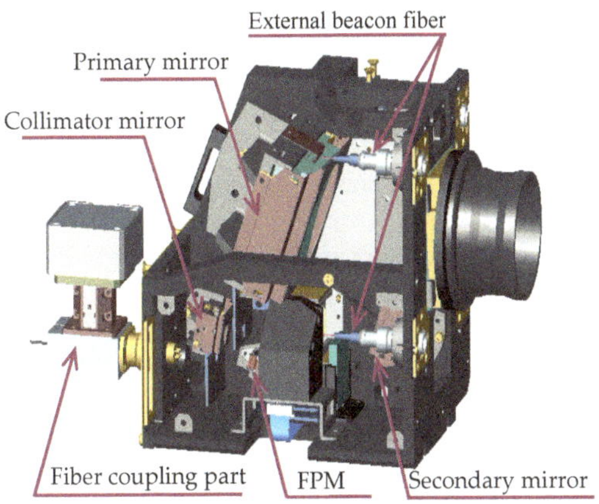

Figure 12. Appearance of prototype of optical antenna module

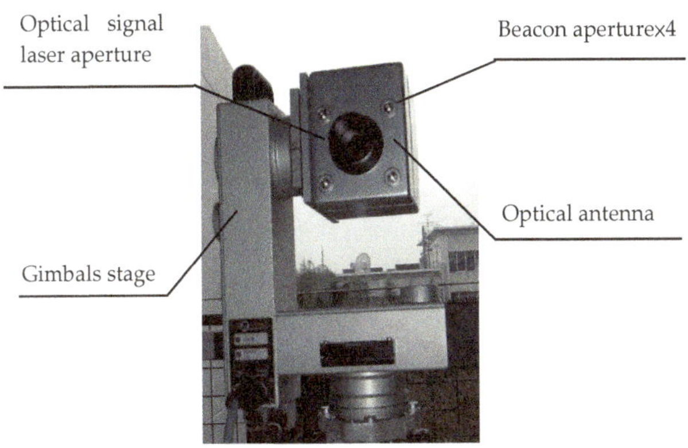

Figure 13. Photograph of prototype optical antenna on the gimbal stage

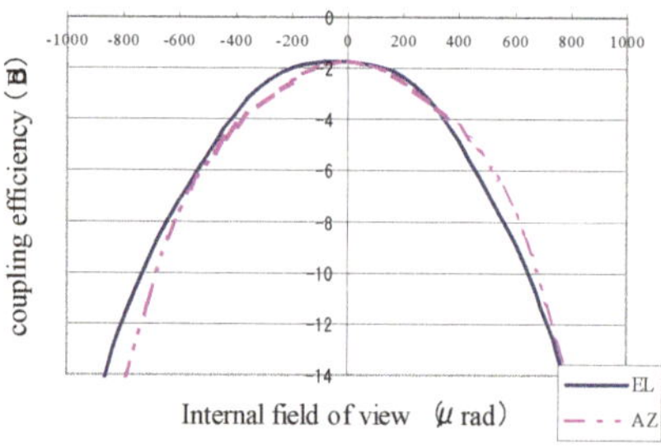

Figure 14. Coupling efficiency depend on internal field of view

5.4. Experimental set up of NG-FSO system

The test beds were put on two campuses of the Waseda University campus area situated in Shinjuku Ward, Tokyo city [Kazaura 2007][Kazaura 2006]. Figure 11 shows a map view of where the FSO systems are deployed. Two antennas were set on the rooftop of a 10 floor building (figure 15) and the corresponding antennas on the rooftop of a 9 floor building with a link spanning 1 km between the transceivers on the top of those two buildings. The antenna shown on the right in figure 16 is the antenna under experiment while the antenna on the left is used for scintillation and optical power attenuation measurement. They are both connected to data acquisition system placed in the experiment setup room which records all the data as shown in figure 17. The FSO beam propagation path travels over low rising residential, office or campus buildings, thus transmitted beam will experience higher probability of scintillation from ground heating and heating from these buildings. The experimental antennas use a 980 nm beacon for alignment and 1550 nm wavelength for communication. A charge coupled device (CCD) camera is used for initial alignment purposes which is done manually adjusting thimble screws located at the base of the antenna. Repeated adjustments are made on both sides until the light is centered on both CCD cameras (see TV monitor in figure16). Fine alignment of the antennas is achieved by using a QD which gets a feedback of the maximum received power from the inbuilt data acquisition system. The 1550 nm transmission beam was driven by an erbium doped fiber amplifier (EDFA) capable of outputting 100 mW total power. The experimental antenna can be controlled from a remote location by a serial computer interface to send control information that is used for remote alignment. For the scintillation and optical power attenuation measurement antennas, a serial computer interface provides the beam output power and receives signal strength information and other information.

The basic configuration of our experimental setup for the developed FSO system by all-optical connection of free-space and SMF is show in Figure 18. Because the signal light is fiber coupled at both ends, booster/post EDFA and other measurement and data collection devices can be conveniently placed inside the building. The fibers are run to the respective rooftops and then coupled directly to the antenna as shown in Figure 16. Separation of transmitting light and receiving light is performed by an optical circulator. The test signal

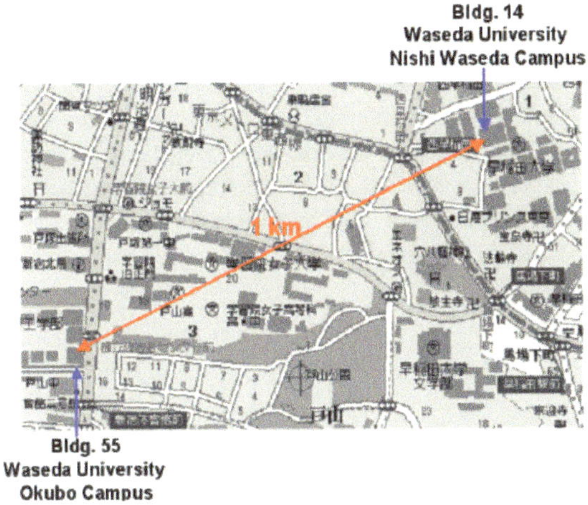

Figure 15. A map view of the 1km test between Waseda University Nishi-Waseda campus and Okubo campus

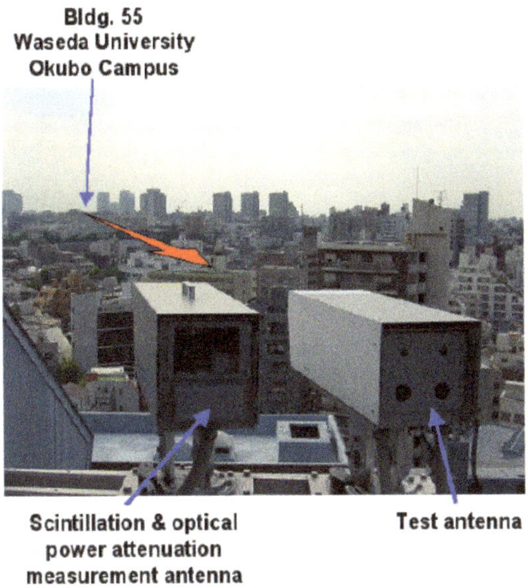

Figure 16. Optical antennas on the rooftop of building 14 in Nishi-Waseda

(pseudo random bit sequence) from bit error rate tester (BERT) is changed into a light signal by an E/O converter, and after being amplified with a booster EDFA, it is sent to the optical antenna of the rooftop. At the other end after the optical signal received with the optical antenna is removed in amplification by post-EDFA and a light filter removes noise, it is changed into an electric signal by an O/E converter, and becomes a receiving bit sequence in a clock data recovery circuit, and calculation of the bit error is carried out by BERT. The light signal which branched by 3 dB-coupler on the way is used for the monitor of receiving intensity, or other measurement. In a WDM experiment, the wavelength of each signal is made to fit the ITU grid (100 GHz channel spacing), and multi/de-multiplex is performed using a DWDM multi/de-multiplex device. Moreover, a weather monitor device and another optical antenna are installed in the rooftop and weather condition (visibility, precipitation and temperature) and link line condition (scintillation and optical power attenuation) are measured simultaneously.

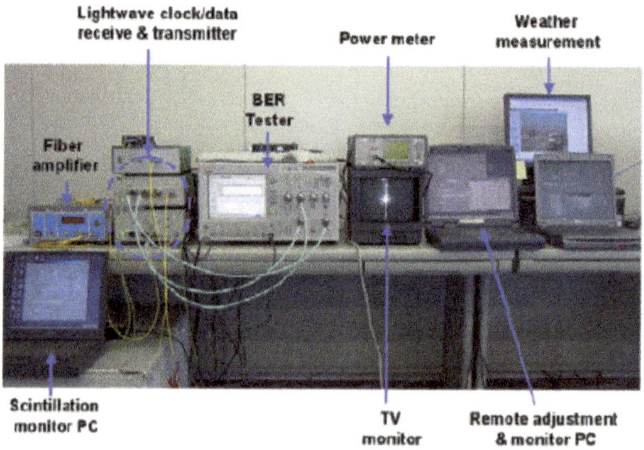

Figure 17. Photograph of experimental hardware setup

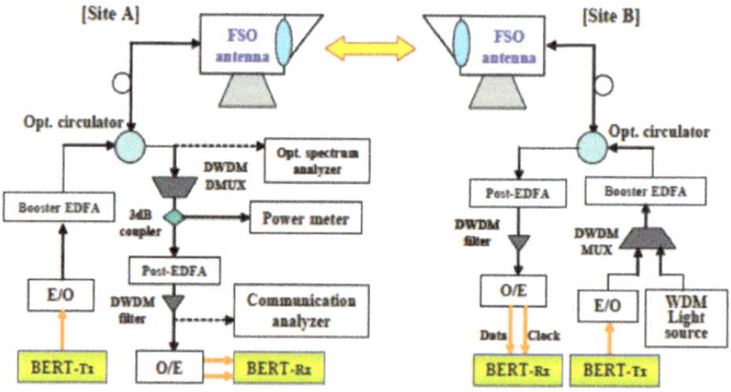

Figure 18. Experimental FSO communication system setup

5.5. Result and analysis

First, in order to evaluate the all-optical connection FSO system characteristic, the result of the 10 Gbps single wavelength light signal transmission experiment [Arimoto 2007] which is not achievable with conventional FSO system is shown in Figure 19. The bit error rate (BER) is measured by accumulated at 1 minute interval is shown in the figure. The data encoded optical signal, is amplified by 100 mW booster EDFA shown in Figure 18. The minimum receiving sensitivity of the optical receiver used in this experiment is -37 dBm. Stable transmission with few errors has been achieved over 12 hours. Figure 20 shows the eye patterns measured continuously for 5 minutes at this time. Although the fluctuation of the amplitude direction becomes large when compared with the eye patterns before transmission, it seems that there are very few increases in a time jitter (horizontal axis), and high-speed transmission is possible. In addition, the error sometimes shown in Figure 19 could be burst error produced by the momentary fall of received light intensity of several milliseconds or less, and this influence can hardly be observed on TCP/IP which is commonly used. An experiment using two optical wavelengths and connection to a Giga-bit Ethernet circuit and a 2.5 Gbps test signal was conducted. Optical and electrical signal conversion were performed by the media converter, and the Ethernet circuit performed huge size file transfer of between PCs connected through a switch, and has calculated the throughput from transfer time. In addition, it is connection by 100 Mbps between the switch and PC. Figure 21 shows the measurement result of the throughput of the file transfer performed for every minute and BER accumulated per minute. The minimum receiving sensitivity is -34 dBm which is almost same as the optical receiving module currently used for the media converter and the optical receiver currently used for BER measurement. Although errors sometime appeared in BER measurement, there is almost no change of a throughput almost similar result as in wired network. Since the great portion of error is short-time burst error, the throughput is hardly affected. Moreover, to the experimental result of BER and Ethernet, change considered to be mutual interference among other light signals cannot be found out at all, but has realized stable WDM transmission to it.

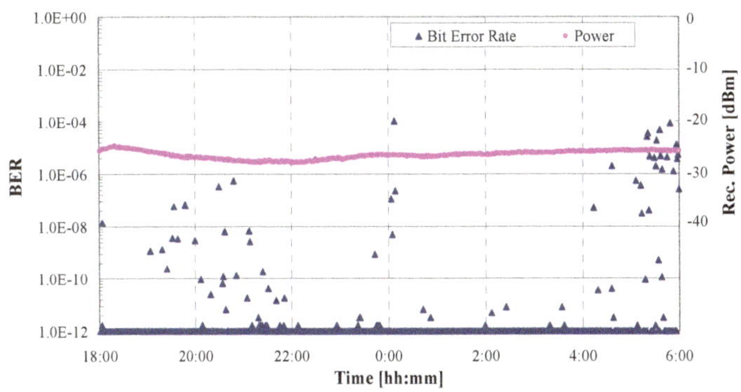

Figure 19. BER measurement result for 10 Gbps transmission over 1 km distance.

From these results, we demonstrated the next-generation FSO system using all-optical connection technology realized 10 Gbps transmission and also possible to apply WDM transmission which were not able to be realized in the conventional FSO system, and showed that the capability near an optical fiber could be offered. However in presence of strong atmospheric turbulence increased burst errors are observed in the system because of the fall in the received signal power. In order to realize stable communication even under strong atmospheric turbulence more improvement in is needed is required in the tracking system performance. Moreover, it is important to clarify the link design technique about the application distance of atmospheric turbulence and a FSO system.

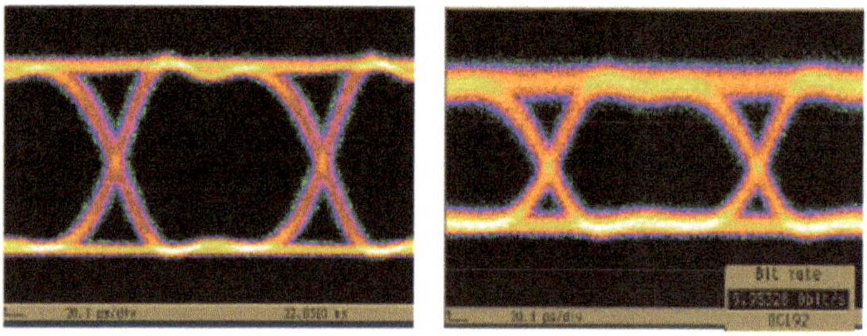

(a) (b)

Figure 20. 10 Gbps transmission test eye pattern. (a) Before transmission (b) After 1 km transmission

5.6. Conclusions of the NG-FSO system

The next-generation optical wireless communication system offered seamless connection of free space and fiber system. The transceiver incorporates a FPM for high-speed beam tracking and control function, therefore, having the capability to mitigate the effects of atmospheric turbulence on the transmitted optical beam. The FSO system performance was verified and error free transmission over an extended period of time was demonstrated. The system performance expressed in terms of BER performance was also evaluated and showed to be consistently above acceptable levels. Stable performance after increasing the system bandwidth using WDM technology was also attained.

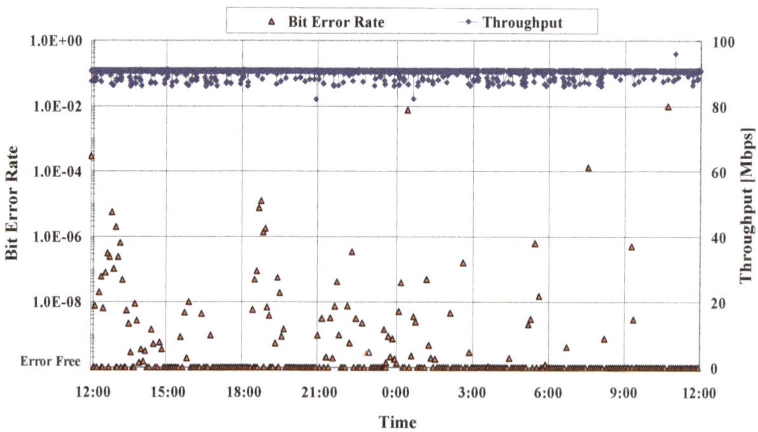

Figure 21. Throughput and BER characteristics in DWDM transmission

6. Research and development on RoFSO system

6.1. Description of RoFSO system

Examination of a RoFSO system which expands RoF technology to a free-space using all-optical connection technology is shown. RoF technologies [Al-Raweshidy 2002][Hai 2006] can realize a cost effective universal platforms for future ubiquitous wireless services. Furthermore, RoF networks can be extended to virtual radio free-space network with layer 1 routing realized [Komaki 2003][Tsukamoto 2005]. By using RoF, architecture for radio access zones easily employs micro or pico cellular systems. However, the optical fiber as an infrastructure is needed for a RoF network. We aim at quick and effective provide of heterogeneous wireless services for not only urban area but rural area, that has a little or no infrastructure for broadband services by the system with which developed next-generation FSO and a RoF system were united.

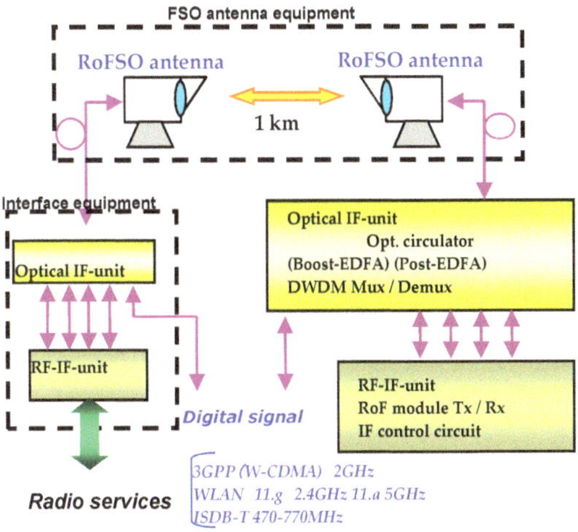

Figure 22. Outline of RoFSO system setup

We developed the RoFSO antenna which is improved the optical antenna, and installed in the same experimental field [Kazaura 2009][Kim 2009][Tsukamoto 2008]. Moreover, RF interface which represents the RoF system, and optical interface which performs multi/de-multiplex and amplification of the optical signal carrying the various services developed for an experiment as indoor equipment. A new optical antenna of 80 mm of the main aperture aimed at better tracking accuracy by feeding back the influence of the atmospheric fluctuation which signal light receives itself by diverting and using a part of signal light for fine tracking. RF interface unit consists of a RoF module which performs the electro/optical conversion of RF signal corresponding to the planned service (3G cellular, WLAN 11g/a, DBV-T), and its control circuit. Each light signal transmitted is matched with the ITU grid wavelength of a 100 GHz space. The optical interface unit consists of a DWDM multi/de-multiplex device, EDFA (booster/post) and optical circulator for separation of transmit/receive signal. A figure showing the whole system just described is depicted in Figure 22. In this experiment, signal generators and analyzers for evaluating the transmission quality of various services are connected to RF interface unit.

6.2. Consideration of optical antenna specification

The average expansion and fluctuation of light intensity of the laser beam at 1.550 μm wavelength with the atmospheric turbulence are calculated by using the above expressions (Section 3). The relation between the beam radius at the transmitter and the average beam expansion at receiver point is shown in Figure 23.

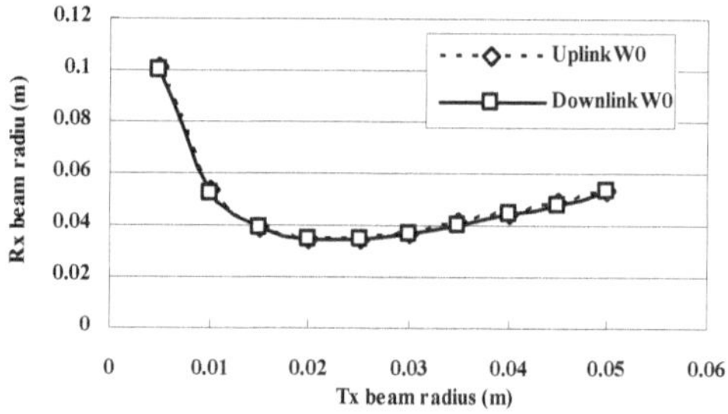

Figure 23. Rx beam radius versus Tx beam radius

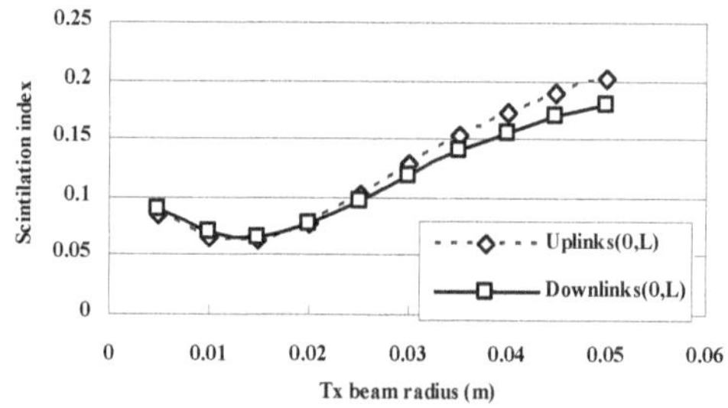

Figure 24. Scintillation index depend on Tx beam radius

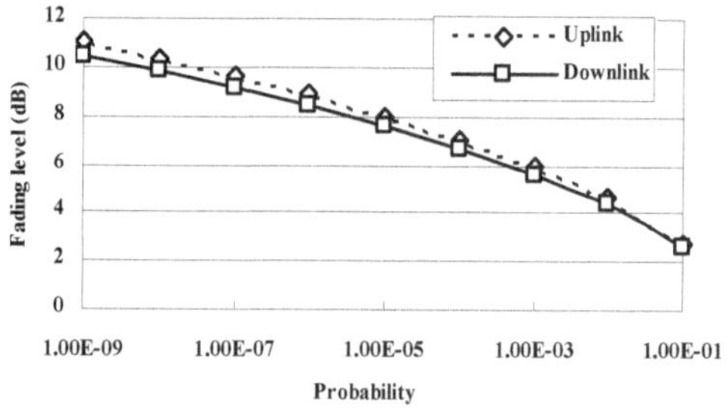

Figure 25. Probability of fading level

According to these results, the radius of received beam is varied slightly and the expansion rate of the radius of received beam is constant when the radius of transmit beam is more than 20 mm. Moreover, the larger the transmission beam diameter is, the smaller the amount of the decrease of center strength of the received beam, and the free space transmission loss is less than 5 dB for a transmit beam diameter of 40 mm or more. We calculated the fluctuation of the received light intensity (scintillation index) by varying the transmitted beam radius. The result is shown in figure 24. The change is small in weak fluctuation case and there is no big received optical power changes for propagation distance of 1~2 km with laser beam transmit and receive aperture diameter of 10~100 mm therefore an optical antenna having this range of aperture beam transmission is possible. Figure 25 shows the cumulative probability distribution when the transmit beam radius is set at 40 mm and pointing error is 5 μradian. We found that the fluctuation of receiving optical intensity is ± 3 dB at 10 % and ± 8 dB at 0.001%. In other words, it is necessary to consider fading margins that are 9.6 dB for the uplink and 9.2 dB for the downlink if inoperable ratio is 10^{-7}.

In addition, the previous NG-FSO system, the beam for fine tracking systems was a beacon light which was different from the signal light. In that case, the propagation path of the beacon light signal and signal beam is not exactly consistent therefore the atmospheric fluctuation behavior is also not similar. Especially, in the case of the data rate is high, such tendency is significantly. Therefore, it was not possible to get enough fine tracking characteristics. That is why in the new system for fine tracking the signal light is used for detection of angle of arrival by utilizing beam splitter.

Based on these calculations, if a diffraction-limited Gaussian beam is transmitted, it is desirable for the pointing error to be less than 5 μradian considering fading. The pointing error of the FPM used for the fine tracking of the RoFSO system is approximately 20 μradian and the angular magnification of the optical antenna is 40. The tracking accuracy of the entire system can be calculated by the division of those two values and it is 0.5 μradian. Thus, it is possible to expect the improvement of the atmosphere fluctuation.

6.3. Results of optical design

A constitution of the prototype of the optical antenna module [Takahashi 2008] which had a QD feedback type built-in fine tracking system by a micro-miniaturized fine pointing mirror with the part of SMF coupling components is shown in figure 26, the photograph of SMF coupling part and tracking system is shown in figure 27 and the photograph of entire prototype optical system is shown in figure 28. Based on this design, results of the calculation of coupling efficiency and spot diagram when an ideal lens is deployed in the radiation side are shown in figure 29. The on-axis SMF coupling loss is -1.02 dB with less the -5 dB loss for up to full size field angle 25 μradian. Optical performance for on-axis and off-axis of the designed optical system is summarized in table 1. It is observed that it is possible to obtain almost twice image formation efficiency of the diffraction limit of the design requirement for field angle ±0.5 degrees.

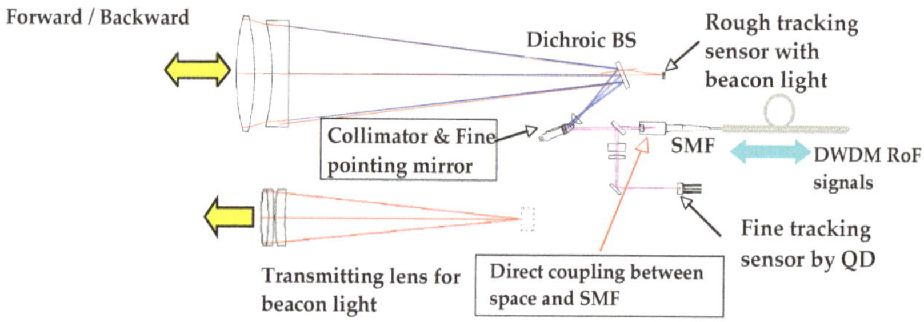

Figure 26. Optical layout of the antenna of RoFSO system

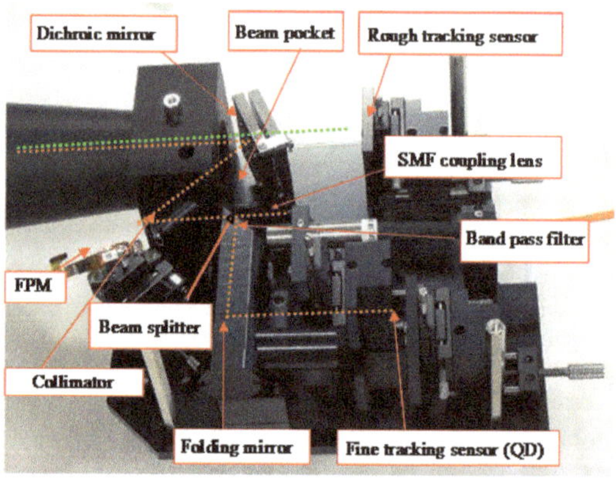

Figure 27. Photograph of SMF coupling part and tracking system

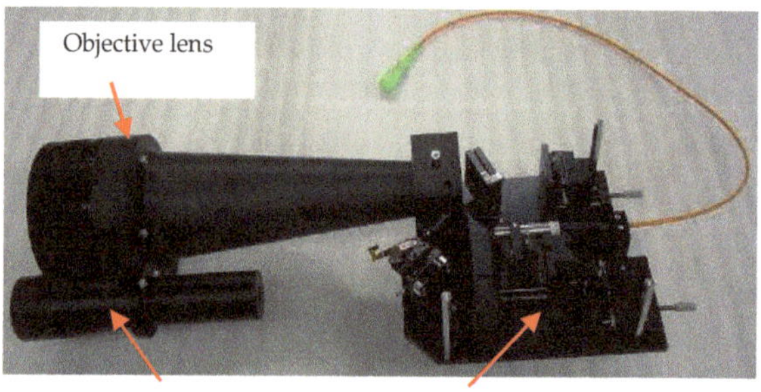

Figure 28. Photograph of entire optical antenna of prototype RoFSO system

To evaluate real performance in the trial product; the coupling efficiency that measured in experiment system deploy a coupling lens after an object lens and a collimator, and to be received light to SMF is approximately - 5dB.

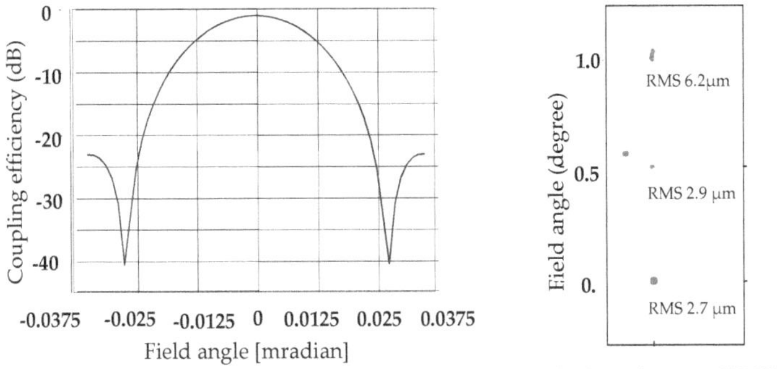

(a) Coupling efficiency

(b) Spot diagram (RMS)

Figure 29. RoFSO antenna optical performance (a) Coupling efficiency (b) Spot diagram

Field angle (deg)	0	0.5	1.0
Spot diameter (μm)	2.7	2.9	6.2
Wave aberration (λ)	0.0037	0.0094	0.0327
Coupling efficiency (dB)	-1.02	-	-

Table 1. Optical performance of the RoFSO optical antenna in design condition

6.4. Evaluation of prototype RoFSO system

6.4.1. Experimental evaluation setup

We installed the developed RoFSO system at a same place of NG-FSO system in Waseda University and performed experiments [Kazaura 2009] [Kim 2009] [Tsukamoto 2008] [Takahashi 2008]. Photographs showing the devices setup on the rooftop as well as the various measurement devices setup in the laboratory is depicted in Figure 30(a) and 30(b) respectively. Three FSO antennas are placed on the rooftop of one building as shown in Figure 30(a), which include the DWDM RoFSO antenna under investigation and two other conventional antennas used for measuring and quantifying the deployment environment characteristics for example atmospheric turbulence induced scintillation.

6.4.2. Basic performance

As a result, the system's stability and tracking properties satisfied the design requirements (tracking response is more than 2 kHz, turbulence suppression characteristics of this feedback system are more than suppression ability 20 dB in frequency less than 100 Hz with $Cn^2 < 1.7\times10^{-14}$), and we confirmed that the system can transmit WDM with sufficient performance regarding not only error-free transmission of digital signals but also RF signal transmission. Figure 31 shows the intensity fluctuation characteristics of received beam when the fine tracking of antenna is set OFF or ON, and the average of turbulence Cn^2 is 2×10^{-14}. If the fine tracking is set OFF, the fluctuation of receiving optical intensity largely and frequently decays because it is not possible to control arrival angle, also because location of beam spot of SMF consistently fluctuates. In addition, the mean of receiving optical intensity is low because it is not possible to control location of light focus accurately. On the other hand, the fine tracking of antenna is set to ON, the entire fluctuation decreases and the mean of receiving optical intensity is improved. Figure 32 shows the receiving optical spectrum when four of wireless service signals are transmitted with 1 km distance and WDM transmission using this RoFSO. It can be found that each service signal is clearly separated and is transmitted without mutual interaction.

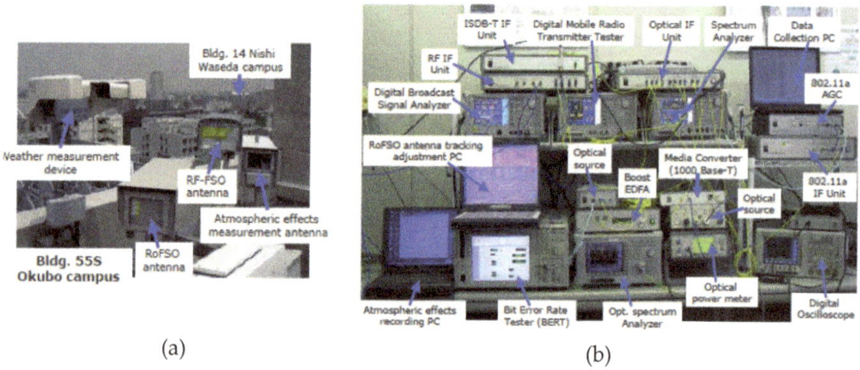

Figure 30. RoFSO system experimental device (a) rooftop setup and (b) devices setup in the laboratory

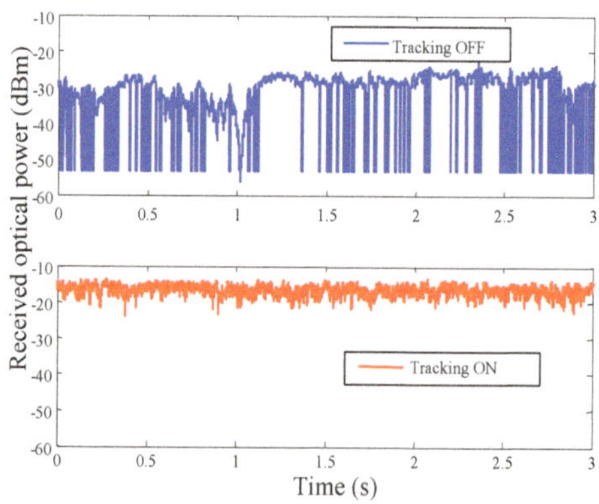

Figure 31. Optical intensity fluctuation characteristics of received beam with tracking OFF (upper) and tracking ON (lower)

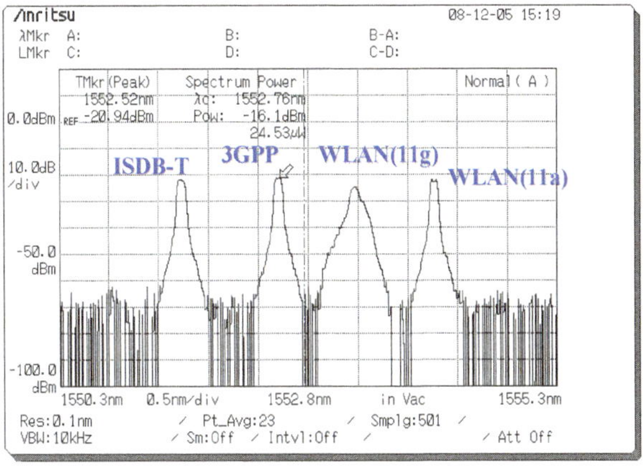

Figure 32. Optical spectrum of four wireless service signals using WDM RoFSO system

6.4.3. W-CDMA (Cellarer phone)

In W-CDMA system, the downlink signal transmitted by the base station is designed to fulfill the specifications set in 3GPP standard [3FPP 2002]. The spectral properties of the signal are measured by he adjacent channel leakage ratio (ACLR) which is considered to be a more stringent quality metric parameter, and is defined as the ratio of the amount of leakage power in an adjacent channel to the total transmitted power in the main channel. The 3GPP specifies one main channel and two adjacent channels. The standard requires the ACLR to be better than 45 dB at 5 MHz offset and 50 dB at 10 MHz offset. In our experimental setup, we use a signal generator to generate a test signal (W-CDMA Test Model 1) with a signal power of -20 dBm which is transmitted over the RoFSO link and at the receiver side a digital mobile radio transmission tester is used to measure and record the quality of the W-CDMA signal. Figure 33(a) shows a received W-CDMA signal ACLR spectrum after transmission over the 1 km RoFSO link. It is observed that the signals' spectral properties shown on Figure 33(a) satisfy the 3GPP specified values of ACLR at the 5 MHz and 10 MHz offsets.

The variation of the measured received optical power and the W-CDMA signal ACLR characteristics is shown in Figure 33(b). Two cases are considered i.e. first case is back-to-back (B-to-B) measurement using the RoF modules, signal generator and analyzer and an optical attenuator for incrementing the attenuation to represent channel losses and in the offsets. The B-to-B actual transmission over the RoFSO system measurements shows almost similar characteristics and the minimum optical received power to satisfy the prescribed 3GPP value at 5 MHz at 10 MHz offsets is about -15 dBm. Using a post EDFA the required received optical power can be even as low as -25 dBm and -20 dBm and still satisfy the 3GPP specification for W-CDMA signal transmission at 5 MHz and 10 MHz offsets respectively.

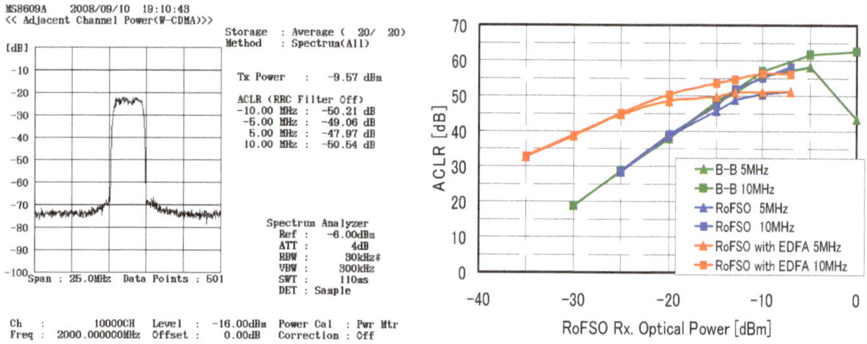

(a) (b)

Figure 33. (a) Received W-CDMA signal ACLR spectrum and (b) variations of ACLR and optical received power

6.4.4. Wireless LAN

In another example the RoFSO system is evaluated by transmitting a WLAN IEEE802.11 based signal. In this experiment, an IEEE802.11g compliant waveform is generated by a vector signal generator at -24 dBm which is applied to the RoF module in the RF interface unit. After transmission through the RoFSO link a spectrum analyzer is utilized to measure and analyze the quality of the received WLAN signal. A pass/fail judgment of the spectrum mask as defined in the IEEE specification 802.11a/b/g is used. As a test signal, IEEE802.11g waveform at 2.4 GHz with 54 Mbps 64QAM is used. Figure 34(a) depicts a WLAN signal with spectrum mask in this case. A constellation graph of the WLAN signal modulation analysis is shown in Figure 34(b).

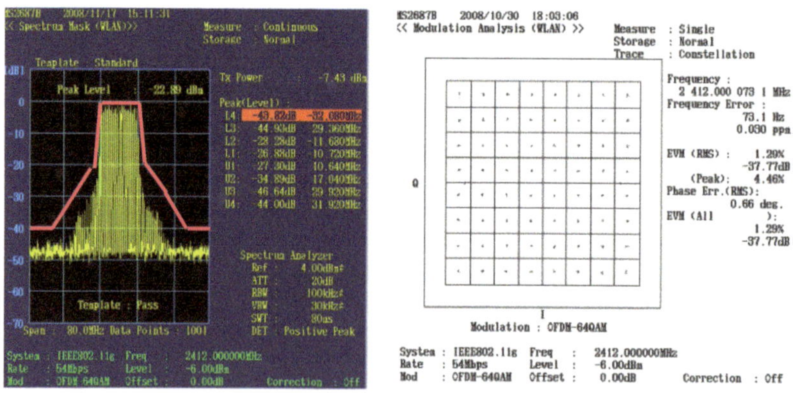

(a) (b)

Figure 34. WLAN (a) spectrum mask and (b) modulation analysis constellation

The recorded RMS of Error Vector Magnitude (EVM) value is within the acceptable tolerance for WLAN signal transition. The result of continuous measurement of the spectral mask test collected over two days is shown in Figure 35. The mask pattern Pass/Fall judgment is measured per second and recorded. This result is accumulated for every 1 minute, and the passed number is plotted in the figure. Moreover, in order to evaluate the variation with respect to the condition of the propagation path, the metrological data such as visibility and rain rate data which is simultaneously collected is plotted.

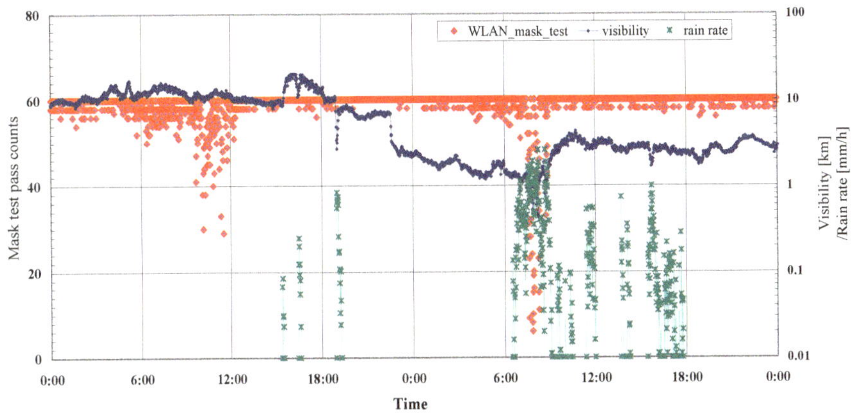

Figure 35. Mask test pass counts per 1 min. vs. weather condition

The measurement data depicted in the figure represent a fine weather condition before noon on the first day which later turned cloudy. On the second day it rained. The drop in visibility because of rain is significant around 8:00am on the second day. Because of the rain there is an increase in attenuation in the propagation path, so the received optical signal power falls therefore increasing the rate of the spectrum mask test failure. It can be observed in the figure that on the first day, there is an increase of spectrum mask failure rate around noon, this is due to the effect of atmospheric turbulence.

6.4.5. *ISDB-T (Digital TV)*

Digital terrestrial television broadcasting, referred to as Integrated Service Digital Broadcasting – Terrestrial (ISDB-T) in Japan, is designed to provide reliable high-quality video, sound and data broadcasting not only for fixed receivers but also for mobile receivers. The system is designed to provide flexibility, expandability and interoperability for multimedia broadcast. The ISDB-T system uses the UHF band at frequencies between 470 MHz and 770 MHz, giving a total bandwidth of 300 MHz. The bandwidth is divided into 50 channels named from 13 to 62. Each channel is further divided into 13 OFDM segments which includes a single segment, (A-Layer or 1seg), for mobile receivers (LDTV, audio and data) and the remainder can be allocated as one 12-segment for high definition television (HDTV) programs [DiBEG]. In this setup, channel 32 is used for ISDB-T signal transmission. A vector signal generator (Anritsu MG3700A) is used to output simple BER data and video waveforms for ISDB-T transmission evaluation. In this example, two signals are set simultaneously with the following waveform patterns (a) ISDBT_16QAM_1_2 (A-Layer: 1seg, 16QAM and B-Layer: 12seg, 64QAM) and (b) ISDBT_2layer_Movie, both at -20 dBm with a 6 MHz frequency offset. The combined signal at -17 dBm is fed into the RoF module. The optical modulation index (OMI) for each channel (at -20 dBm input) is 10%. The signal is subsequently transmitted over the RoFSO link. At the receiving site a digital broadcasting signal analyzer (Anritsu MS8901A) is used to measure the quality of the received ISDB-T signal. A received signal spectrum showing the two transmitted ISDB-T signals is depicted in Figure 36(a). A modulation error ratio (MER) quality metric parameter used to evaluate the modulation signal quality of the digital terrestrial television broadcasting signal directly and quantitatively is measured and analyzed. An example of modulation analysis constellation for the digital terrestrial broadcasting signal made of A-Layer 16QAM and B-Layer 64QAM is shown in Figure 36(b) and 36 (c) respectively which were captured when the recorded average received optical power was -5.92 dBm and -5.88 dBm respectively (i.e. the monitor output measured power adjusted for the respective RF signal). The constellation is very useful for analyzing the condition of the received signal by monitoring the modulation symbol movement. In Figure 36(b) and 36(c) the received signals exhibit little signal distortion (in terms of amplitude or frequency fluctuations) and the signal deterioration is minimal thus confirming the suitability of the RoFSO system for ISDB-T signal transmission conforming to the specified standard [ARIB]. In this example the measurement was made in the evening after 20:00 hrs considered to be weak atmospheric turbulence conditions. The ISDB-T signal transmission using the RoFSO system is also evaluated using a BER quality metric parameter. Figure 37 shows the BER measurement and the mean received optical power characteristics for one-segment (1seg) and 12-segment transmission collected over a 24 hour period on 11 December 2008. Increased bit errors are observed at around midday because of increase in the atmospheric turbulence which affects the received optical power. The variation of the measured mean received optical power can be correlated with the BER in this case. For 1seg transmission (A-Layer) the BER characteristics shows satisfactory performance with most values being below the error correction limit (2×10^{-4}) demonstrating the suitability of the RoFSO system for ISDB-T 1seg transmission. Unfortunately, occasionally the automatic gain control (AGC) is inadequate in the case for 12-segment HDTV (B-Layer) transmission. As an example the ISDB-T transmitted video captured screen shots for 1seg and 12-segment are shown in Figure 38(a) and 38(b) respectively. The 1seg video quality was consistently clear and continuous without any stoppages. However, the 12-segment video is clear and continuous only in the absence of atmospheric turbulence or other effects which contribute to the deterioration of the transmitted RoFSO signal quality.

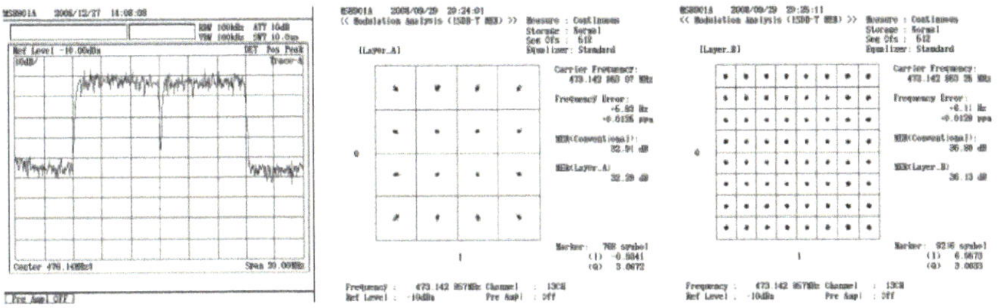

(a) (b) (c)

Figure 36. ISDB-T signal transmission experiment (a) received ISDB-T signal spectrum and modulation analysis constellations for (b) A-Layer 1 Seg (16QAM) and (c) B-Layer (64QAM)

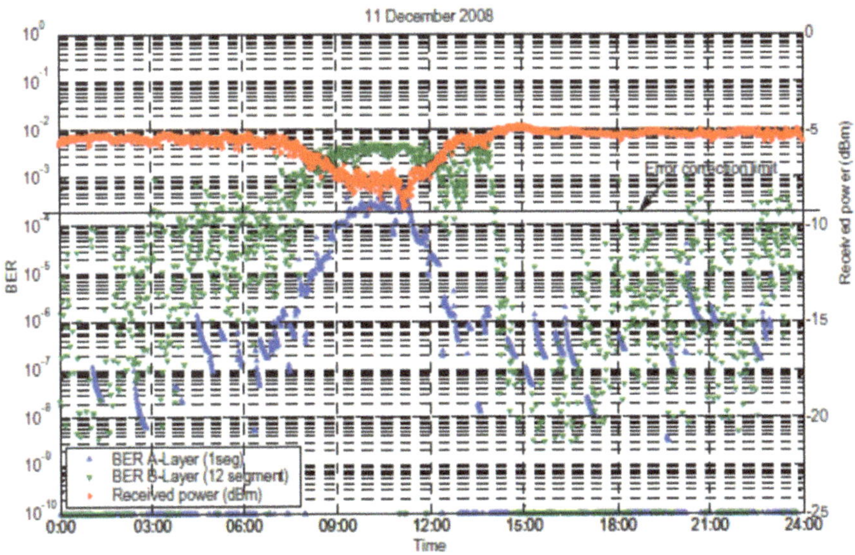

Figure 37. Digital terrestrial television broadcasting BER and received optical power characteristics

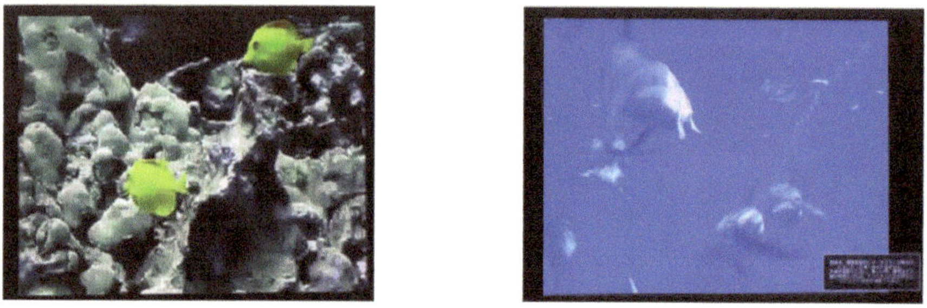

(a) (b)

Figure 38. ISDB-T captured video screen shots (a) A-Layer 1seg and (b) B-Layer 12-segment

6.5. Conclusions of the RoFSO system

Simultaneous transmission of different kinds of wireless services using a newly developed RoFSO system has been presented. As the result, we have been presented that an all-optical connection FSO system could treat not only digital signal transmission but RF signal on a par with an optical fiber. And it is shown that the new generation optical wireless optical communication technology has a possibility of uniting radio environment and cable environment also with a service level, and the thing expectable as a solution of a heterogeneous network is shown.

7. Conclusions

We have described the concept and technology for the next-generation optical wireless communication systems. We explained the optics design method and the design results of the optical antennas considering laser propagation phenomena such as the scintillation and arrival beam angle fluctuation which occurs by atmospheric turbulence. We also mentioned the fine tracking mechanism using the FPM for the optical antennas. The FSO system incorporating high speed and highly precise tracking mechanism in which the influence of angle of arrival change is compensated succeed in maintaining free-space to SMF stably. A free-space optical communication system using specially designed compact antenna for easy, cost effective means of constructing a robust and reliable high-speed link for next generation optical wireless communication system was developed and investigated. The actual proof experiment using the developed NG-FSO system shows that this system enabled offer of a link equivalent to the fiber independent of the bit rate or a transmission protocol. Furthermore, we tried to unite this system with a RoF system and also enabled offer of various wireless services further. RoFSO system which expands RoF technology to a free-space using all-optical connection technology also shows the possibility of RF signal transmission through the field experiment. As these results, a next-generation optical wireless communication system using all-optical connection technology shows a possibility of becoming an effective solution of extension of a next-generation optical fiber system and heterogeneous wireless service.

Author details

Koichi Takahashi
Future Creation Laboratory, Olympus Corporation, Japan

Acknowledgement

I am indebted to Mr. Y. Miyazaki, Mr. T. Sakamoto, Ms. J. Aruga, Mr. T. Suzuki and Dr. Y. Arimoto for the help and support during NG-FSO system development and experiment and to Dr. M. Matumoto, Dr. K. Wakamori, Dr. K. Kazaura, Mr. T. Suzuki, Dr. S. Komaki, Dr. K. Tukamoto, and Dr. T. Higashino for NG-FSO system and RoFSO system experiment and evaluation.

This work is supported by a grant from the National Institute of Information and Communications Technology (NICT) of Japan.

8. References

[Cochennec 2002] J-Y Cochennec, "Activities on Next-Generation Networks Under Global Information Infrastructure in ITUT," IEEE Communications Magazine, vol. 40, pp. 98–101, July 2002.

[ITU-T 2001] ITU-T Rec. Y.2001, "General overview of NGN," Dec. 2004.

[ITU-T 2011] ITU-T Rec. Y.2011, "General principles and general reference model for Next Generation Networks,"Oct. 2004.

[Lee 2007] S-Q. Lee, et al., "The wireless broadband (wibro) system for broadband wireless internet services," IEEE Communications Magazine, vol. 44, pp. 106–112, July 2007.

[Ben 2006] S. J. Ben Yoo, "Optical Packet and Burst Switching Technologies for the Future Photonic Internet, " J. Lightwave Technol. 24, 4468-4492 (2006)

[Zhang 2005] Y. Zhang, O. W. Yang and Y. Zhai, "The optimization issues in an agile all-photonic backbone network, " Proc. SPIE 5626, 1233 (2005)

[Kitayama 2005] K. Kitayama, T. Miki, et al., "Photonic Network R&D Activities in Japan-Current Activities and Future Perspectives," J. Lightwave Technol. 23, 3404- (2005)

[Kim 1999] I. I. Kim, M. Mitchel, and E. Korevaar, "Measurement of Scintillation for Free-space Laser Communication at 785nm and 1550nm," in Optical Wireless Communications II, vol. 3850, (Boston, Massachusetts), pp. 49-62, SPIE, Sept. 1999.

[Bouchet 1999] O. Bouchet, H. Sizun, et al., "Free-Space Optics Propagation and Communication. " ISTE, 2006.assachusetts), pp. 49-62, SPIE, Sept. 1999.

[Kaliski 1999] R. W. Kaliski, S. M. Genco, et al., "Laser communication intersatellite links realized with commercial off-the-shelf technology," in Free-Space Laser

Communication Technologies XI (G. S. Mecherle, ed.), vol. 3615, pp. 170—178, SPIE, Apr. 1999.

[Borocom 2005] D.M.Borocom, C.C.Chan and B.Edwards, "Overview of the Mars laser communications demonstration Project," IEEE LEOS Newslett., 19 No.5(2005) 8-11

[Koyama 2004] Y. Koyama, E. Morikawa, et al., "Optical terminal for NeLS in-orbit demonstration," in Free-Space Laser Communication Technologies XVI (G.S.Mecherle, C. Y. Young, and J. S. Stryjewski, eds.), vol. 5338, (San Jose, CA, USA), pp. 29—36, SPIE, June 2004.

[Infrared] Infrared Communication Systems Association. http://www.icsa.org.jp.

[Free-sSpace] Free-Space Optical Alliance. http://www.wcai.com/fsoalliance.

[Takahashi 2008] K. Takahashi and Y. Arimoto, "Compact optical antennas using free-form surface optics for ultrahigh-speed laser communication systems," SPIE Optical Engineering 47(1), pp. 016002- 1-9, 2008.

[Matsumoto 2008] M. Matsumoto, K. Kazaura, P. Dat, et al., "An alternative access technology for next generation networks based on full-optical wireless communication links," in Kaleidoscope-Innovations in NGN: Future Network and Services (K-INGN 2008), pp. 221—228, May 2008. Proc. of SPIE Vol. 7199 719907-11

[Kazaura 2007] K. Kazaura, K. Omae, T. Suzuki, et al., "Performance Evaluation of Next Generation Free-Space Optical Communication System," IEICE Transactions on Electronics E90-C, pp. 381—388, Feb. 2007.

[Khaleghi 1996] F. Khaleghi, J. Li, M. Kavehrad, and H. Kim, "Increasing Repeater Span in High-speed Bidirectional WDM Transmission Systems Using a New Bidirectional EDFA Configuration," IEEE Photon. Technol. Lett., vol. 8, no. 9, 1996, pp. 1252-1254.

[Luo 1998] G. Luo, J.L. Zyskind, J.A. Nagel, and Mohamed A. Ali, "Experimental and Theoretical Analysis of Relaxation-Oscillations and Spectral Hole Burning Effects in All-Optical Gain-Clamped EDFA's for WDM Networks", IEEE J. Lightwave Technology, vol. 16, no. 4, 1998, pp. 527-529.

[Al-Raweshidy 2002] H. Al-Raweshidy and S. Komaki, eds., "Radio Over Fiber Technologies for Mobile Communications Networks. "Artech House Publishers, 1 ed., 2002.

[Hai 2006] L. H. Hai, et al., "Performance improvement of radio-on-fiber ubiquitous antenna system using sub-carrier resource management, " Proc. SPIE 6390, 63900B (2006)

[Komaki 2003] S. Komaki, "Microwave Technologies for Software Radio Networks", Proceedings of APMC2003, Vol.3, No.FC7-1, pp.1780-1785, (November 2003).

[Tsukamoto 2005] K. Tsukamoto, "Software Definable Radio Networks for the Ubiquitous Networks", XXXIIIth GA of URSI, vol.1,p.205, (October 2005).

[Tatarskii 1997] V.I.Tatarskii, "The Effects of the Turbulent Atmosphere on Wave Propagation, U.S.Dept. of Commerce", National Technical Information Service, Springfield, VA, 1971.

[Dayton 1992] D.Dayton, B.Pierson, B.Spielbusch, J.Gonglewski, "Atomospheric structure function measurements with a Shack-Hartmann wave-front sensor," Opt. lett. 17,1737,1992.

[Hufnagel 1974] R.E.Hufnagel, Proc. Top. Mtg. on Opt., "Propagation through turbulence," Boulder, CO 1974

[Fried 1967] D. L. Fried and J. B. Seidman, "Laser beam scintillations in the atmosphere," J. Opt. Soc. Am. 57, 181–185 (1967)

[Andrew 1998] L. C. Andrews and R. L. Phillips, "Laser Beam Propagation through Random Media, " SPIE Optical Engineering Press, Bellingham, WA, (1998).

[Andrew 1995] L. C. Andrews, R. L. Phillips and P. T. Yu: "Optical scintillations and fade statistics for a satellite-communication system," Applied Optics, Vol. 34, No. 33 (1995).

[Arimoto 2003] Y.Arimoto, M.Nagatsuka, et al., "Laser Communication Experiment Using Stratospheric Platform System," Technical Report of IEICE SANE2003-82 (2003-11).

[Katsuo 2005] F. Katsuo, H. Kikuchi, et al., "Wireless System Demonstration Trial Using a Low-Altitude Station-Keeping Test Airship Optical Communications Experiment– ,"Technical Report of IEICE SANE2004-81(2005-01)

[Takahashi 2006] K. Takahashi and Y. Arimoto, "Development of optical antennas utilizing free form surface optics for the high speed laser communication systems," Free-Space Laser Communication Technologies XVIII, vol.6105, pp.40–49, SPIE, Jan. 2006.

[Takahashi 2011] K. Takahashi, "Development of Ultrawide-Angle Compact Camera Using Free-Form Optics," OPTICAL REVIEW Vol.18,No.1 (2011)55-59

[Kazaura 2006] K. Kazaura, K. Omae, T Suzuki, et al., "Mitigation of atmospheric effects on terrestrial FSO communication systems by using high-speed beam tracking antenna", Free-Space Laser Communication Technologies XVIII, 24 ~ 25 Jan 2006, San Jose, California, USA

[Arimoto 2007] Y. Arimoto, "Multi-gigabit Free-space Optical Communication System with Bidirectional Beacon Tracking," IEEJ Trans. FM, vol. 127, no. 7, pp. 385-390, 2007.

[Kazaura 2009] K. Kazaura, P. Dat, A. Bekkali, et al., "Experimental evaluation of a Radio-on-FSO communication system for multiple RF signal transmission", Paper 7199-6, Photonics West 2009, SPIE, San Jose, 2009

[Kim 2009] K-H. Kim, T. Higasino, et al., "Performance evaluation of WLAN signal transmission by using a new statistical model of RoFSO Channel, " 1st International Conference on Space Optical Systems and Applications (ICSOS 2009), ICSOS2009-33, Miraikan, Tokyo, Japan, 4-6 Feb. 2009.

[Tsukamoto 2008] K. Tsukamoto, H. Onodera, et al., "Project of Radio on Free Space Optic System Development for Heterogeneous Wireless Services," IPGC 2008 IEEE Photonics Global@Singapore, pp: 1-4 8-11 Dec. 2008

[Takahashi 2008] K. Takahashi, T. Higashino, et al., "Design and evaluation of optical antenna module suitable for radio-on free-space optics link system for ubiquitous wireless", Proceedings of SPIE for Photonic West, vol. 6877, pp. 68770H_1-11, San Jose, California USA, Jan 2008.

[Wakamori 2008] K. Wakamori, K. Kazaura, et al., "An Experimental Study of Next Generation FSO Systems, " IEICE TRANS Commu, Vol. J91-C, No. 1, pp. 28-37, Jan. 2008

[3GPP 2002] 3GPP TS 25.141, 3rd Generation Partnership Project. http://www.3gpp.org/, 2002.
[DiBEG] DiBEG. http://www.dibeg.org/.
[ARIB] ARIB STD-B31. http://www.arib.or.jp/english/html/overview/ov/std_b31.html.

Section 4

Optical Communication: Optical Networks

Subwavelength and Nanometer Diameter Optical Polymer Fibers as Building Blocks for Miniaturized Photonics Integration

Xiaobo Xing, Huaqing Yu, Debin Zhu, Jiapeng Zheng, Huang Chen, Wei Chen and Jiye Cai

Additional information is available at the end of the chapter

1. Introduction

The development of nanotechnology in photonics offers significant scientific and technological potentials for miniaturized photonic integrated circuits (PICs) [1,2]. It fosters the substantial efforts for exploring novel materials, developing easy fabrication techniques, reducing the size of photonic components, improving device integration density, and fabricating low-cost nanodevices. One-dimensional (1D) nanomaterials are promising candidates for photonics integration due to their intriguing optical, electronic, and mechanical properties [3-9]. In 2003, L. Tong et al. experimentally demonstrated low-loss optical waveguiding in silica micro/nanofibers (MNFs) with diameters far below the wavelength of the guided light [10], which renewed research interests in optical MNFs as potential building blocks for applications in high density and miniaturized PICs [11-22]. In particular, polymers have been widely regarded as useful materials for manipulating light in optical waveguide applications due to their good processability, biocompatibility, tunable properties, flexibility, and low cost for integration [15-22]. On the other hand, polymeric materials offer a unique opportunity to carry other chemical composition and to readily produce hybrid nanocomposites. 1D nanostructures fabricated from polymers have been the hot subject of recent research with regard to their special physical, chemical, electronic, and photonic properties [19,23-38]. Demonstration of viable polymer MNF based elements and architectures will require the development of reliable methods for the production of such structures with good control over critical parameters such as diameter, length, morphology, and chemical composition. Up to now, a bunch of new materials and new techniques have been developed to fabricate subwavelength and nanometer diameter polymer fibers, including electrospinning, phase separation, self assembling and one-step drawing, etc. [29].

Polymer MNFs with diameter down to several nanometers and length up to hundreds of millimeters have been reported. Besides its low cost and tiny size, polymer MNFs are easy to couple with other general optical devices, enabling them to fulfill their potentials in optical functions. Due to their good flexibility and large tunability, many shapes and structures can be achieved from polymer MNFs, including optical splitters [18,30], resonator [31], couplers [32], Mach–Zehnder Interferometers (MZI) [33], light-emitting polymer nanofibers [34], photodetector [35], organic nanofiber laser [36], sensors [37,38], polymeric nanofiber architecture [38].

In this chapter, we will introduce the fabrication techniques of polymer MNFs. Then we will focus on polymer MNF-based elements and assess their potential used as passive and active components in miniaturized photonic devices. Final is a perspective.

2. Fabrication methods of polymer MNFs

A number of different techniques including physical, chemical, thermal, and electrostatic method have been used to fabricate polymer MNFs. Ten of these fabrication methods are briefly described in the following section [29].

2.1. Electrospinning

Electrospinning is an electrostatically driven method of fabricating polymer nanofibers. Nanofibers are formed from a liquid polymer solution or melt that is feed through a capillary tube into a region of high electric field [39]. The electric field is most commonly generated by connecting a high voltage power source in the kilovolt range to the capillary tip (Fig. 1). As electrostatic forces overcome the surface tension of the liquid, a Taylor cone is formed and a thin jet is rapidly accelerated to a grounded or oppositely charged collecting target. Instabilities in this jet cause violent whipping motions that elongate and thin the jet allowing the evaporation of some of the solvent or cooling of melts to form solid nanofibers on the target site. Nanofiber size and microstructure can be controlled by several processing parameters including: solution viscosity, voltage, feed rate, solution conductivity, capillary to collector distance, and orifice size [40]. The electrospinning technique is very versatile and a wide range of polymer and copolymermaterials with a wide range of fiber diameters (several nanometers to several microns) can be fabricated using this technique. Many different types of molecules can be easily incorporated during the electrospinning fabrication process to produce functionalized nanofibers. Electrospun nanofibers are usually collected from an electrospinning jet as non-woven randomly or uniaxially aligned sheets or arrays.

2.2. Phase separation

Nanofibrous foam materials have been fabricated by a technique called thermally induced liquid-liquid phase separation [41]. This fabrication procedure involves (a) the dissolution of polymer in solvent (b) phase separation and polymer gelatination in low temperature (c) solvent exchange by immersion in water and (d) freezing and freeze-drying (Fig. 2). The morphology of these structures can be controlled by fabrication parameters such as gelatination temperature and polymer concentration. Interconnected porous nanofiber networks have been formed from polymers such as, poly-L-lactide acid (PLLA), poly-lactic-co-glycolic acid (PLGA), and poly-DL-lactic acid (PDLLA) with fiber diameters from 50–500 nm, and porosities up to 98.5%.

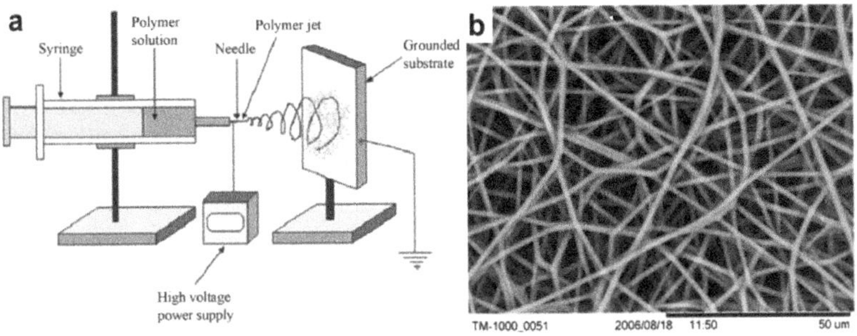

Figure 1. (a) schematic of a standard electrospinning setup [39] and a scanning electron microscope (SEM) image (b) of electrospun polyurethane nanofibers.

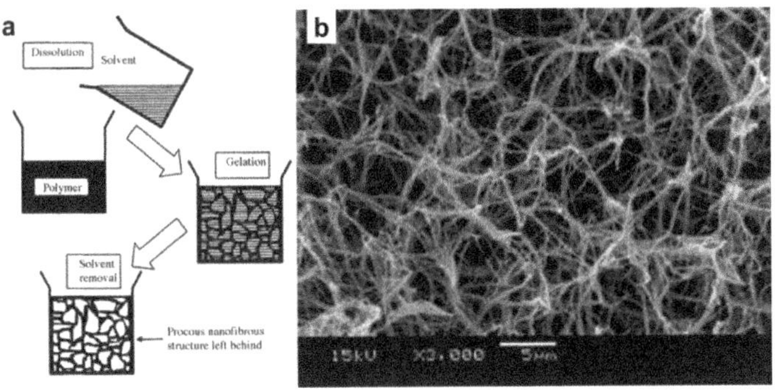

Figure 2. A schematic (a) of nanofiber formation by phase separation [42], and an SEM image (b) of nanofibrous structure fabricated by this technique [41].

2.3. Self assembly

Self assembly is a process by which molecules organize and arrange themselves into patterns or structures through non-covalent forces such as hydrogen bonding, hydrophobic forces, and electrostatic reactions. Dialkyl chain amphiphiles containing peptides were developed to mimic the ECM. These peptide amphiphiles (PA), derived from a collagen ligand, allow for a self assembling system that consists of a hydrophobic tail group and a hydrophilic head group [43]. Figure 3 shows the hierarchy of structure found in the self-assembled PA nanofiber networks [44]. The specific composition of amino acid chains in peptide amphiphile systems determines the assembly, chemical, and biological properties of the system, and therefore PA systems can be tailored to specific applications [45,46]. Nanofibers with diameters around 5–25 nm can be formed by the self assembly process. Cells can be encapsulated in a nanofibrous PA structureif they are added during the self assembly process and PA can also be injected in vivo where they subsequently self assemble into a nanofibrous network. It has been demonstrated that self assembled peptide nanofibers can spontaneously undergo reassembly back to a nanofibers scaffold after destruction by sonication, and after multiple cycles of destruction and reassembly, the peptide nanofibers scaffolds were still indistiquishable from their original structures [47].

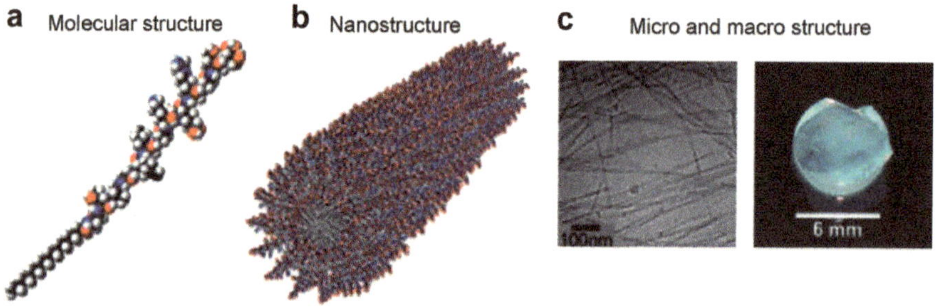

Figure 3. Schematics of the (a) molecular structure and (b) nanostructure, and images of the (c) micro and macro structure of a self assembling peptide amphiphile nanofiber network [44].

2.4. One-step drawing technique

Nanofibers can be mechanically drawn from viscous polymer liquids directly [48]. In one example, nanofibers were drawn directly when a rod was placed in a polymer melt and moved up forming a thin filament that cooled to form a nanofiber [17]. Figure 4 shows the schematic illustration of the drawing process. Fig. 4a shows a vertical direction tip-drawing process. Poly(trimethylene terephthalate) (PTT) pellets (melt temperature T_m = 225°C) was melt by a heating plate and thetemperature was kept at around 250°C during the wire drawing. First, an iron or silicarod/tip with radius of about 125 μm is being approached and its tip is immerged into the molten PTT. Then the rod tip is retracted from the molten PTT with a speed of 0.1–1 m/s, leaving a PTT wire extending between the molten PTT and the tip. The extended PTT wire is quickly quenched in air and finally, a naked amorphous PTT nanowire is formed. PTT nanofibers with diameters as low as 60 nm, and lengths up to 500 mm have been achieved. A SEM image (Fig. 4b) shows part of the coiled nanofiber with a length of about 200 mm and an average diameter of 280 nm. The diameter variation ratio is about 8.4×10^{-8}. Figure 4c demonstrates flexible and elastic connection by pulling the polymer nanofibers with diameters of 140 and 170 nm. To examine surface roughness of the polymer nanofibers, high-magnification transmission electron microscope (TEM) was done. Figure 4d shows a TEM image of a 190-nm-diameter nanowire, indicating no visible defect and irregularity on the surface of the polymer nanofiber. Typical average sidewall root-mean-square roughness of the polymer nanofiber is 0.28 nm. The electron diffraction pattern (inset of Fig. 4d) demonstrates that the obtained PTT MNF is amorphous. The results demonstrate that the obtained polymer nanofibers exhibit high surface smoothness, length uniformity, high mechanical properties, and excellent flexibility.

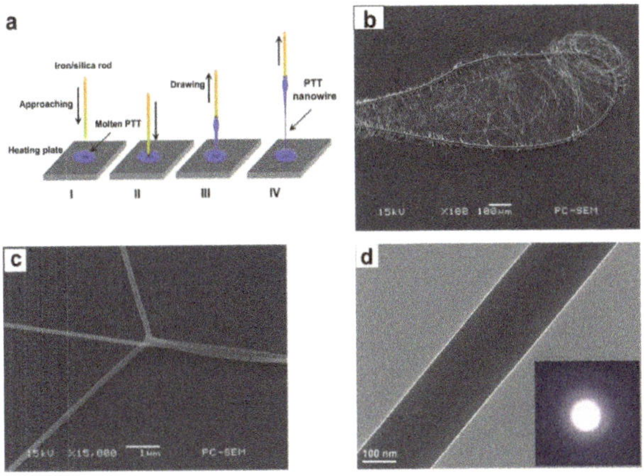

Figure 4. Schematic of nanofiber fabrication by the drawing technique. (b) SEM images of a nanofiber with average diameter of 280 nm coiled on a 12-μm-diameter PTT bending rod, the length of the nanofiber displayed is about 200 mm. (c) Flexible and elastic enough nanofiber connection with diameters of 140 and 170 nm. (d) Transmission electron microscope (TEM) image of a 190-nm-diameter fiber. The insetshows its electron diffraction pattern [17,18].

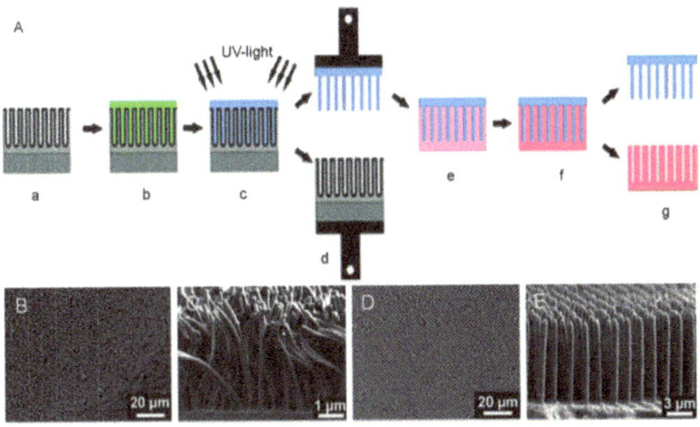

Figure 5. (A) Schematic of the fabrication of polymer nanofibers using a nondestructive templating technique (grey: alumina template, green: resin, blue: polymer nanofibers, pink: silica replica template. (B-E) SEM images of 120 nm (b&c) and 1μm (d&e) polymer fibers fabricated by the above technique [49].

2.5. Templating

Polymer nanofibers can be fabricated using templates such as self-ordered porous alumina. Alumina network templates with pore diameters from 25 to 400 nm, and pore depths from around 100 nm to several 100 μm have been be fabricated. Polymer nanofiber arrays can be released from these molds by destruction of the molds or mechanical detachment (Fig. 5) [49,50]. The length of polycaprolactone (PCL) nanofibers fabricated from alumina templates can be controlled as a function of parameters such as melt time and temperature [51].

2.6. Vapor-phase polymerization

Polymer MNFs have also been fabricated from vapor-phase polymerization. Plasma-induced polymerization of vapor phase vinyltrichlorosilane produced organosiloxane fibers with diameters around 25 nm and typical lengths of 400–600 nm and cyanoacrylate fibers with diameters from 100 to 400 nm and lengths of hundreds of microns (Fig. 6) [52,53].

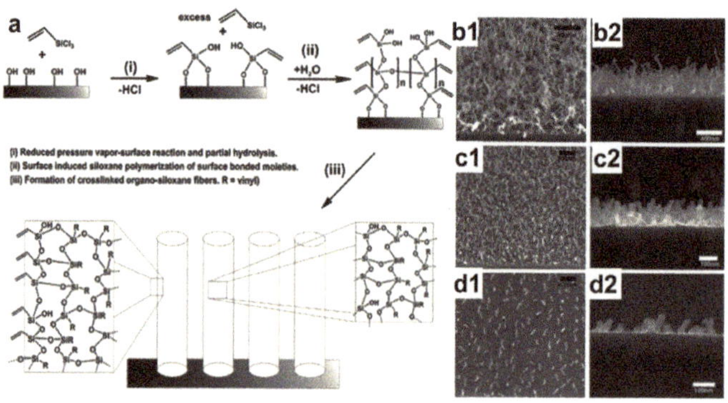

Figure 6. (a) Schematic describing a proposed mechanism for nanofiber formation by vapor-phase polymerization (b) Arial (1) and side views (2) of polymer nanofibers fabricated from vapor-phase polymerization at high (b), intermediate (c) and low (d) packing densities [52].

2.7. Extraction

Nanofibers can be extracted from natural materials using chemical and mechanical treatments. Cellulose fibrils can be disintegrated from plant cell walls. In one example, cellulose nanofibers were extracted from wheat straw and soy hull with diameters ranging from 10 to 120 nm and lengths up to a few thousand nanometers (Fig. 7) [54]. Invertebrates have also been used as a source for the extraction of nanofibers. Chitin nanofibers 3–4 nm in diameter and a few micrometers in length were extracted from squid pen and Poly-N-acetyl glucosamine nanofibers isolated from a marine diatom demonstrated prothrombotic interactions with red blood cells [55,56].

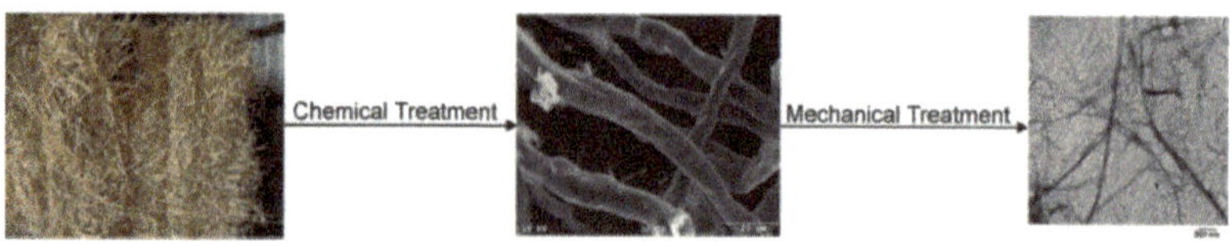

Figure 7. Images of natural wheat straw [57], wheat straw microfibers [54] after chemical treatment and wheat straw nanofibers [54] after chemical treatment.

2.8. Conventional chemical oxidative polymerization of aniline

Chemical oxidative polymerization of aniline is a traditional method for synthesizing polyaniline and during the early stages of this synthesis process polyaniline nanofibers are formed (Fig. 8). Optimization of polymerization conditions such as temperature, mixing speed, and mechanical agitation allows the end stage formation of polyaniline nanofibers with diameters in the range of 30–120 nm [58,59].

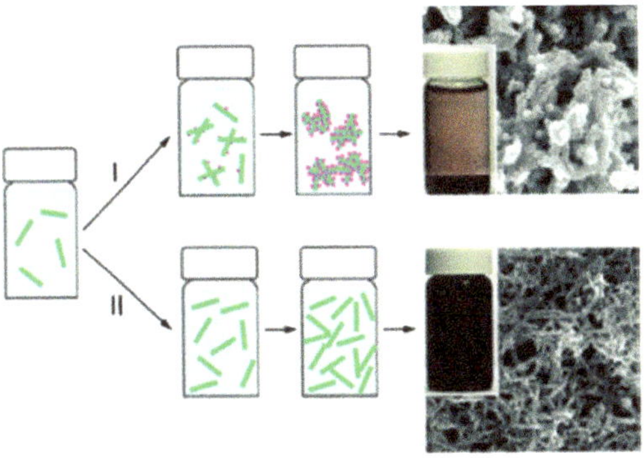

Figure 8. Schematic showing the nucleation of polyaniline nanofibers [59]. (I) Under non-ideal nucleation conditions aggregate formation is present. (II) When ideal nucleation conditions are predominant, well-dispersed polyaniline nanofibers are formed. Typical images of the reaction vials and microstructure are displayed next to the schematic.

2.9. Bacterial cellulose

Cellulose nanofibers produced by bacteria have been long used in a variety of applications, including biomedical applications [60]. Cellulose synthesis by Acetobacter involves the polymerization of glucose residues into chains, followed by the extracellullar secretion, assembly and crystallization of the chains into hierarchically composed ribbons (Fig. 9). Networks of cellulose nanofibers with diameters less than 100nmare readily produced, and fibers with different characteristics may be produced by different strains of bacteria [49]. Copolymers have been produced by adding polymers to the growth media of the cellulose producing bacteria [50,51].

2.10. Kinetically controlled solution synthesis

Nanofibers and nanowires have been fabricated in solution using linear aligned substrates as templating agents such as iron-cation absorbed reverse cylindrical micelles and silver nanoparticles [61]. Poly(vinyl alchohol)-poly(methyl methacrylate) nanofibers were fabricated using silver nanoparticle that were linearly aligned in solution by vigorous magnetic stirring (Fig. 10) [62]. These nanoparticle chain assemblies acted as a template for further polymerization of nanofibers with diameters from 10 to 30 nm and lengths up to 60 μm.

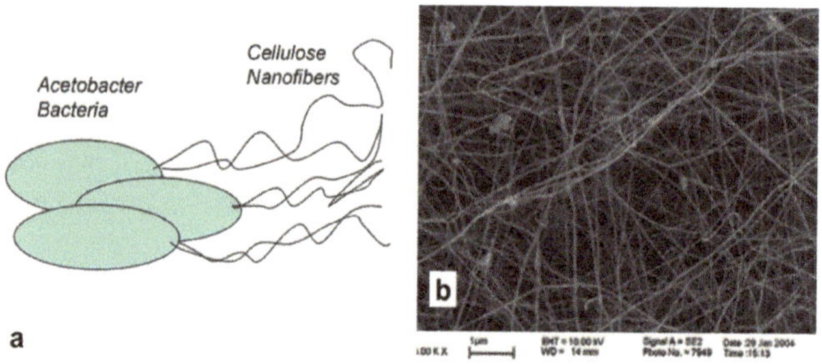

Figure 9. (a) Schematic of Acetobacter cells depositing cellulose nanofibers, (b) an SEM image of a cellulose nanofiber mesh produced by bacteria [60].

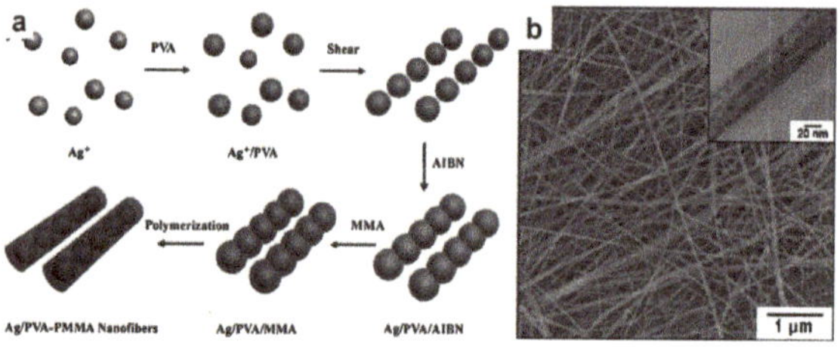

Figure 10. (a) Schematic of silver nanoparticle embedded polymer nanofibers fabrication (b) SEM and TEM images of silver nanoparticle embedded Polymer nanofibers [62].

3. Polymer MNF based photonic components and devices

Optical fiber based components and devices have been very successful in the past 30 years and will surely continue to thrive in a variety of applications including optical communications, optical sensing, power delivery and nonlinear optics [63-65]. With increasing requirements for higher performance, wider applicability and lower energy consumption, there is a strong demand for the miniaturization of fiber-optic components or devices. When operated on a smaller spatial scale, a photonic circuit can circulate, process and respond to optical signals on a smaller time scale. Only at wavelength or subwavelength size does the photonic structure manifest evident near-field features that can be utilized for interlinking and processing optical signals highly efficiently. For example, it was estimated that to reach an optical data transmission rate as high as 10 Tb/s, the size of photonic matrix switching devices should be reduced to 100-nm scale [66]. At the same time, to perform a given function that relies on a certain kind of light-matter interaction, usually less energy is required when smaller quantities of matter are involved. MNFs featured at subwavelength scale, provide a number of interesting properties such as mechanical flexibility, high optical cross-sections, large and ultrafast nonlinear responses, and broad spectral tunability that are highly desirable for functionalizing high density and miniaturized PICs [67]. Besides its low cost and tiny size, MNFs are also easy to couple with other general optical devices, providing excellent compatibility with standard optical fiber systems.This section gives an up-to-date review of polymer MNF based photonic components/devices that have been investigated very recently.

3.1. Polymer MNF-based passive components/devices

3.1.1. M×N coupling photonic splitters and sensors

X. Xing et al. have demonstrated PTT nanofibers with high surface smoothness, diameter uniformity, as well as high mechanical strength and excellent flexibility, make them promising candidates for building blocks to construct ultracompact photonic devices and device arrays [17,18]. Fig.11a shows a 4×4 photonic coupling splitter assembled by twisting four PTT nanofibers with diameters of 450, 450, 510, and 570 nm for branches A to D [18]. The inset of Fig. 11a shows that the coupling section is composed of a 3×4 and a 1×4 couplers, where the total width of the coupling section is 1.98 μm. The maximum length of the coupling regionis about 16.1 μm, and that for the 1×4 splitter is about 8.5 μm. When two red lights (650 nm) are simultaneously launched into the branches A and B, the measured splitting ratio is 29:18:29:24. As shown in Fig. 11b, blue light (532 nm) are coupled into the branchG of an 8×8 coupling splitters and divided into branches 1 to 8.The coupling section of the splitter (Fig. 11b, inset) is 38-μm-long and 2.5-μm-wide. We also launched other visible lights into the devices to observe the splitting phenomenon. Experimental demonstration shows that the properties of the splitters are dependent on the operation wavelength and the input branch which the optical signal launched into. For a fixed operation wavelength and the input branch, desirable splitting ratio can be tuned by controlling input/output branching angle.

As shown in Fig. 11c–e, a tunable refractive index sensor with ultracompact structure in a 2×2 PTT nanofiber coupling splitter assembled by twisting two flexible PTT nanowires with diameter of 400 nm [19]. The sensor consists of two input branches, a twisted coupling region, and two output branches. The changes of optical power caused due to variations in the surrounding medium around the twisted coupling region were measured in the output branches.The highest measured sensitivity of the sensor is 26.96 mW/RIU (refractive index unit) and the maximum detection limit on refractive index change is 1.85×10^{-7}.

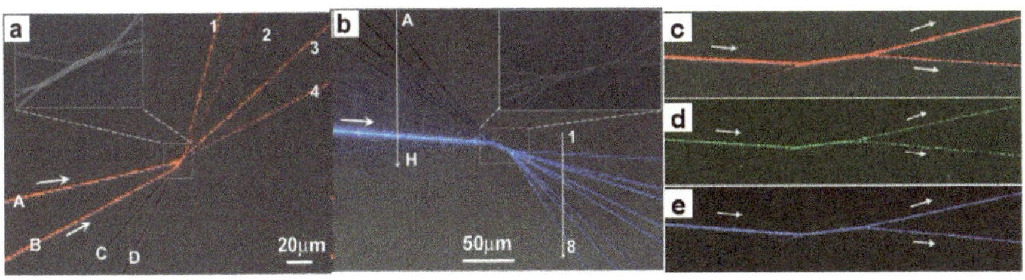

Figure 11. Optical microscope images [18,19]: (a) A 4×4 photonic coupling splitterwith diameters of 450, 450, 510, and 570 nm for branches A to D when red light (650 nm) was launched into branches of A and B. (b) An 8×8 photonic coupling splitter of guided blue light (532 nm) with diameters of 400, 400, 400, 400, 400, 750, 750, and 600 nm for branches A to H. The inset of Fig. a and b shows the coupling section of devices. (c–e) the sensor (400-nm-diameter) taken with (c) red light, (d) green light (532 nm), and (e) blue light (473 nm) (without sample solution).The white arrows show the propagation directions of the launched lights.

3.1.2. Cascaded MZI[21]

Figure12 shows an optical microscope image of the assembled two-cascaded MZI (wire diameter, 900 nm) [21]. The inset (a) showsa scanning electron microscope image of MZI 1 and the inset (b) shows guided red light (650 nm) in the cascaded MZI. Themeasured insertion loss is about 0.94 dB for the red light. The total length of the cascaded MZI is 327 μm. The width and length of each bow-shaped MZI are 32 μm and 121 μm, respectively. According to the analysis, to get coupling ratios of 0.147, 0.501, and 0.147, the lengths of the couplers C_1, C_2, and C_3 are 27, 31, and 27 μm, respectively. The estimated totalpath-length difference is 40 μm. The bright spot in the inset (b) of Fig. 12 is the scattering spot of the input light at the end of the tapered fiber I. The average bandwidth of 3-dB pass-band is measured to 33 nm for the cascaded MZI over the wavelengthsof 1.3 to 1.6 μm. The measured extinction ratio for the cascaded MZI is 16 to 19 dB, and optical insertion loss is 1.1 to 1.8 dB at wavelengths of 1.3 to 1.6 μm.This is good for band-pass filter applications.

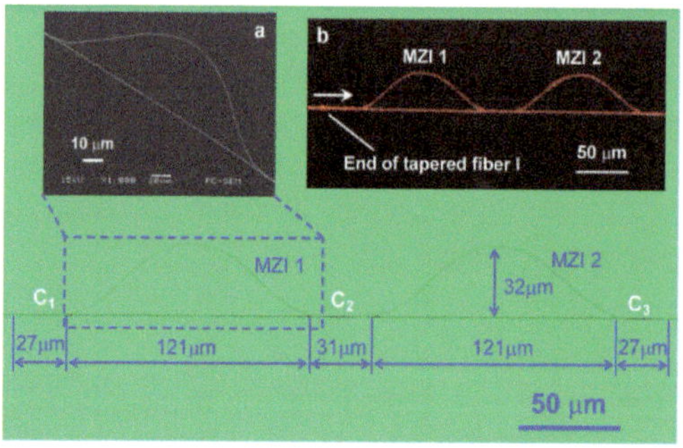

Figure 12. Optical microscope image of the assembled cascaded MZI (wire diameter, 900 nm) [21]. Inset (a) shows the scanning electron microscope image of bow-shaped MZI 1 and inset (b) shows the optical microscope image of guided red light (650 nm) in the cascaded MZI. Estimated path-length difference is 40 μm. The white arrow in the inset (b) indicates propagation direction of the light.

3.1.3. Ring resonator[31]

Owing to relatively high quality (Q) factor and ease off abrication, fiber-based microrings were widely used for resonators. Y. Wang et al. have assembled microring resonators by knotting a uniform PTT wire with micromanipulator assistance under an optical microscope. Subsequently, two ends of the microring resonator were fixed on two branches of a tunable microstage as schematically shown in Fig. 13a. Then an input microtaper and an output microtaper were used to launch and collect optical signals, respectively, by evanescent wave coupling. In Fig. 13a, when we adjust the tunable microstage to the right, the wire will be pulled tightly and finally, the ring radius will become smaller. The red arrows in Fig. 13a show the propagation directions of the optical signals while the yellow one indicates the moving direction of the tunable stage. As an example, Fig. 13b shows a SEM image of the microring with aradius of 70 μm, which was assembled by the PTT wire with adiameter of 3.5 μm.

Three tunable microrings were assembled by PTT wire withdiameters of 3.5, 2.5, and 2.0 μm, respectively. Ring radius and wire diameter dependent optical properties were demonstrated by measuring the FSRs and the Q factors and showed that, the maximum Q factors are 28090, 28071, and 23528 at the ring radii of 417 (wire diameter, 3.5 μm), 258 (wire diameter, 2.5 μm), and 204 (wirediameter, 2.0 μm) μm, respectively. Their corresponding FSRs are 0.55, 0.92, and 1.19 nm. The FSR is mainly dependent on the ringradius and little affected by the wire diameter. By using anappropriate wire diameter with an appropriate ring radius, amaximum Q can be obtained with a desirable FSR. The tunable microring resonators would be useful for optical filters and sensors. Also, the assembly method used in this work could become acandidate for fabricating highly-integrated photonic devices basedon micro/nano-wire rings.

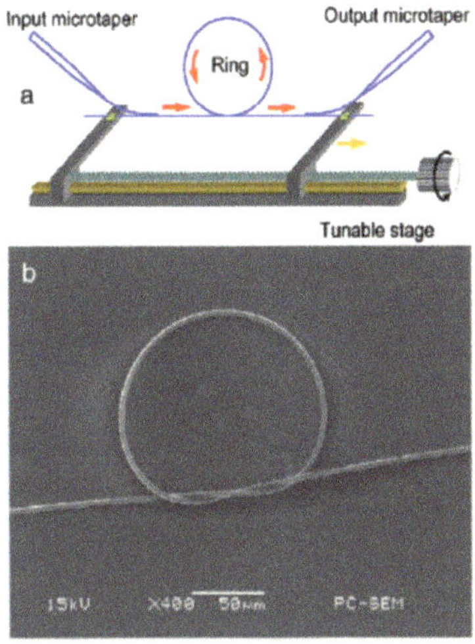

Figure 13. (a) Schematic illustration of the experimental setup. The red arrows show propagation directions of the optical signals whereas the yellow one indicates moving direction of the tunable stage. (b) SEM image of a microring assembled by the PTT wire. The wire diameter is 3.5 μm and the ring radius is 70 μm [31].

3.1.4. Arbitrary and vertical optical couplers using flexible polymer MNFs[32]

The polymer MNFs were directly drawn from a polymethyl methacrylate (PMMA) solution in which acetone was used as a solvent. To investigate the possibility of arbitrary and vertical optical couplers, different topologies were assembled. Figures 14a, b show the optical microscope images of two crossed interconnect structures with different cross angles. The structure of Fig. 14a as assembled by two polymer fibers with diameters of 1.05 μm (wire 1) and 1.00 μm (wire 2). Cross angle between channels C_1 and C_2 is 57° while the cross angle between channels C_3 and C_4 is 42°. Figure 14a shows that a light (wavelength λ=650 nm) was launched into channel C_1 and divided into channels C_2, C_3, and C_4. The output powers were collected using the tapered fibers. For this structure, 4% optical power was divided into channel C_2. The powers divided into channels C_3 and C_4 are 41% and 42%, respectively. The total insertion loss is 0.6 dB. Figure 14b further shows a crossed interconnect structure, which was assembled by two wires with diametersof 1.18 μm (wire 1) and 1.40 μm (wire 2). The cross angle between channels C_1 and C_2 is 55° and that between channels C_3 and C_4 is 89°. For this structure, when the light (λ=650 nm) was launched into channel C_1, the measured powers at channels C_2, C_3, and C_4 are 1%, 39%, and 42%, respectively. The total insertion loss is 0.86 dB. Figure 14c shows the optical microscope image of the star topology network with five channels. It was assembled by twisting three wires with diameters of 1.08 μm (wire 1), 917 nm (wire 2), and 970 nm (wire 3). The cross angle between channels C_1 and C_2 is 29°, that between channels C_2 and C_3 is 57°, between channels C_3 and C_4 is 110°, and between channels C_4 and C_5 is 35°. When the light (λ = 650 nm) was launched into channel C_2, the measured powers in the channels C_3, C_4, and C_5 are 1%, 40%, and 36%, respectively. The total insertion loss is 1.14 dB. Figure 14d shows a tree topology network, which was assembled by three wires with diameters of 1.17 μm (wire 1), 1.37 μm (wire 2), and 1.20 μm (wire 3). The cross angle between channels C_1 and C_3 is 87° and that between channels C_4 and C_5 is 39°. When the light (λ = 650 nm) was launched into channel C_2, it was dividedinto channels C_1 and C_3 by the first node A with a splitting power ratio of 0.41:0.40. At the second node B, the light inchannel C_3 was divided into channels C_4 and C_5 with a splitting power ratio of 0.16:0.17. The total insertion loss for this structure is 1.31 dB. By using the structures shown in Figs. 14a–c, optical channels can be further increased while keeping one node. This is very useful for ultracompact arbitrary and vertical optical interconnect structures. We hope that the arbitrary and vertical coupling structures could find applications in high-density PICs and miniaturized optical networks.

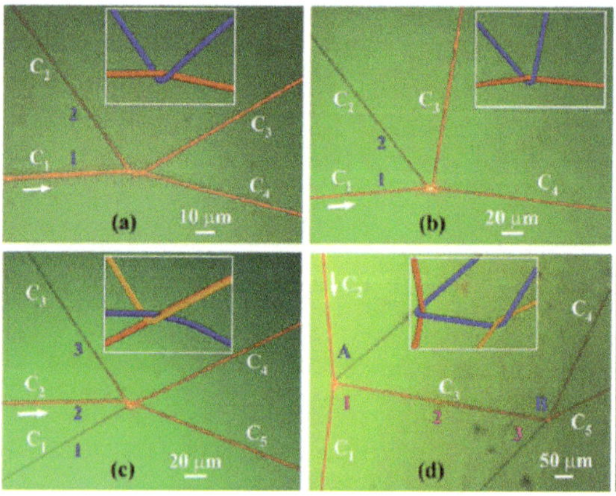

Figure 14. Optical microscope images of different arbitrary and vertical optical couplers [32]. The insets schematically represent respective structures,and the white arrows indicate the transmission direction of the guided light. (a) Optical microscope image of a crossed interconnect structure assembled by wires 1 and 2 with diameters of 1.05 and 1.00 μm, respectively. (b) Optical microscope image of a crossed interconnect structure assembled by wires 1 and 2 with diameters of 1.18 and 1.40 μm,respectively. (c) Optical microscope image of a star topology network assembled by three wires with diameters of 1.08 μm (wire 1), 917 nm (wire 2), and 970 nm(wire 3). (d) Optical microscope image of a tree topology network assembled by three wires with diameters of 1.17 μm (wire 1), 1.37 μm (wire2), and 1.20 μm (wire 3).

3.2. Polymer MNFs as active waveguides

The development of miniaturized PICs are challenging directions in next generation all-optical signal processing, in which light-emitting sources are important elements for integration [68,69]. In the past decades, doped glass optical fibers have been widely used as solid hosts for photonic applications such as tunable lasers and amplifiers for optical communication [70,71]. For most widely used rare-earth-doped glass fibers, the concentrations of rare-earth dopants are limited by the concentration quenching, for example, typically lower than 1% for Er^{3+} in silica fibers, resulting in relatively low energy conversion efficiency within a limited length. Therefore, it is difficult to obtain efficient energy conversion in doped glass fibers within a scale comparable to the compactness of a nanophotonic integrated system. Compared to those of glass fibers, the most attractive prospects of polymer MNFs are as follows: First, the polymer matrix can host functional dopants ranging from metal oxides, quantum dots (QDs), and fluorescent dyes to enzymes that can be used to tailor the properties of the MNFs with greater versatility. Second, the dopants can be easily doped into the solvated polymer and drawn into polymer MNFs at room temperature with higher doping concentration than that in glass fibers. Also, the mechanical flexibility, perm-selective nature to gas molecules, biocompatibility, easy processing, and low cost of the polymer materials [72] offers more opportunities for doped polymer MNFs over doped glass fibers in nanophotonic systems. Especially, high efficient light emission dopants (e.g., fluorescent dyes and QDs), which are much more compatible with polymer MNFs, show great potential to realize compact light-emitting sources with feature sizes acceptable in nanophotonic integrated systems.

3.2.1. Fluorescent dye doped polymer MNFs

In 2010, F. Gu et al. reported light-emitting polymer nanofibers based on waveguiding excitation [34]. By waveguiding excitation light along the polymer nanofiber, the interaction of light with polymer nanofiber is enhanced over 3 orders of magnitude compared with the currently used irradiating excitation. Intriguing advantages such as enhanced excitation efficiency, low excitation power operation down to nW levels, tightly confined excitation with low cross talk, and high photostability of the light-emitting polymer nanofibers are obtained. The waveguiding excitation allows incorporation of various fluorescent dyes into polymer nanofibers to generate multicolor emitting sources covering the entire visible spectrum.

Figure15a shows a photoluminescence (PL) microscope image of a 380-nm-diameter 520-μm-length RhB-PS nanofiber taken with a long-pass emission filter [34]. Upon 473-nm laser (λ_{ex}) launched from the left side with excitation power (P_{ex}) of 70 nW, bright fluorescent emission with a peak (λ_{em}) around 578 nm is generated and guided along the nanofiber. For reference, Fig. 15b shows the PL image taken without the emission filter, in which a small light spot at the output end of the nanofiber and no obvious scattering along the nanofiber are observed, suggesting that the excitation light was efficiently absorbed during its waveguiding excitation. The absorption spectrum of the RhB-PS nanofiber in Fig. 15c (black line) exhibits a peak at 560 nm due to the absorption of the doped RhB molecules. The measured absorption coefficient α of the RhB-PS NFs at 473 nm is ~50 cm^{-1}.

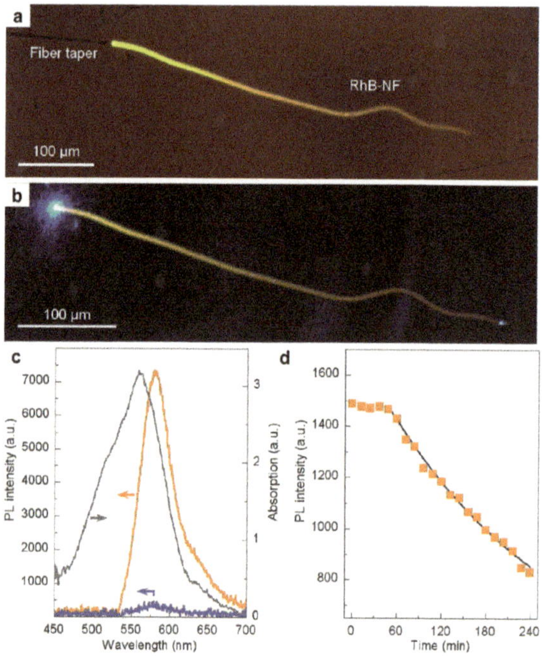

Figure 15. PL microscope images of a 380-nm-diameter 520-μm-length RhB-PS nanofiber excited by 473-nm light from the left side at P_{ex} = 70 nW, taken with (a) and without (b) the long-pass emission filter. (c) PL spectra of the RhB-PS nanofiber under 70-nW waveguiding (orange line) and 3-μW irradiation (blue line) schemes. The absorption spectrum of the NF is also provided (black line). (d) PL intensity of the RhB-PS nanofiber at 579 nm with P_{ex} = 30 nW as a function of time [34].

3.2.2. *Quantum-dot-doped Polymer MNFs*

H. Liu et al. incorporated semiconductor QDs as integrated light sources into the polymer fiber to produce subwavelength-diameter optical waveguides [16]. Fibers of polymeric photoresist were fabricated with diameters down to 50 nm and lengths up to tens of millimeters using an electrospinning method, and QDs consisting of CdSe/ZnS (core/shell) were incorporated into the polymer fibers. A fluorescence optical image (Fig. 16a), taken by uniformly illuminating the fiber by a mercury lamp light source and imaging the 605 nm wavelength fluorescence, shows the distribution of luminescence along the fiber length. The image suggests that the QDs were not uniformly distributed along the fiber. Nanometer-sized QDs were clearly recognizable, and were observed to be concentrated in a constrained region as a strand parallel to the fiber axis (Fig. 16b).

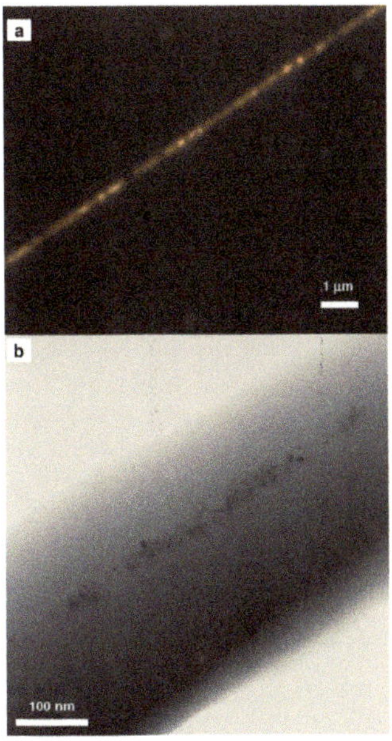

Figure 16. QD distribution in the SU8 nanofiber [16]: a) Fluorescence image of ensembles of QDs embedded in an SU8 nanofiber, showing the distribution of luminescence along the fiber length. b) TEM image of QD distribution inside the SU8 nanofiber.

3.2.3. QD-Decorated Polymer MNFs

H. Yu et al.developed a one-step process to decorate PTT nanofiber with the CdSe/ZnS core/shell QDs [73]. The QDs-decorated nanofibers with diameters of 400–800 nm can be used as active subwavelength waveguides with some advantages such as good photostability, low excitation power operation (less than 0.2 μW), low propagation loss (11.3 dB/cm), low absorption coefficient (down to 2.6 cm^{-1}), and 200 times enhancement in excitation efficiency excited by the evanescent waveguiding excitation than that of by the irradiation excitation.

Figure 17a shows an optical microscope image of adrop of a rugby-ball-like QDs-decorated cross structure, composing of a PTT nanofiber 1 (495 nm) and a PTT nanofiber 2 (590 nm), excited by the EW excitation from laser B at P_{ex} = 0.1 μW. It can be seen that red light was excited in the rugby-ball-like QDs. Figure 17b shows the optical microscope image of the cascaded rugby-ball-like QDs-decorated nanofiber with diameter of 597 nm, in which there are seven rugby-ball-like QDs. The lengths of the QDs B_1 to B_7 are 2.3, 2.2, 2.3, 2.2, 1.8, 2.0, and 2.0 μm, respectively. Their corresponding maximum widths are 1.6, 1.5, 1.6, 1.6, 1.0, 1.1, and 1.1 μm. The average length of the rugby-ball-like QDs is 2.1 μm and the average value of maximum width is 1.4 μm. Figure 17c shows the corresponding optical microscope image with seven red spots excited by the 473 nm blue light fromlaser B at P_{ex} = 1.2 μW.

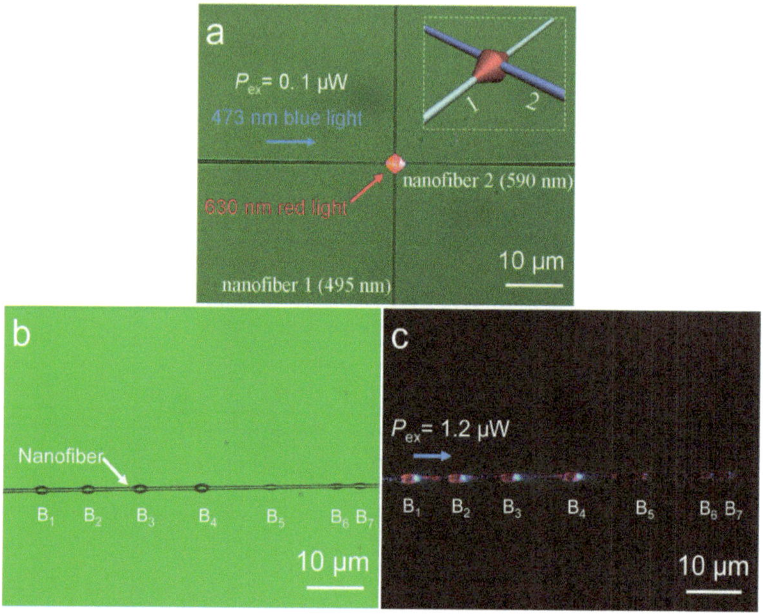

Figure 17. Optical microscope images of nanofibers with a rugby-ball-like QDs-decorated cross structure and a cascaded rugby-ball-like QDs-decorated structure [73]. (a) Crossed structurewith a guided 473 nm blue light in nanofiber 2 and excited 630 nm red light at the cross junction. The inset schematically representsthe cross structures. (b) Cascaded rugby-ball-like QDs-decorated structure. The lengths of the QDs B_1 to $B_7$are 2.3, 2.2, 2.3, 2.2, 1.8, 2.0, and 2.0 μm, respectively. The corresponding maximum widths are 1.6, 1.5, 1.6, 1.6, 1.0, 1.1, and 1.1 μm. (c) Rugby-ball-like QDs-decorated fiber excited by 473 nm blue light at P_{ex} = 1.2 μW. The blue arrow shows the direction of propagation of the input blue light.

3.3. Polymer nanofiber laser

3.3.1. Dye Doped Polymer MNF Knot Resonator Laser

Recently, many kinds of microsized polymer fiber and dye doped polymer fiber were also successfully demonstrated. Q. Song et al. reported the lasing action in a dye doped PNF knot ring resonator [74]. The dye doped polymer nanofiber was fabricated by general fiber drawing technique. Then it was bent to a knot ring resonator and fixed onto glass tubes (the inset of Fig. 18a), which can increase the stability of the knot resonator and fix its diameter [74]. Laser emission was observed. The emission spectrum of knot resonator is shown in Fig. 18b. Here the pump intensity is 4.2 mJ/cm^2. Periodic peaks can be observed in the laser spectrum. This group of peaks is considered to come from the whispering galley mode resonance inside the ring resonator. The linewidth of laser peaks is only 0.07 nm and the mode spacing in Fig. 18 is about 0.206 nm.

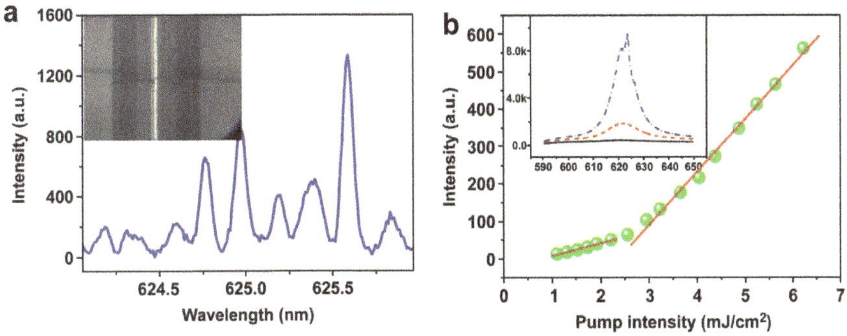

Figure 18. Laser spectrum of dye doped polymer fiber [74].

3.3.2. Random lasering in a singleorganic Nanofiber

One-dimensional light random lasing in individual p-sexiphenyl nanofibers (p-6P) is investigated by F. Quochi et al [36]. The laser action happens in single p-6P nanofibers grown on (001)-oriented muscovite mica. Isolated nanofibers are shown to yield low-threshold random laser emission in the deep blue. Random lasing from isolated nanofibers starts at pump fluences on the order of 10 μJ/cm^2 per pulse. Lasing nanofibers are demonstrated. Figure 19a shows a lasing micrograph taken in imaging mode slightly above threshold. It displays both lasing from vertically aligned nanofibers and spontaneous emission from a set of neighboring nanofibers oriented approximately at 60° with respect to the vertical axis of the detection system. The latter faintly appear in the lower part of the graph. Scattering of the lasing emission into out-of-plane directions does not take place homogeneously along the nanofibers' axis; conversely, scattering is highly spotted, indicating that wave-guiding is interleaved with light scattering and outcoupling at special sites along the fibers. Emission spectra relating to the ~100 μm long nanofiber placed at the center of the imaging field of view in Fig. 19a are reported in Fig. 19b. They refer to the emission spatially integrated over the whole nanofiber length. Below threshold, spontaneous emission exhibits a broad vibronic progression with 0–1 and 0–2 emission bands peaked near 425 and 450 nm, respectively.

3.3.3. Optically pumped lasing in single conjugated polymer nanowires

Conjugated polymers have chemically tuneable opto-electronic properties and are easily processed, making them attractive materials for photonics applications [75,76]. Conjugated polymer lasers, in a variety of resonator geometries such as microcavity, micro-ring, distributed feedback and photonic bandgap structures, have been fabricated using a range of coating and imprinting techniques [77-80]. D. O'Carroll et al. reported the first observation of optically pumped lasing in single conjugated polymer nanowires [81]. The waveguide and resonator properties of each wire are characterized in the far optical field at room temperature. The end faces of the nanowire are optically flat and the nanowire acts as a cylindrical optical cavity, exhibiting axial Fabry–Pérot mode structure in the emission spectrum. Above a threshold incident pump energy, the emission spectrum collapses to a single, sharp peak with an instrument-limited line width that is characteristic of single-mode excitonic laser action. Fluorene-based conjugated polymers are attractive.

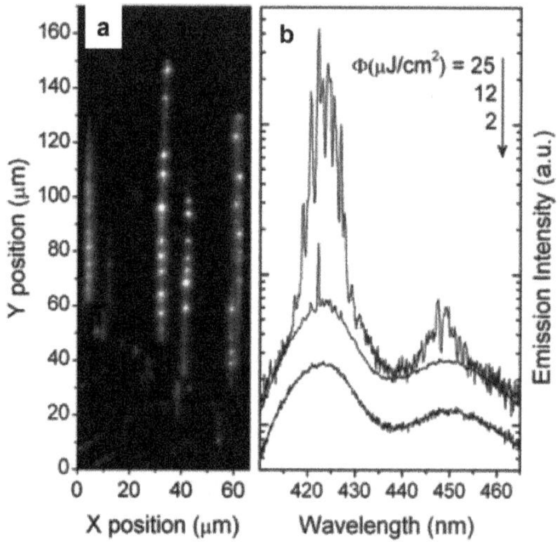

Figure 19. (a) Gray-scale optical emission intensity image of lasing and luminescent *p*-6P nanofibers excited at a pump fluence (Φ) of 15 $\mu J/cm^2$ per pulse. The gray-level scale is logarithmic. The *y* coordinate refers to the position along the vertical direction, which is parallel tothe input slit of the detection system. (b) Emission intensity spectra of the nanofiber positioned at $x \approx 30$ μm in panel a and extending vertically from $y \approx 50$ μm to $y \approx 150$ μm for different values of the pump fluence. Note that the threshold fluence is lower than 12 $\mu J/cm^2$ per pulse [36].

Fluorene-based conjugated polymers are attractive photonic materials because they exhibit high photoluminescence quantum efficiencies, large stimulated emission crosssections and chemically tuneable emission wavelengths [82]. Poly(9,9-dioctylfluorene) (PFO) is a prototypical main-chain liquid-crystalline homopolymer that emits in the blue and exhibits polymorphic behaviour, with striking implications for its photophysical properties [83]. Isolated PFO nanowires were uniformly excited by the 355-nm output of a 0.7 ns, 1.25 kHz pulsed Nd/YVO laser and spatially resolved photoluminescence spectra were acquired from the bodies and tips (Fig. 20a). To confirm that the wires operated as axial Fabry–Pérot microcavities, the mode spacing at 460 nm was plotted versus the inverse nanowire length for 14 wires, and was shown to exhibit a linear dependence (Fig. 20b). The number of Fabry–Pérot modes per guided mode that could propagate in a nanowire microcavity was estimated from $\Delta\lambda_{spont}/\Delta\lambda_m$, where $\Delta\lambda_{spont}$ is the full-width-at-half-maximum, FWHM, of the spontaneous emission and is ~19 nm.

Single isolated PFO nanowire microcavities were then uniformly excited (355 nm, 0.7 ns, 1.25 kHz) and tip emission spectra were collected as a function of pump energy (Fig. 21a).

At lower pump energies, tip spectra exhibited nanowire microcavity emission, with pronounced Fabry–Pérot modes apparent at the 0–1 peak. Above an energy threshold of ~100 nJ (2.8 mJ cm^{-2}), a single spectrally narrowed emission peak developed by preferential gain in a single Fabry–Pérot mode and the onset of lasing. A slight blue shift in peak position with increasing pump energy suggested that stimulated emission occurred on timescales comparable to or faster than exciton energy migration. Concerning the dependence of tip emission intensity on pumpenergy, below the energy threshold, emission increased linearly with excitation energy (Fig. 21b). Above threshold, a kink in emission output was followed by a super-linear increase due to optical gain. Also, the emission peak width narrowed from 19.6 nm to 1.4 nm (instrument-limited) at threshold, indicating the high quality factor of the nanowire cavity.

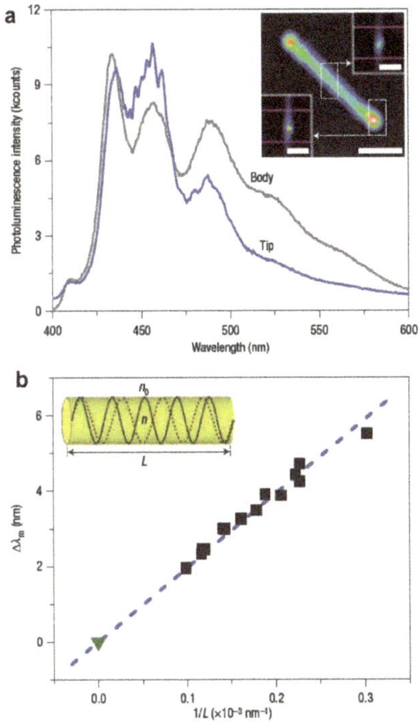

Figure 20. Microcavity effects in single PFO nanowires [81]. a, Emission spectra collected from the tip (blue) and body (grey) of an isolated PFO nanowire under uniform pulsed excitation (1.4 nJ). Inset: emission image of an excited wire (1nJ) with spatially filtered emission images of the tip and body locations from which spectral data were acquired; mauve lines indicate the area over which each spectrum was integrated. Scale bars, 2 μm. b, Plot of mode spacing measured at 460 nm versus inverse nanowire length for 14 different nanowires. Black squares, experimental data points; green triangle, extrapolation to infinite length; dashed blue line, fit of Fabry–Pérot equation to data for λ =460 nm and $[n - \lambda(dn/d\lambda)] = 5.4$. Inset: schematic depiction of a nanowire with well-defined end facets acting as a Fabry–Pérot microcavity.

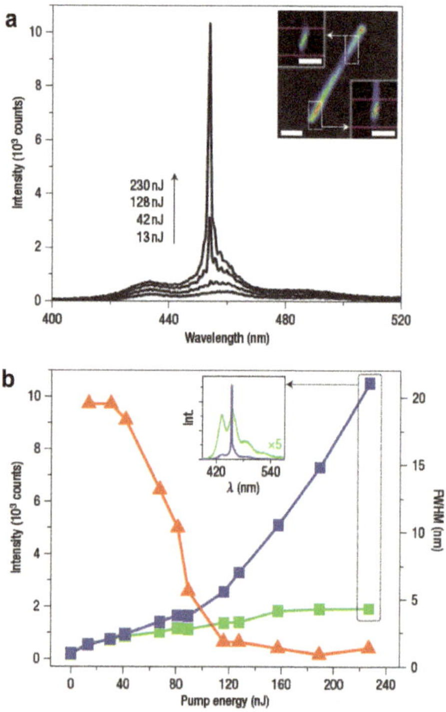

Figure 21. Optically pumped single PFO nanowire laser [81]. a, Emission spectra collected from the tip of an isolated PFO nanowire under uniform excitation, as a function of increasing pump energy at room temperature. Inset: emission image of the wire (1.3 nJ) along with spatially filtered emission images of the tip and body locations from which spectral data were acquired. Scale bars, 2 μm. b, Plot of the tip emission peak intensity (blue squares) and FWHM (red triangles) versus pump energy for the wire shown in a. The intensity of emission from the nanowire body (green squares) is low and almost linear with pump energy. Solid symbols correspond to experimental data points and lines are guides to the eye. Inset: above threshold (230 nJ) emission spectra acquired from the wire tip (blue) and body (green).

3.4. Polymer nanofiber photodetector

Semiconducting polymers are attractive materials due to their chemically tunable optical and electronic properties, as well as their facility for solution processing [84,85]. Garret A. O'Brien et al. have demonstrated that solution-assisted template wetting may be successfully exploited for high-yield controlled synthesis of poly[(9,9-dioctylfluorenyl-2,7-diyl)-co-(bithiophene)] (F8T2) nanofibes and employed single nanowire devices as ultraminiature photodetectors. [35] Discrete nanowires with average lengths of 15 μm and mean diameters of 200 nm has been fabricated. As expected, structural data point to a low degree of crystallinity within the wires. Individual nanowires can be electrically interfaced using either bottom- or top-contact geometries. Top-contacted single-nanowire devices with interelectrode gaps of approximately 5 μm were fabricated on glass substrates using shadow masking and gold evaporation, see

Fig. 22a. The blue curve in Figure 22b shows the measured dark current (I_{dark}) for a typical device. The data show quasilinear characteristics at low bias, similar to measured data for bottom-contacted nanowires, with some asymmetry at higher bias. The red curve in Fig. 22b shows the measured current (I_{illum}) under continuous 405 nm illumination. A marked increase in the measured current is observed across the entire bias range. The green curve in Fig. 22b shows the measured current under manually chopped 405 nm illumination, where the illumination was switched on or off at 10 V intervals during the bias sweep, corresponding to a duty cycle of 50 %. The present photoconductivity measurements of F8T2 nanowire devices, which yield single-nanowire responsivities of approximately 0.4 mAW^{-1} and external quantum efficiencies of approximately 0.1% under monochromatic illumination; these values are comparable with data reported for single-inorganic-nanowire devices. The results demonstrate the promise of these novel nanostructures as ultraminiature photodetectors with the potential for integration into future hybrid nanophotonic devices and systems.

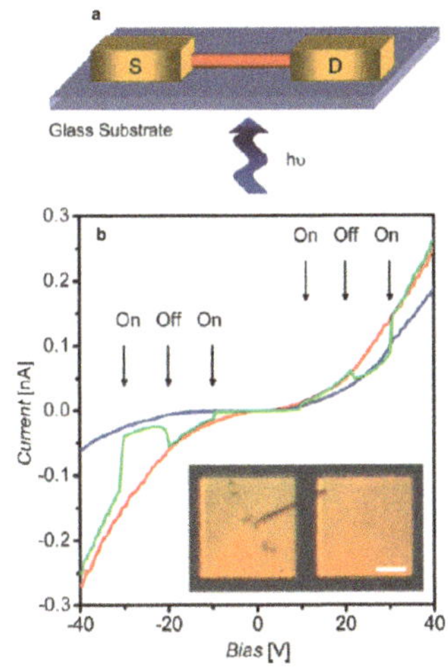

Figure 22. a) Schematic of a top-contact nanowire device. b) I–V characteristicsof a top-contacted F8T2 nanowire acquired in the dark (blue line) and under 405 nm illumination (red line). Reversible switching in I–V data measured under manually chopped 405 nm illumination is alsoshown (green line). Inset: optical microscopy image of a typical top-contacted F8T2 nanowire device. Scale bar: 10 μm.[35]

3.5. Nanopatterning of single light-emitting polymer nanofibres

Polymer nanofibers are compatible with sub-micrometre patterning capability and electromagnetic confinement within subwavelength volumes [2,86], they can offer the benefits of organic light sources to nanoscale optics. F. D. Benedetto et al. reported on the optical properties of fully conjugated, electrospun polymer nanofibres, demonstrated the enhancement of the fibre forward emission through imprinting periodic nanostructures using room-temperature nanoimprint lithography, and investigate the angular dispersion of differently polarized emitted light [87]. Fibers with diameters intentionally produced to be in the range 0.5–5 μm are used to collect reliable fluorescence micrographs and photoluminescence spectra. Confocal microscopy (Fig. 23a) on nanopatterned fibres shows uniformly bright imprinted gratings. The spectra of a typical poly[2-methoxy-5-(2-ethylhexyloxy)-1,4-phenylenevinylene] (MEH-PPV) fibre are displayed in Fig. 23b. in the emission of the untextured fibre the two replicas exhibit almost the same relative intensity, that is, a relative decrease of the high-energy spectral component (λ < 600 nm). Different angular dispersion for light with s- and p-polarization, namely with the electrical field vector parallel or perpendicular to the grating grooves and corresponding to transversal electric (TE) or transversal magnetic (TM) guided modes, is demonstrated in Fig. 23c, respectively. The two modes are concomitantly present in the Bragg outcoupled emission, with TM peaks blueshifted by 18–46 meV with respect to the TE peaks, indicating a lower effective refractive index (n_{eff}) for light polarized perpendicularly to the nanoimprinted features. For an MEH-PPV nanopatterned fibre, neff for TM and TE light monotonously decreases from ~1.9 to 1.2 upon increasing the wavelength from 570 to 575 nm, whereas the difference, $\Delta n_{eff} = n_{eff,TE} - n_{eff,TM}$, slightly increases from 0.01 to 0.04 (Fig. 23d).

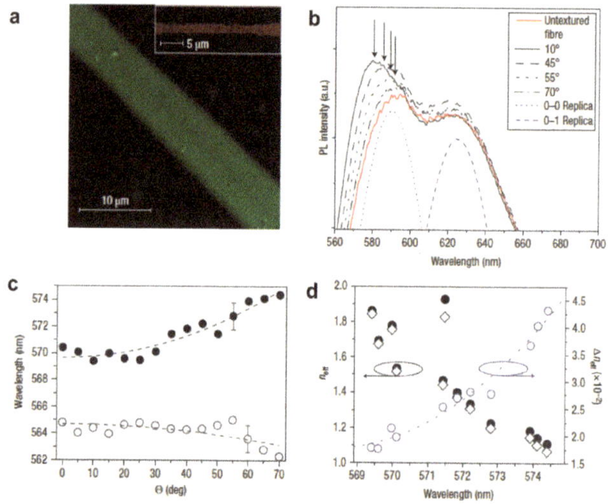

Figure 23. Emission properties of nanopatterned fibres [87]. a, Confocal microscopy pictures of nanoimprinted fibres. The grating period varies from 520 nm (green GE fibre from 1:8 dimethyl formamide:chloroform solution) to 640 nm (red MEH-PPV fibre from 1:5 dimethyl formamide:chloroform solution, inset). The GE fibre was imaged under two-photon excitation at λ = 800 nm. b, Normalized angle-resolved photoluminescence spectra of an unpatterned and nanoimprinted single MEH-PPV fibre, for various collection angles and TE polarization of the guided mode. The 0–0 and 0–1 vibronic replicas of the untextured MEH-PPV fibre, obtained by fitting this spectrum with a Gaussian superposition, are shown as blue curves. The arrows indicate the angular dependence of the emission peaks. c, Angular dispersion of the Bragg-outcoupled modes with *s*- (that is, for TE guided modes, filled circles) and *p*- (transverse magnetic (TM), open circles) polarizations. The error bar shown is the same for all points. The lines are guides for the eye. d, Wavelength dependence of the effective refractive index of the Bragg-outcoupled TE (filled circles) and TM (open diamonds) modes of the nanoimprinted fibre, and of the corresponding difference, $\Delta n_{eff} = \Delta n_{eff,TE} - n_{eff,TM}$ (open circles). The dotted line is a guide for the eye.

3.6. Optical sensors

Single-nanowire detection presents special advantages of high sensitivity and fast response and may offer potentials for highly localized sensing with small footprint and high spatial resolution, as have been recently demonstrated in single-polymer-nanowire electrical sensors [88,89]. And optical sensing offers potentials of high sensitivity, fast response, immunity to electromagnetic interference, and safe operation in explosive or combustive atmosphere, as well as more options for signal retrieval from optical intensity, spectrum, phase, polarization, and fluorescence lifetime [90]. L. Tong group have demonstrated that polymer single-nanowire optical sensors with extraordinary fast response and high sensitivity for humidity and gas sensing [37,91].

When blended or doped with other functional materials,polymer nanowires can be used for optical sensing with high versatilities. For instance, here a 250-nm-diameter PANI/PS nanowire was employed for gas sensing. The nanowire is drawn from a polymer-blend solutionof 2 wt % PANI doped with 10-camphorsulfonic and 5 wt% PS in chloroform and is suspended by a 250-μm-width MgF_2 microchannel and optically connected to fiber tapers at both ends. The sensor is operated by applying a nitrogen-diluted NO_2 gas onto the nanowire witha probing light of 532-nm wavelength. When exposed to NO_2, the increase of the oxidation degree of PANI resultsin spectral absorption at the wavelength of the probing light, in which the absorbance is proportional to the degree of the oxidation that increases with the concentration of NO_2. Figure 24a shows a typical response of a 250-nm-diameter PANI/PS nanowire to 1 ppm NO_2. A clear absorbance is observed,with a response time of about 7s, which is orders ofmagnitude faster than in other types of NO_2 sensors [92]. The time-dependent absorbance of the nanowire at roomtemperature to cyclic NO_2/nitrogen exposure with NO_2 concentration from 0.1 to 4 ppm is given in Figure 24b, indicating good reversibility of the nanowire response. The linear dependence of the absorbance over the NO_2 concentration(see inset) suggests that the PANI/PS nanowire couldfunction as a NO2 optical sensor with a detection limit below 0.1 ppm.

For optical sensing, we suspended a QD/PS nanofiber across a 245-μm-wide MgF_2 microchannel with two ends of the NF coupled to fiber tapers for optical launching and signal collection, as schematically illustrated in Figure 25a. For robustness, the coupling region between the QD/PS NF and fiber tapers was bonded to the substrate using low-index UV-cured fluoropolymer (EFIRON PC-373; Luvantix Co. Ltd.), and the sensing element was sealed inside a glass chamber. To operate the sensor for humidity sensing, 532 nm excitation light was launched from the left-hand fiber taper, and measured the light output from the right-hand fiber taper, while changing the surrounding relative humidity (RH) from 7% to 81% by circulating moisture gases insidethe chamber. The optical response of the nanofiber sensor is shown in Fig. 25b, which works well with the excitation power ofabout 100 pW used here. The RH-dependent PL intensity can be attributed to the passivation of surface trap states of QDs by water molecules [93-95]. The monotonic increase of the PL output with increasing RH can be used for RH sensing, with estimated resolution (calculated from the response curve in Figure 25b) better than 1% RH. Excellent reversibility of the nanofiber sensor was obtained on alternately cycling 19% and 54% RH air, as shown in Figure 25c. The instant response of the sensor was investigated by introducing sudden changes of the humidity in the chamber, with measured response time less than 90 ms (Figure 25d), which is 1–2 orders of magnitude faster than those of RH sensors based on films ormonolayers [96–98]. In addition, the PL intensity of the CdSe/ZnS QD-doped PS matrix was also found to be sensitive to otherspecies, such as CN-ions [99]; the nanofiber sensor proposed here is promising for optical detection of many other types of samples.

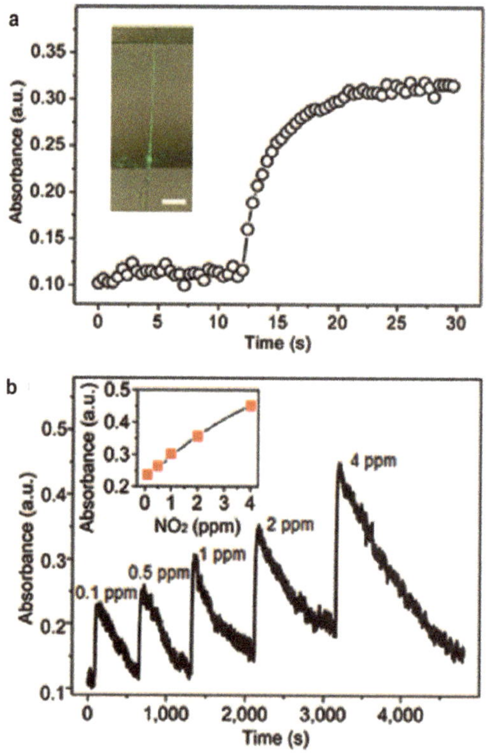

Figure 24. PANI/PS single-nanowire NO_2 sensors [37]. (a) Opticalresponse of a 250-nm-diameter PANI/PS nanowire to 1 ppm NO_2 with a 532-nm-wavelength light. Inset, a close-up optical micrograph of the sensing element with a 532-nm-wavelength probing light guided along the nanowire. Scale bar, 50 μm. (b) Time dependentabsorbance of the nanowire to cyclic NO_2/nitrogen exposure with NO_2 concentration from 0.1 to 4 ppm. Inset, dependence of the absorbance over the NO_2 concentration ranging from 0.1 to 4 ppm.

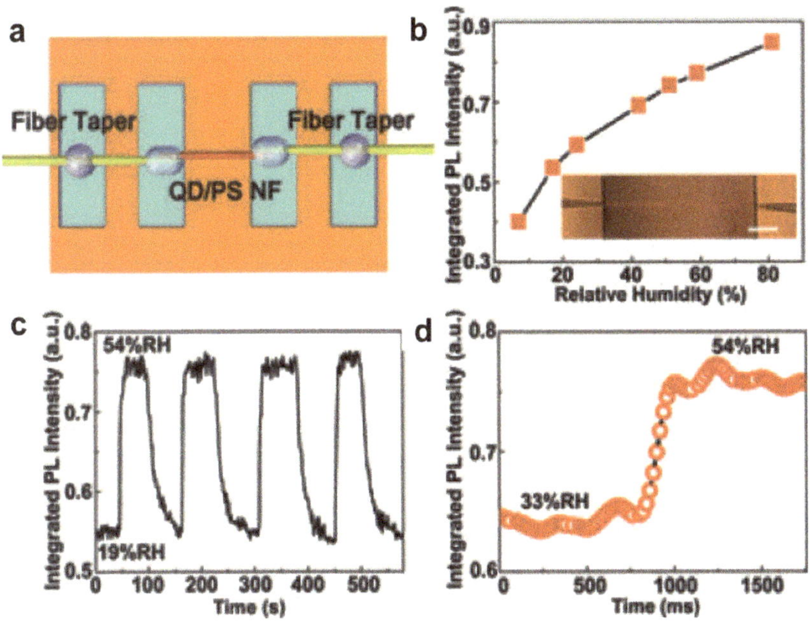

Figure 25. QD/PS single-NF humidity sensors [91]. a) Schematic illustration of the sensor. b) Integrated PL intensity of the nanofiber exposed to ambient RH ranging from 7% to 81%. Inset: Optical microscopy image of the waveguided-light-excited 480-nm-diameter 245- μm-long QD/PS nanofiber used in the sensor. Scale bar: 50 μm. c) Response of the nanofiber sensor to alternately cycled 54% and 19% RH air. d) Typical time-dependent integrated PL intensity of the nanofiber reveals a response time of about 90 ms when RH jumps from 33% to 54%.

4. Polymeric nanowire architecture as full-color micro/nanodisplays

Recently, a number of crossed nanofiber structures for full-color micro/nanodisplays has been reported by H. Yu et al. [22] The crossed array structures were formed by assembling flexible PTT nanofibers under an optical microscope with the assistance of micromanipulators while avoiding the use of color filters and complicated photolithography processing. The color pixels of the displays consist of micro/nanometer sized color spots in a radius of 300–1500 nm. The colors and sizes of the spots were tuned by changing the total power and power ratios of the launched red, green, and blue (RGB) lights.

As shown in Fig. 26a, a 3×3 structure was assembled by using nanofibers 1, 2, and 3 with diameters of 558, 598, and 598 nm, respectively, and RGB lights were launched into the nanofibers 1, 2, and 3, respectively, with a total power of 63.4 μW and a power ratio of about 50:13: 10. Inset of Figure 25b shows that a white colour at the centre of the spot with a radius of about 818 nm was observed at the crossed junction. The colour coordinate was (0.41, 0.35). Its correlated colour temperature (CCT) is 3026 K.

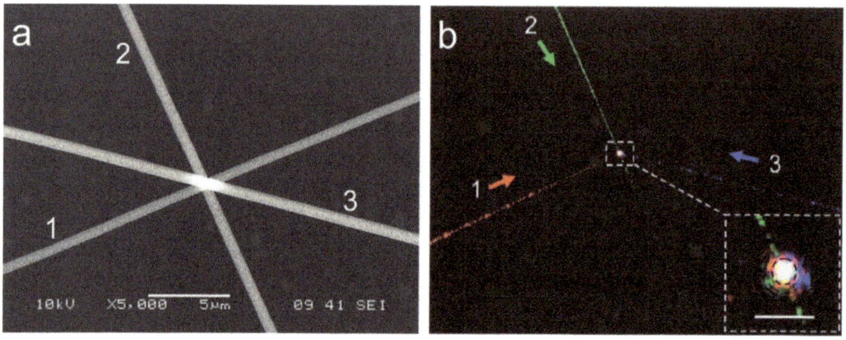

Figure 26. 3×3 crossed structure [22]. (a) SEM image of the device with diameter of 558, 598, and 598 nm for the nanofibers 1, 2, and 3, respectively. The nanofiber 1 is perpendicular to the nanofiber 2, the cross-angle of the nanofibers 2 and 3 is about 50°. (b) Optical microscope image of the guided visible lights in the structure. The inset shows a magnified (×5) view of the white spot at the crossed junction. The arrows show the propagation directions of the launched lights. Scale bar in the inset is 20 μm.

5. Potential to construct integrated optical circuit

The requirement for any photonic device in optical computing, data communications or telecommunications is to be robust, lowcost, high speed and easy to integrate. Polymer MNFs are an ideal material for assembling structures that address both of these requirements, as they offer lowcost processing and are easy to integrate into existing and future networks made either entirely from polymeric or a combination of polymeric with semiconductive. Inaddition, polymer MNFs posses many other advantages such as versatility for biofunctionalization, and the promotion of specificdesired cell behaviors that are elicited by the nanofibers architecture. Polymer MNF shave can be easily produced by current technique, which exhibiting many advantages such as mechanical flexibility, perm-selective nature to gas molecules, biocompatibility, easy processing, and low cost. It is demonstrated that polymer nanofibers offer a unique materials platform for producing key photonic elements including lasing, light emitting diode, detectors, cavity resonator, and passive elements. All these indicate that polymer MNFs is an promising candidate for integrated optical circuit with high density and multifunction. As an excellent research platform merging fiber-optic technology and nanotechnology, there is no doubt that polymer MNFs will continue to open up new opportunities in broad areas including micro- and nanoscale photonics, nonlinear optics and quantum optics in the near future.

Author details

Xiaobo Xing
Education Ministry's Key Laboratory of Laser Life Science & Institute of Laser LifeScience, College of Biophotonics, South China Normal University,Guangzhou, Guangdong, China
Department of Chemistry, Jinan University, Guangdong, Guangzhou, 510632, China

Huaqing Yu
School of Physics and Electronic-information Engineering, Hubei Engineering University, Xiaogan, China

Debin Zhu
Education Ministry's Key Laboratory of Laser Life Science & Institute of Laser LifeScience, College of Biophotonics, South China Normal University,Guangzhou, Guangdong, China

Jiapeng Zheng
School of Physics and Optoelectronic Science and Engineering, South China Normal University, Guangzhou, Guangdong, China

Huang Chen
School of Information and Optoelectronic Science and Engineering, South China Normal University, Guangzhou, Guangdong, China

Wei Chen
School of Information and Optoelectronic Science and Engineering, South China Normal University, Guangzhou, Guangdong, China

Ruibin Xie
School of Information and Optoelectronic Science and Engineering, South China Normal University, Guangzhou, Guangdong, China

Jiye Cai
Department of Chemistry, Jinan University, Guangdong, Guangzhou, 510632, China

6. Acknowledgement

The work thanks for Prof. Baojun Li from State Key Laboratory of Optoelectronic Materials and Technologies, School of Physics and Engineering, Sun Yat-Sen University, for his guidance. This work was supported by China's National Natural Science Foundation (6117707, 81071790, 11104162), China Postdoctoral Science Foundation Funded Project (201003359), Guangdong Excelent Doctoral Dissertation Funded Project (SYBZZM201126), the Key Project of Chinese Ministry of Education (211131), the Science and Technology Project of Guangzhou Nansha District (RG201001003).

7. References

[1] D. Appell. Wired for success. Nature, 419, 553-555 (2002).
[2] M. Law, D. J. Sirbuly, J. C. Johnson, J. Goldberger, R. J. Saykally, and P. Yang. Nanoribbon waveguides for subwavelength photonics integration. Science, 305, 1269-1273 (2004).
[3] Y. Xia, P. Yang, Y. Sun, Y. Wu, B. Mayers, B. Gates, Y. Yin, F. Kim, and Y. Yan. One-dimensional nanostructures: synthesis, characterization, and applications. Adv. Mater., 15, 353-389 (2003).
[4] Y. Huang, X. Duan, Y. Cui, L. Lauhon, K.–H. Kim, and C. Lieber. Logic gates and computation from assembled nanowire building blocks. Science,294, 1313-1317 (2001).

[5] J. Wang, M. Gudiksen, X. Duan, Y. Cui, and C. Lieber. Highly polarized photoluminescence and photodetection from single indium phosphide nanowires. Science, 293, 1455-1457 (2001).

[6] H. Kind, H. Yan, B. Messer, M. Law, and P. Yang. Nanowire ultraviolet photodetectors and optical switches. Adv. Mater.,14, 158-160 (2002).

[7] M. Law, H. Kind, B. Messer, F. Kim, P. Yang. Photochemical sensing of NO_2 with SnO_2 nanoribbon nanosensors at room temperature. Angew. Chem. Int. Ed., 41, 2405-2408 (2002).

[8] J. P. Zhang, D. Y. Chu, S. L. Wu, S. T. Ho, W. G. Bi, C.W. Tu, and R. C. Tiberio. Photonic-wire laser. Phys. Rev. Lett., 75, 2678-2681 (1995).

[9] X. Duan, Y. Huang, R. Agarwal, and C. M. Lieber. Single-nanowire electrically driven lasers. Nature, 421, 241-245 (2003).

[10] L. Tong, R. R. Gattass, J. B. Ashcom, S. He, J. Lou, M. Shen, I. Maxwell, and E. Mazur. Subwavelength-diameter silica wires for low-loss optical wave guiding. Nature, 426, 816-819 (2003).

[11] A. L. Pyayt, B. Wiley, Y. Xia, A. T. Chen and L. Dalton. Integration of photonic and silver nanowire plasmonic waveguides. Nat. Nanotechnol., 3, 660-665 (2008).

[12] G. Brambilla, F. Xu, and X. Feng. Fabrication of optical fiber nanowires and their optical and mechanical characterization. Electron. Lett., 42, 517-518 (2006).

[13] L. Tong, J. Lou, R. R. Gattass, S. He, X. Chen, L. Liu, and E. Mazur. Assembly of silicananowires on silica aerogels for microphotonic devices. Nano. Lett., 5, 259-262(2005).

[14] Y. Li and L. Tong. Mach-Zehnder interferometers assembled with optical microfibers or nanofibers. Opt. Lett., 33, 303-305 (2008).

[15] S. A. Harfenist, S. D. Cambron, E. W. Nelson, S. M. Berry, A. W. Isham, M. M. Crain, K. M. Walsh, R. S. Keynton, and R. W. Cohn. Direct drawing of suspendedfilamentary micro- and nanostructures from liquid polymers. Nano. Lett., 4, 1931-1937 (2004).

[16] H. Liu, J. B. Edel, L. M. Bellan, and H. G. Craighead. Electrospun polymer nanofibers as subwavelength optical waveguides incorporating quantum dots. Small, 2, 495-499 (2006).

[17] X. B. Xing, Y. Q. Wang, H. Zhu, and B. J. Li. Nanofiberdrawing and nanodeviceassembly in poly(trimethylene terephthalate). Opt. Express, 16, 10815-10822 (2008).

[18] X. B. Xing, H. Zhu, Y. Q. Wang, and B. J. Li. Ultra Compact photonic coupling splitters twisted by PTT nanowires. Nano.Lett., 8, 2839-2843 (2008).

[19] H. Zhu, Y. Q. Wang, and B. J. Li. Tunable refractive index sensor with ultracompact structure twisted by poly(trimethylene terephthalate) nanowires. ACS Nano, 3, 3110-3114 (2009).

[20] M. L. Guo, J. C. Shi, and B. J. Li. Polymer-based micro/nanowire structures for three-dimensional photonic integration. Opt. Lett., 33, 2104-2106 (2008).

[21] Y. Q. Wang, H. Zhu, and B. J. Li. Cascaded Mach–Zehnder interferometers assembled by submicron PTT wires. Photon. Technol. Lett., 21, 1115-1117(2009).

[22] H. Yu, D. Liao, M. B. Johnston, and B. Li. All-optical full-color displays using polymer nanofibers. ACS Nano, 5, 2020-2025 (2011).

[23] J. A. Merlo, C. D. Frisbie. Field effect conductance of conducting polymer nanofibers. J. Polym. Sci. Part B: Polym. Phys., 41, 2674-2680 (2003).

[24] J. A. Merlo, C. D. Frisbie. Field effect transport and trapping in regioregular polythiophene nanofibers. J. Phys. Chem. B, 108, 19169-19179 (2004).
[25] H. Liu, C. H. Reccius, and H. G. Craighead. Single electrospun regioregular poly(3-hexylthiophene) nanofiber field-effect transistor. Appl. Phys. Lett., 87, 253106 (2005).
[26] C. R. L. P. N. Jeukens, P. Jonkheijm, F. J. P. Wijnen, J. C. Gielen, P. C. M. Christianen, A. P. H. Schenning, E. W. Meijer, and J. C. Maan. Polarized emission of individual self-assembled oligo(p-phenylenevinylene)-based nanofibers on a solid support. J. Am. Chem. Soc., 127, 8280-8281 (2005).
[27] K. Kim, J. Jin. Preparation of PPV nanotubes and nanorods and carbonized products derived therefrom. Nano Lett., 1, 631-636 (2001).
[28] M. Steinhart, J. H. Wendorff, A. Greiner, R. B. Wehrspohn, K. Nielsch, J. Schilling, J. Choi, and U. Gosele. Polymer nanotubes by wetting of ordered porous templates. Science, 296, 1997 (2002).
[29] V. Beachley, X. Wen. Polymer nanofibrous structures: Fabrication, biofunctionalization, and cell interaction. Prog. Polym. Sci., 35, 868-892 (2010).
[30] X. Xing, H. Zhu, Y. Wang, and B. Li. Ultrasmall optical beam splitters assembled by polymer nanowires. Phys. Express, 1, 89-92 (2011).
[31] Y. Wang, H. Zhu, B. Li. Optical characterization of mechanically tunable microwire based resonators bychanging ring radius and wire diameter. Opt. Commun., 284, 3276-3279 (2011).
[32] J. Shi, M. Guo, and B. Li. Assembly of arbitrary and vertical optical couplers using flexible polymer micro/nanowires. Appl. Phys. Lett., 93, 121101 (2008).
[33] Y. Wang, H. Zhu, and B. Li. Cascaded Mach–Zehnder interferometers assembled by submicrometer PTT wires. IEEE. Photon. Technol. Lett., 21, 1115-1117 (2009).
[34] F. Gu, H. Yu, P. Wang, Z. Yang, and L. Tong. Light–emitting polymer single nanofibers via waveguiding excitation. ACS Nano, 4, 5332-5338 (2010).
[35] G. A. O'Brien, A. J. Quinn, D. A. Tanner, and G. Redmond.A single polymer nanowire photodetector. Adv. Mater., 18, 2379-2383 (2006).
[36] F. Quochi, F. Cordella, A. Mura, G. Bongiovanni, F. Balzer, and H.–G. Rubahn. One–dimensional random lasing in a single organic nanofiber. J. Phys. Chem. B, 109, 21690-21693 (2005).
[37] F. Gu, L. Zhang, X. Yin, and L. Tong. Polymer single-nanowire optical sensors. Nano Lett., 8, 2757-2761 (2008).
[38] J. kameoka, D. Czaplewski, H. Liu, and H. G. Craighead. Polymeric nanowire architecture. J. Mater. Chem., 14, 1503-1505 (2004).
[39] R. Vasita, D. S. Katti. Nanofibers and their applications in tissue engineering. Int. J. Nanomed., 1, 15-30 (2006).
[40] V. Beachley, X. Wen. Fabrication of nanofiber reinforced protein structures for tissue engineering. Mater. Sci. Eng. C Biol. Appl., 29, 2448-2453 (2009).
[41] P. X. Ma, R. Zhang. Synthetic nano–scale fibrous extracellular matrix. J. Biomed. Mater. Res., 46, 60-72 (1999).
[42] S. Ramakrishna, K. Fujuhara, W. E. Teo, T. C. Lim, and Z. Ma. An introduction to electrospinning and nanofibers. Singapore: World Scientific Publishing Company, 2005, 396pp.

[43] P. Berndt, G. Fields, and M. Tirrell. Synthetic lipidation of peptides and amino acids: monolayer structure and properties. J. Am. Chem. Soc., 117, 9515-9522 (1995).
[44] H. Jun, V. Yuwono, S. Paramonov, and J. Hartgerink. Enzyme mediateddegradation of peptide-amphiphile nanofiber network. Adv. Mater.,17, 2612-2617 (2005).
[45] J. D. Hartgerink, E. Beniash, and S. I. Stupp. Self–assembly and mineralization of peptide–amphiphile nanofibers. Science, 294, 1684-1688 (2001).
[46] G. A. Silva, C. Czeisler, K.L. Niece, E. Beniash, D. A. Harrington, J. A. Kessler, and S. I. Stupp. Selective differentiation of neural progenitor cells by high–epitope density nanofibers. Science, 303, 1352-1355 (2004).
[47] H. Yokoi, T. Kinoshita, and S. Zhang. Dynamic reassembly of peptide RADA16 nanofiber scaffold. Proc. Natl. Acad. Sci., USA, 102, 8414-8419 (2005).
[48] H, Jeong, S. Lee, P. Kim, and K. Suh. Stretched polymer nanohairs bynanodrawing. Nano Lett, 6, 1508-1513 (2006).
[49] S. Grimm, R. Giesa, K. Sklarek, A. Langner, U. Gosele, H. W. Schmidt, and M. Steinhart. Nondestructive replication of self–ordered nanoporous alumina membranes via cross–linked polyacrylate nanofiber arrays. Nano Lett., 8, 1954-1959 (2008).
[50] J. R. Porter, A. Henson, and K. C. Popat. Biodegradable poly(epsiloncaprolactone) nanowires for bone tissue engineering applications. Biomaterials, 30, 780-788 (2009).
[51] S. L. Tao, T. A. Desai. Aligned arrays of biodegradable poly(epsilon-caprolactone) nanowires and nanofibers by template synthesis. Nano Lett., 7, 1463-1468 (2007).
[52] D. A. Rollings, S. Tsoi, J. C. Sit, and J. G.C. Veinot. Formation and aqueous surface wettability of polysiloxane nanofibers prepared via surface initiated, vapor–phase polymerization of organotrichlorosilanes. Langmuir, 23, 5275-5278 (2007).
[53] P. Mankidy, R. Rajagopalan, and H. Foley. Facile catalytic growth of cyanoacrylate nanofibers. Chem. Commun., 1139-1141 (2006).
[54] A. Alemdar, M. Sain. Isolation and characterization of nanofibers from agricultural residues: wheat straw and soy hulls. Bioresour. Technol., 99, 1664-1671 (2008).
[55] T. H. Fischer, C. R. Valeri, C. J. Smith, C. M. Scull, E. P. Merricks, T. C. Nichols, M. Demcheva, and J. N. Vournakis. Non–classical processes in surface hemostasis: mechanisms for the poly–N–acetyl glucosamine–induced alteration of red blood cell morphology and surface prothrombogenicity. Biomed. Mater., 3, 015009/1–9(2008).
[56] Y. Fan, T. Saito, A. Isogai. Preparation of chitin nanofibers from squid pen β–chitin by simple mechanical treatment under acid conditions. Biomacromolecules, 9, 1919-1923 (2008).
[57] Aden Brook Farms Website. Large 3×3 bales of brightwheat straw; 2009 http://adenbrookfarms.com/order/index.php?main page=index&cPath=2.
[58] X. Jing, Y. Wang, D. Wu, J. Qiang. Sonochemical synthesis of polyaniline nanofibers. Ultrason. Sonochem., 14, 75-80 (2007).
[59] D. Li, J. Huang, R. B. Kaner. Polyaniline nanofibers: a unique polymer nanostructure for versatile applications. Acc. Chem. Res., 42, 135-145 (2009).
[60] W. K. Czaja, D. J. Young, M. Kawecki,and J. R. M. Brown. The future prospects of microbial cellulose in biomedical applications. Biomacromolecules, 8, 1-12 (2007).
[61] J. Jang, M. Chang, and H. Yoon. Chemical sensors based on highly conductive Poly(3,4–ethylene–dioxythiophene) nanorods. Adv. Mater., 17, 1616-1620 (2005).

[62] H. Kong, J. Jang. One–step fabrication of silver nanoparticle embedded polymer nanofibers by radical–mediated dispersion polymerization. Chem. Commun., 3010-3012 (2006).
[63] P. P. Bishnu. Fundamentals of fibre optics in telecommunication and sensor systems. John Wiley & Sons, New York, 1992.
[64] D. K. Mynbaev, and L. L. Scheiner. Fiber-optic communications technology. PrenticeHall, New York, 2001.
[65] A. Mendez, T. F. Morse. Specialty optical fibers handbook. Academic Press, Burlington, 2007.
[66] T. Kawazoe, T. Yatsui, and M. Ohtsu. Nanophotonics using optical near fields. J. Non-Cryst. Solids, 352, 2492-2495 (2006).
[67] L. Tong, M. Sumetsky. Subwavelength and nanometer diameter optical fibers. Springer, Zhejiang University Press, 2009.
[68] R.Yan, D.Gargas, and P.Yang. Nanowire photonics. Nat. Photon, 3, 569-576 (2009).
[69] Y. Hung, X.Duan, and C. M. Lieber. Nanowires for integrated multicolor nanophotonics. Small, 1, 142-147 (2005).
[70] D. K. Mynbaev, and L. L. Scheiner. Fiber-optic communications technology. Prentice Hall: Upper Saddle River, N.J., 2002.
[71] P. Urquhart. Review of rare earth doped fiber lasers and amplifiers. Proc. IEE, Part J, 135, 385-407 (1988).
[72] H. Ma, A. K.-Y. Jen, L. R. Dalton. Polymer-based optical waveguides: materials, processing, and devices. Adv. Mater., 14, 1339-1365 (2002).
[73] H. Yu, R. Zhang, and B. Li. Optical properties of quantum-dot-decorated polymer nanofibers. Nanotechnology, 22, 335202 (2011).
[74] Q. Song, L. Liu, and L. Xu. Lasing action in dye doped polymer nanofiber knot resonator. J. lightwave Technol., 27, 4374-4376(2009).
[75] H. Sirringhaus, N. Tessler, and R. H. Friend. Integrated optoelectronic devices based on conjugated polymers. Science, 280, 1741-1744 (1998).
[76] F. Carpi, D. De Rossi. Colours from electroactive polymers: electrochromic, electroluminescent and laser devices based on organic materials. Opt. Laser Technol., 38, 292-305 (2006).
[77] N. Tessler, G. J. Denton, and R. H. Friend. Lasing from conjugated–polymer microcavities. Nature, 382, 695-697 (1996).
[78] S. V. Frolov, M. Shkunov, Z. V. Vardeny, and K. Yoshino. Ring microlasers from conducting polymers. Phys. Rev. B, 56, R4363-R4366 (1997).
[79] G. Heliotis, R. Xia, G. A. Turnbull, P. Andrew, W. L. Barnes, I. D. W. Samuel, and D. C. Bradley. Emission characteristics and performance comparison of polyfluorene lasers with one– and two–dimensional distributed feedback. Adv. Funct. Mater., 14, 91-97 (2004).
[80] S. Riechel, C. Kallinger, U. Lemmer, J. Feldmann, A. Gombert, V. Wittwer, and U. Scherf. A nearly diffraction limited surface emitting conjugated polymer laser utilizing a two–dimensional photonic band structure. Appl. Phys. Lett., 77, 2310-2312 (2000).
[81] D. O'carroll, I. Liberwirth, and G. Redmond. Microcavity effects and optically pumped lasing in single conjugated polymer nanowires. Nat. Nanotechnol., 2, 180-184 (2007).
[82] R. Xia, G. Heliotis, Y. Hou, and D. D. C. Bradley. Fluorene–based conjugated polymer optical gain media. Org. Electron., 4, 165-177 (2003).

[83] M. Ariu, D. G. Lidzey, M. Sims, A. J. Cadby, P. A. Lane, and D. D. C. Bradley. The effect of morphology on the temperature-dependent photoluminescence quantum efficiency of the conjugated polymer poly(9,9–dioctylfluorene). J. Phys.: Condens. Matter., 14, 9975-9986 (2002).
[84] S. R. Forrest. The path to ubiquitous and low-cost organic electronic appliances on plastic. Nature, 428, 911-918 (2004).
[85] G. Malliaras, R. Friend. An organic electronics primer. Phys. Today, 58, 53-58 (2005).
[86] E. Menard, M. A. Meitl, Y. Sun, J.-U. Park, D. J.-L. Shir, Y.-S. Nam, S. Jeon, and J. A. Rogers. Micro and nanopatterning techniques for organic electronic and optoelectronic systems. Chem. Rev., 107, 1117-1160 (2007).
[87] F. D. Benedetto, A. Camposeo, S. Pagliara, E. Mele, L. Persano, R. Stabile, R. Cingolani, and D. Pisignano. Patterning of light-emitting conjugated polymer nanofibres. Nat. Nanotechnol, 3, 614-619, (2008).
[88] H. Q.Liu, J.Kameoka, D. A.Czaplewski, and H. G. Craighead. Polymeric nanowire chemical sensor. Nano Lett., 4, 671-675 (2004).
[89] K. Ramanathan, M. A. Bangar, M. Yun, W. Chen, N. V. Myung, and A. Mulchandani. Bioaffinity sensing using biologically functionalized conducting-polymer nanowire. J. Am. Chem. Soc., 127, 496-497(2005).
[90] William B. Spillman Jr. Fiber optical sensors. An introduction for engineers and scientists. John Wiley&Sons. Inc., Hobokn, New Jersey, 2001.
[91] C. Meng, Y. Xiao, P. Wang, L. Zhang, Y. Liu, and L. Tong. Quantum-dot-doped polymer nanofibers for optical sensing. Adv. Mater., 23, 1514-1544 (2011).
[92] M. C. McAlpine, H. Ahmad, D. W. Wang, and J. R. Heath. Highly ordered nanowire arrays on plastic substrates for ultrasensitive flexible chemical sensors. Nat. Mater., 6, 379-384 (2007).
[93] S. R. Cordero, P. J. Carson, R. A. Estabrook, G. F. Strouse, and S. K. Buratto. Photo-activated luminescence of CdSe quantum dot monolayers. J. Phys. Chem. B, 104, 12137-12142 (2000).
[94] B. Alperson, I. Rubinstein, and G. Hodes. Identification of surface states on individual CdSe quantum dots by room-temperature conductance spectroscopy. Phys. Rev. B, 63, 081303(2001).
[95] M. Jones, J. Nedeljkovic, R. J. Ellingson, A. J. Nozik, and G. Rumbles. Photoenhancement of luminescence in colloidal CdSe quantum dot solutions. J. Phys. Chem. B, 107, 11346-11352 (2003).
[96] S. K. Khijwania, K. L. Srinivasan, and J. P. Singh. An evanescent-wave optical fiber relative humiditysensor with enhanced sensitivity. Sens. Actuators, B, 104 , 217-222 (2005).
[97] Y. Zhang, K. Yu, D. Jiang, Z. Zhu, H. Geng, and L. Luo. Zinc oxide nanorod and nanowire for humidity sensor. Appl. Surf. Sci., 242, 212-217 (2005).
[98] R. A. Barry, P. Wiltzius. Humidity-sensing inverse opal hydrogels. Langmuir, 22, 1369-1374 (2006).
[99] A. Touceda-Varela, E. I. Stevenson, J. A. Galve-Gasion, D. T. F. Dryden, J. C. Mareque-Rivas. Selective turn-on fluorescence detection of cyanide in water using hydrophobic CdSe quantum dots. Chem. Commun., 1998-2000 (2008).

Chapter 12

Efficient Forward Error Correction Decoder Design for High-Speed Optical Networking

Bo Yuan, Li Li and Zhongfeng Wang

Additional information is available at the end of the chapter

1. Introduction

Due to the fast development of Internet, the traffic load of data network has increased dramatically in past decades. Accordingly, optical network, as the major carrier for data transmission, needs to increase its capacity to meet the increasing data rate requirement. As stated in the white paper of Optical Internetworking Forum (OIF) (see [1]), the rapid growth of data flow demands optical network to double its capacity every 12-18 months. As a result, this critical requirement pushes various types of optical transmission systems to improve their delivered data rate at the same time.

Generally, in high-speed optical communication, the increase on data rate usually comes with the increase on signal bandwidth and sampling rate. In this case, due to the sensitiveness of optical and electronic devices, the additive transmission noise will inevitably increase as well. Therefore, how to increase the throughput of optical network without loss of robustness is an essential task when designing modern high-speed optical network.

To date, various techniques have been employed to enhance the quality of data transmission in optical network. Among those approaches, Forward error correction (FEC), or commonly called error correction coding (ECC), is viewed as the most cost-effective solution, and has been widely adopted in many industrial optical transmission systems. Many specific FEC codes, including Reed-Solomon (RS), BCH, and LDPC codes, are proposed in different industrial standards for error correction in physical layer. Among them, RS code is the earliest and most widely used FEC code in optical communication. For example, RS (255, 239) was the first generation FEC code for submarine fiber-optical transmission in [2], and its code rate is still the standard parameter for frame design. In addition, for Ethernet network such as 10GBase-LR in [3], Reed-Solomon (255, 239) code is also the standard FEC code.

There are several reasons for the wide application of RS code. First, modern long-distance optical network, especially long-haul network, is very high-speed system (10Gbps and beyond). For other promising FEC codes, such as LDPC or Turbo code, the corresponding decoding throughput usually can not meet such stringent requirement on data rate, or with the penalty of very high hardware complexity. Instead, RS decoder can achieve such high throughput with affordable hardware resource. Second, for local optical network, such as Ethernet network, the real-time response is an important metric for system design. Compared with its counterpart, RS code has the particular advantage on low decoding latency. Therefore, RS codes are widely employed in modern optical transmission system and are believed to play an important role in next generation optical networks.

Considering the importance of RS code, its efficient implementation is quite important for the optical transmission system. A low-complexity high-speed RS encoding/decoding system will improve the overall performance significantly. Particularly, since RS decoding is the most complex procedure in the RS-based FEC system, efficient RS decoder design should be well-studied. Therefore, targeted to different level of optical communication ranging from short-distance Ethernet network to long-haul backbone system, this chapter fully introduces efficient VLSI design of RS decoder. In addition, to meet the requirement of 100Gbps era, this chapter also discusses some new FEC schemes for ultra high-speed application (beyond 100Gbps).

The chapter is organized as follows. Section 2 reviews the RS decoding. The low-complexity high-speed RS decoders for short-distance network are discussed in Section 3. Section 4 analyzes performance-improved RS burst-error decoder for medium-distance system. Some recent FEC schemes targeted to 100Gbps long-haul network are introduced in Section 5. Section 6 draws the conclusion.

2. Review of RS decoding

According to the coding theory in [4], the procedure for decoding RS code contains three main steps: **syndrome computation** (SC), **key equation solving** (KES) and **Chien search & error evaluation** (CSEE). Therefore, the decoding procedure of RS code is summarized as below:

Step 1. (Syndrome computation): For an (n, k) RS code defined over GF(2^m) whose primitive element is α in reference [4], let $\mathbf{C}(x)$ and $\mathbf{R}(x)$ be the transmitted and received codeword polynomial respectively, and then assumes $\mathbf{R}(x) = \mathbf{C}(x) + \mathbf{E}(x)$, where $\mathbf{E}(x)$ is the error polynomial which reflects the errors induced by transmission channel noise. Then, the syndrome polynomial $\mathbf{S}(x)$ is computed as follows:

$$\mathbf{S}(x)=s_0+s_1x+s_2x^2+\ldots+s_{2t-1}x^{2t-1}, \text{ where } s_i=R(\alpha^{i+1}) \text{ and } t=(n-k)/2. \quad (1)$$

The architecture of SC block of an example RS (255, 239) decoder is shown in Fig. 1. Here $\mathbf{R}(x)=r_{n-1}x^{n-1}+r_{n-2}x^{n-2}+\ldots+r_1x+r_0$ is serially transmitted to SC block with the sequence of r_{n-1}, r_{n-2}, $\ldots$, r_0. Every partial syndrome is calculated with shown multiply-accumulate circuits (MAC) in every clock cycle. After n=255 clock cycles, the $2t$=16 syndromes are computed and serially transmitted to the next KES block.

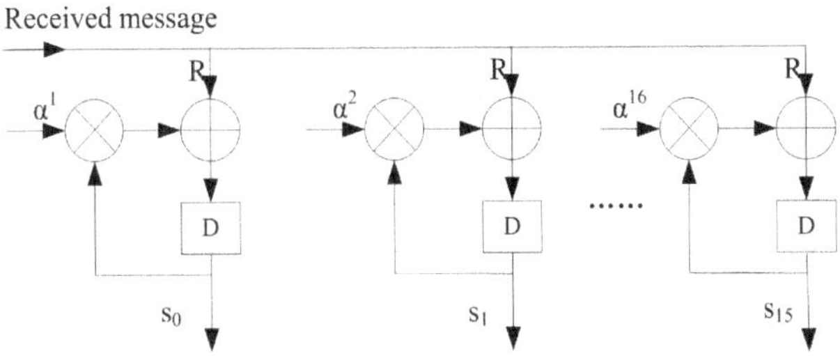

Figure 1. The block diagram of syndrome computation for example RS (255, 239) code.

Step 2. (Key equation solving): With the help of inputted $\mathbf{S}(x)$, in this step, Key equation solver (KES) block will calculate error evaluator polynomial $\mathbf{\Omega}(x)$ and error locator polynomial $\mathbf{\Lambda}(x)$ by solving key equation: $\mathbf{\Lambda}(x)\mathbf{S}(x)\equiv\mathbf{\Omega}(x) \bmod x^{2t}$. This part is the most important step in the whole RS decoding procedure, which usually dominates the performance of the overall decoder. Therefore, in this chapter, we focus on the algorithm and architecture optimization of KES block.

Generally, Berlekamp-Massey (BM) algorithm or modified Euclidean (ME) algorithm can be employed to solve key equation. To data, many efforts have addressed for efficient VLSI implementation of the above two algorithms. In [5], BM algorithm was reformulated as RiBM with the same regular architecture format compared with conventional ME algorithm in [6] and [7], and a folded BM algorithm based on RiBM was introduced in [8]. Reference [6] and [7] implemented conventional ME algorithm with systolic and recursive architecture. In Section 3 and Section 4, based on the above efforts, some improved KES algorithms and their corresponding hardware implementations will be discussed for efficient RS decoder design.

Step 3. (Chien search & error evaluation): After KES block finishes its computation for the current codeword, the calculated error locator polynomial $\mathbf{\Lambda}(x)$ and the error evaluator polynomial $\mathbf{\Omega}(x)$ will be outputted to CSEE block to generate the error positions and magnitudes.

Chien search is a widely employed approach to look for error position. Its basic idea is simple but efficient: If $\Lambda(\alpha^{-i})$=0 for current i, it indicates that the i-th symbol of the received codeword is wrong and needs to be corrected. After obtaining the position of error, the following Forney algorithm is applied to determine the error value:

$$Y_i = -\left.\frac{\Omega(x)}{x\Lambda'(x)}\right|_{x=\alpha^{-i}} \tag{2}$$

where Y_i is the error magnitude for the i-th erroneous symbol.

Based on the above described Chien search and Forney algorithm, the architecture of CSEE block for example RS (255, 239) code is illustrated in Fig. 2. It consists of several unit cells (shown in Fig. 2(a)). Both of the sub-blocks that carry out Chien search and Forney algorithm consist of these basic cells. In the beginning, λ_i, and ω_i, (represented by U_i in the figure), as the coefficients of $\mathbf{\Lambda}(x)$ and $\mathbf{\Omega}(x)$, are parallel loaded into these basic cells (enable=1). Then, during the next 255 cycles, those basic cells will carry out multiply iteratively. Fig. 2(b) is the overall architecture for CSEE block. Once a zero is detected in Chien search, the corresponding error magnitude will be computed via executing the above Forney algorithm.

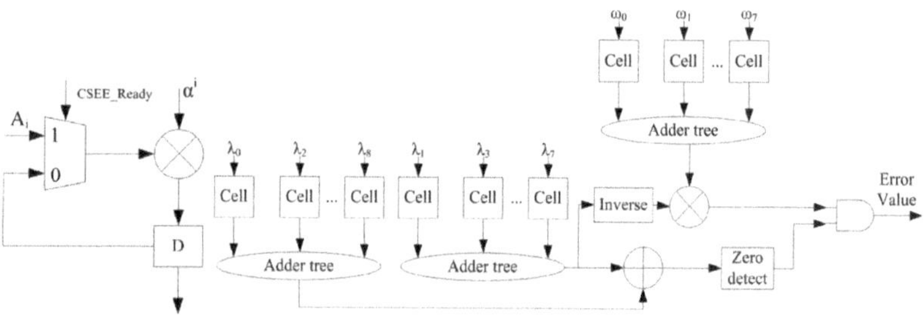

(a) (b)

Figure 2. (a) The diagram of CSEE cell. (b) The block diagram of CSEE.

The overall architecture of RS decoding is summarized in Fig. 3.

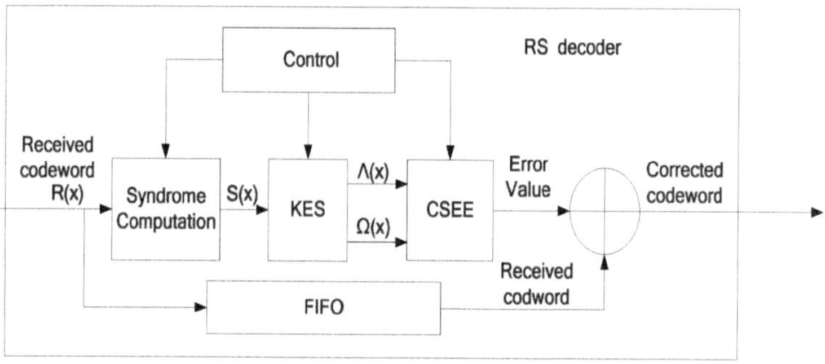

Figure 3. The overall architecture of RS decoder.

As mentioned in previous paragraph, since KES is the dominating step in the whole RS decoding, Section 3 and 4 will focus on the algorithm and architecture optimization of KES block.

3. Low-complexity high-speed RS decoders for short-distance network

For short-distance optical transmission, such as 10GBase-LR, since the noise rendered from transmission distance is quite limited, the requirement of coding gain is not as strict as long-distance backbone network (which will be discussed later). Therefore, as discussed in Section 1, Reed-Solomon (255, 239) code is widely used in this kind of network due to its high code rate and good error correction capability.

Although coding gain is not the major concern in this scenario, because of limited hardware resource, in order to implement efficient RS decoder, the designers have to consider the challenge of achieving high data rate with low hardware complexity. Accordingly, optimization of RS decoding architectures is necessary for high-efficiency hardware implementation.

In this section, based on the two main RS decoding algorithms, the improved ME-based and BM-based decoders are introduced.

3.1. rDCME-based RS decoder

In traditional ME algorithm, the inherent degree computation and systolic architecture renders large consumption of area and power (see [6] and [7]), which is not suitable for the discussed application. To reduce the unnecessary degree computation, DCME algorithm was introduced in [9]. By generating internal switch and shift signals, the DCME algorithm can achieve the same function as ME algorithm without degree computation.

DCME algorithm

The initial conditions:
$R_0 = xS(x)$ $Q_0 = x^{2t}$ $L_0 = x$ $U_0 = 0$
$i = 0$
Iteration phase:
while $i < 2t - 1$
generate leading coefficients of R_i and Q_i: a_i and b_i
generate switch (*sw*) and polynomial computation (*pc*) signals by control FSM
if (*sw*=1) then switch R_i and Q_i;
switch L_i and U_i;
switch a_i and b_i;
if (*pc*=1) then
$R_{i+1} = b_i R_i + a_i Q_i$
$Q_{i+1} = Q_i$ (delay a clock cycle)
$L_{i+1} = b_i L_i + a_i U_i$
$U_{i+1} = U_i$ (delay a clock cycle)
else polynomial of R_i is shifted by one degree
$i = i + 1$
Output L_i as error locator polynomial and R_i as error value polynomial

It is needed to point out that the initialization of DCME and ME is different due to the consideration of the design of the following introduced FSM.

Fig. 4 shows the FSM for generating control signals. In each iteration, there are two possible states: S0 and S1. S0 represents the case when both of a_i and b_i are nonzero; otherwise the state of FSM is S1. The different combinations of the current and previous states will determine control signals in the current iteration.

When the leading coefficients of R_i and Q_i are both nonzero, it denotes that polynomial computation can be carried out (*pc*=1), otherwise shift operation would be performed (*pc*=0) to reduce leading coefficient (and in this case the leading coefficient must be zero, the details will be shown in next paragraph). In each iteration, the possible shift operation would be executed once at most. The whole shift process would not stop until both of a_i and b_i are nonzero, which means the degrees of R_i and Q_i are equal again. And in that case KES block starts executing polynomial computation. So it is clear that in each iteration the algorithm would perform only shift operation or only polynomial computation operation.

It should be pointed out that after every polynomial computation, if being carried out, the original leading coefficient of R_{i+1} must be zero due to the arithmetic character of $R_{i+1} = b_iR_i + a_iQ_i$. Different from the leading coefficients referred in above paragraph, this kind of leading zero is a "false" leading coefficient which will cause logic errors in next iteration. (For example, after polynomial computation if R_{i+1} is represented by 0, 0, 0, α^2, α^3, the "false" leading coefficient is the first zero, and the real representation of R_{i+1} should be0, 0, α^2, α^3.) So in every possible polynomial computation process, the designed rDCME KES block has automatically eliminated this kind of leading zero with the aid of "*start*" signal in hardware design (Fig.5): the coefficients which arrive simultaneously with "*start*" signal are selected as the leading coefficients. So once polynomial computations are finished, by delaying Q_{i+1}, U_{i+1} and *start* signal one more clock cycle, the "false" leading zero is eliminated, and degR_{i+1} is one less than deg R_i or equal to it (this condition happens when the previous iteration's actual input is xR_i brought by initial input $R_0=xS(x)$). Then a_i and b_i represent the real leading coefficients respectively.

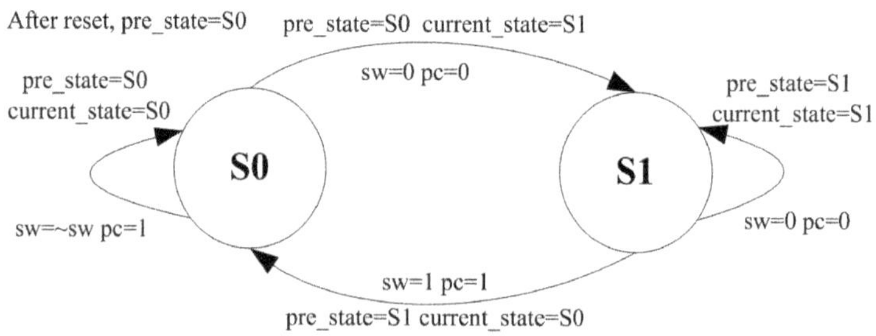

Figure 4. The FSM for generating control signals.

If the previous state and the current state are both S0, it indicates that the polynomial computation is able to be executed in the two successive iterations. So *pc*=1 since the current operation is polynomial computation. Due to the fact the previous state is S0, after the previous polynomial computation and the degree reduction, degR_i is one smaller than deg

R_{i-1} or equal to it. So deg R_i is the same with degQ_i or one smaller than degQ_i. These two possible conditions occur successively and the switch signal (*sw*) alternates successively (*sw*=~*sw*).

If the previous state is S0 and the current state is S1, KES block would process shift operation to eliminate leading zero (*pc* is set to 0) in the current iteration, because S1 shows leading coefficient is zero. *sw* is also 0 because switch operation always be carried out with polynomial operation.

If the previous state and the current state are both S1, it indicates that the two successive iterations are both in shifting operations. Similar with the above condition, *sw* and *pc* are both set to 0.

If the previous state is S1 and the current state is S0, the polynomial computation would be executed (*pc*=1) in the current iteration. Since in the previous iteration R_i is in shift operation (Q_i is never in shift operation because of its character in polynomial computation, it is also guaranteed by rDCME's initial conditions), actual degree of R_i must be smaller than Q_i, so *sw* is set to 1.

After $2t$=16 iterations, the rDCME KES block stops and outputs error value polynomial $R(x)=\Omega(x)$ and error locator polynomial $L(x)=\Lambda(x)$.

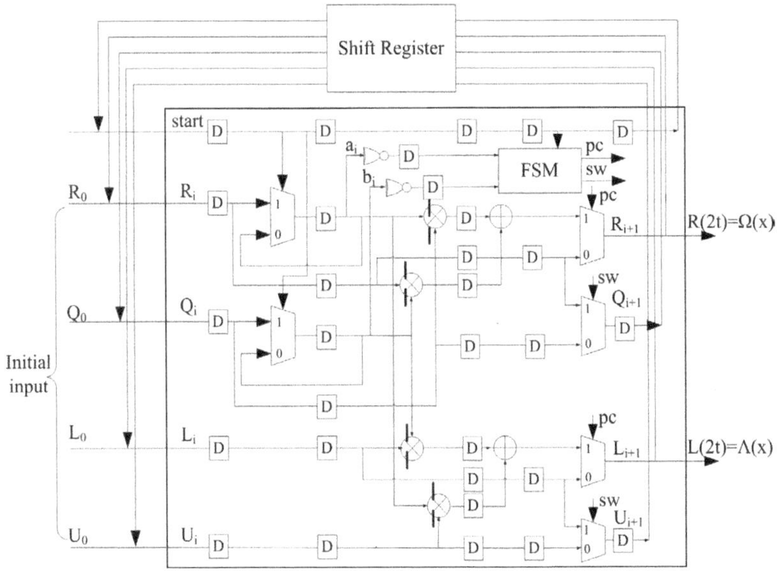

Figure 5. The block diagram of KES.

Fig. 5 shows the detailed architecture of rDCME algorithm. The KES block is designed with single PE. It is commonly known that recursive architecture usually can not be pipelined due to data dependency. And recursive architecture is always a bottleneck for high-speed. However, in the rDCME architecture, these disadvantages can be avoided. A 11-stages ($2t$-5=11) shifter registers are used to store the last iteration results and feedback to the next iteration for avoiding dependency between successive iterations: At the end of each iteration, the leading coefficients of five updated inputs (R, Q, L, U and *start*) are just stored back into the leftmost registers of shift registers and ready to be updated in the next iteration. Because during the computation procedure the whole iteration process of KES block is a close loop, the property of leading coefficients' in-time arrival makes dependency between iterations be avoided and logical validity guaranteed. Furthermore, because the former SC block takes n clock cycles to output one codeword, the PEs in conventional systolic DCME architecture in [10] and [11] are idle in the most of processing time and at the same time it occupies a large amount of chip area. So the multi-stages pipeline can be employed in the area efficient recursive KES block with valid logic and only a little data processing rate degradation. Note that in Fig. 5 the multipliers are pipelined.

Architect.	**rDCME**	**pDCME in [10]**	**DCME in [11]**	**PrME in [7]**
Tech.(μm)	0.18	0.13	0.25	0.13
PE	1	2t	3t+2	1
SC	2900	2900	2900	2900
KES	11400	46200	21760	17000
CSEE	4100	4100	4100	4100
Total gates	18400	53200	28760	24000
fmax(MHz)	640	660	200	625
Throughput (Gb/s)	5.1	5.3	1.6	5

Table 1. Implementation results and comparisons

Table 1 presents performance comparisons between the rDCME RS decoder and other existing RS decoders. It can be observed that the rDCME decoder has very low hardware complexity and high throughput. Compared with the existing ME architectures, the total gate count of the rDCME architecture is reduced by at least 30.4%. Therefore, the hardware efficiency is improved at least 1.84 times, which means under the same technology condition our design would be much more area-efficient compared with other existing RS decoder designs for multi-Gb/s optical communication systems.

3.2. PI-iBM-based RS decoder

Besides ME algorithm, BM algorithm is another main decoding approach for RS codes. An important and inevitable disadvantage of traditional iBM/RiBM algorithms is the high cost of area or iteration time for computing error value polynomial $\Omega(x)$. In iBM architecture stated in [12], one third of total iteration time or half of hardware complexity is employed to compute $\Omega(x)$; in RiBM architecture stated in [5], one third of processing elements (PE) are utilized to calculate and store $\mathbf{\Omega}(x)$. Therefore, the calculation of $\mathbf{\Omega}(x)$ impedes further performance improvement of current BM architectures.

The PI-iBM algorithm employs simplified Forney algorithm to compute error values. Simplified Forney algorithm, presented in [13] and [14], replaces $\Omega(x)$ with scratch polynomial B(x) as follows:

$$Y_i = \left. \frac{\lambda_0 \delta}{xB(x)\Lambda'(x)} \right|_{x=\alpha^{-i}}$$

In each iteration, scratch polynomial $\mathbf{B}(x)$, discrepancy δ, error locator polynomial $\mathbf{\Lambda}(x)$ and its coefficient λ_0 are simultaneously updated. After completing iteration, KES block outputs them to CSEE block for calculating error values Y_i. So the computation of $\mathbf{\Omega}(x)$ is completely eliminated, which enables KES block to reduce a large amount of extra computation circuitry and iteration time.

Furthermore, in order to reduce hardware complexity significantly without sacrificing throughput per unit area, pipeline interleaving techniques in [15] is employed in the PI-iBM algorithm and architecture proposed in [16].

As depicted in the following PI-iBM algorithm, interleaving factor g is a crucial factor to design overall architecture. In practical RS (n, k, t) codes, such as (255, 239, 8) code, t=8 is a common value. So in this paper we set both p and g as 3 for demonstrating PI-iBM architecture.

The PI-iBM architecture consists of two blocks: pipeline interleaving error locator update (PI-ELU) block and pipeline interleaving discrepancy computation (PI-DC) block. As it is illustrated in Fig. 6, PI-ELU block is designed to execute Step3 for updating polynomials. Fig. 6(a) shows the internal architecture of the i-th PE. Initial values of upper and leftmost registers are shown in the figure and other registers are initialized to zero. For the i-th PE, in each iteration 10 cycles are required to update the stored coefficients of $\mathbf{\Lambda}(x)$ and $\mathbf{B}(x)$, meanwhile "ctrl" signal is set to be "1 0 0 0 0 0 0 0 0 0". At the beginning of r-th iteration, $b_{3i}(r)$, $b_{3i+1}(r)$, $b_{3i+2}(r)$ are stored in the leftmost three registers with $\lambda_{3i}(r\text{-}1)\gamma(r\text{-}1)$, $\lambda_{3i+1}(r\text{-}1)\gamma(r\text{-}1)$, $\lambda_{3i+2}(r\text{-}1)\gamma(r\text{-}1)$ in the upper three registers, then they are shifted in the upper and lower loops to be updated. During the first 3 cycles, $\lambda_{3i}(r)$, $\lambda_{3i+1}(r)$, $\lambda_{3i+2}(r)$ are successively computed and outputted to PI-DC block for calculating discrepancy $\delta(r)$ (Step 1). After current iteration is completed, $b_{3i}(r+1)$, $b_{3i+1}(r+1)$, $b_{3i+2}(r+1)$ and $\lambda_{3i}(r)\gamma(r)$, $\lambda_{3i+1}(r)\gamma(r)$, $\lambda_{3i+2}(r)\gamma(r)$ are just fed back to the initial registers which stored them at the beginning. The two dashed rectangles indicate that the critical path between lower multiplier and adder has been 3-stage fine-grain pipelined; the path between upper multiplier and adder is tackled in the same way.

In addition, PI-DC block mainly implements the function of updating discrepancy $\delta(r)$ (Step1). A low-complexity and high-speed architecture of PI-DC block is shown in Fig. 7. As shown in Fig. 7, 2t-1 syndromes are serially sent to PI-DC block and shifted in the upper t+1 registers **every 10 cycles.** The initialization of leftmost register is S_0 while other registers are initialized to zero. In each iteration "ctrl 1"signal is set to be "0 0 0 0 0 0 1 0 0 0". In the first 5 cycles of each iteration, input λ_j, λ_{3+j}, λ_{6+j} and corresponding syndromes selected by multiplexers are multiplied by three 3-stage pipelined multipliers (shown by dashed lines). At the end of 6-th cycle accumulator circuit computes $\delta(r)$ and outputs it to control block for updating $\gamma(r)$ and SEL(r). Passing another register which cuts path between PI-ELU and PI-DC in control block, the three signals are fed back to PI-ELU block. In the overall architecture of PI-iBM (Fig. 8), it takes 7 cycles to calculate and output $\delta(r)$ (PI-DC block), and another 3 cycles is the cost for calculating new coefficients (PI-ELU block), so the total time for one iteration is 10 cycles.

The PI-iBM Algorithm

Initialization and Input:

Let $t+1 = p \times g$, where g is the coefficient of pipeline interleaving;

$\lambda_0(0) = b_0(0) = 1$,

$\lambda_l(0) = b_l(0) = 0$ for l = 1, 2, ..., t;

$k(0) = 0,\ \gamma(0) = 1$;

$S(x) = S_0 + S_1 x + S_2 x^2 + \ldots + S_{2t-1} x^{2t-1}$.

Iteration Process:

for $r = 0$ step 1 until $2t$ - 1 do

Begin

Step1: $\delta(r) = S_r \lambda_0(r) + S_{r-1}\lambda_1(r) + \ldots + S_{r-t}\lambda_t(r)$;

Step2: If $\delta(r) \neq 0$ and $k(r) \geq 0$ then $a = 1$; else $a = 0$;

Step3: for $j = 0$ step 1 until $g-1$ do

Begin

Step3.1: $\lambda_{gi+j}(r+1) = \gamma(r)\lambda_{gi+j}(r) - \delta(r) b_{gi+j-1}(r)$;

Step3.2: $b_{gi+j}(r+1) = \begin{bmatrix} \lambda_{gi+j}(r) & b_{gi+j-1}(r) \end{bmatrix} \begin{bmatrix} a \\ \bar{a} \end{bmatrix}$ for $i = 0,1,\ldots p-1$;

End

Step4: $\begin{bmatrix} \gamma(r+1) \\ k(r+1) \end{bmatrix} = \begin{bmatrix} \delta(r) & \gamma(r) \\ -k(r)-1 & k(r)+1 \end{bmatrix} \begin{bmatrix} a \\ \bar{a} \end{bmatrix}$;

End

Output: $\Lambda(x) = \lambda_0(2t) + \lambda_1(2t)x + \lambda_2(2t)x^2 + \ldots + \lambda_t(2t)x^t$;

$B(x) = b_0(2t) + b_1(2t)x + b_2(2t)x^2 + \ldots + b_t(2t)x^t$.

Table 2 gives the implementation results of PI-iBM decoder and also lists some other designs. From this table we can find that the PI-iBM architecture deliver very high throughput with relatively low hardware complexity: the total throughput rate and throughput per unit area in the PI-iBM design are at least 200% more than those existing works. To achieve data rates from 10 Gb/s to 100 Gb/s, PI-iBM decoder has the lowest hardware complexity. If 65 nm CMOS technology is used in the implementation, the throughput of our design can be increased significantly. Thus the current designs can fit well for 10 Gb/s-40 Gb/s optical communication systems. For 100 Gb/s applications, we may need two to three independent hardware copies of the designs. However, the PI-iBM architecture will remain to have the lowest hardware complexity compared with existing designs. In short, the PI-iBM decoder is very area-efficient for very high-speed optical applications.

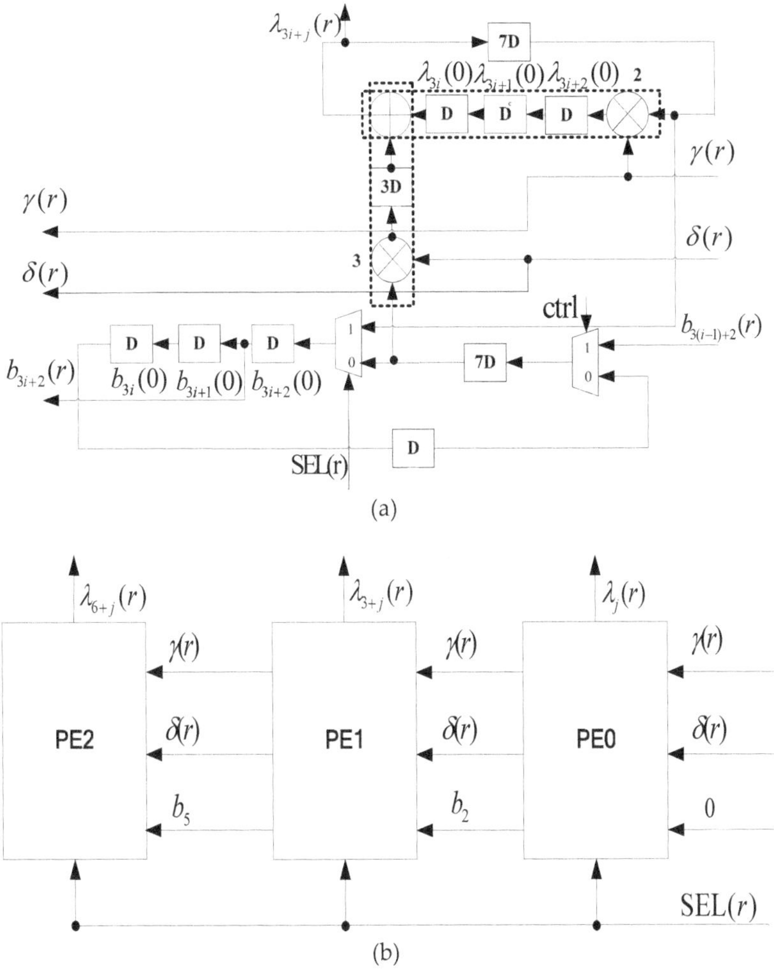

Figure 6. The diagram of PI-ELU block. (a) The internal architecture of the *i*-th PE. (b) The overall architecture of PI-ELU block.

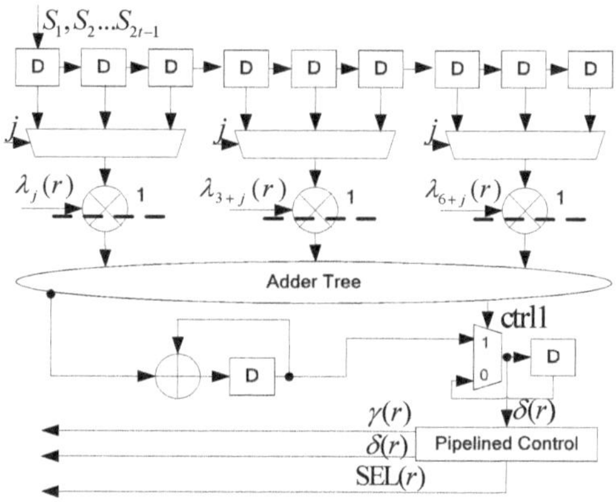

Figure 7. The diagram of PI-DC block.

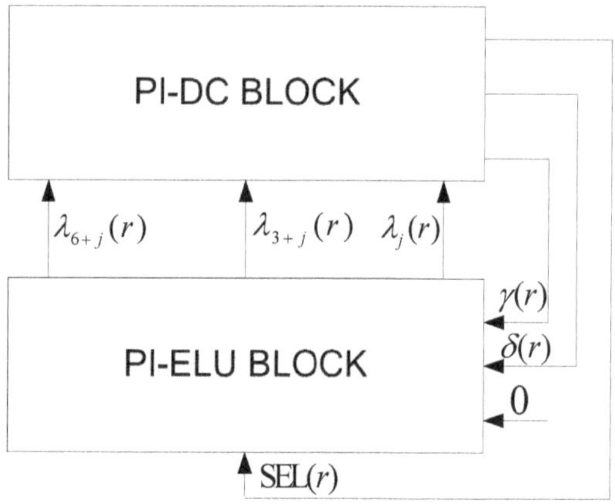

Figure 8. The diagram of overall PI-iBM architecture.

Design	Tech. (μm)	Total of gates	fmax (MHz)	Throughput (Gb/s)	Latency (cycles)
Retimed iBM in [17]	0.18	8423	654	5.23	288
Multi-mode RiBM in [18]	0.18	9566	400	3.20	128
Systolic ME in [19]	0.13	102500	770	6.16	80
Folded ME in [7]	0.13	17000	625	5.00	256
PI-iBM	0.18	10951	980	12.5	160

Table 2. Implementation results and comparisons

4. High-performance low-complexity RS burst-error decoders for mediate distance network

For mediate distance optical network, such as Metro Ethernet network, traditional RS decoding can not provide enough coding gain for the data transmission in this scenario. Instead, enhanced FEC scheme should be employed for improved error-correcting capability. Notice that in this kind of systems, long burst error is the major error pattern in transmission procedure; therefore, burst-error decoding algorithm and architecture are attractive solutions for this case. In this chapter, we introduce efficient burst-error-correcting RS decoder to meet the requirement in this type of application.

4.1. Reformulated inversionless burst-error correcting (RiBC) algorithm

As excellent Maximum Distance Separable (MDS) code [20], RS code is very effective in correcting long burst errors. However, previous RS burst decoding algorithms in [21] and [22] are infeasible for hardware implementation due to their high computation complexity. In [20], Wu proposed a new approach to track the position of burst of errors. By introducing a new polynomial that is a special linear function of syndromes, this approach can correct a long burst of errors with length up to $2t$-1-2β plus a maximum of β random errors. Here β is a pre-chosen parameter that determines the specific error correcting capability. In this case, the miscorrection probability is upper bounded by $(n-2f)(n-f)^{\beta}2^{m(\beta+f-2t)}$.

Although the approach in [20] has reduced computation complexity, it still contains inversion operation and long data path, which impedes its efficient VLSI implementation, therefore, the algorithm in [20] was reformulated to the RiBC algorithm. The RiBC algorithm is a kind of list decoding algorithm. 8 polynomials are updated simultaneously in each iteration. After every 2β inner iterations, $\tilde{\Lambda}^{(2\beta)}(x)$, as the candidate of the error locator polynomial of the random errors, is computed for current l-th outer iteration. When l reaches n, we track the $\tilde{\Lambda}^{(2\beta)}(x)$ that is identical for longest consecutive l, and record the last element l^* of the consecutive l's. Then the corresponding $\Lambda^{(2\beta)}(x)$ and $\tilde{\Delta}^{(2\beta)}(x)$ at the l^*-th loop are marked as overall error locator polynomial $\Lambda^*(x)$ and error evaluator polynomial $\Omega^*(x)$ respectively. Finally Forney algorithm is used to calculate the error value in each error position with the miscorrection probability up to $(n-2f)(n-f)^{\beta}2^{m(\beta+f-2t}$.

The RiBC algorithm is targeted for correcting burst error plus some random errors. By observing step2.3 and step 2.4, it can be founded that both of them are quite similar to the essential update equations in RiBM algorithm (see [5]). Therefore, it inspires us that both of the RiBC algorithm for burst-error correction and RiBM for random error correction can be implemented one the same hardware. Furthermore, considering single burst error correcting algorithm in [20] is a specific instance of RiBC algorithm with β=0, so it can also be implemented on the RiBC architecture. Accordingly, a unified hybrid RS decoder, which can be configured to the above three types of error correcting mode, is introduced in the next subsection.

Reformulated Inversionless Burst-Error Correcting (RiBC) Algorithm

---

f-length ($f < 2t - 2\beta$) burst of errors plus maximum β random errors):

Input: Syndromes $S_0, S_1, S_2, \ldots, S_{2t-1}$;

step1: Compute $\Xi(x) = (1-\alpha^{1-(2t-2\beta)}x)(1-\alpha^{2-(2t-2\beta)}x)\ldots(1-\alpha^{-1}x)(1-x)$
$= 1 + \xi_1 x + \xi_2 x^2 + \ldots + \xi_{2t-2\beta}x^{2t-2\beta}$;

step2: For $l = 0$ Step 1 Until $n-1$ do

step2.1: Compute $\Phi(x) = \Xi(\alpha^l x) = 1 + \phi_1 x + \phi_2 x^2 + \ldots + \phi_{2t-2\beta}x^{2t-2\beta}$;

step2.2: Compute $\Psi(x) = \psi_0 + \psi_1 x + \ldots + \psi_{2t-1}x^{2t-1}$, where $\psi_i = \sum_{j=0}^{2t-2\beta} \phi_j S_{i+2t-2\beta-j}$;

step2.3: Initialize $\Lambda^{(0)}(x) = \lambda_0^{(0)} + \lambda_1^{(0)}x + \ldots + \lambda_{2t-2}^{(0)}x^{2t-2} = \Phi(x)$;

$B^{(0)}(x) = b_0^{(0)} + b_1^{(0)}x + \ldots + b_{2t-2}^{(0)}x^{2t-2} = \Phi(x)$;

$\tilde{\Lambda}^{(0)}(x) = \tilde{\lambda}_0^{(0)} + \tilde{\lambda}_1^{(0)}x + \ldots + \tilde{\lambda}_{2t-2}^{(0)}x^{2t-2} = 1$;

$\tilde{B}^{(0)}(x) = \tilde{b}_0^{(0)} + \tilde{b}_1^{(0)}x + \ldots + \tilde{b}_{2t-2}^{(0)}x^{2t-2} = 1$;

$\tilde{\Delta}^{(0)}(x) = \tilde{\delta}_0^{(0)} + \tilde{\delta}_1^{(0)}x + \ldots + \tilde{\delta}_{2t-1}^{(0)}x^{2t-1} = \Psi(x)$;

$\tilde{\Theta}^{(0)}(x) = \tilde{\theta}_0^{(0)} + \tilde{\theta}_1^{(0)}x + \ldots + \tilde{\theta}_{2t-1}^{(0)}x^{2t-1} = \Psi(x)$;

$\tilde{\Delta}^{*(0)}(x) = \tilde{\delta}_0^{*(0)} + \tilde{\delta}_1^{*(0)}x + \ldots + \tilde{\delta}_{2t-1}^{*(0)}x^{2t-1} = S(x)$;

$\tilde{\Theta}^{*(0)}(x) = \tilde{\theta}_0^{*(0)} + \tilde{\theta}_1^{*(0)}x + \ldots \tilde{\theta}_{2t-1}^{*(0)}x^{2t-1} = S(x)$;

$\gamma^{(0)} = 1, k^{(0)} = 0$;

step2.4: For $r = 0$ Step 1 Until $2\beta - 1$ do

step2.4.1: Compute $\tilde{\delta}_i^{(r+1)} = \gamma^{(r)}\tilde{\delta}_{i+1}^{(r)} - \tilde{\delta}_0^{(r)}\tilde{\theta}_i^{(r)}$;

$\tilde{\delta}_i^{*(r+1)} = \gamma^{(r)}\tilde{\delta}_{i+1}^{*(r)} - \tilde{\delta}_0^{(r)}\tilde{\theta}_i^{*(r)}$;

$\lambda_i^{(r+1)} = \gamma^{(r)}\lambda_i^{(r)} - \tilde{\delta}_0^{(r)}b_{i-1}^{(r)}$;

$\tilde{\lambda}_i^{(r+1)} = \gamma^{(r)}\tilde{\lambda}_i^{(r)} - \tilde{\delta}_0^{(r)}\tilde{b}_{i-1}^{(r)}$;

step2.4.2: If $\tilde{\delta}_0^{(r)} \neq 0$ and $k^{(r)} \geq 0$ then $a = 1$; else $a = 0$;

step2.4.3:
$$\begin{bmatrix} b_i^{(r+1)} \\ \tilde{b}_i^{(r+1)} \\ \tilde{\theta}_i^{(r+1)} \\ \tilde{\theta}_i^{*(r+1)} \\ \gamma^{(r+1)} \\ k^{(r+1)} \end{bmatrix} = \begin{bmatrix} \lambda_i^{(r)} & b_{i-1}^{(r)} \\ \tilde{\lambda}_i^{(r)} & \tilde{b}_{i-1}^{(r)} \\ \tilde{\delta}_{i+1}^{(r)} & \tilde{\theta}_i^{(r)} \\ \tilde{\delta}_{i+1}^{*(r)} & \tilde{\theta}_i^{*(r)} \\ \tilde{\delta}_0^{(r)} & \gamma^{(r)} \\ -k^{(r)}-1 & k^{(r)}+1 \end{bmatrix} \begin{bmatrix} a \\ \bar{a} \end{bmatrix}$$

step3: Track the longest consecutive $\tilde{\Lambda}^{(2\beta)}(x)$ that are identical, recorded the last element l^* of the consectuive l's, then the overall error locator polynoimal $\Lambda^*(x) = \Lambda^{(2\beta)}(x)$ at the l^*-th outer loop.

The overall evaluator polynomial $\Omega^*(x)$ is corresponding $\tilde{\Delta}^{(2\beta)}(x)$ at he l^*-th outer loop.

Output: $\Lambda^*(x)$, $\Omega^*(x)$

4.2 Unified hybrid decoding (UHD) architecture

The overall UHD architecture is shown in Fig. 9. Here different blocks are used to process different steps in algorithm. Since excluding KES and PT blocks, other blocks are quite straightforward to be implemented; in this section we only introduce the architectures of KES and PT blocks and focus the discussion for the case of RiBC work mode. Interested readers can refer to [23] for the introduction of other blocks and other modes.

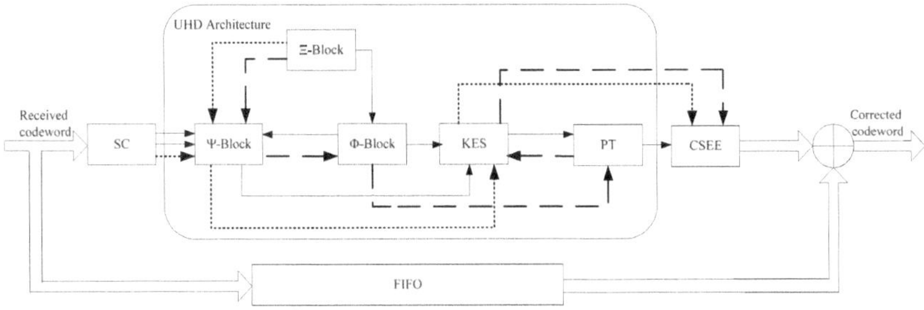

Figure 9. The overall architecture of the UHD decoder. Three types of lines illustrate data flows for different work modes: solid line (mode-1) for burst combined with random error correction RiBC algorithm, dashed line (mode-2) for only burst-error correction and dotted line (mode-3) for only random error correction.

4.2.1. KES block architecture

For RiBC algorithm, KES block is employed to carry out steps 2.4. Fig. 10 presents the overall architecture of KES block and the internal structure of its two types of processing elements (PE): PE0 and PE1. As shown in Fig. 10(a), the KES block consists of 2*t*-1 PE0's and 2*t* PE1's. In the *r*-th iteration, each register in $PE0_i/PE1_i$ stores the corresponding coefficients of different polynomials (Fig. 10(b) (c)). For each outer iteration, it takes 2β cycles to compute $\lambda_i^{(2\beta)}$ and $\tilde{\delta}_i^{(2\beta)}$ as the coefficients of $\Lambda^{(2\beta)}(x)$ and $\tilde{\Delta}^{(2\beta)}(x)$. Meanwhile, $\tilde{\lambda}_i^{(2\beta)}$ will also be computed and outputted into PT block to track the longest consecutive $\tilde{\Lambda}^{(2\beta)}(x)$ that are identical.

4.2.2. Position track (PT) block architecture

PT block is used to track the longest consecutive polynomials that are identical (step 3). Fig. 11 illustrates the architecture of PT block. The input $\tilde{\lambda}_i^{(2\beta)}$, $\lambda_i^{(2\beta)}$ and $\tilde{\delta}_i^{(2\beta)}$ from KES block at the *l*-th outer iteration are denoted as $\tilde{\lambda}_i(l)$, $\lambda_i(l)$ and $\tilde{\delta}_i(l)$. In addition, $\tilde{\lambda}_i(temp)$ represents $\tilde{\lambda}_i(l-1)$, while $\tilde{\lambda}_i(store)$ are the coefficients of current continuously identical $\tilde{\Lambda}^{(2\beta)}(x)$. Moreover, $\tilde{\lambda}_i(longest)$ stores the coefficients of current *longest* continuously identical $\tilde{\Lambda}^{(2\beta)}(x)$. Control signals *shift* and *equal* are generated from the signal generation schedule. After *l* reaches n, $\lambda_i(longest)$ and $\tilde{\delta}_i(longest)$ are outputted as the coefficients of overall error locator polynomial $\Lambda^*(x)$ and overall error evaluator polynomial $\Omega^*(x)$.

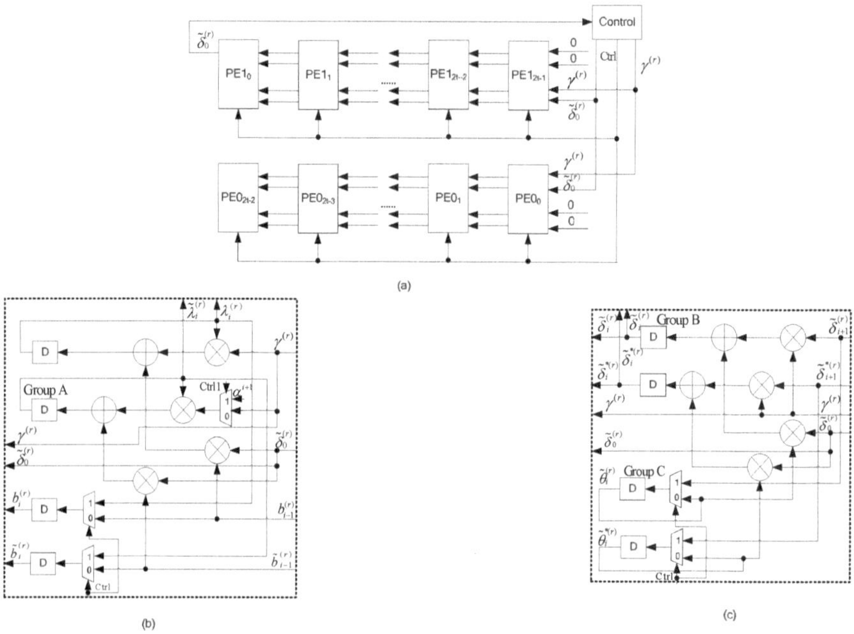

Figure 10. (a) The overall architecture of KES block. (b) The block diagram of PE0*i*. (c) The block diagram of PE1*i*.

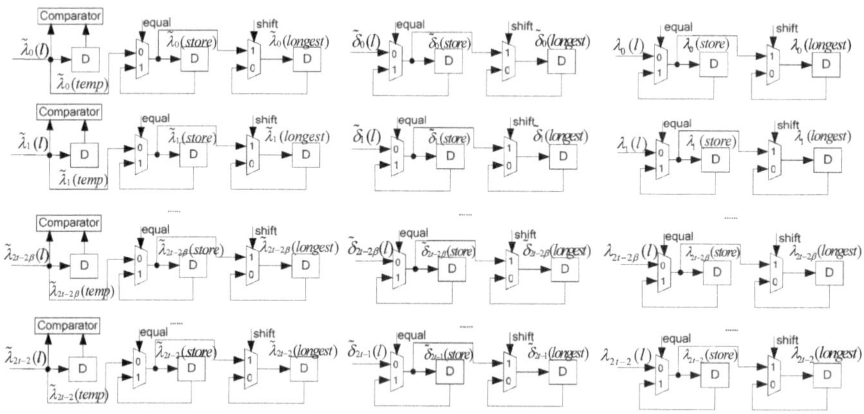

Figure 11. The architecture of PT block for mode-1.

Table 3 presents the comparison between UHD and RiBM decoder. Here for the example RS (255, 239) code, n=255, t=8 and m=8. The hardware complexity is estimated based on the work in [24]. Although the area requirement of the UHD decoder is about 1.7 times of that of the RiBM decoder, the UHD decoder can achieve significantly enhanced burst-error

---

Track Control Signal Generation Schedule for RiBC Algorithm

Initiliazation: $l_1 = 0,\ l_2 = 1$;

Step S1: Input: $\tilde{\lambda}_i^{(2\beta)}$ for current l, denote them as $\tilde{\lambda}_i\,(l)$ for $i = 0,1 \ldots 2t - 2\beta$;

Step S2: If $\tilde{\lambda}_i\,(l) = \tilde{\lambda}_i\,(temp)$ for all $i = 0,1 \ldots 2t - 2\beta$ then $equal = 1,\ l_2 = l_2 + 1$;
  else $equal = 0,\ l_2 = l_1$;

Step S3: If $l_2 > l_1$ then $shift = 1$;
  else $shift = 0$;

Step S4: If $shift = 1$ then $l_1 = l_2$;
  else l_1 remains;

Step S5: Output $equal,\ shift$;

Step S6: Goto Step S1

Architecture		UHD	RiBM
Total gates(# of XOR gates)		34308	18968
		44392 (two codewords)	
Critical path (# of gates)		10	8
(Mode-1) (f=11, β=1)	**Latency**	4846	Unavailable to decode
	Throughput (Normalized)	1	
	Miscorrection Probability	$(n\text{-}2f)(n\text{-}f)^{\beta}2^{m(\beta+f-2t)}$ $=1.32*10^{-5}$	
(Mode-2) (f=12)	**Latency**	542 (the worst case)	Unavailable to decode
	Throughput (Normalized)	8.9~16.8	
	Miscorrection Probability	$(n\text{-}2f)2^{m(f-2t)}$ $=5.38*10^{-8}$	
(Mode-3) ($t \leq 8$, 2t+ρ≤16)	**Latency**	255	255
	Throughput (Normalized)	19	23.8
		38 (two codewords)	
	Miscorrection Probability	0	0

Table 3. Comparisons of performance on hardware and error correction capability.

correcting capability. In the channel environments that likely generate long burst of errors (f>8), the traditional RiBM decoder fails to decode the codewords for its limited error correcting capability, while UHD decoder can be still effective. In short, the UHD design provides an efficient and attractive unified solution for multi-mode RS decoding in optical applications that demands enhanced error correcting capability.

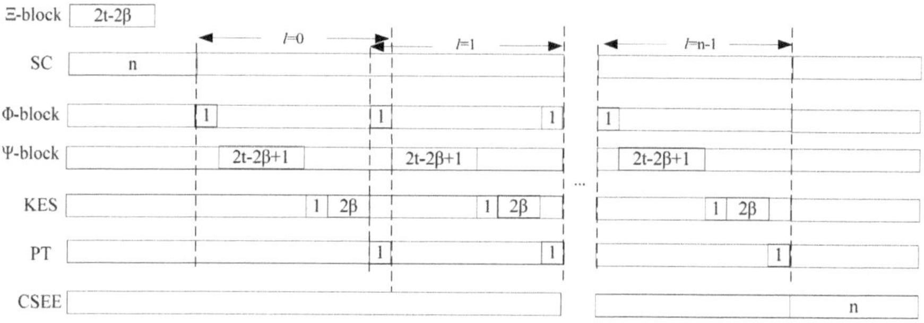

Figure 12. The timing charts for RiBC architecture.

5. Ultra high performance FEC schemes for long-haul network

For long-haul optical transport networks (OTN), because the performance loss mainly results from long distance transmission, the requirement on coding gain is very strict. This requirement even gets more and more strict when optical backbone networks enter 100Gbps era. Based on OIF whitepaper, the new FEC schemes applied in 100G long-haul systems should achieve waterfall performance at very low BER region. Meanwhile the other requirements for OTN FEC such as capable of achieving very high speed and having very low error floor still remain. Therefore, 100Gbps era puts more challenges on FEC schemes. In this section, some recent FEC schemes targeted to 100Gbps applications are introduced.

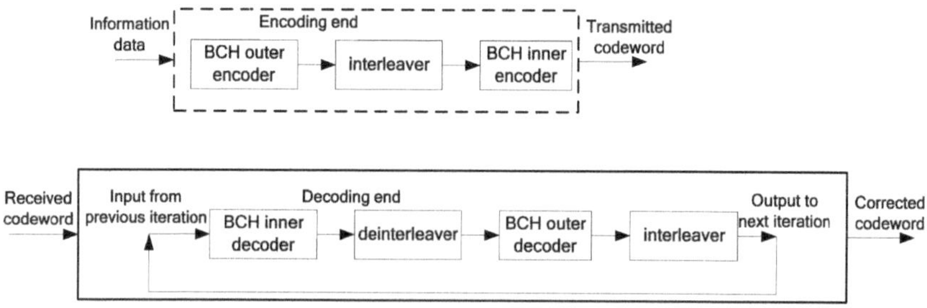

Figure 13. The hard-decision product BCH scheme.

5.1. Hard-decision BCH product codes

One candidate for 100Gbps application is BCH-based binary product codes such as the one presented in [25]. The component BCH (992, 960) and (987, 956) codes are constructed carefully over $GF(2^{10})$ for hardware amenity, which have the 3-error correction capability, therefore its decoder design can be developed based on simple PGZ algorithm in [4]. The simulation results show the performance of this FEC scheme can be very close to the Shannon limit.

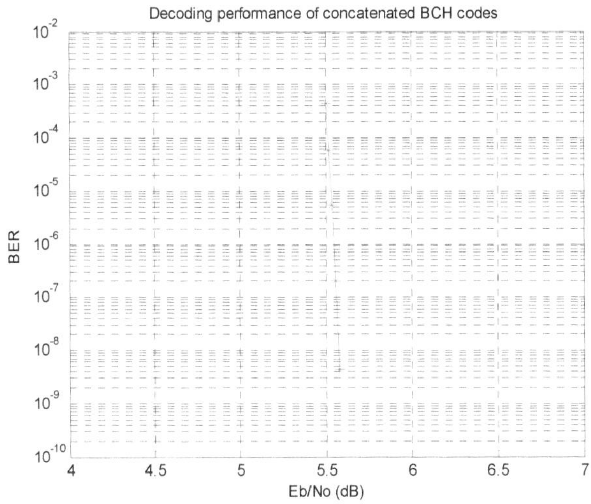

Figure 14. Decoding performance of product BCH codes based on maximum 7 iterations.

5.2. Some other soft-decision based concatenated codes

The above binary product scheme is based on hard-decision. If soft information is available in the system, soft-decision decoding approach can work with the product codes to enhance the overall decoding performance. Fig. 15 illustrates a LDPC code concatenated with BCH-based product code for long-haul network systems in [26]. In this scheme, LDPC code is used as inner code and BCH-based product code is used as outer code. Some other soft-decision based concatenated FEC scheme such as RS code concatenated with LDPC coding system in [27] can also provide significant coding gain for targeted ultra high-speed optical communication.

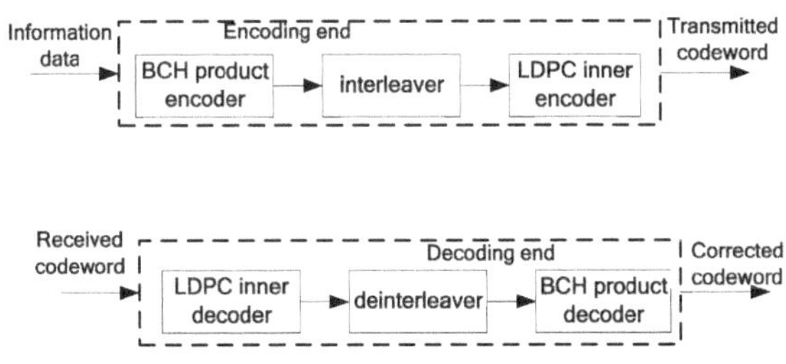

Figure 15. Product BCH-LDPC concatenated scheme in [26].

6. Conclusion

With the evolution of optical network, the employed FEC scheme has been developed in several generations. The requirement on high data rata and large coding gain is always challenging the design for efficient FEC decoder. In this chapter, targeted to different types of optical transmission networks, ranging from local Ethernet to long-haul backbone system, different FEC solutions with efficient VLSI implementations are discussed. For short-distance networks, two kinds of area-efficient high-speed RS decoders are analyzed for the scenario. For mediate distance networks, which require some tradeoff between decoding performance and hardware efficiency, the introduced RS burst-error decoder can be employed to meet such requirement. For long-haul systems, which have stringent requirement on decoding performance, some candidate FEC schemes targeted to the future 100Gbps era are discussed. In summary, these various FEC architectures and schemes are good candidates for their specific targeted optical transmission applications.

Author details

Bo Yuan
University of Minnesota-Twin Cities, USA

Li Li
Nanjing University, China

Zhongfeng Wang
Broadcom Corporation, USA

7. References

[1] 100G Forward Error Correction White Paper, OIF, OIF-FEC-100G-01.0, May. 2010.
[2] Forward Error Correction for high bit-rate DWDM Submarine System, Telecommunication Standardization Section, International Telecom Union, ITU-T G. 975.1, Feb. 2004.
[3] H. -Y. Hsu, A.-Y. Wu and J. -C. Yeo. Area-efficient VLSI design of Reed-Solomon decoder for 10Gbase-LX4 optical communication systems. IEEE Transactions on Circuits and Systems II 2006; 53(11) 1245-1249.
[4] E. R. Berlekamp. Algebraic coding theory. Newyork: McGraw-Hill; 1968.
[5] D. V. Sarwate and N. R. Shanbhag. High-speed architectures for Reed-Solomon decoders. IEEE Transactions on VLSI Systems 2001; 9(5) 641-655.
[6] H. M. Shao, T. K. Truong, L. J. Deutsch, J. H. Yuen and I. S. Reed. A VLSI design of a pipeline Reed-Solomon decoder. IEEE Transactions on Computer 1985; 34(5) 393-403.
[7] H. Lee. A High-speed low-complexity Reed-Solomon decoder for optical communications. IEEE Transactions on Circuits and Systems II 2005; 52(8) 461-465.

[8] K. Seth, K. N. Viswajith, S. Srinivasan, and V. Kamakoti. Ultra folded high-speed architectures for Reed-Solomon decoders: proceeding of International. Conference on VLSI Design, 3-7 Jan. 2006.
[9] B. Yuan, Z. Wang, L. Li, M. Gao, J. Sha and C. Zhang. Area-efficient Reed-Solomon decoder design for optical communications. IEEE Transactions on Circuits and Systems II 2009; 56(6) 469-473.
[10] S. Lee, H. Lee, J Shin and Je-Soo Ko. A high-speed pipelined degree-computationless modified Euclidean algorithm architecture for Reed-Solomon decoders: proceeding of IEEE International. Sympousium on. Circuits and System, 27-30 May 2007.
[11] J. H. Baek and M. H. Sunwoo. New degree computationless modified Euclid's algorithm and architecture for Reed-Solomon decoder. IEEE Transactions on VLSI Systems 2006; 14(8) 915-920.
[12] I. S. Reed, M. T. Shih, and T. K. Truong. VLSI design of inverse-free Berlekamp-Massey algorithm. proceeding of IEE, part. E 1991; 138(5) 295-298.
[13] T. Horiguchi, High-speed Decoding of BCH Codes Using a New Error-evaluation Algorithm, Electronics and Communications in Japan Part 3 1989; 72(12) 63-71.
[14] Z. Yu and W. Feng. Efficient high-speed Reed-Solomon dcoder, U.S. Patent 7 322 004, Jan 22, 2008.
[15] K. K. Parhi. VLSI digital signal processing systems: Design and implementation. MA: Wiley; 1999.
[16] B. Yuan, L. Li, J. Sha, and Z. Wang. Area-efficient RS decoder design for 10-100 Gb/s applications: proceeding of IEEE International. Sympousium on. Circuits and System, 24-27 May 2009.
[17] S. Rizwan. Retimed decomposed serial Berlekamp-Massey architecture for high-speed Reed-Solomon decoding: proceeding of International Conference on VLSI Design, 4-8 Jan. 2008.
[18] M. –D. Shieh, Y. –K. Lu, S. –M. Chung, and J. –H. Chen. Design and implementation of efficient Reed-Solomon decoders for multi-mode applications: proceeding of IEEE International. Sympousium on. Circuits and System, 21-24 May 2006.
[19] H. Lee. High-speed VLSI architecture for parallel Reed-Solomon decoder. IEEE Transactions on VLSI Systems 2003; 11(2) 288-294.
[20] Y. Wu. Novel burst error correcting algorithms for Reed-Solomon codes. proceeding of IEEE Allerton Conference on Communication, Control and Computing, Sep. 30 - Oct. 2 2009.
[21] E. Dawson and A. Khodkar. Burst error-correcting algorithm for Reed-Solomon codes. Electronics Letters 1995; 31(11) 848-849.
[22] L. Yin, J. Lu, K. B. Letaief and Y. Wu. Burst-error-correcting algorithm for Reed-Solomon codes. Electronics Letters 2001; 37(11) 695-697.
[23] L. Li, B. Yuan, Z. Wang, J. Sha, H. Pan and W. Zheng. Unified Architecture for Reed-Solomon Decoder Combined with Burst-Error Correction. IEEE Transactions on VLSI Systems, to appear.

[24] X. Zhang and J. Zhu. High-throughput interpolation architecture for algebraic soft-decision Reed-Solomon decoding. IEEE Transactions on Circuits and Systems-I 2010; 57(3) 581-591.

[25] Z. Wang, C-J. Chen, K. Xiao, H. Jiang, J. R. Fife, and S. Bhoja. Communication device employing binary product coding with selective additional cyclic redundancy check(CRC) therein. U.S. Patent 12 726 062, Mar. 17, 2010.

[26] Z. Wang, B. Shen, K. Xiao and J. R. Fife. Communication device employing LDPC coding with Reed-Solomon and/or binary product coding. U.S. Patent 12 726 219, Mar. 17, 2010.

[27] T. Mizuochi, Y. Konishi, Y. Miyata, T. Inoue, K. Onohara, S. Kametani, T. Sugihara, K. Kubo, H. Yoshida, T. Kobayashi, and T. Ichikawa. Experimental demonstration of concaenated LDPC and RS codes by FPGAs emulation. IEEE photonics technology letter 2009; 21(18) 1302 – 1304.

Rate-Adaptive Free-Space Optical Links Over Atmospheric Turbulence and Misalignment Fading Channels

Antonio García-Zambrana, Carmen Castillo-Vázquez and Beatriz Castillo-Vázquez

Additional information is available at the end of the chapter

1. Introduction

Atmospheric free-space optical (FSO) transmission using intensity modulation and direct detection (IM/DD) can provide high-speed links for a variety of applications, being an interesting alternative to consider for next generation broadband in order to support large bandwidth, unlicensed spectrum, excellent security, and quick and inexpensive setup [9]. Recently, the use of FSO transmission is being specially interesting to solve the *"last mile" problem*, as well as a supplement to radio-frequency (RF) links [35, 41]. However, atmospheric turbulence produces fluctuations in the irradiance of the transmitted optical beam, which is known as *atmospheric scintillation*, severely degrading the link performance [3, 50]. Additionally, since FSO systems are usually installed on high buildings, building sway causes vibrations in the transmitted beam, leading to an unsuitable alignment between transmitter and receiver and, hence, a greater deterioration in performance.

Error control coding as well as diversity techniques can be used over FSO links to mitigate turbulence-induced fading [6, 7, 13, 33, 44]. In [20, 24, 26], selection transmit diversity is proposed for FSO links over strong turbulence channels, where the transmit diversity technique based on the selection of the optical path with a greater value of irradiance has shown to be able to extract full diversity as well as providing better performance compared to general FSO space-time codes (STCs) designs, such as conventional orthogonal space-time block codes (OSTBCs) and repetition codes (RCs). The combined effect of atmospheric and misalignment fading is analyzed in the case of single-input/single-output (SISO) FSO channels in [5]. In [16], the effects of atmospheric turbulence and misalignment considering aperture average effect were considered to study the outage capacity for SISO links. In [38] the error rate performance for coded FSO links over strong turbulence and misalignment fading channels is studied. The capacity calculation and the analysis of error rate performance for FSO links over log-normal and gamma-gamma turbulence and misalignment fading channels is presented in [8] using a wave optics based approach. In [18, 39], a wide range of turbulence conditions with gamma-gamma atmospheric turbulence and pointing errors is also considered on terrestrial FSO links, deriving closed-form expressions for the error-rate performance in terms of Meijer's G-functions. In [36], average capacity optimization is considered in this same context by numerically solving the derivative of the corresponding capacity expression, also mathematically treated as a Meijer's G-function. In [17], the study of the outage probability and diversity gain has been considered for multiple-input/multiple-output (MIMO) FSO communication systems impaired by log-normal atmospheric turbulence and misalignment fading, assuming repetition coding on the transmitter side and equal gain combining on the receiver side and showing that the diversity gain is conditioned only by misalignment parameters. In [21, 25], comparing different diversity techniques, a significant improvement in terms of outage and error-rate performance is demonstrated when MIMO FSO links based on transmit laser selection are adopted in the context of wide range of turbulence conditions (weak to strong), showing that better performance is achieved when increasing the number of transmit apertures instead of the number of receive apertures in order to guarantee a same diversity order.

An alternative approach to improving the performance in this turbulence FSO scenario is the employment of rate-adaptive transmission in order to make suitable the communication to the adverse channel conditions, depending on the available signal-to-noise ratio (SNR) until a sufficiently low error probability can be attained. Various FSO systems using adaptive modulation have been proposed [10, 12, 14, 34]. In [34], a variable rate FSO system employing adaptive Turbo-based coding schemes with on-off keying (OOK) formats was investigated. In [12, 14], an adaptive transmission scheme that varied both the power and the modulation order of a FSO system with M-ary pulse amplitude modulation (MPAM) has been studied. In [10], an adaptive transmission technique employing subcarrier phase shift keying (S-PSK) intensity modulation has been proposed. Another solution is the employment of adaptive transmission based on repetition coding, a well known technique employed in RF systems [2]. Based on the concept of temporal-domain diversity reception (TD-DR), this idea has been applied for FSO links in [32, 42, 43], where two separate channels over the same transmit and receive path are implemented. Both channels carry the same data, but one of the channels is delayed by the expected fade duration. In [23], a new and simple rate-adaptive transmission scheme for FSO communication systems with intensity modulation and direct detection over atmospheric turbulence channels is analyzed. This scheme is based on the joint use of repetition coding and variable silence periods, exploiting the potential time-diversity order (TDO) available in the turbulent channel as well as allowing the increase of the peak-to-average optical power ratio (PAOPR), which has shown to be a favorable characteristic in IM/DD FSO links [19, 24, 26].

In this chapter, this approach is extended to FSO communication systems using IM/DD over atmospheric turbulence channels with pointing errors. Novel closed-form asymptotic expressions are derived when the irradiance of the transmitted optical beam is susceptible to either a wide range of turbulence conditions (weak to strong), following a gamma-gamma distribution of parameters α and β, or pointing errors, following a misalignment fading model, as in [16–18, 38], where the effect of beam width, detector size and jitter variance is considered. Obtained results provide significant insight into the impact of various system and channel parameters, showing that the time-diversity order that can be fully exploited is independent of the pointing error when the equivalent beam radius at the receiver is at least $2(\min\{\alpha, \beta\})^{1/2}$ times the value of the pointing error displacement standard deviation at the receiver. In this fashion, the same slope of the bit-error rate (BER) performance versus average SNR as in a similar FSO scenario without pointing errors is corroborated. However, different coding gain, i.e. the horizontal shift in the BER performance in the limit of large SNR, is achieved as a consequence of the severity of the pointing error effects and turbulence conditions. Here, not only rectangular pulses are considered but also on-off keying (OOK) formats with any pulse shape, corroborating the advantage of using pulses with high PAOPR. Moreover, since proper FSO transmission requires transmitters with accurate control of their beamwidth, asymptotic expressions are used to find the optimum beamwidth that minimizes the BER at different turbulence conditions. Simulation results are further demonstrated to confirm the accuracy and usefulness of the derived results, showing that asymptotic expressions here obtained lead to simple bounds on the bit error probability that get tighter over a wider range of SNR as the turbulence strength increases.

2. System and channel model

The use of infrared technologies based on IM/DD links is considered, where the instantaneous current in the receiving photodetector, $y(t)$, assumed to be ideal noncoherent (direct-detection) receiver can be written as

$$y(t) = \eta i(t)x(t) + z(t) \tag{1}$$

where η is the detector responsivity, assumed hereinafter to be the unity, $X \triangleq x(t)$ represents the optical power supplied by the source, assumed to be intensity-modulated, and $I \triangleq i(t)$ the equivalent real-valued fading gain (irradiance) through the optical channel between the laser and the receive aperture; $Z \triangleq z(t)$ is assumed to include any front-end receiver thermal noise as well as shot noise caused by ambient light much stronger than the desired signal at detector. In this case, the noise can usually be modeled to high accuracy as AWGN with zero mean and variance $N_0/2$, i.e. $Z \sim N(0, N_0/2)$, independent of the on/off state of the received bit [31]. Since the transmitted signal is an intensity, X must satisfy $\forall t\, x(t) \geq 0$. Due to eye and skin safety regulations, the average optical power is limited and, hence, the average amplitude of X is limited. Although limits are placed on both the average and peak optical power transmitted, in the case of most practical modulated optical sources, it is the average optical power constraint that dominates [30]. The received electrical signal $Y \triangleq y(t)$, however, can assume negative amplitude values. We use Y, X, I and Z to denote random variables and $y(t)$, $x(t)$, $i(t)$ and $z(t)$ their corresponding realizations. In this fashion, the atmospheric turbulence channel model consists of a multiplicative noise model, where the optical signal is multiplied by the channel irradiance.

We consider OOK formats with any pulse shape and reduced duty cycle, allowing the increase of the PAOPR parameter [19, 26]. In spite of the atmosphere can cause pulse distortion and broadening during propagation at extremely high signalling rates and, especially, at high levels of turbulence strength when increasing either C_n^2 or the path length L, or both, pulse shape distortion is assumed to be negligible for the typical FSO scenario here analyzed [15, 49]. A new basis function $\phi(t)$ is defined as $\phi(t) = g(t)/\sqrt{E_g}$ where $g(t)$ represents any normalized pulse shape satisfying the non-negativity constraint, with $0 \leq g(t) \leq 1$ in the bit period and 0 otherwise, and $E_g = \int_{-\infty}^{\infty} g^2(t)dt$ is the electrical energy. In this way, an expression for the optical intensity can be written as

$$x(t) = \sum_{k=-\infty}^{\infty} a_k \frac{2T_b P}{G(f=0)} g\left(t - kT_b\right) \tag{2}$$

where $G(f = 0)$ represents the Fourier transform of $g(t)$ evaluated at frequency $f = 0$, i.e. the area of the employed pulse shape, and T_b parameter is the bit period. The random variable (RV) a_k follows a Bernoulli distribution with parameter $p = 1/2$, taking the values of 0 for the bit "0" (off pulse) and 1 for the bit "1" (on pulse). From this expression, it is easy to deduce that the average optical power transmitted is P, defining a constellation of two equiprobable points in a one-dimensional space with an Euclidean distance of $d = 2P\sqrt{T_b\xi}$ where $\xi = T_bE_g/G^2(f = 0)$ represents the square of the increment in Euclidean distance due to the use of a pulse shape of high PAOPR, alternative to the classical rectangular pulse.

Since atmospheric scintillation is a slow time varying process relative to typical symbol rates of an FSO system, having a coherence time on the order of milliseconds, we consider the time variations according to the theoretical block-fading model, where the channel fade remains constant during a block (corresponding to the channel coherence interval) and changes to a new independent value from one block to next. In other words, channel fades are assumed to be independent and identically distributed (i.i.d.). This temporal correlation can be overcome by means of long interleavers, being usually assumed both in the analysis from the point of view of information theory and error rate performance analysis of coded FSO links [4, 37, 45]. However, as in [48], we here assume that the interleaver depth can not be infinite and, hence, we can potentially benefit from a degree of time diversity limited equal to TDO. This consideration is justified from the fact that the latency introduced by the interleaver is not an inconvenience for the required application. For example, for repetition coding and a time diversity order available of TDO=2, i.e. two channel fades i_1 and i_2 per frame, perfect interleaving can be done by simply sending the same information delayed by the expected fade duration, as shown experimentally in [32] for a rate reduction of 2. In this system model, shown in Fig. 1, it is assumed that channel state information (CSI) is known

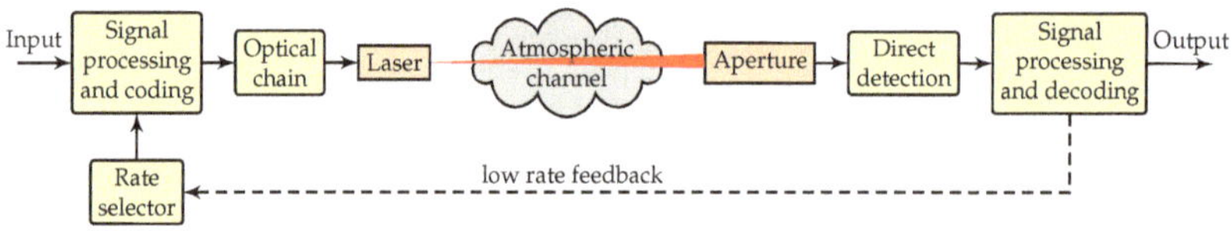

Figure 1. FSO system with rate-adaptive transmission.

not only at the receiver but also at the transmitter (CSIT). The knowledge of CSIT is feasible for FSO channels given that scintillation is a slow time varying process relative to the large symbol rate. Adaptive transmission here considered is based on reducing the initial bit rate, $R_b = 1/T_b$, for a rate reduction (RR) parameter as R_b/RR in order to satisfy a predefined BER requirement.

2.1. Proposed rate-adaptive transmission scheme

In this subsection, we firstly explain the case corresponding to the rate-adaptive transmission only based on variable silence periods; secondly, we present the case corresponding to the rate-adaptive transmission only based on repetition coding and, finally, we consider the adaptive transmission scheme here proposed and based on the joint use of repetition coding and variable silence periods, exploiting the potential time-diversity order available in the turbulent channel as well as allowing the increase of the PAOPR, which has shown to be a favorable characteristic in IM/DD links.

A rate-adaptive transmission scheme based on variable silence periods has already been proposed in [28] for indoor optical wireless communication systems as a consequence of the good performance of signaling techniques having a reduced duty cycle [27] and, hence, providing a high PAOPR. The advantage of using pulses with high PAOPR has also been corroborated in FSO links over atmospheric turbulence channels by closed-form expressions corresponding to the average BER performance [19, 24, 26], suggesting its application to the turbulent FSO scenario. A similar expression to (2) for the optical intensity corresponding to this adaptive technique only based on variable silence periods can be written as

$$x_{RR_s}(t) = \sum_{k=-\infty}^{\infty} a_k \frac{RR_s 2T_b P}{G(f=0)} g\left(t - kRR_s T_b\right) \tag{3}$$

where $RR_s - 1$ is the number of silence bit periods added in order to accomodate the transmission rate to the channel conditions, so that the higher rate reduction ($RR = RR_s$), the larger silence time. From this expression, it is easy to deduce that an increase of the PAOPR is required in order to maintain the average optical power at the same constant level of P.

In relation to the rate-adaptive transmission only based on repetition coding, the expression for the optical intensity can be written as

$$x_{RR_{rc}}(t) = \sum_{k=-\infty}^{\infty} a_k \frac{2T_b P}{G(f=0)} \sum_{l=0}^{RR_{rc}-1} g\left(t - lT_b - kRR_{rc}T_b\right) \tag{4}$$

where RR_{rc} represents the repetition of each information bit and, hence, a rate reduction of $RR = RR_{rc}$ is considered. From this expression, it is easy to deduce that an increase of the PAOPR is not achieved; however, unlike rate-adaptive transmission only based on variable silence periods, a potential diversity gain can be exploited based on the concept of temporal-domain diversity reception [32]. To the best of the authors' knowledge, closed-form expressions for the error rate performance corresponding to this rate-adaptive scheme in the context proper to FSO systems over atmospheric turbulence channels with pointing errors where a limited time-diversity order is only available have not been reported in the open literature.

Next, we combine the best of previous approaches, proposing a novel adaptive transmission scheme based on the joint use of repetition coding and variable silence periods and, this way, exploiting the potential time-diversity order available in the turbulent channel as well as allowing the increase of the PAOPR. In this fashion, the expression for the optical intensity can be written as

$$x_{RR_{rcs}}(t) = \sum_{k=-\infty}^{\infty} a_k \frac{RR_s 2T_b P}{G(f=0)} \sum_{l=0}^{RR_{rc}-1} g\left(t - lRR_s T_b - kRRT_b\right) \tag{5}$$

where the final rate reduction $RR = RR_{rc} \cdot RR_s$. In this adaptive transmission technique, repetition coding is firstly used in order to accomodate the transmission rate to the channel conditions until the whole time diversity order available in the turbulent channel by interleaving is exploited, i.e. $RR_{rc} \leq$ TDO and $RR_s = 1$. Then, once no more diversity gain is available, the rate reduction can be increased by using variable silence periods in order to increase the PAOPR, i.e. $RR_{rc} =$ TDO and $RR_s > 1$. For the sake of simplicity, we here consider values multiples of 2 for RR when rate reduction is applied, i.e $RR = \{1, 2, 4, 8, \cdots\}$

as well as for the time-diversity order effective TDO which is provided by interleaving, i.e. TDO $\bullet$ $\{1,2,4\}$, in this fashion, allowing to satisfy different latency requirements in the system [48].

2.2. Atmospheric turbulence channel

The irradiance is susceptible to either atmospheric turbulence conditions and pointing error effects. In this case, it is considered to be a product of two independent random variables, i.e. $I \bullet I^{\bullet a \bullet} I^{\bullet p \bullet}$, representing $I^{\bullet a \bullet}$ and $I^{\bullet p \bullet}$ the attenuation due to atmospheric turbulence and the attenuation due to geometric spread and pointing errors, respectively, between transmitter and receiver. Although the effects of turbulence and pointing are not strictly independent, for smaller jitter values they can be approximated as independent [8]. To consider a wide range of turbulence conditions (weak to strong), the gamma-gamma turbulence model proposed in [1, 3] is here assumed, whose probability density function (PDF) is given by

$$f_{I^{\bullet a \bullet}} \bullet i \bullet \bullet \frac{2 \bullet \alpha\beta \bullet^{\bullet \alpha \bullet \beta \bullet /2}}{\Gamma \bullet \alpha \bullet \Gamma \bullet \beta \bullet} i^{\bullet \bullet \alpha \bullet \beta \bullet /2 \bullet -1} K_{\alpha-\beta}\left(2\sqrt{\alpha\beta i}\right), \quad i \geq 0 \tag{6}$$

where $\Gamma \bullet \cdot \bullet$ is the well-known Gamma function and $K_{\nu} \bullet \cdot \bullet$ is the νth-order modified Bessel function of the second kind [29, eqn. (8.43)]. The parameters α and β can be selected to achieve a good agreement between Eq. •6• and measurement data [1]. Alternatively, assuming spherical wave propagation, α and β can be directly linked to physical parameters through the following expresions [1, 7, 45]:

$$\alpha \bullet \left[\exp\left(\frac{0.49\chi^2}{\bullet 1 \bullet \ 0.18d^2 \bullet \ 0.56\chi^{12/5 \bullet 7/6}}\right) - 1\right]^{-1} \tag{7}$$

$$\beta \bullet \left[\exp\left(\frac{0.51\chi^2 \bullet 1 \bullet \ 0.69\chi^{12/5 \bullet -5/6}}{\bullet 1 \bullet \ 0.9d^2 \bullet \ 0.62d^2\chi^{12/5 \bullet 7/6}}\right) - 1\right]^{-1} \tag{8}$$

where $\chi^2 \bullet 0.5C_n^2 k^{7/6} L^{11/6}$ and $d \bullet \bullet kD^2/4L \bullet^{1/2}$. Here, $k \bullet 2\pi/\lambda$ is the optical wave number, λ is the wavelength, D is the diameter of the receiver collecting lens aperture and L is the link distance in meters. C_n^2 stands for the altitude-dependent index of the refractive structure parameter and varies from $10^{-13}\ m^{-2/3}$ for strong turbulence to $10^{-17}\ m^{-2/3}$ for weak turbulence [3, 45]. Since the mean value of this turbulence model here considered is normalized and the second moment is given by $E \bullet I^2 \bullet \bullet \bullet 1 \bullet 1/\alpha \bullet \bullet 1 \bullet 1/\beta \bullet$, the scintillation index (SI), a parameter of interest used to describe the strength of atmospheric fading, is defined as

$$SI \bullet \frac{E \bullet I^2 \bullet}{\bullet E \bullet I \bullet \bullet^2} - 1 \bullet \frac{1}{\alpha} \bullet \frac{1}{\beta} \bullet \frac{1}{\alpha\beta}. \tag{9}$$

It must be noted that PDF in Eq. •6• contains other turbulence models adopted in strong turbulence FSO scenarios such as the K-distribution ($\beta \bullet 1$ and $\alpha > 0$) and the negative exponential distribution ($\beta \bullet 1$ and $\alpha \to \infty$) [20, 38, 44]. From the point of view of scintillation index, it is easy to deduce the fact that the strength of atmospheric fading represented by the gamma-gamma distributed turbulence model with channel parameters $\beta \bullet 1$ and increasing α tends to be closer and closer to 1, SI corresponding to the negative exponential distributed turbulence model. Regarding to the impact of pointing errors, we use the general model of misalignment fading given in [16] by Farid and Hranilovic, wherein the effect of beam width, detector size and jitter variance is considered. Assuming a Gaussian spatial intensity profile of beam waist radius, ω_z, on the receiver plane at distance z from the transmitter and a circular receive aperture of radius r, the PDF of $I^{(p)}$ is given by

$$f_{I^{(p)}}(i) = \frac{\varphi^2}{A_0^{\varphi^2}} i^{\varphi^2-1}, \quad 0 \leq i \leq A_0 \tag{10}$$

where $\varphi = \omega_{z_{eq}}/2\sigma_s$ is the ratio between the equivalent beam radius at the receiver and the pointing error displacement standard deviation (jitter) at the receiver, $v = \sqrt{\pi}r/\sqrt{2}\omega_z$, $\omega_{z_{eq}}^2 = \omega_z^2\sqrt{\pi}\mathrm{erf}(v)/2v\exp(-v^2)$, $A_0 = [\mathrm{erf}(v)]^2$ and $\mathrm{erf}(\cdot)$ is the error function [29, eqn. (8.250)]. Here, independent identical Gaussian distributions for the elevation and the horizontal displacement (sway) are considered, being σ_s^2 the jitter variance at the receiver. Using the previous PDFs for turbulence and misalignment fading, a closed-form expression of the combined PDF of I was derived in [39] as

$$f_I(i) = \frac{\alpha\beta\varphi^2}{A_0\Gamma(\alpha)\Gamma(\beta)} G_{1,3}^{3,0}\left(\frac{\alpha\beta}{A_0} i \,\middle|\, \begin{matrix} \varphi^2 \\ \varphi^2-1, \alpha-1, \beta-1 \end{matrix}\right), \quad i \geq 0 \tag{11}$$

where $G_{p,q}^{m,n}[\cdot]$ is the Meijer's G-function [29, eqn. (9.301)]. Even though Meijer's G-function can be expressed in terms of more familiar generalized hypergeometric functions, this PDF appears to be cumbersome to use in order to obtain simple closed-form expressions in the analysis of rate-adaptive FSO systems, leading to numerical solutions that obscure the impact of the basic system and channel parameters on performance. To overcome this inconvenience, the PDF in Eq. (11) is approximated by a single polynomial term as $f_I(i) \approx ai^b$, based on the fact that the asymptotic behavior of the system performance is dominated by the behavior of the PDF near the origin, i.e. $f_I(i)$ at $i \to 0$ determines high SNR performance [46]. Hence, using the series expansion corresponding to the Meijer's G-function [47, eqn. (07.34.06.0006.01)] and considering the fact that the two parameters α and β related to the atmospheric conditions are greater than each other, different asympotic expressions for Eq. (11), depending on the relation between the values of φ^2 and $\min\{\alpha,\beta\}$, can be written as

$$f_I(i) \approx \frac{\varphi^2(\alpha\beta)^{\min\{\alpha,\beta\}}\Gamma(|\alpha-\beta|)}{A_0^{\min\{\alpha,\beta\}}\Gamma(\alpha)\Gamma(\beta)\left(\varphi^2-\min\{\alpha,\beta\}\right)} i^{\min\{\alpha,\beta\}-1}, \quad \varphi^2 > \min\{\alpha,\beta\} \tag{12a}$$

$$f_I(i) \approx \frac{\varphi^2(\alpha\beta)^{\varphi^2}\Gamma\left(\alpha-\varphi^2\right)\Gamma\left(\beta-\varphi^2\right)}{A_0^{\varphi^2}\Gamma(\alpha)\Gamma(\beta)} i^{\varphi^2-1}, \quad \varphi^2 < \min\{\alpha,\beta\} \tag{12b}$$

It must be noted that the assumption $\alpha \neq \beta$ can be assumed under a wide variety of simulated turbulence conditions [1].

3. Error-rate performance analysis

In this section, using the previous asymptotic expressions for the combined PDF, we reveal some insights into the BER performance of rate-adaptive FSO links over atmospheric turbulence and misalignment fading channels. We can take advantage of these simpler expressions in order to quantify the bit error-rate probability at high SNR, showing that the asymptotic performance of this metric as a function of the average SNR is characterized by two parameters: the diversity and coding gains. Assuming channel side information at the receiver, we present closed-form expressions for the average BER when the scintillation follows a gamma-gamma distribution, which cover a wide range of atmospheric turbulence conditions.

For the adaptive scheme in (3) wherein only variable silence periods are used for implementing the rate reduction and the information is detected each RR_s bit periods, the conditional BER is given by

$$P_{b_s}(E|\, I) = Q\left(\sqrt{RR_s^2 d^2 i^2/2N_0}\right) \tag{13}$$

where $Q(\cdot)$ is the Gaussian-Q function defined as $Q(x) = \frac{1}{\sqrt{2\pi}}\int_x^\infty e^{-\frac{t^2}{2}}dt$. Substituting the value of the Euclidean distance d gives $P_{b_s}(E|\, I) = Q\left(\sqrt{2\gamma\xi RR_s i}\right) = Q\left(\sqrt{2\gamma\xi\Phi_c^s i}\right)$ where $\gamma = P^2T_b/N_0$ is the average receiver SNR in the presence of the turbulence [50], knowing that PDF in (6) is normalized. It can be noted that the parameter $\Phi_c^s = RR_s$ is related to the coding gain corresponding to this rate-adaptive scheme. Hence, the average BER, $P_{b_s}(E)$, can be obtained by averaging $P_{b_s}(E|\, I)$ over the turbulence PDF as follows

$$P_{b_s}(E) = \int_0^\infty Q\left(\sqrt{2\gamma\xi\Phi_c^s i}\right) f_I(i)di. \tag{14}$$

In relation to the adaptive scheme in (4) wherein only repetition coding is used for implementing the rate reduction, the information is detected each bit period, combining with the same weight RR_{rc} noisy faded signals in a similar manner to a single-input multiple-output (SIMO) FSO scheme with equal gain combining (EGC) [44] and, this way, achieving a diversity gain, never greater than TDO. When TDO $> RR_{rc}$, considering that the variance of the noise of the combined signal is $RR_{rc}N_0/2$ and substituting the value of d, the conditional BER is given by

$$P_{b_{rc}}\left(E|\, \{I_k\}_{k=1}^{RR_{rc}}\right) = Q\left(\sqrt{\frac{d^2}{2RR_{rc}N_0}}\sum_{k=1}^{RR_{rc}} i_k\right) = Q\left(\sqrt{\frac{2\gamma\xi}{RR_{rc}}}\sum_{k=1}^{RR_{rc}} i_k\right). \tag{15}$$

When TDO $< RR_{rc}$, assuming for simplicity in this analysis that RR_{rc} is multiple of TDO, the conditional BER can be written as

$$P_{b_{rc}}\left(E|\, \{I_k\}_{k=1}^{\mathrm{TDO}}\right) = Q\left(\sqrt{\frac{2\gamma\xi}{RR_{rc}}}\frac{RR_{rc}}{\mathrm{TDO}}\sum_{k=1}^{\mathrm{TDO}} i_k\right). \tag{16}$$

Both expressions can be presented in a unified way as

$$P_{b_{rc}}\left(E|\, \{I_k\}_{k=1}^{\Phi_d^{rc}}\right) = Q\left(\sqrt{2\gamma\xi\Phi_c^{rc}}\sum_{k=1}^{\Phi_d^{rc}} i_k\right) \tag{17}$$

where $\Phi_d^{rc} = \min\{\text{TDO}, RR_{rc}\}$, representing the diversity gain corresponding to the adaptive transmission scheme and Φ_c^{rc}, the parameter related to the coding gain corresponding to this rate-adaptive scheme, will be $\Phi_c^{rc} = 1/\sqrt{RR_{rc}}$ when $\text{TDO} \geq RR_{rc}$ and $\Phi_c^{rc} = \sqrt{RR_{rc}}/\text{TDO}$ when $\text{TDO} < RR_{rc}$. The average BER, $P_{b_{rc}}(E)$, can be obtained by averaging $P_{b_{rc}}\left(E|\{I_k\}_{k=1}^{\Phi_d^{rc}}\right)$ over the turbulence PDF as follows

$$P_{b_{rc}}(E) = \underbrace{\int_0^\infty \int_0^\infty \cdots \int_0^\infty}_{\Phi_d^{rc}\text{-fold}} Q\left(\sqrt{2\gamma\xi}\Phi_c^{rc} \sum_{k=1}^{\Phi_d^{rc}} i_k\right) \prod_{k=1}^{\Phi_d^{rc}} f_{I_k}(i_k) di_1 di_2 \ldots di_{\Phi_d^{rc}}. \tag{18}$$

From the expressions in (14) and (18), it can be deduced that a greater value for the parameter related to the coding gain using silence periods can be achieved once the whole time diversity order available in the turbulent channel by interleaving is exploited in order to achieve as much diversity gain as possible. In this way, repetition coding is firstly used in the rate-adaptive transmission proposed in (5) and then, once no more diversity gain is available, RR is increased by using variable silence periods in order to improve the coding gain. A better performance can be obtained when $\text{TDO} < RR$, being $RR = \text{TDO} \cdot RR_s$. In this way, the conditional BER is given by

$$P_{b_{rcs}}\left(E|\{I_k\}_{k=1}^{\Phi_d^{rcs}}\right) = Q\left(\sqrt{2\gamma\xi}\Phi_c^{rcs} \sum_{k=1}^{\Phi_d^{rcs}} i_k\right) \tag{19}$$

where $\Phi_d^{rcs} = \min\{\text{TDO}, RR\}$, representing the diversity gain corresponding to the rate-adaptive transmission and Φ_c^{rcs}, the parameter related to the coding gain corresponding to this rate-adaptive scheme, will be $\Phi_c^{rcs} = 1/\sqrt{RR}$ when $\text{TDO} \geq RR$, as in repetition coding, and $\Phi_c^{rcs} = RR_s/\sqrt{\text{TDO}}$ when $\text{TDO} < RR$, a greater value than Φ_c^{rc}. Hence, the average BER, $P_{b_{rcs}}(E)$, can be obtained by averaging $P_{b_{rcs}}\left(E|\{I_k\}_{k=1}^{\Phi_d^{rcs}}\right)$ over the turbulence PDF as follows

$$P_{b_{rcs}}(E) = \underbrace{\int_0^\infty \int_0^\infty \cdots \int_0^\infty}_{\Phi_d^{rcs}\text{-fold}} Q\left(\sqrt{2\gamma\xi}\Phi_c^{rcs} \sum_{k=1}^{\Phi_d^{rcs}} i_k\right) \prod_{k=1}^{\Phi_d^{rcs}} f_{I_k}(i_k) di_1 di_2 \ldots di_{\Phi_d^{rcs}}. \tag{20}$$

Finally, the average BER corresponding to the three previous rate-adaptive transmission schemes can be written in a unified way as

$$P_b(E) = \int_0^\infty Q\left(\sqrt{2\gamma\xi}\Phi_c i\right) f_{I_T}(i) di \tag{21}$$

where I_T represents the sum of variates $I_T = \sum_{k=1}^{\Phi_d} I_k$ whose PDF is given by $f_{I_T}(i)$, Φ_c represents the parameter related to the coding gain, being equal to Φ_c^s, Φ_c^{rc} and Φ_c^{rcs}, respectively, and Φ_d represents the diversity gain, i.e. the slope of the BER versus average SNR, being equal to 1, Φ_d^{rc} and Φ_d^{rcs}, respectively. Since the variates are independent, knowing that the resulting PDF of their sum is the convolution of their corresponding PDFs, obtained via single-sided Laplace and its inverse transforms, approximate expressions for the PDF, $f_{I_T}(i)$, of the combined variate can be easily derived as

$$f_{I_T}(i) \approx \frac{1}{\Gamma(\Phi_d \Omega_{min})} \left(\frac{(\alpha\beta)^{\Omega_{min}} \varphi^2 \Gamma(|\alpha-\beta|)}{A_0^{\Omega_{min}} \Gamma(\Omega_{max}) (\varphi^2 - \Omega_{min})} \right)^{\Phi_d} i^{\Phi_d \Omega_{min} - 1}, \quad \varphi^2 > \Omega_{min} \tag{22a}$$

$$f_{I_T}(i) \approx \frac{1}{\Gamma(\Phi_d \varphi^2)} \left(\frac{(\alpha\beta)^{\varphi^2} \Gamma(\alpha-\varphi^2)\Gamma(\beta-\varphi^2)\Gamma(1+\varphi^2)}{A_0^{\varphi^2} \Gamma(\alpha)\Gamma(\beta)} \right)^{\Phi_d} i^{\Phi_d \varphi^2 - 1}, \quad \varphi^2 < \Omega_{min} \tag{22b}$$

where $\Omega_{min} = \min\{\alpha, \beta\}$ and $\Omega_{max} = \max\{\alpha, \beta\}$ are used for readability. To evaluate the integral in Eq. (21), we can use that the Q-function is related to the complementary error function $\text{erfc}(\cdot)$ by $\text{erfc}(x) = 2Q(\sqrt{2}x)$ [29, eqn. (6.287)] and the fact that $\int_0^\infty \text{erfc}(x) x^{a-1} dx = \Gamma((1+a)/2)/(\pi^{1/2} a)$ [29, eqn. (6.281)], obtaining the corresponding closed-form asymptotic solutions for the BER as can be seen in

$$P_b(E) \doteq \left(\left(\frac{A_0^{-\Omega_{min}} (\alpha\beta)^{\Omega_{min}} \varphi^2 \Gamma(|\alpha-\beta|) (\Gamma(\Omega_{max}))^{-1}}{\left(2\Gamma\left(1+\frac{\Phi_d \Omega_{min}}{2}\right)\right)^{1/\Phi_d} (-\Omega_{min}+\varphi^2)} \right)^{-\frac{2}{\Omega_{min}}} 4\Phi_c^2 \gamma \xi \right)^{-\frac{\Phi_d \Omega_{min}}{2}}, \quad \varphi^2 > \Omega_{min} \tag{23a}$$

$$P_b(E) \doteq \left(\left(\frac{(\alpha\beta)^{\varphi^2} \Gamma(\alpha-\varphi^2) \Gamma(\beta-\varphi^2) \Gamma(1+\varphi^2)}{A_0^{\varphi^2} \left(2\Gamma\left(1+\frac{\Phi_d \varphi^2}{2}\right)\right)^{1/\Phi_d} \Gamma(\alpha)\Gamma(\beta)} \right)^{-\frac{2}{\varphi^2}} 4\Phi_c^2 \gamma \xi \right)^{-\frac{\Phi_d \varphi^2}{2}}, \quad \varphi^2 < \Omega_{min} \tag{23b}$$

The results corresponding to this FSO scenario are illustrated in the Fig. 2, when different levels of turbulence strength of $(\alpha, \beta) = (4, 2)$ and $(\alpha, \beta) = (2, 1)$ are assumed together with values of normalized jitter of $\sigma_s/r = \{1, 3\}$ and a normalized beamwidth of $\omega_z/r = \{5, 10\}$, corresponding to values of scintillation index of $SI = 0.875$ and $SI = 2$, respectively. Here, rectangular pulse shapes with $\xi = 1$ are used for values of TDO $= \{1, 2, 4\}$, rate reductions of $RR = \{1, 2, 4, 8\}$ and the three rate-adaptive transmission schemes presented in the previous section, based on: the only use of silence periods (Sil), the only use of repetition coding (Rep) and, finally, the joint use of repetition codes and variable silence periods (Rep&Sil). Monte Carlo simulation results are furthermore included as a reference, confirming the accuracy and usefulness of the derived results. Due to the long simulation time involved, simulation results only up to BER=10^{-8} are included. Simulation results corroborate that asymptotic expressions here obtained lead to simple bounds on the bit error probability that get tighter over a wider range of SNR as the turbulence strength increases. In particular, Eq. (23*a*) is corroborated in this figure since $\varphi^2 > \Omega_{min}$ is satisfied when values of $(\omega_z/r, \sigma_s/r) = (5, 1)$ and $(\omega_z/r, \sigma_s/r) = (10, 3)$ are assumed. It is straightforward to show that the average BER behaves asymptotically as $(G_c \gamma \xi)^{-G_d}$, where G_d and G_c denote diversity order and coding gain, respectively [40, 46]. At high SNR, if asymptotically the error probability behaves as $(G_c \gamma \xi)^{-G_d}$, the diversity order G_d determines the slope of the BER versus average SNR curve in a log-log scale and the coding gain G_c (in decibels) determines the shift of the curve in SNR.

In the same way, Eq. (23*b*) is corroborated in Fig. 3 when $(\alpha, \beta) = (4, 2)$ is assumed together with values of normalized jitter of $\sigma_s/r = \{1, 4\}$ and a normalized beamwidth of $\omega_z/r = 5$.

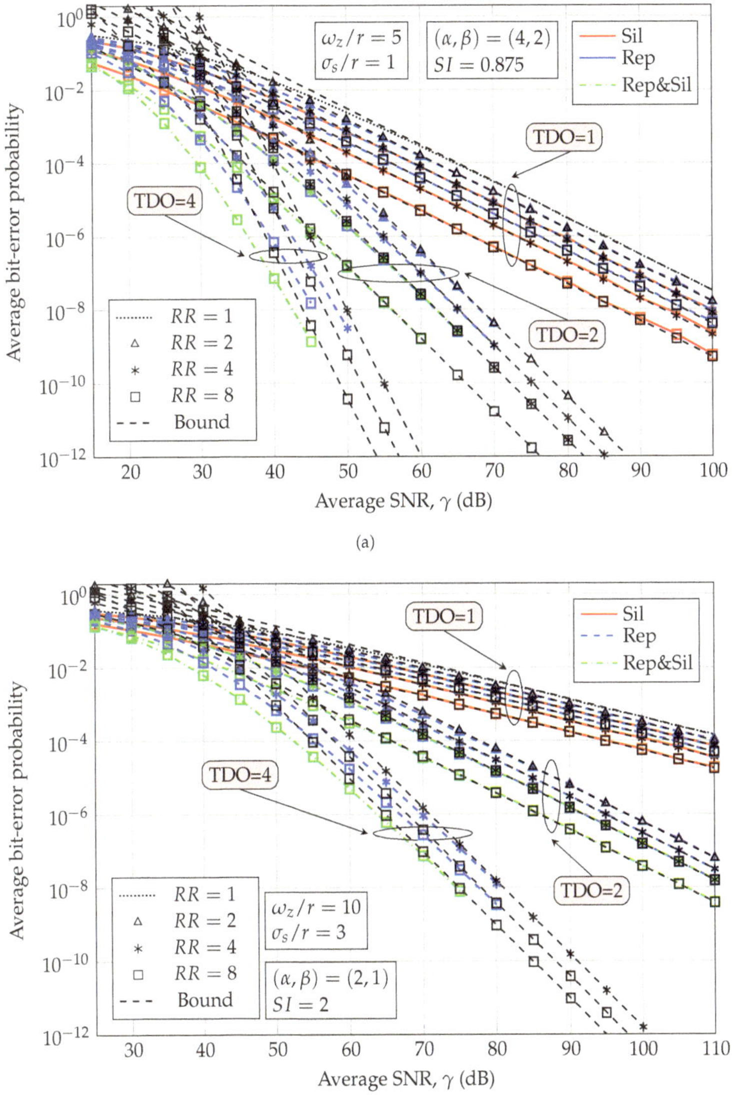

(a)

(b)

Figure 2. BER performance in FSO IM/DD links using different rate-adaptive transmission schemes over gamma-gamma atmospheric and misalignment fading channels, when different levels of turbulence (a) $(\alpha, \beta) = (5, 1)$, (b) $(\alpha, \beta) = (10, 3)$ are assumed together with values of normalized jitter of $\sigma_s/r = \{1, 3\}$ and a normalized beamwidth of $\omega_z/r = \{5, 10\}$.

Here, rectangular pulse shapes with $\xi = 1$ are used for values of TDO $= \{2,4\}$, rate reductions of $RR = \{2,4,8\}$ and rate-adaptive transmission schemes presented in the previous section, based on: the only use of repetition coding (Rep) and the joint use of repetition codes and variable silence periods (Rep&Sil). It can be noted that the bound corresponding to Eq. (23b) is displayed in this figure when $\sigma_s/r = 4$ is considered since $\varphi^2 < \Omega_{min}$ is satisfied, showing lower performance due to the diversity order is determined by the pointing errors. At this

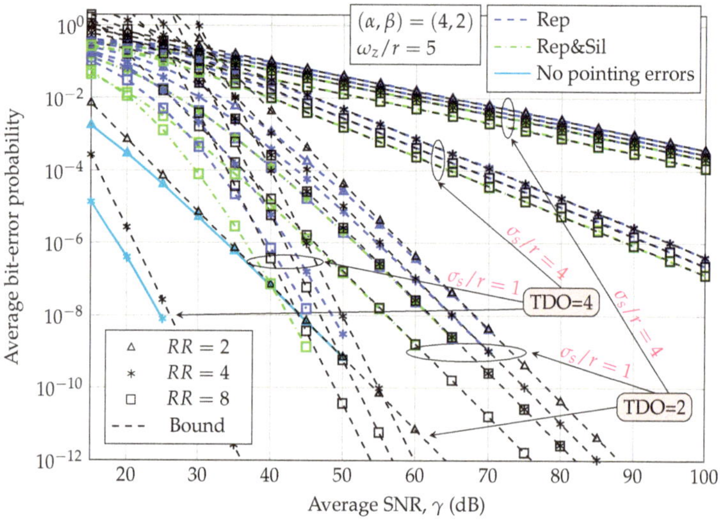

Figure 3. BER performance in FSO IM/DD links using different rate-adaptive transmission schemes over gamma-gamma atmospheric channels with/without pointing errors, assuming TDO= $\{2,4\}$ and values of normalized jitter of $\sigma_s/r = \{1,4\}$ and normalized beamwidth of $\omega_z/r = 5$.

point, it can be convenient to compare with the BER performance obtained in a similar context when misalignment fading is not present. Knowing that the impact of pointing errors in our analysis can be suppressed by assuming $A_0 \to 1$ and $\varphi^2 \to \infty$ [16], the corresponding asymptotic expression can be easily derived from Eq. (23a) as follows

$$P_b(E) \doteq \left(\left(\frac{(\alpha\beta)^{\Omega_{min}} \Gamma\left(|\alpha-\beta|\right)\left(\Gamma(\Omega_{max})\right)^{-1}}{\left(2\Gamma\left(1+\frac{\Phi_d \Omega_{min}}{2}\right)\right)^{1/\Phi_d}} \right)^{-\frac{2}{\Omega_{min}}} 4\Phi_c^2 \gamma \xi \right)^{-\frac{\Phi_d \Omega_{min}}{2}} \tag{24}$$

In Fig. 3, BER performance in FSO links using different rate-adaptive transmission schemes over gamma-gamma atmospheric channels with/without pointing errors is displayed, exploiting a potential time-diversity order available in the turbulent channel of 2 and 4, and values of σ_s/r of 1 and 4 with normalized beamwidth of $\omega_z/r = 5$, where a turbulence strength of $(\alpha,\beta) = (4,2)$ has been assumed. From this asymptotic analysis, it can be deduced that the main aspect to consider in order to optimize the error-rate performance is the relation between the values of φ^2 and $\min\{\alpha,\beta\}$. So, it is shown that the diversity order is independent of the pointing errors when the equivalent beam radius at the receiver is at least

$2(\min\{\alpha,\beta\})^{1/2}$ times the value of the pointing error displacement standard deviation at the receiver. Once this condition is satisfied, taking into account the coding gain in Eq. (23a), the impact of the pointing error effects translates into a coding gain disadvantage, $D[dB]$, relative to gamma-gamma atmospheric turbulence without misalignment fading given by

$$D[dB] \triangleq \frac{20}{\Omega_{min}} \log_{10}\left(\frac{\varphi^2}{A_0^{\Omega_{min}}(\varphi^2 - \Omega_{min})}\right). \tag{25}$$

In a similar way as in the outage and error rate performance analysis of MIMO FSO links [21, 25], it is also here corroborated that the impact of pointing errors is not related to the rate-adaptive scheme, obtaining the same coding gain disadvantage regardless the rate reduction. According to this expression, it can be observed in Fig. 3 that a coding gain disadvantage of 23.9 decibels is achieved for a value of $(\omega_z/r, \sigma_s/r) = (5,1)$, for both a rate reduction of $RR = 2$ and $RR = 4$.

Additionally, as previously reported by the authors [25, 26], a relevant improvement in performance must be noted as a consequence of the pulse shape used, providing an increment in the average SNR of $10\log_{10}\xi$ decibels. This superiority has been previously reported by the authors in terms of closed-form expressions corresponding to the average BER performance in several FSO scenarios [19, 24, 26], as well as from the point of view of information theory [22]. Following the approach here presented, this improvement can be viewed as an additional coding gain, regardless of the rate-adaptive transmission scheme employed. So, for instance, when a rectangular pulse shape of duration κT_b, with $0 < \kappa \leq 1$, is adopted, a value of $\xi = 1/\kappa$ can be easily shown. Nonetheless, a significantly higher value of $\xi = 4/\kappa\sqrt{\pi}$ is obtained when a Gaussian pulse of duration κT_b as $g(t) = \exp\left(-t^2/2\rho^2\right)$ $\forall |t| < \kappa T_b/2$ is adopted, where $\rho = \kappa T_b/8$ and the reduction of duty cycle is also here controlled by the parameter κ. In this fashion, 99.99% of the average optical power of a Gaussian pulse shape is being considered. Ultrashort pulses from mode-locked lasers often have a temporal shape which can be described with a Gaussian function. This improvement in performance due to the pulse shape can be even increased when a squared hyperbolic secant (sech^2) function is employed which is another temporal shape proper to ultrashort pulses from mode-locked lasers [11]. In this way, a value of $\xi = 8/3\kappa$, greater than that of the Gaussian pulse, is obtained when a sech^2 pulse of duration κT_b as $g(t) = \text{sech}^2(t/\rho)$ $\forall |t| < \kappa T_b/2$ is adopted, where $\rho = \kappa T_b/8$ and the reduction of duty cycle is also here controlled by the parameter κ. As in previous pulse shape, 99.99% of the average optical power of a sech^2 pulse shape is being considered. Obtained results corresponding to the sech^2 pulse shape with $\kappa = 0.25$ for the rate-adaptive transmission (Rep&Sil) here proposed with TDO $= 2$ and rate reductions of $RR = \{1,2,4,8\}$ are displayed in Fig. 4 together with results previously displayed in Fig. 2(a) where rectangular pulse shapes with $\kappa = 1$ are used. From this figure, it can be observed that a significant horizontal shift in the BER performance above 10 decibels is achieved in any case, regardless of the rate reduction assumed.

4. Achievable information rate performance analysis

In this section, the analysis of the achievable information rate corresponding to the three rate-adaptive transmission schemes here presented is considered. From the mode of operation based on reducing the initial rate, $R_b = 1/T_b$, for a RR parameter as R_b/RR in order to achieve

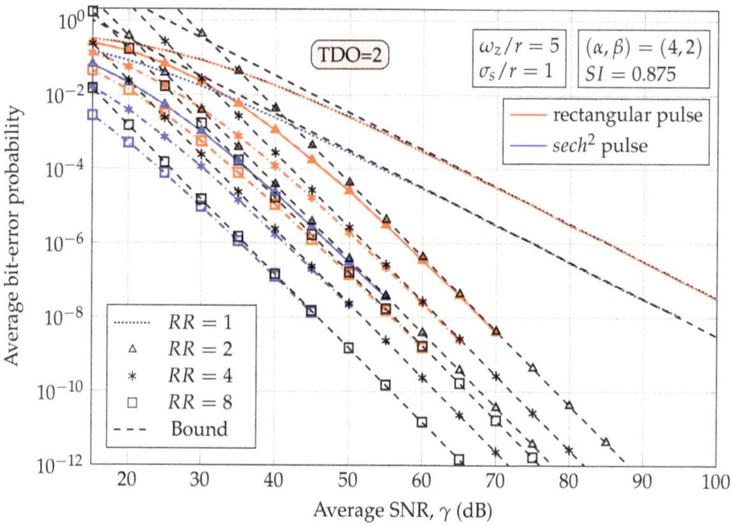

Figure 4. Performance of the sech^2 pulse shape with $\kappa = 0.25$ for the rate-adaptive scheme (Rep&Sil) with TDO = 2 and $RR = \{1, 2, 4, 8\}$ in FSO IM/DD links over gamma-gamma atmospheric channels with pointing errors, assuming a value of normalized jitter of $\sigma_s/r = 1$ and normalized beamwidth of $\omega_z/r = 5$.

a target BER requirement P_b, the achievable information rate, R, in bits/channel use, can be defined as $R = 1/RR$. While the required value of RR to satisfy a target BER requirement can be numerically solved from analytical results in (21), we can significantly simplify this analysis when target BER requirements and levels of turbulence imply a sufficiently tight performance for the corresponding asymptotic BER in (23). From results displayed in Fig. 2, it can be deduced that the asymptotic BER expressions are closer and closer upper bounds for the BER performance as the level of turbulence is increased. This is an expected conclusion since a greater value of average SNR is required to satisfy the same predefined target BER requirement. In this sense, making use of the asymptotic BER in (23a), the achievable information rate can be written as

$$R_{Sil}(\gamma) = \left(\frac{2^{\Omega_{min}} A_0^{\Omega_{min}} P_b \Omega_{min} (\xi\gamma)^{\Omega_{min}/2} \left(\varphi^2 - \Omega_{min}\right) \Gamma(\Omega_{max}) \Gamma\left(\Omega_{min}/2\right)}{\varphi^2 (\alpha\beta)^{\Omega_{min}} \Gamma(|\alpha - \beta|)} \right)^{1/\Omega_{min}} \tag{26}$$

$$R_{Rep}(\gamma) = \left(\frac{\mathrm{TDO}^{-\Omega_{min}} A_0^{\Omega_{min}} \Gamma(\Omega_{max}) \left(\varphi^2 - \Omega_{min}\right) (4\gamma\xi)^{\Omega_{min}/2}}{\varphi^2 (\alpha\beta)^{\Omega_{min}} \Gamma(|\alpha - \beta|) \left(2P_b \Gamma\left(\frac{\mathrm{TDO}\Omega_{min}}{2} + 1\right)\right)^{-1/\mathrm{TDO}}} \right)^{2/\Omega_{min}} \tag{27}$$

$$R_{Rep\&Sil}(\gamma) = \left(\frac{\mathrm{TDO}^{-3\Omega_{min}/2} A_0^{\Omega_{min}} \Gamma(\Omega_{max}) \left(\varphi^2 - \Omega_{min}\right) (4\gamma\xi)^{\Omega_{min}/2}}{\varphi^2 (\alpha\beta)^{\Omega_{min}} \Gamma(|\alpha - \beta|) \left(2P_b \Gamma\left(\frac{\mathrm{TDO}\Omega_{min}}{2} + 1\right)\right)^{-1/\mathrm{TDO}}} \right)^{1/\Omega_{min}} \tag{28}$$

for the rate-adaptive transmission schemes Sil, Rep and Rep&Sil, respectively. It can

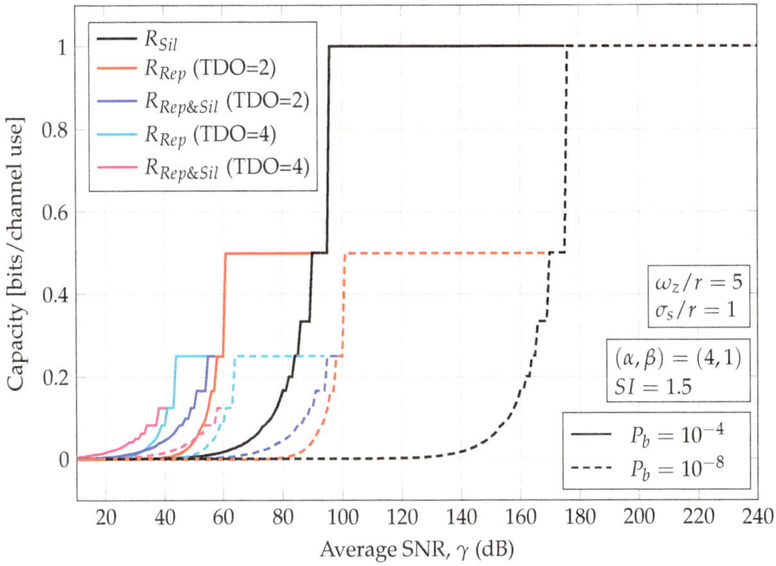

(a)

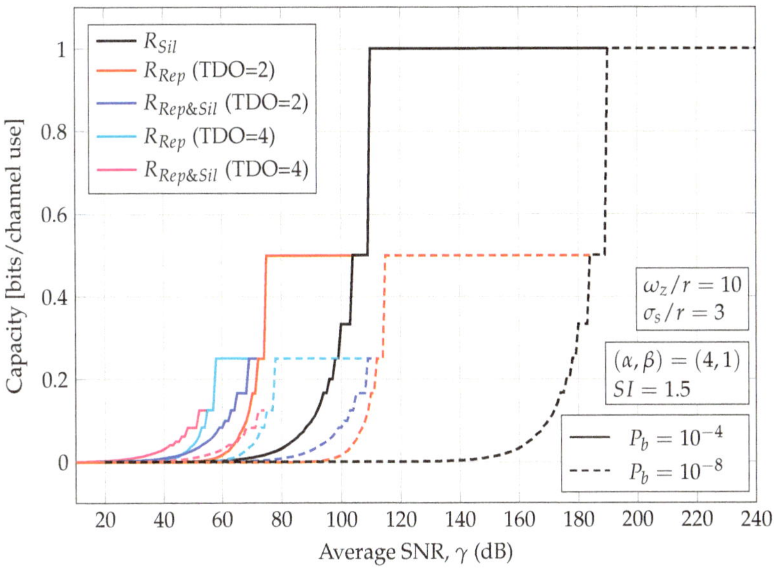

(b)

Figure 5. Achievable information rate in FSO IM/DD links corresponding to the rate-adaptive transmission schemes Sil, Rep and Rep&Sil over gamma-gamma atmospheric and misalignment fading channels for two different target BER requirements, $P_b = 10^{-4}$ and $P_b = 10^{-8}$, and time-diversity orders of TDO=$\{2, 4\}$. Here, $(\alpha, \beta) = (4, 1)$ is assumed together with values of normalized jitter and normalized beamwidth of (a) $(\omega_z/r, \sigma_s/r) = (5, 1)$ and (b) $(\omega_z/r, \sigma_s/r) = (10, 3)$.

be observed that the rate-adaptive scheme Sil does not take advantage of the potential time-diversity order available in the turbulent channel. The achievable information rate in (26), (27) and (28) is depicted in Fig. 5, corresponding to the rate-adaptive transmission schemes Sil, Rep and Rep&Sil, respectively, for two different target BER requirements, $P_b = 10^{-4}$ and $P_b = 10^{-8}$, and TDO=$\{2, 4\}$. It can be noted that obtained results are in excellent agreement with previous results presented in terms of BER performance in Fig. 2. In this sense, it can be observed that the required value of SNR to satisfy a target BER is greater when the impact of pointing errors is more severe, even when the same level of atmospheric turbulence strength is considered.

From this figure, it can be deduced not only the superiority of the rate-adaptive transmission scheme based on the joint use of repetition coding and variable silence periods but also that an adaptive transmission design approach based on taking advantage out of the potential time-diversity order available in the turbulent channel is required, corroborating the fact that the rate-adaptive transmission scheme only based on variable silence periods implies a remarkably inefficient performance from the point of view of information theory. In this way, it can be observed that even when the available time-diversity order is low (TDO=2) a relevant improvement in achievable information rate is obtained, especially when a lower target BER is demanded. In spite of high values of TDO cannot be possible because of the latency introduced by the interleaver, to achieve a time diversity order available of TDO=2, perfect interleaving can be done by simply sending the same information delayed by the expected fade duration, as shown experimentally in [32] for a rate reduction of 2. It must be commented that the rate-adaptive transmission scheme proposed is not based on an adaptive signal constellation where more complicated modulation techniques can be defined, being assumed the use of OOK signaling due to its simplicity and low implementation cost, and, hence, an achievable information rate not higher than 1 bit/channel use can be achieved, since this is determined by the signal constellation and how the coding technique is able to take advantage of it. At the expense of a greater simplicity in hardware implementation, lower values of capacity are achieved if compared to rate-adaptive transmission schemes based on adaptive modulation or coding techniques more sophisticated than repetition coding and the inclusion of variable silence periods [10, 12, 14].

5. Conclusions

In this chapter, a simple rate-adaptive transmission scheme for FSO communication systems with intensity modulation and direct detection over atmospheric turbulence channels with pointing errors is analyzed. This scheme is based on the joint use of repetition coding and variable silence periods, exploiting the potential time-diversity order available in the turbulent channel as well as allowing the increase of the PAOPR, which has shown to be a favorable characteristic in IM/DD FSO links [19, 24, 26]. Here, repetition coding is firstly used in order to accomodate the transmission rate to the channel conditions until the whole time diversity order available in the turbulent channel by interleaving is exploited. Then, once no more diversity gain is available, the rate reduction can be increased by using variable silence periods in order to increase the PAOPR. Novel closed-form asymptotic expressions are derived when the irradiance of the transmitted optical beam is susceptible to either a wide range of turbulence conditions (weak to strong), following a gamma-gamma distribution of parameters α and β, or pointing errors, following a misalignment fading model, as in [16–18, 38], where the effect of beam width, detector size and jitter variance is considered.

Furthermore, we extend the concepts of diversity and coding gain, which are well known from the RF communication literature [40], to the rate-adaptive FSO systems under study, allowing us to provide simple, insightful, and accurate closed-form approximations for the BER performance at high SNR. Simulation results are further demonstrated to confirm the analytical results. Here, not only rectangular pulses are considered but also OOK formats with any pulse shape, corroborating the advantage of using pulses with high PAOPR, such as Gaussian pulses or squared hyperbolic secant pulses, and concluding the fact that this improvement can be viewed as an additional coding gain, regardless of the rate-adaptive scheme employed or even in absence of rate-adaptive transmission.

In terms of pros and cons of using the adaptive transmission scheme Rep&Sil here analyzed and proposed, we can conclude that one of the pros of this adaptive transmission scheme is the greater simplicity, requiring a lower implementation complexity if compared to alternative rate-adaptive transmission schemes proposed for use in FSO systems [10, 12, 14, 34]. In this sense, the use of OOK signaling with repetition coding implies a hardware implementation of a significant lower complexity if compared to that required by the variable rate turbo-coding scheme as proposed in [34], the adaptive LDPC-coded modulation and transmission scheme that varied both the power and the modulation order of a FSO system with M-ary pulse amplitude modulation as proposed in [12, 14] and the adaptive transmission techniques employing subcarrier phase shift keying intensity modulation as proposed in [10]. Additionally, together with the use of repetition coding, the adaptive scheme Rep&Sil here proposed considers the inclusion of variable silence periods once no more diversity gain is available, providing a higher level of PAOPR and, hence, a better performance, without the need of increasing complexity in hardware implementation. On the contrary, one of the cons of the rate-adaptive transmission scheme here proposed in relation to previous rate-adaptive transmission schemes is the lower capacity achieved since no adaptive modulation is considered, proposing the use of OOK signaling due to its simplicity and low implementation cost and, hence, providing an achievable information rate not higher than 1 bit/channel use.

Acknowledgments

The authors are grateful for financial support from the Junta de Andalucía (research group "Communications Engineering (TIC-0102)").

Author details

Antonio García-Zambrana and Beatriz Castillo-Vázquez
Department of Communications Engineering, University of Málaga, Spain

Carmen Castillo-Vázquez
Department of Statistics and Operations Research, University of Málaga, E-29071 Málaga, Spain

6. References

[1] Al-Habash, M. A., Andrews, L. C. & Phillips, R. L. [2001]. Mathematical model for the irradiance probability density function of a laser beam propagating through turbulent media, *Opt. Eng.* 40: 8.

[2] Ali, A. A. & Al-Kadi, I. A. [1989]. On the use of repetition coding with binary digital modulations on mobile channels, *IEEE Trans. Veh. Technol.* 38(1): 14–18.

[3] Andrews, L., Phillips, R. & Hopen, C. [2001]. *Laser beam scintillation with applications*, Bellingham, WA: SPIE Press.

[4] Anguita, J., Djordjevic, I., Neifeld, M. & Vasic, B. [2005]. Shannon capacities and error-correction codes for optical atmospheric turbulent channels, *J. Opt. Netw.* 4(9): 586–601.

[5] Arnon, S. [2003]. Effects of atmospheric turbulence and building sway on optical wireless-communication systems, *Opt. Lett.* 28(2): 129–131.

[6] Bayaki, E. & Schober, R. [2010]. On space-time coding for free-space optical systems, *IEEE Trans. Commun.* 58(1): 58–62.

[7] Bayaki, E., Schober, R. & Mallik, R. K. [2009]. Performance analysis of mimo free-space optical systems in gamma-gamma fading, *IEEE Trans. Commun.* 57(11): 3415–3424.

[8] Borah, D. K. & Voelz, D. G. [2009]. Pointing error effects on free-space optical communication links in the presence of atmospheric turbulence, *J. Lightwave Technol.* 27(18): 3965–3973.

[9] Chan, V. W. S. [2006]. Free-space optical communications, *J. Lightwave Technol.* 24(12): 4750–4762.

[10] Chatzidiamantis, N. D., Lioumpas, A. S., Karagiannidis, G. K. & Arnon, S. [2011]. Adaptive subcarrier PSK intensity modulation in free space optical systems, *IEEE Trans. Commun.* 59(5): 1368–1377.

[11] Diels, J. C. & Rudolph, W. [2006]. *Ultrashort Laser Pulse Phenomena*, Optics and Photonics Series, second edn, Academic Press.

[12] Djordjevic, I. B. [2010]. Adaptive modulation and coding for free-space optical channels, *IEEE/OSA Journal of Optical Communications and Networking* 2(5): 221–229.

[13] Djordjevic, I. B., Denic, S., Anguita, J., Vasic, B. & Neifeld, M. [2008]. Ldpc-coded mimo optical communication over the atmospheric turbulence channel, *J. Lightwave Technol.* 26(5): 478–487.

[14] Djordjevic, I. B. & Djordjevic, G. T. [2009]. On the communication over strong atmospheric turbulence channels by adaptive modulation and coding, *Opt. Express* 17(20): 18250–18262.

[15] Fante, R. L. [1975]. Electromagnetic beam propagation in turbulent media, *Proc. IEEE* 63(12): 1669–1692.

[16] Farid, A. A. & Hranilovic, S. [2007]. Outage capacity optimization for free-space optical links with pointing errors, *J. Lightwave Technol.* 25(7): 1702–1710.

[17] Farid, A. A. & Hranilovic, S. [2010]. Diversity gains for mimo wireless optical intensity channels with atmospheric fading and misalignment, *Proc. IEEE GLOBECOM Workshops (GC Wkshps)*, pp. 1015–1019.

[18] Gappmair, W., Hranilovic, S. & Leitgeb, E. [2010]. Performance of ppm on terrestrial fso links with turbulence and pointing errors, *IEEE Commun. Lett.* 14(5): 468–470.

[19] García-Zambrana, A. [2007]. Error rate performance for STBC in free-space optical communications through strong atmospheric turbulence, *IEEE Commun. Lett.* 11(5): 390–392.

[20] García-Zambrana, A., Castillo-Vázquez, B. & Castillo-Vázquez, C. [2010a]. Average capacity of FSO links with transmit laser selection using non-uniform OOK signaling over exponential atmospheric turbulence channels, *Opt. Express* 18(19): 20445–20454.

[21] García-Zambrana, A., Castillo-Vázquez, B. & Castillo-Vázquez, C. [2012]. Asymptotic error-rate analysis of fso links using transmit laser selection over gamma-gamma atmospheric turbulence channels with pointing errors, *Opt. Express* 20(3): 2096–2109.
[22] García-Zambrana, A., Castillo-Vázquez, C. & Castillo-Vázquez, B. [2010b]. On the capacity of FSO links over gamma-gamma atmospheric turbulence channels using OOK signaling, *EURASIP Journal on Wireless Communications and Networking* . Article ID 127657, 9 pages, 2010. doi:10.1155/2010/127657.
[23] García-Zambrana, A., Castillo-Vázquez, C. & Castillo-Vázquez, B. [2010c]. Rate-adaptive FSO links over atmospheric turbulence channels by jointly using repetition coding and silence periods, *Opt. Express* 18(24): 25422–25440.
[24] García-Zambrana, A., Castillo-Vázquez, C. & Castillo-Vázquez, B. [2010d]. Space-time trellis coding with transmit laser selection for FSO links over strong atmospheric turbulence channels, *Opt. Express* 18(6): 5356–5366.
[25] García-Zambrana, A., Castillo-Vázquez, C. & Castillo-Vázquez, B. [2011]. Outage performance of MIMO FSO links over strong turbulence and misalignment fading channels, *Opt. Express* 19(14): 13480–13496.
[26] Garcia-Zambrana, A., Castillo-Vazquez, C., Castillo-Vazquez, B. & Hiniesta-Gomez, A. [2009]. Selection transmit diversity for FSO links over strong atmospheric turbulence channels, *IEEE Photon. Technol. Lett.* 21(14): 1017–1019.
[27] García-Zambrana, A. & Puerta-Notario, A. [1999]. RZ-Gaussian pulses reduce the receiver complexity in wireless infrared links at high bit rates, *IEE Electronics Letters* 35(13): 1059–1061.
[28] García-Zambrana, A. & Puerta-Notario, A. [2001]. Large change rate-adaptive indoor wireless infrared links using variable silence periods, *IEE Electronics Letters* 37(8): 524–525.
[29] Gradshteyn, I. S. & Ryzhik, I. M. [2007]. *Table of integrals, series and products*, 7th edn, Academic Press Inc.
[30] Hranilovic, S. & Kschischang, F. R. [2003]. Optical intensity-modulated direct detection channels: signal space and lattice codes, *IEEE Trans. Inf. Theory* 49(6): 1385–1399.
[31] Kahn, J. M. & Barry, J. R. [1997]. Wireless infrared communications, *Proc. IEEE* 85: 265–298.
[32] Kwok, C. H., Penty, R. V. & White, I. H. [2008]. Link reliability improvement for optical wireless communication systems with temporal-domain diversity reception, *IEEE Photon. Technol. Lett.* 20(9): 700–702.
[33] Lee, E. J. & Chan, V. W. S. [2004]. Part 1: optical communication over the clear turbulent atmospheric channel using diversity, *IEEE J. Sel. Areas Commun.* 22(9): 1896–1906.
[34] Li, J. & Uysal, M. [2003]. Achievable information rate for outdoor free space optical communication with intensity modulation and direct detection, *Proc. IEEE Global Telecommunications Conference GLOBECOM '03*, Vol. 5, pp. 2654–2658.
[35] Lim, W., Yun, C. & Kim, K. [2009]. Ber performance analysis of radio over free-space optical systems considering laser phase noise under gamma-gamma turbulence channels, *Opt. Express* 17(6): 4479–4484.
[36] Liu, C., Yao, Y., Sun, Y. X., Xiao, J. J. & Zhao, X. H. [2010]. Average capacity optimization in free-space optical communication system over atmospheric turbulence channels with pointing errors, *Opt. Lett.* 35(19): 3171–3173.

[37] Nistazakis, H. E., Karagianni, E. A., Tsigopoulos, A. D., Fafalios, M. E. & Tombras, G. S. [2009]. Average capacity of optical wireless communication systems over atmospheric turbulence channels, *IEEE/OSA Journal of Lightwave Technology* 27(8): 974–979.

[38] Sandalidis, H. G. [2011]. Coded free-space optical links over strong turbulence and misalignment fading channels, *IEEE Trans. Commun.* 59(3): 669–674.

[39] Sandalidis, H. G., Tsiftsis, T. A. & Karagiannidis, G. K. [2009]. Optical wireless communications with heterodyne detection over turbulence channels with pointing errors, *J. Lightwave Technol.* 27(20): 4440–4445.

[40] Simon, M. K. & Alouini, M.-S. [2005]. *Digital communications over fading channels*, second edn, Wiley-IEEE Press, New Jersey.

[41] Stotts, L. B., Andrews, L. C., Cherry, P. C., Foshee, J. J., Kolodzy, P. J., McIntire, W. K., Northcott, M., Phillips, R. L., Pike, H. A., Stadler, B. & Young, D. W. [2009]. Hybrid optical rf airborne communications, *Proc. IEEE* 97(6): 1109–1127.

[42] Trisno, S., Smolyaninov, I. I., Milner, S. D. & Davis, C. C. [2004]. Delayed diversity for fade resistance in optical wireless communication system through simulated turbulence, *Proc. SPIE*, pp. 385–394.

[43] Trisno, S., Smolyaninov, I. I., Milner, S. D. & Davis, C. C. [2005]. Characterization of delayed diversity optical wireless system to mitigate atmospheric turbulence induced fading, *Proc. SPIE*, pp. 589215.1–589215.10.

[44] Tsiftsis, T. A., Sandalidis, H. G., Karagiannidis, G. K. & Uysal, M. [2009]. Optical wireless links with spatial diversity over strong atmospheric turbulence channels, *IEEE Trans. Wireless Commun.* 8(2): 951–957.

[45] Uysal, M., Li, J. & Yu, M. [2006]. Error rate performance analysis of coded free-space optical links over gamma-gamma atmospheric turbulence channels, *IEEE Trans. Wireless Commun.* 5(6): 1229–1233.

[46] Wang, Z. & Giannakis, G. B. [2003]. A simple and general parameterization quantifying performance in fading channels, *IEEE Trans. Commun.* 51(8): 1389–1398.

[47] Wolfram Research, Inc. [n.d.]. The Wolfram functions site.
URL: *http://functions.wolfram.com*

[48] Xu, F., Khalighi, A., Caussé, P. & Bourennane, S. [2009]. Channel coding and time-diversity for optical wireless links, *Opt. Express* 17(2): 872–887.

[49] Young, C. Y., Andrews, L. C. & Ishimaru, A. [1998]. Time-of-arrival fluctuations of a space–time gaussian pulse in weak optical turbulence: an analytic solution, *Appl. Opt.* 37(33): 7655–7660.

[50] Zhu, X. & Kahn, J. M. [2002]. Free-space optical communication through atmospheric turbulence channels, *IEEE Trans. Commun.* 50(8): 1293–1300.

Section 5

Optical Communication: Signal Processing

Chapter 14

All-Optical Signal Processing for High Spectral Efficiency (SE) Optical Communication

Y. Ben Ezra, B.I. Lembrikov, Avi Zadok, Ran Halifa and D. Brodeski

Additional information is available at the end of the chapter

1. Introduction

All-optical signal processing is essential for optical communication systems transmitting high speed data signals. All-optical wavelength conversion (WC), pulse generation, orthogonal frequency-division multiplexing (OFDM), demultiplexing, regeneration, modulation format conversion are the important signal processing functions. The limiting factors in the existing ultra long haul optical communication systems are spectral efficiency (SE), fiber attenuation, insertion losses, chromatic dispersion, polarization mode dispersion (PMD), and the optical fiber nonlinearity [1]-[4]. Optical communication systems can be divided into two groups. In the first group, an electrical signal modulates the intensity of an optical carrier inside the optical transmitter, the modulated optical signal is transmitted through the optical fiber and converted to the original electrical signal by an optical receiver [1]. Such a scheme is called an intensity modulation with direct detection (IM/DD) [1], [5]. In the second group, information is transmitted by modulation of the optical carrier frequency or phase, and then the modulated optical signal can be linearly down-converted to a baseband electrical signal by heterodyne or homodyne detection [1], [5]. The phase coherence of the optical carrier is essential for the realization of the second group called coherent optical communication systems [1].

Coherent optical communication systems possess the following advantages with respect to IM/DD systems: (i) the shot-noise limited receiver can be achieved with a sufficient local oscillator (LO) power; (ii) the frequency resolution at the intermediate frequency (IF) or baseband stage is high enough in order to separate close wavelength-division multiplexed (WDM) channels in the electric domain; (iii) the phase detection improves the receiver sensitivity compared with IM/DD systems; (iv) the multilevel modulation formats can be introduced into optical communications by using phase modulation [5].

The rapid development of digital communication and digital signal processing (DSP) caused by the necessity of high SE and, at the same time, by the advance of the high-speed electronics required advanced modulation, coding and digital equalization [6]. Recently, coherent optical communications attracted a large interest due to the feasibility of the high SE, the large bandwidth and multilevel modulation formats [5]. Coherent optical systems using multilevel modulation formats can increase SE up to M b/s/Hz where M is the number of bits per symbol for a given modulation format [5].

Coherent optical OFDM (CO-OFDM) has been recently proposed in order to increase receiver sensitivity, SE, and especially, provide the dispersion compensation at high data transmission rates tending to 100Gb/s [3], [4], [7], [8]. Generally, OFDM is a kind of multicarrier modulation (MCM) in which the data information is carried over many lower rate subcarriers [4]. In OFDM spectra of individual subcarriers overlap, but due to the orthogonality, the subcarriers can be demodulated without interference and without analog filtering for the received subcarrier separation [7]. The signal processing in the OFDM system can be carried out by using the Fast Fourier Transform (FFT)/Inverse Fast Fourier Transform (IFFT) [3], [4], [7]. In CO-OFDM systems the digital signal processing is used in order to mitigate the channel dispersion, nonlinearity and different types of noise.

CO-OFDM system combines the following advantages of coherent detection and OFDM modulation essential for the high-speed optical fiber communication systems: (i) in CO-OFDM systems, chromatic dispersion and PMD can be mitigated; (ii) the SE of CO-OFDM systems is high due to the partial overlapping of OFDM sub-carrier spectra; (iii) linearity in radio frequency (RF)-to-optical (RTO) up-conversion and optical-to-RF (OTR) down-conversion; (iv) the electrical bandwidth requirements for the CO-OFDM transceiver can be greatly reduced by using direct up/down conversion which results in the low cost of the high-speed electronic circuits [3], [4], [8]. Recently, all-optical FFT scheme enabling Tbit/s real-time signal processing has been proposed [9]. The method based on only passive optical components realizes the highest speed signal processing without the power consumption where electronics cannot be used. This approach combines the advantages of the electronic high precision processing of the low bit rates and the optical processing of high bit rates [9].

However, OFDM is characterized by the inter-symbol-interference (ISI) and inter-carrier-interference (ICI) caused by a large number of subcarriers [4]. In the RF systems ISI is mainly due to multipath channel delay spread [10], [11] and ICI is mainly due to the carrier frequency offset [12]. In the case of CO-OFDM, ISI and ICI are caused by channel chromatic dispersion and PMD [3], [4]. A so-called cyclic prefix (CP), i.e. the cyclic extension of the OFDM waveform into the guard interval (GI) Δ_G, has been proposed in order to prevent ISI and ICI [4]. If the GI is long enough to contain the intersymbol transition, then the remaining part of the OFDM symbol satisfies the orthogonality condition and receiver cross-talk occurs only within GI [9]. The addition of CP requires an increase of a bandwidth and sampling rate of analog-to-digital converter (ADC) and digital-to-analog converter (DAC). CP appeared to be an easily recognizable feature of an OFDM system making the signal vulnerable to interception by surveillance receiver [10]. The elimination of CP reduces the probability of interception and improves SE [10].

The need for CP can be avoided if the wavelet packet transform (WPT) is used in CO-OFDM systems instead of discrete Fourier Transform (DFT) and inverse DFT (IDFT) [13]. The sinusoidal functions are infinitely long in the time domain while wavelets have finite length being localized in time and in frequency domains [13]. Wavelet signal analysis can be a base for an effective computational algorithm which is faster and simpler than FFT [14]. Wavelets have been used in optical communications for time-frequency multiplexing and ultrafast image transmission [14]. A signal may be expanded in an orthogonal set of wavelet packets (WPs) as the basis functions, each channel occupies a wavelet packet (WP), and IDWPT/DWPT are used at the transmitter and receiver, respectively [13].

In this chapter, we consider the CO-OFDM based on WPT and its influence on the optical communication network performance. The chapter is constructed as follows. In Section 2, we review the coherent optical communication systems. In Section 3, we discuss high SE CO-OFDM system. In Section 4, we discuss the OFDM based on WPT and present the original results for the WPT-OFDM system performance. In Section 5, we present the original results concerning the simulations of the structure and operation mode of the novel passive components for all-optical signal processing based on Si-on-insulator (SOI) structure, and a novel hierarchical architecture of the 1Tb/s transmission system based on WPT-OFDM [15]. In Section 6, conclusions are presented.

2. Coherent optical communication systems

The number of publications concerning the coherent optical communications is enormous. In this section, we can only relate to a limited number of the fundamental works concerning the concept of the coherent detection and the modulation formats used in digital communication system since these topics are related to high SE CO-OFDM systems.

The most advanced detection method is coherent detection based on the recovery of the full electric field containing both amplitude and phase information [16]. The concept of the coherent detection is to combine in a receiver the modulated optical signal with a continuous wave (CW) optical field generated by a narrow linewidth laser, or local oscillator (LO) before the photodetector (PD) [1]. Coherent detection requires the carrier synchronization with respect to LO that serves as an absolute phase reference [16]. For this purpose, optical systems can use two types of the phase-locked loops (PLLs): (i) an optical PLL (OPLL) that synchronizes the frequency and phase of the LO laser with the transmitter laser; (ii) an electrical PLL where down-conversion with a free-running LO laser is followed by a second-stage demodulation by an electrical voltage controlled oscillator (VCO) with the synchronized frequency and phase [16].

The basic component of coherent optical systems is a coherent optical receiver [1], [5]. Its block diagram is shown in Figure 1 [5].

The electric fields $E_s(t)$ and $E_{LO}(t)$ of the received optical signal and LO, respectively, are given by [5]

$$E_s(t) = A_s(t)\exp\left[j\left(\omega_0 t + \phi_s\right)\right]; E_{LO}(t) = A_{LO}\exp\left[j\left(\omega_{LO} t + \phi_{LO}\right)\right] \quad (1)$$

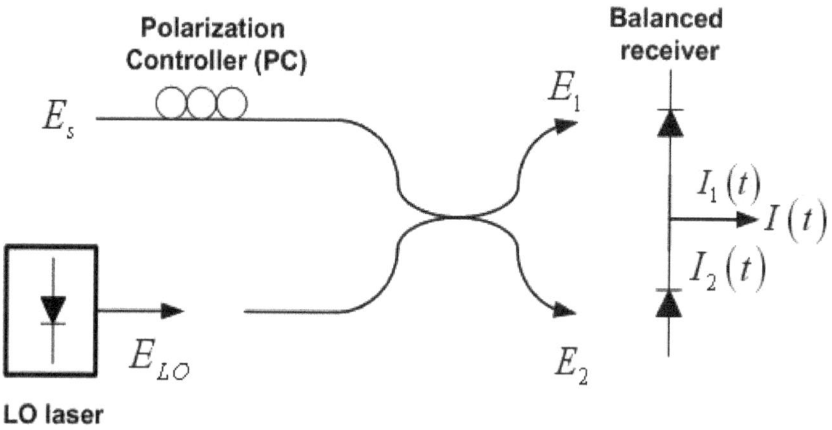

Figure 1. Block diagram of the coherent receiver

where $A_s(t), A_{LO}, \omega_0, \omega_{LO}, \phi_s, \phi_{LO}$ are the amplitudes, frequencies and phases of the received optical signal and LO, respectively. A 3db optical coupler (OC) is used that adds a π phase shift to either the signal field or the LO field between the two output ports. When the signal and LO fields are co-polarized, the electric fields $E_{1,2}$ incident on the upper and lower diodes are given by, respectively.

$$E_{1,2} = \frac{1}{\sqrt{2}}\left[E_s(t) \pm E_{LO}(t)\right] \tag{2}$$

The output photocurrents $I_{1,2}(t)$ are given by, respectively [5]

$$I_{1,2}(t) = \frac{R}{2}\left\{P_s + P_{LO} \pm 2\sqrt{P_s P_{LO}}\cos\left(\omega_{IF}t + \phi_s - \phi_{LO}\right)\right\} \tag{3}$$

where $P_s = A_s^2(t)/2, P_{LO} = A_{LO}^2/2, \omega_{IF} = \omega_0 - \omega_{LO}$ is IF, $R = e\eta_q/(\hbar\omega_0)$ is the detector responsivity, e is the electron charge, η_q is the PD quantum efficiency, $\hbar = h/2\pi, h$ is the Planck's constant. The sum frequency component is neglected since it is averaged out to zero over the bandwidth of PD. Balanced detection is used in order to suppress the DC component and maximize the signal photocurrent $I(t)$ at the balanced detector output given by [5]

$$I(t) = 2R\sqrt{P_s(t)P_{LO}}\cos\left(\omega_{IF}t + \phi_s - \phi_{LO}\right) \tag{4}$$

Eq. (4) demonstrates the main advantage of the coherent detection as compared to the direct detection. The photocurrent $I(t)$ contains explicitly the signal phase ϕ_s making possible to transmit information by modulating the phase or frequency of the carrier signal [1], [5]. There are two cases of the coherent detection: (i) the heterodyne detection when the signal carrier frequency ω_0 and the LO frequency ω_{LO} are different, and $f_{IF} = \omega_{IF} / (2\pi) \sim 1GHz$; (ii) the homodyne detection when the signal carrier frequency ω_0 and the LO frequency ω_{LO} coincide, and IF $\omega_{IF} = 0$ [1]. In the heterodyne detection case, eq. (4) describes the output photocurrent. Typically, the LO power is much larger than the received signal power: $P_{LO} >> P_s$, and LO amplifies the received signal improving signal-to-noise ratio (SNR) [1]. In the case of the heterodyne detection, the optical signal is demodulated in two stages: it is first down-converted to IF and then to the baseband [1]. The output photocurrent $I(t)$ (4) then takes the form.

$$I(t) = 2R\sqrt{P_s(t)P_{LO}}\cos(\phi_s - \phi_{LO}) \tag{5}$$

Similarly, the increase of the average electrical power up to 20dB can occur in the case of the homodyne detection. If additionally, the LO phase is locked to the signal phase so that $\phi_s - \phi_{LO} = 0$ eq. (5) takes the form [1].

$$I(t) = 2R\sqrt{P_s(t)P_{LO}} \tag{6}$$

The phase difference $\phi_s - \phi_{LO}$ can be kept constant by using an OPLL. However, the implementation of OPLL makes the design of optical homodyne receivers a comparatively complicated problem [1], [5].

The coherent detection allows the greatest flexibility in modulation formats, since information can be encoded by modulating the amplitude, the phase, or the frequency of an optical carrier as it is seen from equations (1)-(6) or in both in-phase (I) and quadrature (Q) components of the carrier [1], [16]. In the case of the digital communication systems, these methods correspond to three modulation formats: amplitude-shift keying (ASK), phase--shift keying (PSK), and frequency-shift keying (FSK) [1]. The increased performance, speed, and reliability, and the reduced size and cost of integrated circuits permit to use DSP for the information recovery from the baseband signal [5]. Typically, the M-ary PSK modulation is used in SE high-speed CO-OFDM systems such as quaternary PSK (QPSK) (M=4), 8-PSK (M=8), as well as quadrature amplitude modulation (QAM) such as 4-QAM, 16-QAM, 64-QAM in single or dual polarization [4], [5]. The digital coherent receiver linearly detects incoming signal including phase and polarization diversities and converts this information to digital data by using ADCs while the digital information is processed by DSP circuits [5].

3. High-speed and high SE CO-OFDM system

In this section we present a brief review of the operation principle and architecture of CO-OFDM system. Detailed analysis of CO-OFDM in optical communication systems may be found in the book [4].

A generic optical OFDM system consists of five functional blocks: the RF OFDM transmitter, the RTO up-converter, the optical channel, the OTR down-converter, and the RF OFDM receiver [3], [4]. In such a system the following chain of events occurs [3]. The input data bits are mapped onto corresponding information symbols of the subcarriers within one OFDM symbol. The digital time domain signal is obtained by using IDFT. It is inserted with the GI Δ_G in order to prevent ISI caused by channel dispersion and converted into the real time waveform through DAC [3]. The baseband signal is up-converted to an appropriate RF band with an IQ mixer/modulator. A linear RTO up-conversion can be achieved by using Mach-Zehnder modulator (MZM) [3]. MZM is mainly used for the bit rates of 40GB/s and higher due to its high modulation performance and the possibility of independent modulation of the electric field intensity and phase [6]. At the receiver, the OFDM signal is down-converted to baseband with an IQ demodulator, sampled by an ADC and demodulated by DFT and baseband DSP to recover the data [3]. A linear OTR down-conversion is provided by a coherent detection described in section 2. The high performance of the CO-OFDM transmission systems has been shown both theoretically and experimentally [8], [17]. A single-channel 1Tb/s CO-OFDM signal consisting of 4104 spectrally-overlapped subcarriers with SE of 3.3bit/s/Hz has been generated, transmitted over 600km standard single mode fiber (SSMF) without amplification and dispersion compensation, and successfully received [17]. However, CO-OFDM system is extremely sensitive to nonlinearity and channel dispersion. The dispersion mitigation with the dispersion compensation fiber (DCF) results in the additional noise and nonlinear effects decreasing the system performance [8].

Consider now the analytical expressions describing the CO-OFDM signals. The MCM transmitted signal $s(t)$ is given by [3]

$$s(t)=\sum_{i=-\infty}^{\infty}\sum_{k=1}^{N_{SC}} c_{ki}s_k\left(t-iT_s\right) \tag{7}$$

$$s_k(t)=\Pi(t)\exp\left(j2\pi f_k t\right);\Pi(t)=\begin{cases}1, & 0<t\leq T_s\\ 0, & t\leq 0, t>T_s\end{cases} \tag{8}$$

where c_{ki} is the i th information symbol at the k th subcarrier, s_k, f_k are the waveform and the frequency of the k th subcarrier, respectively, N_{SC} is the number of subcarriers, and T_s is the symbol period. The optimum detector for each subcarrier could use a filter matched to the subcarrier waveform, or a correlator matched to subcarrier [3]. Eq. (7) shows that the modulation can be performed by IDFT of the input information signal c_{ki} .

The detected information signal c'_{ik} at the output of the correlator has the form [3].

$$c'_{ik}=\int_0^{T_s} r\left(t-iT_s\right)\exp\left(-j2\pi f_k t\right)dt \tag{9}$$

where $r\left(t-iT_s\right)$ is the received time-domain signal. Eq. (9) shows that the demodulation is provided by DFT of the sampled received signal $r(t)$ [3]. The classical MCM uses non-overlapped band-limited signals. In order to prevent overlapping of the band-limited signals, a bank of a large number of oscillators and filters is necessary at the transmitter and the receiver [3]. The cost-efficient design of the filters and oscillators requires that the channel spacing should be multiple of the symbol rate. As a result, SE reduces and the required bandwidth increases [3].

The OFDM technique permits to use the overlapped signals under the condition that they are orthogonal [7]. The orthogonality condition for any two subcarriers $s_k(t)$ and $s_l(t)$ is given by [3].

$$\frac{1}{T_s}\int_0^{T_s} s_k(t)s_l^*(t)dt = \delta_{kl} = \begin{cases} 1, & k=l \\ 0, & k \neq l \end{cases} \tag{10}$$

Substituting expression (8) into condition (10) we obtain.

$$\exp\left[j\pi\left(f_k - f_l\right)T_s\right]\frac{\sin\left[\pi\left(f_k - f_l\right)T_s\right]}{\pi\left(f_k - f_l\right)T_s} = \delta_{kl} \tag{11}$$

The left-hand side of (11) vanishes when

$$f_k - f_l = \frac{m}{T_s}; m = 1,2,... \tag{12}$$

Then, the two subcarriers $s_k(t)$ and $s_l(t)$ are orthogonal and can be recovered with the matched filters according to (9) without ICI despite the signal spectral overlapping [3].

In the high speed CO-OFDM systems the problem of ISI and ICI caused by the channel dispersion is critical. ISI is caused by the interference between "slow" and "fast" subcarriers. ICI is due to the breaking of the orthogonality condition (12) for the subcarriers [3]. In order to prevent ISI and ICI, CP was proposed that is realized by cyclic extension of the OFDM waveform into GI [3]. The waveform in GI is essentially an identical copy of that in the DFT window [3]. The condition for ISI-free OFDM transmission requires that the dispersive channel time delay spread $t_d < \Delta_G$ [3].

SE is defined as the ratio of net per-channel information data rate B to WDM channel spacing Δf and measured in b/s/Hz [1], [18]. SE of CO-OFDM η is given by [3]

$$\eta = 2\frac{R_s}{B_{OFDM}} \tag{13}$$

where $R_s = N_{SC} / T_s$ is the total symbol rate, $B_{OFDM} = (2/T_s) + (N_{SC} - 1)/t_s$ is the OFDM bandwidth, t_s is the observation period, and the factor of 2 is taking into account two polarizations of the optical fiber modes. Typically, the subcarriers number is large: $N_{SC} \gg 1$. Then eq. (13) takes the form: $\eta \approx 2t_s / T_s$. The optical SE of 3.6bit/Hz can be achieved for QPSK modulation of subcarriers, and can be improved by using higher-order QAM modulation format [3]. However, the addition of CP requires an increase of a bandwidth B_{OFDM} and sampling rate of ADC and DAC. The need for CP can be avoided if WPT is used in CO-OFDM systems instead of DFT and IDFT [13]. This approach will be discussed in the next section.

4. WPT based CO-OFDM

WPT can be used in CO-OFDM instead of the IDFT/DFT since it improves the system performance, and in particular, mitigates the channel chromatic dispersion without CP [13]. In this section we briefly discuss the main features of WPT and its applications to CO-OFDM. The theory and applications of continuous wavelet transform (CWT) and discrete WT (DWT) can be found in a large number of books and articles (see, for example, [13], [14], [19]-[22] and references therein).

CWT $W_T(a,\tau)$ of a given function f(t) with respect to a mother wavelet (MW) ψ(t) is defined as follows [19], [20]

$$W_T(a,\tau) = \frac{1}{\sqrt{|a|}} \int_{-\infty}^{\infty} \psi^*\left(\frac{t-\tau}{a}\right) f(t)\,dt \tag{14}$$

where the real numbers a and τ are the scaling and shifting, or translation parameters, respectively, and asterisk means complex conjugation. Note that in many practically important cases MW ψ(t) is real. The functions $\psi^{a,\tau}(s) = |a|^{-1/2}\psi((s-\tau)/a)$ are called wavelets [20]. The set of wavelets is orthogonal and can be used as a basis instead of sinusoidal functions [13]. It is possible to localize the events described by f(t) in time and frequency domains simultaneously by means of WT choosing the appropriate values of the parameters a and τ [19]. For this reason, wavelets are used in the multiresolution analysis (MRA) which decomposes a signal at different scales, or resolutions, using a basis whose elements are localized both in time and in frequency domains [14].

DWT is given by [19], [20]

$$W_T^{m,n}(a,\tau) = a_0^{-m/2} \int_{-\infty}^{\infty} \psi^*\left(a_0^{-m}t - n\tau_0\right) f(t)\,dt \tag{15}$$

where m,n∈Z, Z is the set of all integers, and the constants $a_0 > 1, \tau_0 > 1$. Comparison of eqs. (14) and (15) shows that $a = a_0^m$ and $\tau = n\tau_0 a_0^m$ [20]. The orthogonal wavelet series expansions can be successfully used in DSP and multiplexing when the scaling and translation parameters are discrete [14]. In such a case, a signal s(t) $\in V_0$ can be represented by a smooth approximation at resolution 2^M, obtained by combining translated versions of the basic scaling function $\phi(t)$, and M details at the dyadic scales $a = 2^l, (l = 1,2,...,M-1)$ obtained by combining shifted and dilated versions of the MW ψ(t) as follows [14].

$$s(t) = \sum_k 2^{-M/2} c_M[k]\phi\left(2^{-M}t - k\Delta\tau\right) + \sum_{l=1}^{M}\sum_k 2^{-l/2} d_l[k]\psi\left(2^{-l}t - k\Delta\tau\right) \tag{16}$$

Here a subspace $V_0 \in L^2(R)$, $L^2(R)$ is a the linear vector space of square integrable functions, $2^{-l/2}\phi\left(2^{-l}t - k\Delta\tau\right)$ and $2^{-l/2}\psi\left(2^{-l}t - k\Delta\tau\right)$ are the orthonormal bases for the subspaces

$V_l \in L^2(R)$ and $W_l \in L^2(R)$, respectively, $V_l \perp W_l$, (l,k)∈Z, $c_l[k]$ and $d_l[k]$ are the scaling and detail coefficients, respectively, at resolution 2^l, Δτ is the time interval coinciding with the inverse of the free spectral range (FSR). The scaling function $\phi(t)$ and wavelet function ψ(t) satisfy the dilation equations [14], [19], [21]

$$\phi(t) = \sqrt{2}\sum_k h[k]\phi(2t - k\Delta\tau);\psi(t) = \sqrt{2}\sum_k g[k]\phi(2t - k\Delta\tau) \tag{17}$$

where h[k] and g[k] are the coefficients of two half-band (HB) quadrature mirror filters (QMFs) described by the following functions H(ω) and G(ω)

$$H(\omega) = \frac{1}{\sqrt{2}}\sum_k h[k]\exp(-j\omega k\Delta\tau);G(\omega) = \frac{1}{\sqrt{2}}\sum_k g[k]\exp(-j\omega k\Delta\tau) \tag{18}$$

and Δτ is the inverse of their FSR. The functions H(ω) and G(ω) (18) satisfy the conditions [14], [22].

$$|H(\omega)|^2 + |G(\omega)|^2 = 1;G(\omega) = \exp(-j\omega k\Delta\tau)H^*\left(\omega + \frac{\pi}{\Delta\tau}\right) \tag{19}$$

The evaluation of the discrete wavelet coefficients is equivalent to filtering the signal s(t) by a cascade of mutually orthogonal bandpass filters [21]. An optical HB filter can be realized by using Mach-Zehnder interferometers (MZIs) [14], [22].

The DWT decomposition procedure is described by the following recursive expressions for the scaling and detail coefficients $c_l[n], d_l[n]$ [14], [22]

$$c_l[n] = \sum_k c_{l-1}[k]h[2n-k];d_l[n] = \sum_k c_{l-1}[k]g[2n-k] \tag{20}$$

where

$$c_0 = \int s(t)\phi(t - n\Delta\tau)dt \tag{21}$$

In the DWT case only the scaling coefficients $c_l[n]$ are recursively filtered, while the detail coefficients $d_l[n]$ are not reanalyzed [14]. In the case of the WP decomposition both the scaling coefficients $c_l[n]$ and the detail coefficients $d_l[n]$ are recursively decomposed following the same filtering and subsampling scheme, and consequently, all outputs have the same number of samples span over the same frequency bandwidth [14]. The WP decomposition based on the wavelet atom functions $w_{l,m}(t)$ is performed as follows [14]

$$w_{l+1,2m}(t) = \sum_k h[k]w_{l,m}(t - 2^l k\Delta\tau) \tag{22}$$

$$w_{l+1,2m+1}(t) = \sum_k g[k]w_{l,m}(t - 2^l k\Delta\tau) \tag{23}$$

where l is the decomposition level, $0 \le m \le 2^l - 1$ is the wavelet atom position in the tree, $w_{0,0} = \phi(t)$ and

$$w_{l,m}(t) = \sum_k f_{l,m}[k]\phi(t - k\Delta\tau) \tag{24}$$

and $f_{l,m}[k]$ is the equivalent filter from the root to the (l,m) th terminal recursively evaluated using eqs. (22), (23). The orthogonality condition for WP atoms has the form [14]

$$\int w_{l,m}(t - 2^l n\tau) w_{\lambda,\mu}(t - 2^\lambda k\tau) dt = \delta(l - \lambda)\delta(m - \mu)\delta(n - k) \tag{25}$$

where $(l,\lambda) \in Z$, $0 \le m \le 2^l - 1, 0 \le \mu \le 2^\lambda - 1$, $(n,k) \in Z$. The waveform orthogonality is used in WPT based OFDM in order to transmit multiple message signals overlapping in time and frequency domains [14]. The time and frequency localization of wavelets can mitigate the optical channel chromatic dispersion which affects only the detail coefficients, or the highpass-filtered versions of the original signal. Then, a selective reconstruction of the wavelet coefficients is necessary [14].

WPTs can provide orthogonality between OFDM subcarriers similarly to DFT, and consequently, DWPT can replace DFT in the CO-OFDM system [13]. The all-optical WPT based CO-OFDM (WPDM) system has been proposed where the digital sequences are encoded by a set of orthogonal waveforms [13], [14]. The system performance is improved due to the orthogonal properties (25) of the wavelet atoms (22)–(24) and their overlapping in time and frequency [13], [14]. Each optical pulse is transformed into the corresponding wavelet atom function at the device output under the conditions that the input bit duration $t_{bit} = \Delta\tau$ and the processing gain $F = 2^l$ is equal to the number of simultaneous users [14]. In the WPT-OFDM system each channel occupies a WP [13]. At the transmitter, IDWPT reconstructs the time domain signal from WPs; at the receiver DWPT is used in order to decompose the time domain signal into different WPs by using successive low-pass and high-pass filtering [13]. Unlike IDFT/DFT system, in the IDWPT/DWPT OFDM system the basis function wavelets are finite in time, the inter-symbol orthogonality in WT is maintained due to the shift orthogonal property of waveforms, and symbols are overlapped in time domain [13]. As a result, the symbol duration increases, providing the tolerance with respect to the chromatic dispersion and eliminating the need of CP [13].

Consider the computational complexity C_{WPT} of WPT-OFDM defined as the total required number of complex multiplications [23]. It depends on the specific type of wavelets and system configuration. The complexity of one basic block C_{BB} determined by the convolution between complex input data and real QMFs, and the total complexity C_{WPT} are given by, respectively [23]

$$C_{BB} = L_{QMF}; C_{WPT} = (N - 1)L_{QMF} \tag{26}$$

where L_{QMF} is QMF length, N is the number of subcarriers. WTP-OFDM reduces the complexity by a factor of 6 to 10 for different wavelets in the range of moderate accumulated dispersion as compared to FFT based CO-OFDM without CP [23].

The performance of a digital optical communication systems is characterized by the bit error rate (BER) defined as the average probability of incorrect bit identification [1]. The simulations of the BER for WPT-OFDM and FFT based OFDM have been carried out using different wavelets, optical SNR of 20dB , chromatic dispersion parameter of 17ps/(nm·km), and forward error correction code (FEC) threshold of 10^{-3} [13]. The results show the chromatic dispersion tolerance of 5600 ps/nm and the longest distance of 330km for SSMF for the Johnston64 (E) wavelet [13].

We have carried out the numerical simulations of BER dependence on the transmission distance in the single polarization regime for the WPT-OFDM system without CP, with GI length of 5% of the symbol interval, and for generic IDFT/DFT systems with values of CP length from 5% up to 30% of the symbol interval. We used the single-polarization signal with the optical carrier frequency $f_{opt} = 193.1THz$, with 128 subcarriers. An optical fiber is characterized by the attenuation of 0.2dB/km and chromatic dispersion parameter of 17ps/(nm·km). We assumed that the efficient transmission can be realized with the BER less than the FEC threshold of $2 \cdot 10^{-2}$. PMD has not been taken into account. At the receiver, we used window synchronization Schmidl - Cox algorithm and 1 tap equalizer in frequency domain.

The BER dependence on the distance for the Haar WPT-OFDM and FFT CO-OFDM with different CPs is shown in Figure 2.

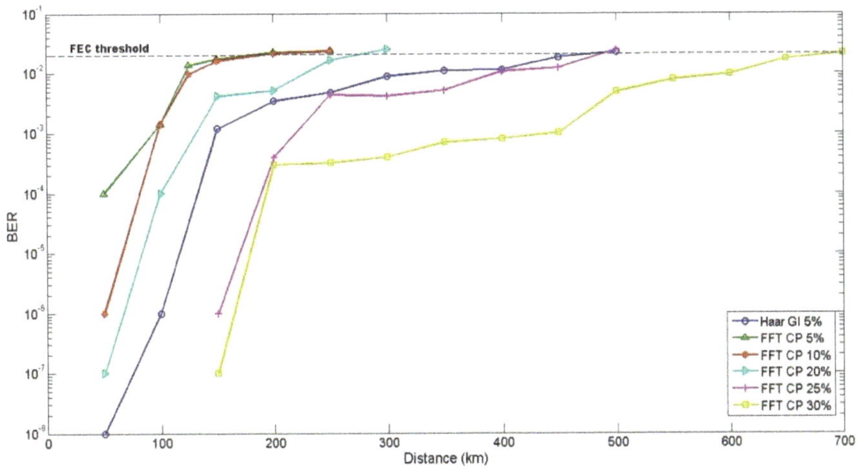

Figure 2. BER dependence on the transmission distance for FFT CO-OFDM with different CPs and WPT-OFDM without CP, with GI 5%

The results clearly show that WPT-OFDM provides the efficient transmission up to 500km without CP with 5% GI, while the FFT based CO-OFDM may achieve the same distance with the CP length of 25% of the symbol interval which substantially reduces SE of the communication system.

5. Passive Si-photonic components for all-optical signal processing

In this section we discuss the implementation of passive WPT-OFDM system components based on the Si photonics and a novel hierarchical architecture of the 1Tb/s WPT-OFDM transmission system that can be realized by using these components.

5.1. SOI optical components

The practical implementation of all-optical signal processing would require some extent of device integration. Much effort is dedicated over the last two decades to the development of photonic integrated circuits (PICs), which bring together multiple discrete devices on a single substrate. Integration helps to minimize the losses associated with the coupling of light in and out of devices, to enhance functionality, and to reduce cost and footprint. Numerous material platforms are prevalent in PICs, such as $LiNbO_3$, GaAs, InP and SiO_2. Among those platforms, the SOI wafer structure stands out as an advantageous choice for the realization of passive devices, such as couplers, interferometers, arrayed-waveguide gratings etc [24], [25]. Silicon is a low-cost material with an excellent crystalline quality, high thermal conductivity and high optical damage threshold. It is transparent over a broad range of wavelengths of 1.1-7 μm, including the telecommunication wavelengths. The silica SiO_2 lower cladding of SOI wafers provides a large contrast in refractive index with respect to silicon, which allows for the tight confinement of light into sub-micron structures. The fabrication of photonic devices in SOI can benefit from the advanced manufacturing technology of electronic integrated circuits. Silicon photonic devices may lead to a true merger of optics alongside electronics in unified devices. The realization of modulation of light on the silicon material platform is more challenging. The concentration of free charges in silicon changes the real and imaginary parts of the refractive index, and this effect in pure silicon is more strongly pronounced than the Pockels effect, the Kerr effect and the Franz-Keldysh effect [26]. Most of the fast modulators integrated on Si are based on free-carrier concentration variations [27]. Optical modulation using SiGe/Si and all-silicon phase shifters based on carrier depletion has been investigated theoretically and experimentally [27]. An all-silicon phase-shifter based on carrier depletion in a doped layer inserted into a PIN diode has been demonstrated [28]. SiGe/Si and all-silicon modulators can be integrated in rib waveguides and in MZIs [27]. Another modulation technique for SOI optical devices is based on the thermo-optic effect, in which the refractive index n of silicon is varied by applying heat to the material [24]. The thermo-optic coefficient in silicon is given by $dn/dT = 1.86\times10^{-4}K^{-1}$, and the refractive coefficient variation of $\Delta n = 1.1\times10^{-3}$ for the controllable temperature increase of 6K [24]. It has been shown experimentally that a 500μm length device thermally isolated from the substrate can provide a phase shift of π radians for an applied power of 10mW [24].

5.2. Example of SOI MZM for all-optical signal processing

In this section we present an example of the SOI based MZI which can realize the WPT operation. The most basic family of wavelet shapes is the Haar transform, proposed initially by Alfred Haar in 1910 [19]. The Haar wavelet and scaling functions $\psi(t),\phi(t)$ and the filter coefficients h[n], g[n] have the form, respectively [13], [19].

$$\psi(t)=\begin{cases}1,0\le t<\frac{1}{2}\\ -1,\frac{1}{2}\le t<1;\ \phi(t)=\begin{cases}1, & 0\le t<1\\ 0, & t<0,t\ge 1\end{cases}\\ 0,t<0,t\ge 1\end{cases} \tag{27}$$

$$h[n]=\frac{1}{\sqrt{2}}(-1,1),g[n]=\frac{1}{\sqrt{2}}(1,1) \tag{28}$$

Note that the equivalent definitions $h[n]=(1,1)/\sqrt{2},g[n]=(1,-1)/\sqrt{2}$ also may be used [21]. Since it is the simplest to implement, we adapt it in the proposed realization of the WPT based CO-OFDM photonic integrated circuit. An n-points signal is decomposed into two groups of n/2 samples. The first group is the sum of pairs of c[n] of the original signal, and can be described as the output of a discrete low-pass filter (LPF) followed by a down-sampling operation by a factor of two. The second group describes the differences between pairs d[n], and can be represented as the output of a discrete high-pass filter (HPF) followed by factor of two down-sampling operation [14]. The Haar WPT can be described by the scheme shown in Figure 3. Here s[n] is the input signal, g[n] and h[n] are the discrete HPF and LPF impulse responses.

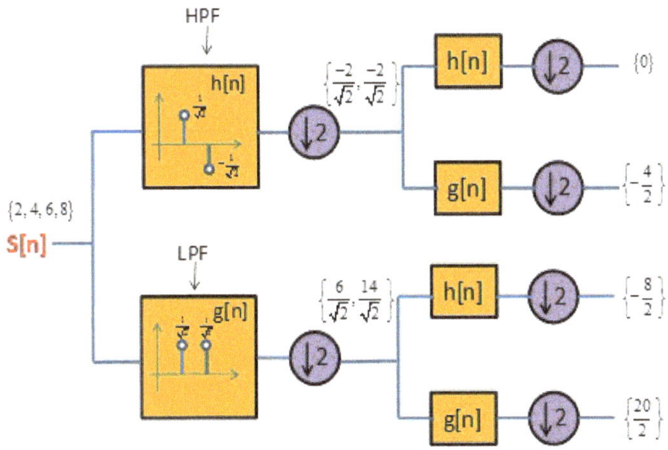

Figure 3. Two levels Haar wavelet-packet decomposition (WPD)

The inverse operation recovers the original signal from its decomposition coefficients. Its scheme is shown in Figure 4. Here S[n] is the output signal, g[n] and h[n] are the discrete HPF and LPF impulse responses, c[n] and d[n] are the approximation and detail coefficients respectively mentioned above.

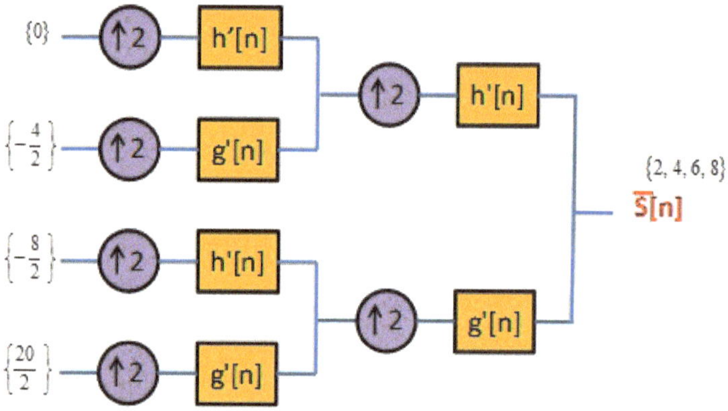

Figure 4. Haar inverse wavelet-packet decomposition (IWPD) transform

The realization of Haar wavelet packet decomposition (WPD) transform and the corresponding inverse wavelet packet decomposition (IWPD) in an optical integrated circuit was theoretically suggested by Gabriella Cincotti and co-workers [14], [22], [29], [30]. The method is based on the following MZI delay line architecture shown in Figure 5.

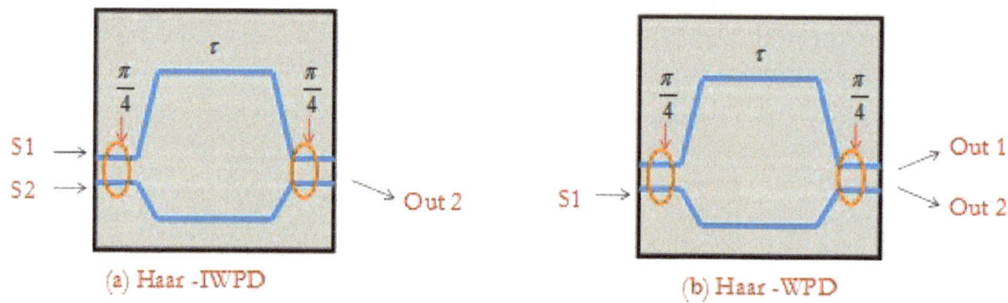

Figure 5. Optical implementation of Haar WPD / IWPD based on MZI. Left: Haar-IWPD used for the transmitter unit. Right: Haar-WPD used for the receiver unit.

The IWPD is represented by the optical field $E_{out2}(t)$ at the lower output of a MZI that is driven by two inputs $S_{1,2}(t)$.

$$E_{out2}(t)=\frac{1}{2}\left[-jS_1(t)+S_2(t)\right]+\frac{1}{2}\left[-jS_1(t-\tau)-S_2(t-\tau)\right] \tag{29}$$

Expression (29) shows that a single MZI could provide the sum and the difference of its two input fields, in series, in one of its output ports. The operation is equivalent to the LPF and

HPF operation of the inverse Haar IWPD. Similarly, the same MZI can generate the sum of successive values in one of its input ports at one output $E_{out2}(t)$, and the difference of the same values at the other output $E_{out1}(t)$, in parallel.

$$E_{out1}(t)=\frac{1}{2}\left[S_1(t)-S_1(t-\tau)\right];E_{out2}(t)=-j\frac{1}{2}\left[S_1(t)+S_1(t-\tau)\right] \quad (30)$$

The latter configuration described by expression (30) can realize the Haar WPD. Hence, MZIs can function as a basic building block of a discrete Haar WPD and IWPD. As can be seen in equations (29) and (30), the MZI realization of the WPD includes an additional relative phase shift of 90° in between the two inputs/outputs, which is not part of the Haar formalism. This additional phase must be compensated for. Furthermore, the optical path lengths connecting between cascaded MZIs cannot be controlled at the fabrication stage to a sub-wavelength precision. Hence, metallic resistors must be deposited in proximity to the waveguides [31], [32]. The driving of currents through the resistors would locally heat the nearby silicon structure, and modify its refractive index through the thermo-optic effect mentioned above [24]. A schematic drawing of a single MZI with the thermo-optic phase-shifters is shown in Figure 6.

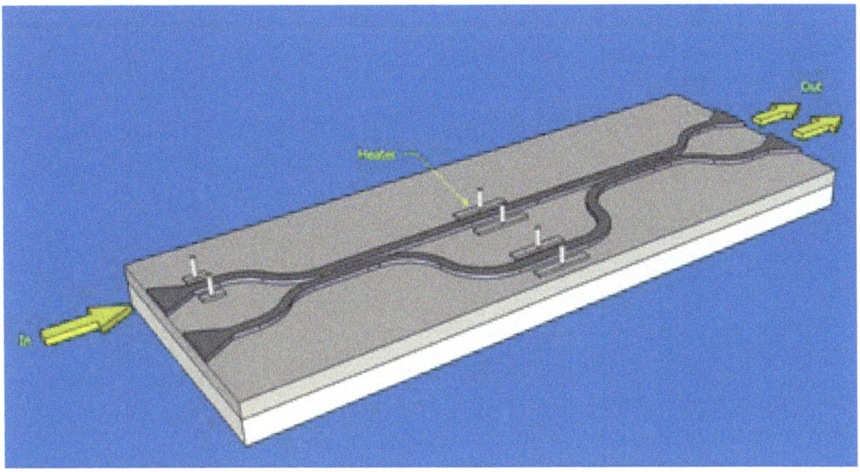

Figure 6. A schematic drawing of a single MZI stage used in a Haar WPD receiver including three thermo-optic phase-shifters

Three-stage MZI-based photonic integrated circuits for the realization of Haar WPT-OFDM encoding and decoding based on the single MZI stage are shown in Figures 7, 8.

In the Haar WPT-OFDM encoder presented in Figure 7, S_1-S_8 are low rate input data channels, with a seven-bits zero padding. The output is the multiplexed Haar transform signal.

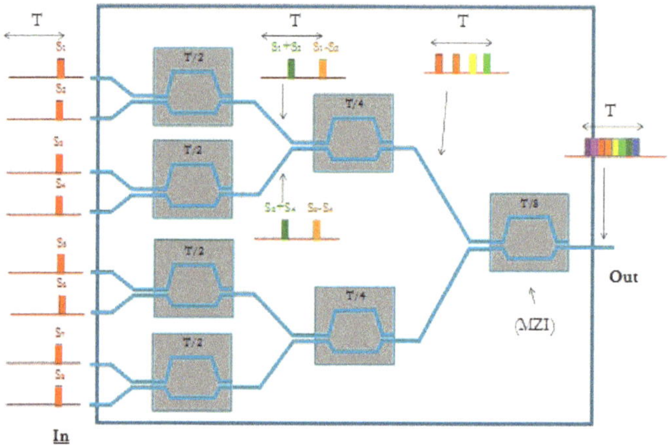

Figure 7. All-optical Haar WP Encoder used as optical transmitter

In the all-optical Haar WP Decoder shown in Figure 8, the input signal is constructed from eight data channels, which are recovered individually at the eight outputs. The output channels must be down sampled by factor of 8.

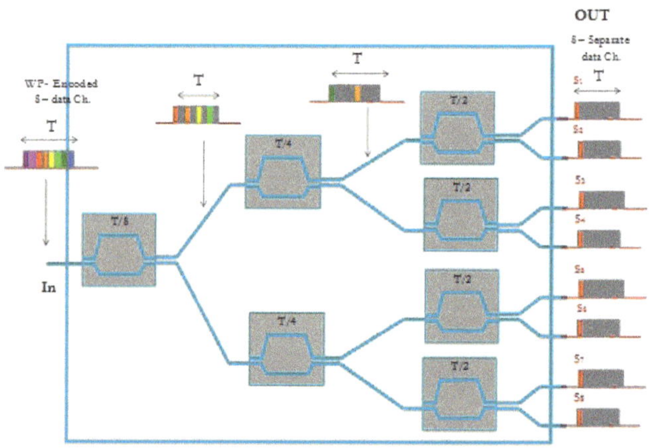

Figure 8. All-optical Haar WP Decoder used as an optical receiver

The all-optical WP encoder calculates the Haar IWPD of eight coefficients, incoming from eight parallel input values. The reconstructed signal appears in series at the output of the circuit. Note that padding by seven zeros is necessary between successive bits at each input, so that the transform coefficients of one input parallel word do not overrun those of the next word at the output [9], [33], [34]. The zero padding is the optical-domain equivalent of the up-sampling that is part of a digital IWPD. Similarly, a proper gating is necessary at the each of the eight outputs of the decoder circuit, since the original data is only reconstructed at specific time slots within the symbol duration [33]. The remainder of the symbol duration is occupied by noise-like ISI.

A WPT based CO-OFDM communication network, employing the encoding and decoding PICs, is shown in Figure 9. Light from a CW laser diode is split in eight paths. Light in each path is individually modulated by a separate stream of data, which are prepared with the necessary zero padding as described above. The eight channels are multiplexed by the WPT-OFDM PIC. At the other end of the link, each of the eight outputs of the WPT-OFDM decoder PIC is separately gated by an electro-optic switch and detected.

The SOI waveguide is a basic component of the Si photonic systems. We calculated the modal profile of such a waveguide in a single mode regime for each polarization of the optical wave. The SOI waveguide cross-section and the modal profile are shown in Figure 10. The analysis of such waveguides can be carried out only by numerical methods [35]. We used the commercial software modeling (COMSOL). The modal field distribution (Figure 10b) clearly shows the electric field confinement in the waveguide core.

A basic building block of a MZI is a directional coupler. Couplers are realized by bringing two SOI waveguides in close proximity for a certain length Z_0. The degrees of freedom in the design are the length and gap between the SOI waveguide cores. A relatively large gap of the order of magnitude of 300nm is advantageous with respect to fabrication imperfections. COMSOL simulations were used to calculate the coupling coefficient κ_{ab} between two waveguides separated by a chosen gap. An even splitting ratio of incoming optical power between the two outputs is obtained when the two waveguides remain in proximity over a length $L = \pi / (4\kappa_{ab})$. The simulation results are shown in Figure 11.

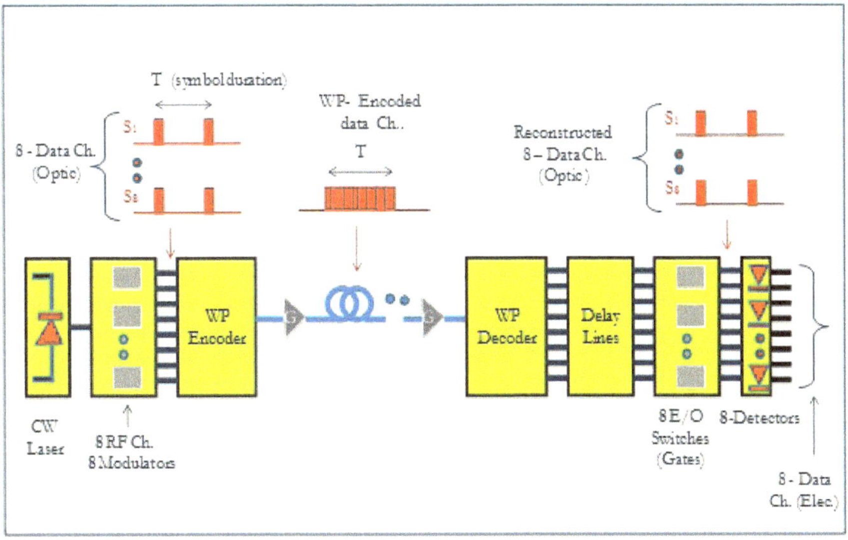

Figure 9. WPT based CO-OFDM data channel based on transmitter and receiver PICs

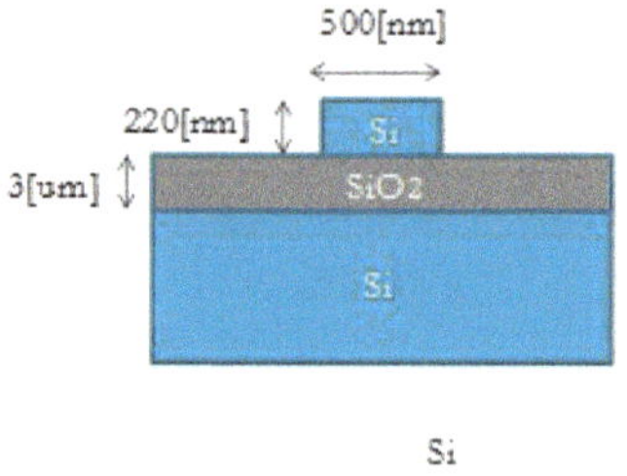

(a)

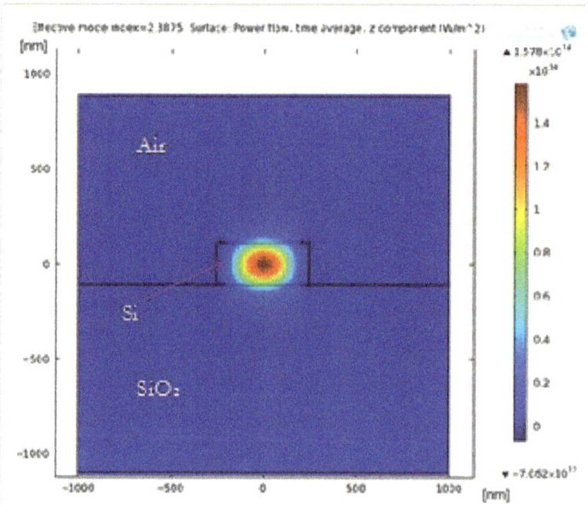

(b)

Figure 10. A single mode SOI-based waveguide; (a) schematic diagram; (b) COMSOL simulation of the transverse profile for the EM mode field super-imposed on the waveguide cross section

Consider now the differential delays of the MZIs. As discussed earlier, different stages in the cascaded MZI PIC require different delays. The basic delay unit is T/8, where T is the symbol duration. For a data rate of 2.5 GSymbols/s for each of the eight multiplexed channels, the fundamental delay unit is 50psec, which corresponds to a physical length of about 3.5mm in SOI waveguides. The heat dissipation from aluminium heaters in proximity of the SOI waveguides was simulated, once more using COMSOL. Figure 12 shows the resulting temperature profile. The Al heaters are heated by an external current up to 60°C. Simulation results show that a temperature in the Si core of the SOI waveguide is 40°C compared to 20°C in the unheated regions. This temperature difference stems from the attachment of the back end of the PIC to a 20°C heat sink.

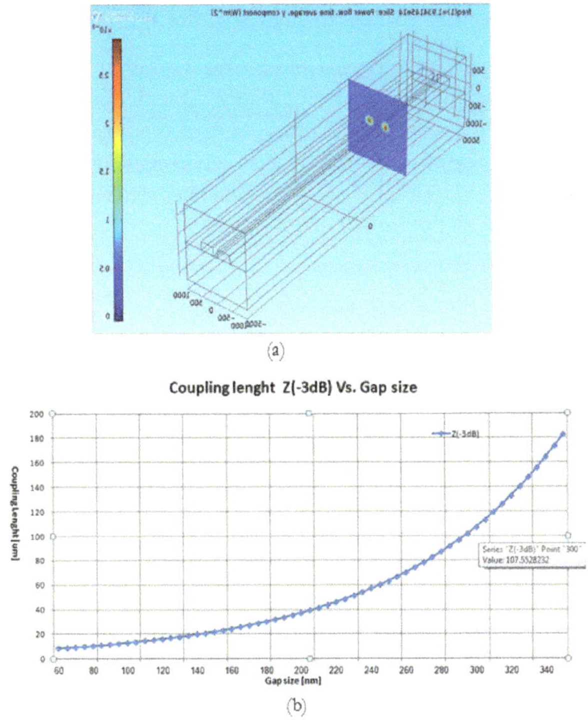

Figure 11. Coupler dimensions design with COMSOL software simulations: (a) an example of a three-dimensional modelling of a directional coupler; (b) calculated coupling length that is required between two parallel waveguides as a function of the gap size. The coupling length for the chosen gap size of 300 nm is approximately 110μm

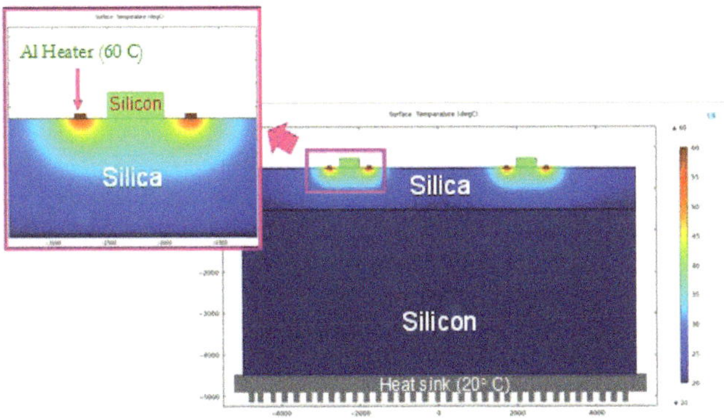

Figure 12. Cross section of heat dissipation in an SOI waveguide with the aluminium heaters located in both sides of the SOI waveguide

5.3. Hierarchical architecture and performance of the WPT based OFDM system

The DSP in CO-OFDM systems is carried out by the algorithms realized with the field programmable gate array (FPGA) and application specific integrated circuit (ASIC) electronic processors. Their computational power is limited with the operation rate of the VLSI electronic elements. For this reason, the electronic "bottleneck" can be eliminated and the system operation rate can be improved if high data rates signal processing is realized by all-optical methods such as all-optical WPT-OFDM.

We proposed a novel hierarchical architecture of the 1Tb/s transmission system based on WPT-OFDM in order to reduce the computational complexity of the DSP algorithms [15]. The hierarchical architecture concept is based on the separation of low bit rate and high bit rate signal channels, unlike the system discussed in Ref. [13]. We used an IDWPT/DWPT system based on the Haar WPT with the wavelet function $\psi(t)$, scaling function $\phi(t)$, and filter coefficients g[n] and h[n] given by eqs. (27), (28) [13], [19].

The WPT-OFDM hierarchical transmitter and receiver are shown in Figures 13 and 14, respectively.

The high data rate bands are multiplexed using all-optical IDWPT. The transmitter includes IQ modulator. QAM 16, QAM 4 and other multilevel modulation formats can be used for each subband. Subbands are multiplexed in electrical domain also by utilizing the IDWPT. At the receiver side, the multiband signal is demultiplexed into the 8 bands using all-optical

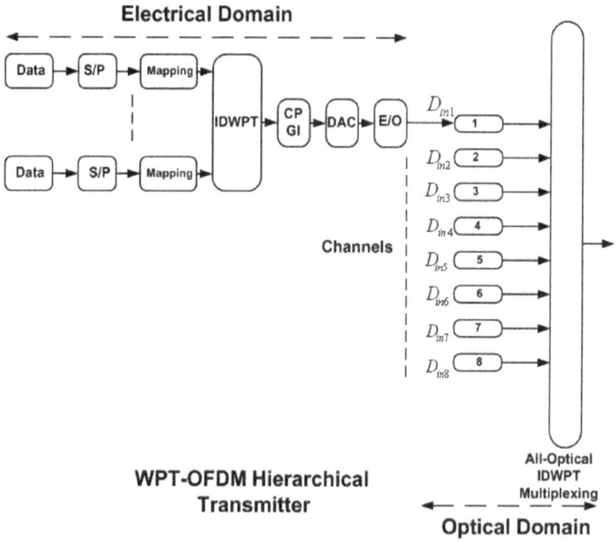

Figure 13. Hierarchical architecture of the WPT-OFDM transmitter (S/P - serial/parallel; E/O - electrical/optical)

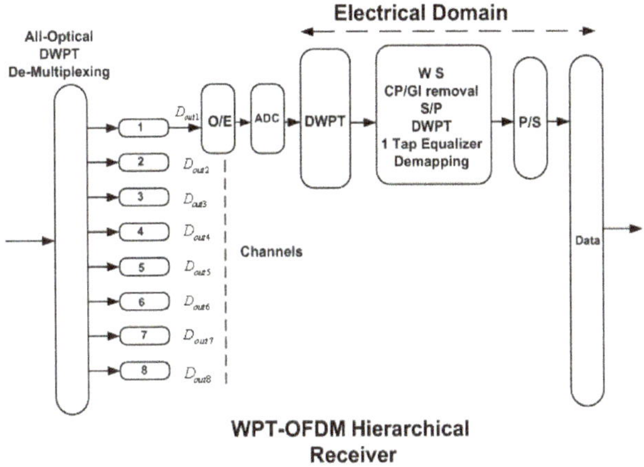

Figure 14. Hierarchical architecture of the WPT-OFDM receiver (O/E - optical/electrical, WS-window synchronization, S/P-serial/parallel, P/S-parallel/serial)

filters and consequently demultiplexed into the subbands in the electrical domain by the DWPT. The performance of the WPT-OFDM communication system based on the hierarchical architecture has been investigated theoretically taking into account the 10% non-ideality of the system devices and the white Gaussian noise model. We used the modulation format QAM 16, 8 level decomposition, the bit rate of 500Gb/s. The 500Gb/s multiple band consists of eight 62.5Gb/s bands. The simulated constellation of the transmitter-receiver "back-to-back" system, i.e. without optical fiber link is shown in Figure 15.

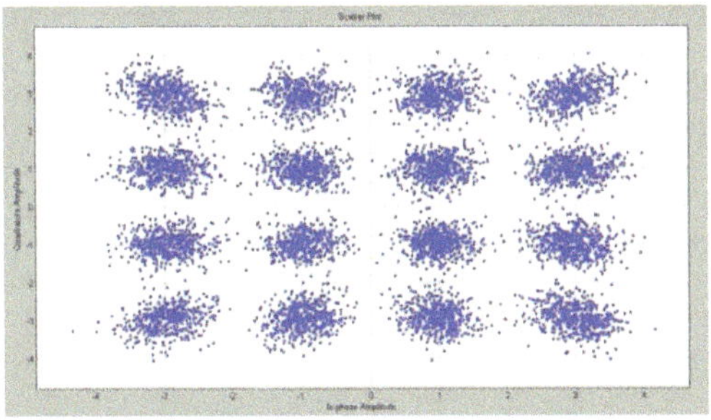

Figure 15. Constellation diagram for all-optical WPT-OFDM with a 500Gb/s bit rate, QAM-16 modulation format and 8 level decomposition

The constellation clearly shows the high performance of the transmitter-receiver system.

6. Conclusions

We discussed in this chapter the structure, operation principle and basic properties of the all-optical high SE CO-OFDM systems. The CO-OFDM is the most promising modulation method in the modern optical systems combining the advantages of the coherent detection and OFDM modulation. However, the high data rate transmission is strongly influenced by the optical channel chromatic dispersion and PMD. As a result, the transmission system performance significantly deteriorates due to ISI and ICI. The dispersion influence can be mitigated by an appropriately chosen CP. Unfortunately, a long enough CP would decrease the system SE. The problem can be solved and the necessity of CP may be eliminated if a generic DFFT based CO-OFDM is replaced with a WPT-OFDM since WPs are localized both in time and frequency domains. The simulation results show that WPT-OFDM system provides a 500 km transmission distance at the FEC level of 2×10^{-2} without CP, with a small 5% GI. We proposed a novel hierarchical architecture of the WPT-OFDM system based on the separation of the low data rate and high data rate signal processing. The former ones are processed in the electrical domain, while the latter ones undergo the all-optical processing. The numerical simulations show that such an approach improves the WPT-OFDM system performance which is demonstrated by the constellation of a signal with QAM 16 modulation. The WPT-OFDM all-optical signal processing units can be implemented by using the passive SOI waveguide components. The passive components of this architecture are the wavelet filters realized by the SOI waveguide based MZIs. The change of the Si refraction index is realized by using the thermo-optic effect. We presented the numerical simulations of the Haar wavelet filters for all-optical signal processing based on such MZIs. The modal profile of a SOI waveguide in a single mode regime for each polarization of the optical wave has been calculated.

Author details

Y. Ben Ezra
Holon Institute of Technology (HIT), Holon, Israel;
Optiway Integrated Solutions LTD, Rosh Haayin, Israel

B.I. Lembrikov
Holon Institute of Technology (HIT), Holon, Israel

Avi Zadok, Ran Halifa
School of Engineering and Institute for Nano-Technology and Advanced Materials, Bar-Ilan University, Israel

D. Brodeski
Optiway Integrated Solutions LTD, Rosh Haayin, Israel

Acknowledgement

This work was supported in part by the Israeli Science Foundation (ISF), and by the Chief Scientist Office of the Israeli Ministry of Industry, Trade and Labor within "Tera Santa" consortium.

7. References

[1] Agrawal, G.P. Fiber-Optic Communication Systems, Wiley, New York, 2002.

[2] Ran, M; Ben Ezra, Y. and Lembrikov B.I. Ultra-wideband Radio-over-optical-fibre Technologies, In: Kraemer, R. & Katz, M. D. (Eds.) Short-Range Wireless Communications, Chichester, England: Wiley; 2009, p271-327.

[3] Shieh, W.; Bao, H. and Tang, Y. Coherent optical OFDM: theory and design. Optics Express, Vol.16, No. 2, January 2008, 841-859.

[4] Shieh, W. and Djordjevic, Ivan. Orthogonal Frequency Division Multiplexing for Optical Communications, Academic Press, 2010.

[5] Kikuchi, K. Coherent optical communication systems, In: Kaminov, I. P.; Li, T. & Willner, A. E. (Eds.) Optical Fiber Telecommunications VB: Systems and Networks, Amsterdam, London, New York: Academic Press; 2008. p91-129.

[6] Winzer, J. P. and Essiambre, R.-J. Advanced Modulation Formats for High-Capacity Optical Transport Networks, IEEE Journal of Lightwave Technology December 2006; 24(12) 4711-4728.

[7] Armstrong, J. OFDM for Optical Communications, IEEE Journal of Lightwave Technology, February 2009; 27(3) 189-204.

[8] Tang, Y. and Shieh, W. Coherent optical OFDM transmission up to 1 Tb/s per channel. Journal of Linghtwave Technology, August 15, 2009; 27(16) 3511-3517.

[9] Hillerkuss, D.et al. Simple all-optical FFT scheme enabling Tbit/s real-time signal processing, Optics Express, April 2010; 18(9) 9324-9340.

[10] Wang, X.; Ho, P., and Wu, Y. Robust Channel Estimation and ISI Cancellation for OFDM Systems with Suppressed Features, IEEE Journal on Selected Areas in Communications, May, 2005; 23(5) 963-972.

[11] Lai, Hung-Quoc; Siriwongpairat, W. Pam, and Liu, K.J. Ray. Performance Analysis of Multiband OFDM UWB Systems with Imperfect Synchronization and Intersymbol Interference. IEEE Journal of Selected Topics in Signal Processing, October 2007; 1(3) 521-534.

[12] Armstrong, J. Analysis of New and Existing Methods of Reducing Intercarrier Interference Due to Carrier Frequency Offset in OFDM, IEEE Transactions on Communications, March 1999; 47(3) 365-369.

[13] Li, An; Shieh, W. and Tucker, Rodney S. Wavelet packet transform-based OFDM for optical communications, Journal of Linghtwave Technology, December 15, 2010; 28 (24) 3519-3528.

[14] Cincotti, G.; Moreolo, M.S. and Neri, A. Optical Wavelet Signals Processing and Multiplexing, EURASIP Journal on Applied Signal Processing, 2005;10, 1574-1583.

[15] Ben-Ezra, Y.; Brodeski, D.; Zadok, A.; Califa, R.; Lembrikov, B.I. 1Tbps Transmission System Based on Hierarchical Approach to Wavelet Packet Transform OFDM. Proceedings of the 13th International Conference on Transparent Optical Networks (ICTON 2011), Stockholm, Sweden, June 26-30 2011, Tu.A5.2 (1-4).

[16] Ip, Ezra; Lau, A.P.T.; Barros, D.J.F.; Kahn, J.M. Coherent detection in optical fiber systems. Optics Express, January 2008; 16(2) 753-791.

[17] Ma, Yiran; Yang, Qi; Tang, Y.; Chen, Simin & Shieh, W. 1 Tb/s single-channel coherent optical OFDM transmission over 600-km SSMF fiber with subwavelength bandwidth access, Optics Express, May 2009; 17(11) 9421-9427.
[18] Bigo, S. Multiterabit DWDM terrestrial transmission with bandwidth-limiting optical filtering. IEEE Journal of Selected Topics in Quantum Electronics, March/April 2004; 10(2) 329-340.
[19] Rao, R.M.& A.S. Bopardikar, A. S. Wavelet Transforms. Introduction to Theory and Applications. Reading, Massachusetts: Addison-Wesley; 1998.
[20] Daubechies, I Ten Lectures on Wavelets. Philadelphia, Pennsylvania: Society for Industrial and Applied Mathematics; 2006.
[21] Sarkar, K.T.; Salazar-Palma, M.; Wicks, M.C. Wavelet Applications in Engineering Electromagnetics. Boston: Artech House; 2002.
[22] Moreolo, M. S.; Cincotti, G. and Neri A. Synthesis of optical wavelet filters, IEEE Photonics Technology Letters. July 2004, 16 (7) 1679-1681.
[23] Bulakci, Ö.; Schuster, M.; Bunge, C.-A.; Spinkler, B.; Hanik, N. Wavelet Transform Based Optical OFDM. In: Optical Fiber Communication (OFC), collocated National Fiber Optic Engineers Conference, 2009 Conference on (OFC/NFOEC), 2009, pp. 1-3.
[24] Reed, G. T. and Knights, A. P. Silicon photonics. An introduction. Chichester, England: Wiley; 2004.
[25] Reed, G. T., editor. Silicon photonics. The state of the art. Chichester, England: Wiley; 2008.
[26] Reed, G. T.; Mashanovich, G.; Gardes, F.Y, & Thomson, D.J. Silicon optical modulators, Nature Photonics. August 2010; 4, 518-526.
[27] Marris-Morini, D. et al. Recent progress in high-speed silicon-based optical modulators. Proceedings of the IEEE, July 2009; 97(7) 1199-1215.
[28] Marris-Morini, D. et al. Optical modulation by carrier depletion in a silicon PIN diode, Optics Express, October 2006; 14(22) 10838-10843.
[29] G. Cincotti, G. Full optical encoders/decoders for photonic IP routers, Journal of Lightwave Technology, vol. 22, no. 2, (February 2004) 337- 342.
[30] Moreolo, M. S. and Cincotti, G. Compact low-loss planar architectures for all-optical wavelet signal processing, In: Transparent Optical Networks, 2005, Proceedings of 2005 7th International Conference, 2005, vol. 1, pp. 319- 322.
[31] Densmore, A. et al. Compact and low power thermo-optic switch using folded silicon waveguides, Optics Express, June 2009; 17(13) 10457-10465.
[32] Harjanne, M.; Kapulainen, M.; Aalto, T. and Heimala, P. Sub-μs switching time in silicon-on-insulator Mach-Zehnder thermooptic switch. IEEE Photonics Technology Letters, September 2004; 16(9) 2039-2041.
[33] Hillerkuss, D. et al. Single source optical OFDM transmitter and optical FFT receiver demonstrated at line rates of 5.4 and 10.8 Tbit/s, In: Optical Fiber Communication (OFC), collocated National Fiber Optic Engineers Conference, 2010 Conference on (OFC/NFOEC), 2010, pp. 1-3.
[34] Hillerkuss, D. et al. 26 Tbit s-1 line-rate super-channel transmission utilizing all-optical fast Fourier transform processing, Nature Photonics, June 2011; 5(6) 364-371.
[35] Okamoto, K. Fundamentals of Optical Waveguides. Academic Press, San-Diego, USA: Academic Press; 2000.

Chapter 15

Stable Higher-Charged Vortex Solitons in Optically-Induced Lattices

Liangwei Dong and Huijun Li

Additional information is available at the end of the chapter

1. Introduction

Vortices are fundamental objects which appear in many branches of physics [1] such as optics [2, 3] and Bose-Einstein condensates [4]. In nonlinear optics, vortex solitons are associated with the phase dislocations (or phase singularities) carried by the nondiffracting optical beams [5], and share many common properties with the vortices observed in other systems, e.g., superfluids and Bose-Einstein condensates [6, 7]. In a homogeneous medium, stable vortex solitons were proposed to exist in the so-called cubic-quintic or other similar nonlinear media, for example, combination of $\chi^{(2)}$ and $\chi^{(3)}$ nonlinear media, based on competing self-focusing and self-defocusing nonlinearities [8–11]. However, the experimental realization of vortex solitons in such media is hard, as the requirement of very high energy flow of light usually excites other higher-order nonlinearities, which may be dominant and suppress the occurrence of competing nonlinearities.

Successful alternatives are confined systems, such as graded-index optical fibers [12], nonlinear photonic crystals with defects [13], linear and nonlinear optical lattices [14–19], or optical lattices with defects [20], where the azimuthal instability of vortices can be suppressed by the corresponding confining potentials. Stable vortices with charges lower than two are possible within certain ranges of lattice (transverse refractive index modulation) parameters [14, 17, 21–28]. Different types of vortex solitons, such as discrete vortices [21], vortex-ring "discrete" solitons [28], and second-band vortices [22] were observed in experiments. For a review of early works, see [5, 29, 30] and references therein.

Thus far, dynamics of higher-charged vortex solitons are still poorly understood. The optical settings allowing stable higher-charged vortex solitons are rare. Main efforts in bulk or lattice-modulated nonlinear media were devoted to the analysis of vortex solitons with charges less than or equal to two. In the following sections, three different schemes for the realization of vortex solitons with higher charges will be addressed.

In section 2, the dynamics of vortex solitons in a radial lattice with a lower-index defect covering several lattice rings is revealed. The defect scale can be utilized to control the energy flow of vortices. Vortex solitons with various charges are stable in a region near the upper cutoffs of propagation constant. Although higher-charged vortices at higher energy flow suffer oscillatory instability, they can survive very long distances without visible distortions. Vortex solitons at lower or moderate energy flow are completely stable under appropriate conditions. The variation of topological charges slightly influences the existence and stability domains of vortex solitons. This property provides an effective way for the experimental realization of vortex solitons with higher charges in an optical setting with fixed parameters.

In section 3, the existence, stability and propagation dynamics of vortex solitons in a defocusing Kerr medium with an imprinted azimuthally modulated Bessel lattice are investigated. Since the special amplitude distribution of vortex soliton resembles to the azimuthons stated in [31] and the phase distribution is also a staircase function of the polar angle, such vortex solitons can also be termed as "azimuthons". The azimuthal refractive index modulation admits stable vortex solitons with lower or higher topological charges. The "stability rule" of azimuthons in defocusing cubic media is exactly opposite to that of vortex solitons in focusing media with the same transverse refractive index modulation. It is the first example of stable azimuthons in local nonlinear media.

In section 4, the stability of vortex solitons supported by a circular waveguide array with out-of-phase modulation of linear and nonlinear refractive indices is studied. The out-of-phase competition between two effects substantially modifies the stability properties of vortex solitons. Vortex solitons undergo remarkable power-dependent shape transformations. They expand or shrink radially with the propagation constant, depending on the phase difference between the neighboring lobes. In particular, we revealed that increasing waveguide number of circular array can stabilize vortex solitons with higher topological charges.

2. Higher-charged vortex solitons in defected radial lattices

2.1. Introduction

Defects and defect states exist in a variety of linear and nonlinear systems, including solid state physics, photonic crystals, and Bose-Einstein condensates. When lights propagate in an optical lattice with a local defect, the band-gap guidance results in the formation of linear or nonlinear defect modes [32, 33]. Recently, defect guiding phenomena of light in diverse settings, such as photonic crystals [13], fabricated waveguide arrays [34, 35], and optically induced photonic lattices [36–42], have been predicted theoretically and observed experimentally. Ye and his coworkers proposed that stable nonlinear modes can be trapped in a lower-index defect sandwiched between two optical lattices, or in the cylindrical core of a radial lattice [43]. The variation of defect scales, depths and shapes can be used to stabilize and reshape the fundamental, dipole and vortex solitons [20].

In this section, we reveal that the defocusing media with an imprinted radially symmetric lattice with a lower-index defect covering several lattice rings can support stable vortex solitons with higher charges under appropriate conditions. In contrast to the cases in competing media, vortex solitons can propagate stably at lower or moderate energy flow. In lattices with fixed depth and defect scale, vortex solitons are completely stable provided that the propagation constant exceeds a critical value. In particular, we illustrate that the variation of topological charges slightly influences the existence and stability domains of vortex solitons. This is in contrast to all previous studies and allows the experimental realization of vortex solitons with higher charges without changing the parameters of an optical setting.

2.2. Theoretical model

We consider light propagation along the z axis of a defocusing Kerr medium with an imprinted transverse modulation of the refractive index. Dynamics of the beam can be described by the nonlinear Schrödinger equation for the dimensionless complex field amplitude A:

$$i\frac{\partial A}{\partial z} = -\frac{1}{2}\left(\frac{\partial^2 A}{\partial x^2} + \frac{\partial^2 A}{\partial y^2}\right) + |A|^2 A - pR(x,y)A. \tag{1}$$

Here, the longitudinal z and transverse x, y coordinates are scaled in the terms of diffraction length and beam width, respectively; p denotes the lattice depth; the refractive-index profile is given by $R(x,y) = \cos^2(\Omega r)$ for $r \geq (2N-1)\pi/(2\Omega)$ and $R(x,y) = 0$ otherwise, where $r = (x^2+y^2)^{1/2}$ is the radial distance, Ω is the frequency, and $N = 1, 2...$ is the number of rings removed from the lattice and characterizes the defect scale. Thus, the transverse modulation of refractive index features a lower-index guiding core. By comparing the defocusing bulk media without external potentials, the radial lattices with defects can confine the beams in a local region. An example of such refractive-index landscapes is shown in Fig. 1.

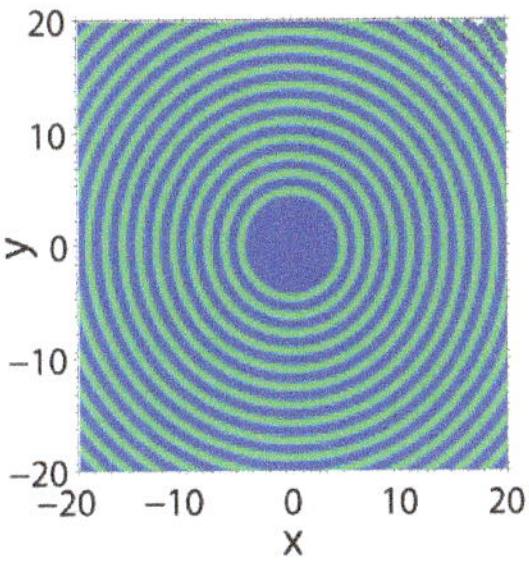

Figure 1. Radial lattice with a defect.

Although there are defects in radial lattices, the wings of nonlinear modes still penetrate into the bulk of lattices. Thus, the existence of nonlinear modes strongly depends on the transverse lattices. Since the term $1/rd/dr$ in Laplacian can be neglected at $r \to \infty$, the band-gap structure of a radially symmetric lattice is slightly different from that of 1D periodic lattice [43]. Thus, it is convenient to use the band-gap structure of 1D periodic lattice to approximately analyze the existence of solitons. Due to the fact that nonlinear modes in defocusing Kerr media can only be found in the finite gaps, we are interested in the solitons residing in the first finite gap. Equation (1) conserves several quantities, including the energy flow $U = \int_{-\infty}^{\infty}\int_{-\infty}^{\infty} |A(x,y)|^2 dxdy$ and the Hamiltonian $H = \frac{1}{2}\int_{-\infty}^{\infty}\int_{-\infty}^{\infty}(\left|\frac{\partial A}{\partial x}\right|^2 + \left|\frac{\partial A}{\partial y}\right|^2 - 2pR|A|^2 + |A|^4)dxdy$:

We search for stationary solutions of Eq. (1) by assuming $A(x,y,z) = q(r)\exp(ibz + im\phi)$, where q is a r-dependent real function depicting the profile of stationary solution, b is a propagation constant associating with the energy flow, and m is an integer known as the topological charge of vortex soliton. The nonlinear mode degenerates to a fundamental radially symmetric mode when $m = 0$. The substitution of the light field into Eq. (1) yields:

$$\frac{d^2 q}{dr^2} + \frac{1}{r}\frac{dq}{dr} - \frac{m^2}{r^2}q - 2bq - 2q^3 + 2pRq = 0, \tag{2}$$

which can be solved numerically by means of a Newton iterative method. Mathematically, various families of stationary solutions are determined by the propagation constant b, lattice depth p, modulation frequency Ω and defect scale N. We vary b, p, N and fix $\Omega \equiv 2$ in this section.

The stability of solitons can be analyzed by considering the perturbed solution in the form of: $A\bullet x, y, z\bullet \;\bullet\; \bullet q\bullet r\bullet\bullet\; u\bullet r\bullet \exp\bullet\lambda z \bullet\; in\phi\bullet\bullet\; v^{*}\bullet r\bullet \exp\bullet\lambda^{*}z - in\phi\bullet\bullet \exp\bullet ibz \bullet\; im\phi\bullet$, here the perturbation components u, v could grow with a complex rate λ during propagation, and n is an integer representing the angle dependence of the perturbation and is termed as an azimuthal index. The substitution of the perturbed solution into Eq. (1) results in a system of eigenvalue equations:

$$i\lambda u \bullet\; -\frac{1}{2}\bullet\frac{d^2}{dr^2} \bullet\; \frac{1}{r}\frac{d}{dr} - \frac{\bullet m \bullet\; n\bullet^2}{r^2}\bullet u \bullet\; bu \bullet\; \bullet v \bullet\; 2u\bullet q^2 - pRu$$

$$-i\lambda v \bullet\; -\frac{1}{2}\bullet\frac{d^2}{dr^2} \bullet\; \frac{1}{r}\frac{d}{dr} - \frac{\bullet m - n\bullet^2}{r^2}\bullet v \bullet\; bv \bullet\; \bullet u \bullet\; 2v\bullet q^2 - pRv. \quad (3)$$

The coupled equations can be solved by a finite-difference method. In Cartesian coordinates, the square of the above linearization operator is self-adjoint if the stationary solutions are angle independent (fundamental solitons). Thus, the discrete eigenvalue is either purely real or purely imaginary. The instability growth rates with purely real parts correspond to the Vakhitov-Kolokolov (V-K) instability [44]. When the stationary solutions are angle dependent (vortex solitons), the eigenvalues may have both real and imaginary parts associating with an oscillatory instability. Stationary solutions are completely stable provided that all real parts of eigenvalues equal zero.

2.3. Discussions

First, we consider vortex solitons with unit charge supported by the defocusing Kerr media with an imprinted radial lattice with a defect. Without loss of generality, we set the defect scale $N \bullet\; 10$ in the following discussions. In contrast to the fundamental solitons, the energy flow U of vortices is always a monotonically decreasing function of propagation constant b [Fig. 2(a)]. We stress that the vortex solitons here are bright or ring-profile ones whose amplitudes decay to zero at infinity. Such vortices cannot be bifurcated from the dark vortices supported by the bulk defocusing media in the vanishing lattice. In other words, vortex soliton only exists when the lattice depth exceeds a critical value. For example, as shown in Fig. 2(b), the threshold value of lattice depth for the appearances of vortices with unit charge is $p_{th} \approx 1.18$, below which no vortex solutions can be found. Therefore, the vortex in the present lattice system is not a continuum of dark vortex in the vanishing lattice case, and it belongs to a different soliton family. For $p \leq 2.57$, the existence domain expands with the lattice depth. It shrinks with the growth of lattice depth if $p \in \bullet 2.57, 7.05\bullet$, due to the restriction of the ascending lower edge of the first band gap. It is the restriction of the first gap which accounts for the hoofed existence domain [Fig. 2(b)].

Figures 2(c) and 2(d) display two typical profiles of vortex solitons with unit charge at different energy flow. The vortex at higher energy flow looks like a flat-topped beam embedded with a dark core at which the amplitude is zero and the phase is undefined. Vortices become more localized with the growth of propagation constant. Vortices at higher energy flow penetrate deeply into the bulk of lattice, which leads to the multi-ring structures of beam intensity distributions.

To examine the stability of vortex solitons with unit charge, we conduct linear-stability analysis on the stationary solutions according to Eqs. (3). Typical spectra of the linearization operator for vortex solitons at $b = -0.8$ and $b = -0.1$ in lattice with defect scale $N = 10$ at $p = 4.2$ are shown in Figs. 3(a) and 3(b), respectively. Vortices at higher energy flow suffer an oscillatory instability with complex growth rates [$\mathrm{Re}(\lambda) \ll \mathrm{Im}(\lambda)$], while vortices at lower or moderate energy flow are completely stable [$\mathrm{Re}(\lambda) = 0$]. To confirm the stability analysis results, we numerically integrate Eq. (1) with a standard beam propagation method code, using the stationary solutions as the initial inputs. Representative unstable and stable propagation examples are illustrated in Figs. 3(c) and 3(d). Obviously, unstable vortex solitons can survive large distances (hundreds of diffraction lengths), greatly exceeding the present experimentally feasible sample lengths.

We summarize the linear-stability analysis results in Fig. 2(b). We show the critical value of propagation constant $b_{cr}^{n=1}$ above which no perturbations with the azimuthal index n and nonzero real part of growth rate were found. Vortex solitons are dynamically stable in a broad region near the upper cutoffs of propagation constant. It is the combination of defocusing nonlinearity and confining potential who affords the stability of vortex solitons. The precise structure of instability regions (patched) is rather complicated. There may exist multiple narrow stability windows.

Now, we focus on the vortex solitons with higher topological charges in radially lattices with defects imprinted in a defocusing Kerr medium. Figure 4(a) shows the hoofed existence domain of vortex solitons with $m = 3$. Vortex solitons can be found in lattices with $p \in [1.18, 6.97]$. By comparing the existence domains of vortex solitons with $m = 1$ [Fig. 2(b)]

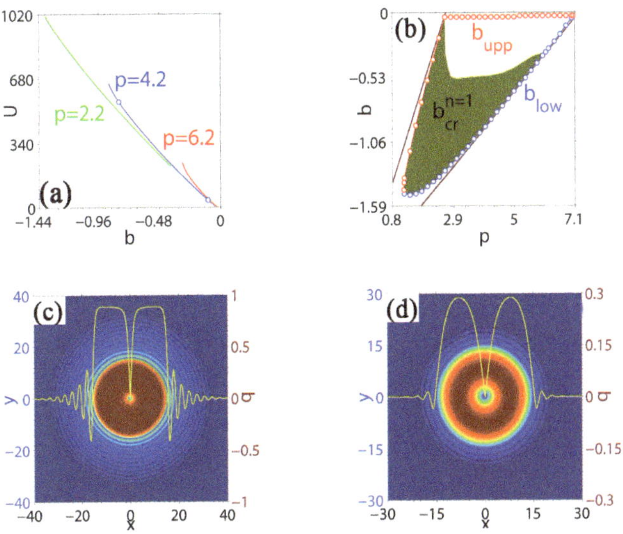

Figure 2. (a) Energy flow of vortex solitons with $m = 1$ vs b for different p. (b) Areas of existence and instability (patched) on the (p, b) plane. Solid lines denote the edges of the first gap of 1D periodic lattice. (c, d) Contour and amplitude profiles of vortices at $p = 4.2$, $b = -0.8$, $U = 556.2$ and $p = 4.2$, $b = -0.1$, $U = 37.9$, respectively. In all cases $N = 10$.

and $m = 3$, one can find that the upper cutoff of propagation constant drops from ~ -0.04 to ~ -0.08, which leads to the decrease of the upper threshold value of lattice depth p (from 7.05 to 6.97) and thus the shrinkage of the existence domain. Yet, the existence area of vortex solitons with $m = 3$ still occupies almost the whole of the first gap of undefected lattice (for $b < 0$).

The energy flow of vortex solitons with $m = 3$ also decreases with the propagation constant. The pronounced decaying oscillations of such modes in the bulk of lattice become stronger with the decrease of propagation constant [Fig. 4(b)]. The maxima of intensity distribution around the phase dislocation move towards the center of the defect core with the growth of propagation constant. Figure 4(c) displays an example of screw-type phase distribution of vortex soliton with $m = 3$.

The instability of vortex solitons with higher charges usually depends on the azimuthal index n [17, 26, 45]. Linear-stability analysis results reveal that for vortices with $m = 3$, the instability area associating with $n = 2$ is always dominant. For vortex solitons with $m = 3$ in a lattice with $p = 5$, the widths of instability windows associating with azimuthal indices $n = 1, 2$ and 3 occupy $\approx 26.77\%$, $\approx 38.69\%$ and $\approx 16.93\%$ of the width of the whole existence domain, respectively. An example of instability growth rate corresponding to azimuthal index $n = 2$ versus propagation constant is illustrated in Fig. 4(d). It indicates that vortex solitons suffer weak azimuthal instability, which allows them to propagate without obvious shape distortion over large propagation distances. Vortex solitons will be completely stable provided that the propagation constant exceeds a critical value.

By comparing the stability areas of vortex solitons with $m = 1$ [Fig. 2(b)] and $m = 3$ [Fig. 4(a)], one finds that the stability area of vortices with $m = 3$ is slightly narrower than that of

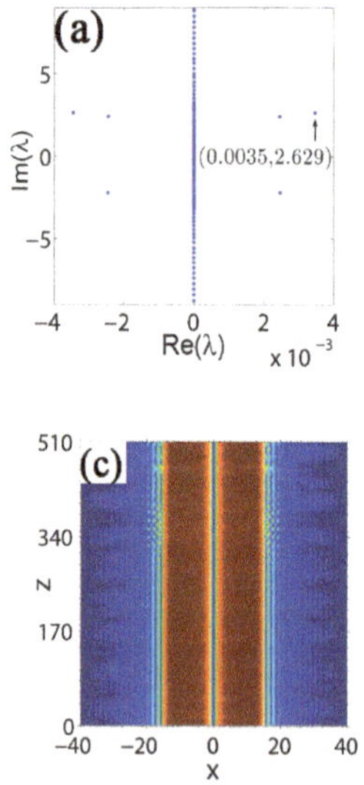

Figure 3. Spectra of the linearization operator (a, b) and unstable and stable propagations (c, d) of vortex solitons shown in Figs. 2(c) and 2(d).

vortices with $m = 1$, which constitutes one of our central results. That is to say, the stability area is slightly affected by the growth of topological charge, which allows one to realize stable vortex solitons with even higher charges. Since vortices with different charges share a collective stability area, one can input beams with different charges to excite vortex solitons with corresponding charges in certain parameter windows without changing the lattice depth, defect scale, modulation frequency, etc. It should be noted again that vortex solitons with different charges can propagate stably at lower or moderate energy flow, which is in sharp contrast to the cases in competing media, where very high energy flow is needed to stabilize the vortices [8, 10]. Thus, in addition to the Bessel lattice [17], the radial lattice with a defect is another effective alternative for the realization of stable vortex solitons at lower or moderate energy flow, especially for vortices with higher charges.

To confirm the above conclusions, we investigate the dynamics of vortices with $m = 4, 5...10$. The existence and stability domains shrink slowly with the topological charge due to the slow decrease of the upper cutoff of propagation constant. The energy flow decreases with the growth of topological charge when the lattice parameters are fixed [Fig. 5(a)]. This can be explained by the law of conservation of energy. For linearly polarized vortex beam, the total energy includes two parts. The first part is the energy carried by the photons, and the second part is the rotational energy associating with the orbital angular momentum. From the definition of the energy flow of vortex solitons, one finds that the expression of energy flow only defines the energy carried by the photons. Yet, the rotational energy of vortex solitons is proportion to the square of topological charge and effectively rotational radius. Given that the allowed energy of different modes in a fixed system is a constant, the energy flow carried by

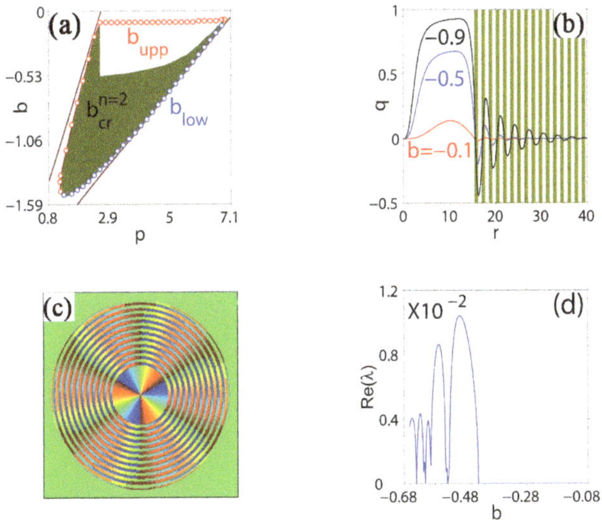

Figure 4. (a) Areas of existence and instability (patched) of vortex solitons with $m = 3$ on the (p, b) plane. Solid lines denote the edges of the first gap of 1D periodic lattice. (b) Profiles of vortices at $p = 4.2$. (c) Phase distribution at $b = -0.5$, $p = 4.2$. (d) Real part of instability growth rate associating with $n = 2$ vs b at $p = 5$. In all cases $N = 10$.

photons decreases with the increase of rotational energy. The conclusion may be generalized to vortex solitons with continuous intensity distributions in other models.

The above discussions can also explain the decrease of the thickness of vortex solitons shown in Fig. 5(b). With the growth of topological charge, the decrease of effective mass of the beam is in companion with the increase of effectively rotational radius and angular velocity. For fixed propagation constant, the delocalization of vortex soliton weakens with the growth of topological charge. A representative propagation example of unstable vortex solitons is illustrated in Fig. 5(c). The vortex can propagate without visible shape distortion over hundreds of diffraction lengths. Figures 5(d) and 5(e) show two instances of stable propagations of vortex solitons at $b = -0.4$ with topological charges $m = 6$ and 9, respectively. The phase structure of vortex soliton at $z = 1024$ with $m = 9$ is displayed in Fig. 5(f).

Finally, we briefly discuss the influence of lattice parameters on the existence of vortex solitons. Vortex solutions cannot be found in radial lattices without defects. The existence domain expands with the growth of defect scale and approaches an ultimate at $N = 4$. It shrinks with the increase of modulation frequency Ω. The existence domain shrinks slowly with the topological charge if other parameters are fixed. No matter what topological charge or lattice parameters are, the stable area always occupies a region near the upper cutoffs of propagation constant. We stress that although the vortex solitons residing in the patched areas shown in Figs. 2(b) and 4(a) are unstable, they can survive large propagation distances. Unstable vortex solitons with higher charges exhibit a similar behavior. Thus, we expect that all vortices in radial lattices with defects can be observed in experiments.

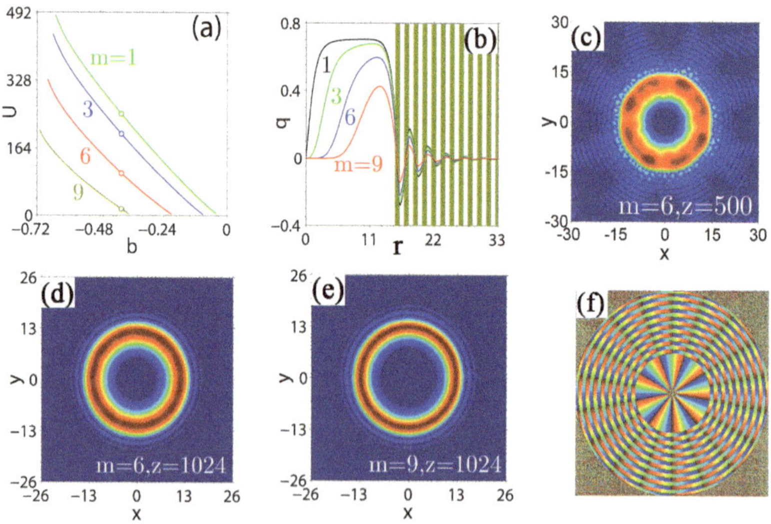

Figure 5. (a) Energy flow of vortex solitons with different m vs b. (b) Profiles of vortex solitons with different m at $b = -0.4$. (c, d) Unstable and stable propagations of vortex solitons with $m = 6$ at $b = -0.67$ and $b = -0.4$, respectively. (e, f) Field modulus and phase distribution of vortex soliton with $m = 9$ at $b = -0.4, z = 1024$. In all panels, $p = 5$.

3. Higher charged vortex solitons in azimuthally modulated Bessel lattice

3.1. Introduction

Besides the harmonic lattice, there is another important optical lattice with unique symmetry, the Bessel lattice, which can be created by nondiffracting Bessel beams with cylindrical symmetry. Kartashov *et al.* systematically investigated the dynamics of various types of solitons supported by the Bessel lattices, including multipole-mode solitons [46], ring-profile vortex solitons [17], spatiotemporal solitons [27] etc. Necklace [47], broken ring solitons [48] can also be trapped stably in different order Bessel lattices. For a review of the early works, see [23].

Interestingly, Bessel lattices with azimuthal modulation are also possible [25, 49]. Such lattices resemble highly nonlinear micro-structured fibres [50] and may be realized in experiment by several incoherent Bessel beams with different intensities and orders [49]. The complex lattices can also be created in photorefractive crystals by the phase-imprinting technique [50, 51]. The azimuthally modulated lattices exhibit several discrete guiding channels of linear refractive index. Stable soliton complexes and azimuthal switching in focusing cubic media with modulated Bessel lattices were reported in [49]. Neighboring components in soliton complex are out-of-phase. Ring-shaped and single-site solitons were observed in azimuthally modulated lattices [50, 51]. Especially, by using group-theory techniques, Kartashov *et al.* derived a general "charge/stability rule" for vortex solitons supported by the azimuthal Bessel lattice [25].

In Ref. [31], Desyathikov and his coworkers introduced a novel class of spatially localized self-trapped ringlike singular optical beams in focusing cubic and saturable media, the so-called "azimuthons". The amplitude of such states is a spatially localized ring modulated azimuthally, and the phase of the azimuthon is a staircase function of the polar angle. This concept provided an important missing link between the radially symmetric vortices and rotating soliton clusters [52]. Following this work, stable azimuthons in nonlocal nonlinear media were found when the nonlocality parameter exceeds a certain threshold value [53, 54].

However, stable azimuthons are only found in media with nonlocal responses. Azimuthons in local nonlinear media unavoidably experience azimuthal modulation instability upon propagation. In this section, we elucidate the existence and stability properties of azimuthons (vortex solitons with special amplitude distribution) supported by the azimuthally modulated Bessel lattices. It is the combination of nontrivial phase and lattice confinement who affords the existence of azimuthons. Thus, the azimuthons we obtained provide a missing link between the radially symmetric vortices and nonrotating soliton clusters although they break the radial symmetry due to the potential we used. Similar to the "azimuthons" stated in [31], the nonlinear localized modes we discussed can also be attributed to the two contributions induced by the internal energy flow and the modulated beam respectively. In sharp contrast to the cases in focusing cubic media [25], we reveal that the "stability rule" in defocusing cubic media is quite the reverse. The result is in good agreement with the conclusion given by [55] where the stability of discrete vortex solitons supported by hexagonal photonic lattices in focusing media is opposite to the stability in the defocusing one, though the discussions were limited to the single-charged and double-charged vortex solitons.

3.2. Theoretical model

We consider beam propagation along the z axis in defocusing cubic media with an imprinted transverse refractive index modulation. The dynamics of the nonlinear modes supported by such a scheme can be described by Eq. (1)

Here, the parameter p describes the lattice depth. The profile of the modulated lattice is given by $R \bullet x, y \bullet \ \bullet \ J_{n_J}^2 \bullet\bullet 2b_{lin} \bullet^{1/2} r \bullet \cos^2 \bullet n\phi \bullet$, where n_J denotes the order of the Bessel function, ϕ is the azimuthal angle, n stands for the azimuthal index and b_{lin} defines the transverse lattice scale. Typical transverse linear refractive index modulation induced by the first order Bessel lattice with azimuthal index $n \ \bullet \ 2$ is shown in Fig. 6(a). The local lattice maxima situated closer to the lattice center are more pronounced than others. The number of guiding channels in the main ring is given by $2n$.

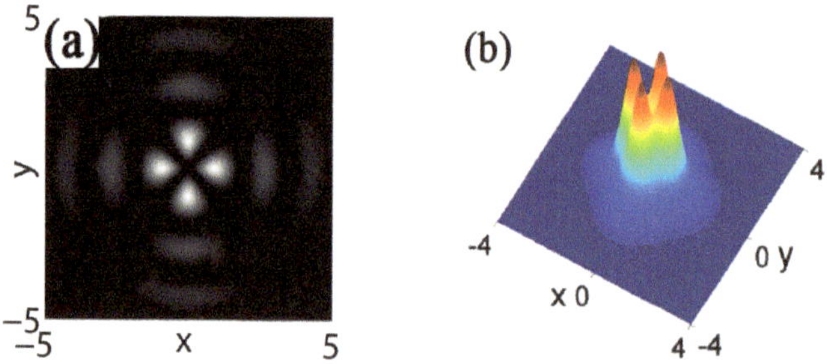

Figure 6. (a) First-order Bessel lattice with azimuthal index $n \ \bullet \ 2$. (b) Profile of the first linear modes with $n \ \bullet \ 2$.

Experimentally, Eq. (1) can be realized by launching a modulated Bessel beam into a photorefractive crystal in the ordinary polarization direction and a soliton beam in the extraordinary polarization direction [46]. In the particular case of optical lattice induction in SBN crystal biased with dc electric field $\sim 10^5$ V/m, for laser beams with 10μm the propagation distance $z \sim 1$ corresponds to 1mm of actual crystal length, while amplitude $q \sim 1$ corresponds to peak intensity about 50 mW/cm^2 [25]. Note that Eq. (1) can also be treated as Gross-Pitaevskii equation for a 2D Bose-Einstein condensate with repulsive interatomic interactions trapped in an optical lattice created by an azimuthally modulated Bessel beam.

We search for stationary solutions of Eq. (1) in the form of $A \bullet x, y, z \bullet \ \bullet \ \bullet q_r \bullet x, y \bullet \ \bullet \ iq_i \bullet x, y \bullet\bullet \exp \bullet ibz \bullet$, where q_r and q_i are real and imaginary parts of the solution profiles and b is a nonlinear propagation constant. The twisted phase structure of the stationary solutions can be defined by $m \ \bullet \ \arctan \bullet q_i / q_r \bullet / 2\pi$, where m is the so-called "topological charge" of vortex solitons. Substituting the expression into Eq. (1), we obtain:

$$\frac{1}{2}\left(\frac{\partial^2}{\partial x^2} \bullet \ \frac{\partial^2}{\partial y^2}\right) q_{r,i} - bq_{r,i} - \bullet q_r^2 \bullet \ q_i^2 \bullet q_{r,i} \bullet \ pRq_{r,i} \bullet \ 0 \qquad (4)$$

We fix $b_{lin} \equiv 2$ and vary b, p and n without loss of generality.

To elucidate the stability properties of solitons, we search for perturbed solutions of Eq. (1) in the form $A \bullet x, y, z \bullet \ \bullet \ \bullet q_r \bullet \ iq_i \bullet \ \bullet u_r \bullet \ iu_i \bullet \exp \bullet \lambda z \bullet\bullet \exp \bullet ibz \bullet$, where u_r and u_i are the real and imaginary parts of the perturbations, respectively. Substituting the perturbed solution into Eq. (1) and linearizing $u_{r,i}$ around $q_{r,i}$ yield a system of coupled Schrödinger-type equations for perturbation components $u_{r,i}$:

$$\pm \lambda u_{i,r} \bullet \ \frac{1}{2}\left(\frac{\partial^2}{\partial x^2} \bullet \ \frac{\partial^2}{\partial y^2}\right) u_{r,i} - b u_{r,i} \bullet \ pRu_{r,i} - \bullet 3q_{r,i}^2 \bullet \ q_{i,r}^2 \bullet u_{r,i} - 2q_r q_i u_{i,r} \qquad (5)$$

where $u_{r,i}$ may grow with a complex rate λ during the propagation of solitons. The eigenfunctions $u_{r,i}$ and eigenvalues λ can be solved numerically. The solitons are stable only when all real parts of λ equal zero.

3.3. Discussions

Before we discuss the dynamics of vortex solitons (azimuthons), it is important to understand the origin of such nonlinear modes. After removing the nonlinear term in Eq. (4), the linear equation has infinite eigenvalues and the corresponding linear eigen-modes. Nonlinear modes bifurcate from these linear modes while the nonlinearity cannot be ignored. Fundamental solitons always bifurcate from the first linear modes and higher order solitons associate with the other linear modes. Corresponding to the azimuthal Bessel lattice shown in Fig. 6(a), we plot the first eigen-mode of the linearized equation of Eq. (4) in Fig. 6(b). The profile of linear mode possesses several amplitude peaks covering on a constitutive ring-like substrate, thus it looks like azimuthons. Such linear modes intuitively indicate the possible profiles of nonlinear modes in the nonlinear system.

Mathematically, the refractive index modulation contributed by the modulated Bessel lattice increases linearly with the growth of lattice depth. However, this relationship cannot hold for practical crystal when the lattice is modulated very deep. Thus, the practical realization of stable vortex solitons with higher topological charges becomes infeasible by solely increasing the lattice depth of the first-order lattice to a very large value. Fortunately, the higher-order Bessel lattice can suppress the azimuthal instability of vortex solitons effectively [26]. To study the properties of vortex solitons with higher charges, one must consider the higher order modulated Bessel lattices.

The following discussion will focus on azimuthons (vortex solitons) carrying different topological charges supported by azimuthally modulated Bessel lattices imprinted in defocusing cubic media. For the convenience of comparing with the results of Ref. [25], we assume $R\bullet x, y\bullet \ \bullet \ J_n^2\bullet\bullet 2b_{lin}\bullet^{1/2} r\bullet \cos^2\bullet n\phi\bullet$, where the order of lattice equals to the azimuthal index. We also search for stationary solutions of azimuthons by a relaxation method. A Gauss beam multiplying a phase dislocation with charge m was selected as an initial iterative guess solution. Figure 7 displays some instances of azimuthons supported by the azimuthally modulated lattices with $n \bullet \ 4$ and 6. The azimuthons exhibit spatially modulated patterns which are in contrast to the vortices in unmodulated Bessel lattices [17], where the vortices are ring-shaped. The number of amplitude peaks is determined by the azimuthal index n. Like the vortices in focusing media [25], azimuthons with similar amplitude distributions allow different topological charges. In the fourth-order Bessel lattices with azimuthal index $n \bullet \ 4$, azimuthons can be found only for $m \bullet \ 1, 2$ and 3. In sufficiently deep lattices with fixed b and p, the discreteness of azimuthons increases with the growth of the topological charge m, while the "radii" of the azimuthons are almost the same. Such properties are similar to the vortex solitons in focusing media [25]. For fixed p and n, the azimuthons will expand to the outer lattice rings at small b and shrink to the main guiding lattice ring at larger b. The local minima of azimuthons around the lattice ring approaches to zero when $b \to b_{co}$.

We also find azimuthons supported by lattices with different azimuthal indices. Numerical study reveals that azimuthon solutions can be found only when the relation $0 < m < n$ is satisfied. The relation also holds for the vortex solitons in focusing cubic media [25]. The phase difference between the neighboring components is $m\pi/n$, which differs from the vortex solitons in harmonic lattices [56] or necklace solitons in Bessel lattices [47].

The properties of azimuthons supported by azimuthally modulated Bessel lattice are summarized in Fig. 8. The power of azimuthons is a descending curve due to the defocusing nonlinearity [Fig. 8(a)]. Azimuthon solutions cannot be found when the propagation constant exceeds a certain value which corresponds to an eigen-value of the linearized equation of Eq. (4). The upper propagation constant cutoffs of vortex solutions with $m = 1$ and 3 are displayed in Fig. 8(b) and 8(c). The existence areas expand with the growth of lattice depth for fixed topological charge and shrink with the growth of topological charges for fixed lattice depth. There is a lower threshold lattice depth for supporting azimuthons. Comparing the points of $b \to 0$ in Fig. 8(b) and 8(c), we find that the threshold lattice depth grows with the increase of topological charge m.

To comprehensively understand the stability properties of azimuthons supported by lattices with different depths and azimuthal indices, we performed the linear stability analysis on azimuthons in lattices with order (azimuthal index) n up to 10 and lattice depths $p \leq 80$. We numerically derived an important "stability rule" for azimuthons in the azimuthally modulated Bessel lattice imprinted in defocusing media. That is, azimuthons might be stable only when the topological charge satisfies the condition:

$$0 < m \leq n/2 \tag{6}$$

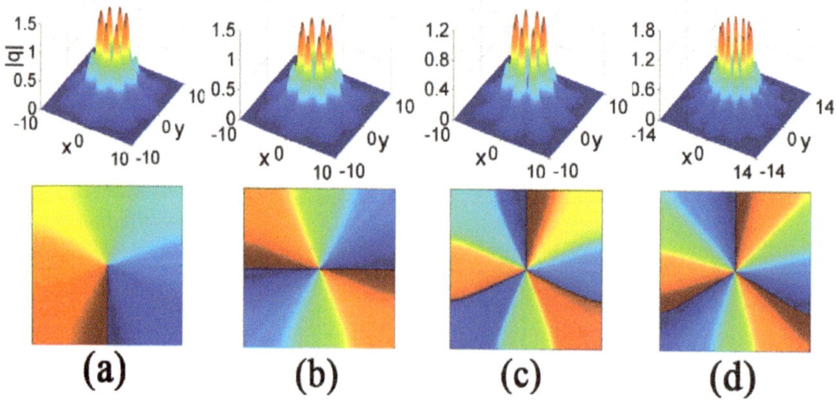

Figure 7. Amplitude distributions of azimuthons with $m = 1$ (a), 2 (b) and 3 (c, d). Parameters $n = 4, p = 30, b = 0.5$ in (a-c) and $n = 6, p = 45, b = 0.5$ in (d). Bottom row: the corresponding phase structures.

Lattice order	Available charges and stability status				
n=2	m=1 unstable				
n=3	m=1 stable	m=2 unstable			
n=4	m=1 stable	m=2 stable	m=3 unstable		
n=5	m=1 stable	m=2 stable	m=3 unstable	m=4 unstable	
n=6	m=1 stable	m=2 stable	m=3 stable	m=4 unstable	m=5 unstable

Table 1. The stability status of azimuthons for different lattice orders.

where $n > 2$. Azimuthons with $m = n/2$ for even n may be stable or unstable depending on the lattice parameters. There exists a narrow instability area near $b \to 0$ when the lattice is modulated shallow (near its lower threshold value). For deeper lattices, completely stable azimuthons are possible. A summary of "stability rule" is presented in Table 1. The table shows the stability status of azimuthons for different lattice orders. It is exactly opposite to the Table I in [25], which was derived by the group-theory and is valid in the focusing cubic media. This finding also verifies the very recent reports [55, 57] in which the stability of discrete vortex solitons supported by hexagonal photonic lattices in focusing media is proved to be opposite to the stability in the defocusing ones. We note that our conclusion is more general since the above two studies are restricted to the single- and double-charge discrete vortex solitons.

Linear instability analysis results of some unstable azimuthons supported by the fourth-, fifth-, and sixth-order lattices are shown in Fig. 8(d)-8(f). Note the relation between the azimuthal index and topological charge does not satisfy the condition Eq. (6). The instability domain vanishes only when propagation constant approaches to the upper cutoff. Direct propagation simulation examples of stable and unstable propagation of azimuthons are presented in Fig. 9.

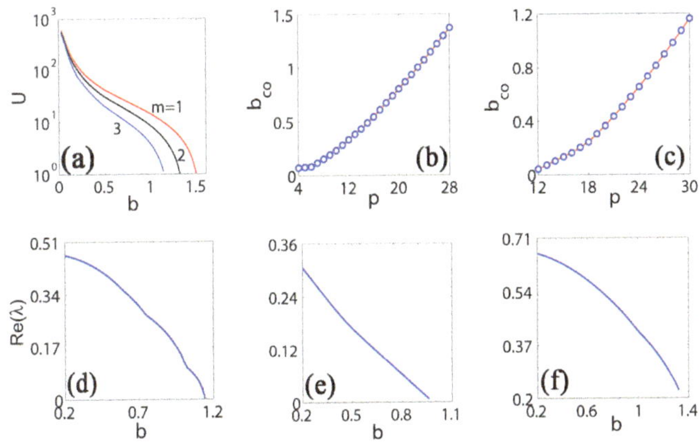

Figure 8. (a) Power of azimuthons with $m = 1,2$ and 3 vs propagation constant, $n = 4, p = 30$. Propagation constant cutoff b_{co} vs lattice depth p for azimuthons with $m = 1$ (b) and 3 (c). Real parts of instability growth rate λ vs propagation constant for azimuthons supported by the fourth- (d), fifth- (e), and sixth- (f) order lattices with $p = 30, m = 3$, $p = 35, m = 3$, $p = 35, m = 5$, respectively. Azimuthal index $n = 4$ in (a-d), 5 in (e) and 6 in (f).

The azimuthons aforementioned are restricted to the particular cases of $n = n_J$. In fact, azimuthon solutions can also be found when $n \neq n_J$. Contrary to the intuition and the cases in nonlocal media [53], the charge m of available azimuthon solutions is independent of the lattice order n_J but less than the azimuthal index n. The initial input guess solutions with $m \geq n$ may converge to the nonlinear modes of the following three different categories: 1. an azimuthon with charge $m' < n$; 2. a multipole mode or necklace soliton with neighboring components out-of phase; 3. a multipole mode or necklace soliton embedded into a global skew phase whose charge $m'' = m - n$. Thus, we conclude that azimuthon solutions can only be found for $m < n$. The reason may be attributed to the Kerr media with a local nonlinear response in our model [31].

4. Higher-charged vortices in mixed linear-nonlinear circular arrays

4.1. Introduction

Recent studies demonstrated that the transverse nonlinearity modulation of optical materials can substantially affect the existence conditions and stability properties of spatial solitons [58]. Current technologies allow one to realize both the linear refractive index and the nonlinearity modulation of materials, for example, in photonic crystals with holes infiltrated with highly nonlinear materials [59]. Nonlinearity can also be modulated by changing the concentration of dopants upon fabrication [60] or be achieved in arrays written in glass by high-intensity femtosecond laser pulses [61].

Very recently, various types of one-dimensional solitons in the form of odd, even, dipole, triple, vector solitons, and two-dimensional solitons in the form of fundamental, multi-pole, vortex, surface solitons, and three-dimensional solitons in the form of light-bullets and optical tandems were predicted in competing linear and nonlinear lattices [18, 62–64] or in purely nonlinear lattices [19, 65–67]. On the other hand, spatially modulated nonlinearity may result in controllable soliton shape transformations [18, 62]. Bound states with an arbitrary number of solitons can be found in media with spatially inhomogeneous nonlinearities [68]. The power-dependent location of stationary solitons and their stability in linear-nonlinear lattices were analyzed in Refs. [69], where the position and stability of the solitons become functions of the power.

Besides models with periodically modulated nonlinearity, settings with local nonlinearity modulation were also suggested [70]. In particular, Kartashov *et al.* demonstrated that necklace solitons can be supported by circular waveguide arrays with out-of-phase modulation of nonlinearity and linear refractive index [63]. They revealed that the stability domain of necklace solitons expands with the increase of the number of necklace spots.

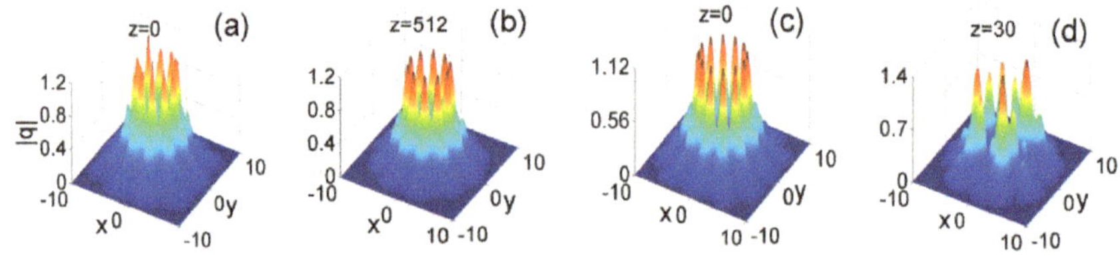

Figure 9. Stable and unstable propagation of azimuthons. (a, b) $n = 5, m = 2, p = 30, b = 0.4$. (c, d) $n = 6, m = 4, p = 35, b = 0.4$. White noise was added into (a).

Although vortex solitons in linear-nonlinear or in purely nonlinear lattices were studied in [18, 19], the discussions were limited to the vortices with unit charge. In fact, stable localized vortex solitons with charges higher than two were only predicted in media modulated by Bessel-like lattices [25, 26]. In addition, the stability domain of vortex solitons in all previous studies shrinks rapidly with the growth of topological charge.

In this section, following the theoretical model proposed by Kartashov and his coworkers [63], we address the existence and stability properties of vortex solitons with higher charges supported by a circular array with mixed out-of-phase linear-nonlinear refractive index modulation. We find that vortex solution can be found only when its topological charge is less than half of the number of waveguides. Vortex solitons expand radially with the propagation constant when the phase difference between neighboring lobes is greater than $\pi/2$ and shrink radially when the phase difference between neighboring lobes is smaller than $\pi/2$. For vortex solitons with fixed charges, the stability weakens with the growth of waveguide sites. In particular, we demonstrate that vortex solitons with higher charges tend to be more stable than those with lower charges, which in sharp contrast to the cases in uniform or periodical lattice modulated media.

4.2. Theoretical model

We consider a beam propagation along the z axis in a circular waveguide array with out-of-phase modulation of linear refractive index and nonlinearity coefficient. Evolution of the nonlinear waves is governed by the nonlinear Schröinger equation for the dimensionless field amplitude A:

$$i\frac{\partial A}{\partial z} \bullet -\frac{1}{2}\left(\frac{\partial^2 A}{\partial x^2} \bullet \frac{\partial^2 A}{\partial y^2}\right) - \bullet 1 - \delta R \bullet x, y \bullet\bullet |A|^2 A - pR\bullet x, y\bullet A \bullet 0 \qquad (7)$$

where p and δ are the depths of modulation of linear refractive index and nonlinearity. We adopt modulation function of the linear refractive index and nonlinearity as $R\bullet x, y\bullet \bullet \Sigma_{k\bullet 1}^{n} \exp\bullet - \bullet x - x_k \bullet^2/d^2 - \bullet y - y_k\bullet^2/d^2\bullet$, which means a circular array composed by n Gaussian waveguides with widths $d \bullet 1/2$ and centers (x_k, y_k) located on a ring of radius $nr_0/2$ [63]. To guarantee that waveguides do not overlap and the soliton components residing in neighboring waveguides can interact with each other, we select $r_0 \bullet 0.6$. The nonlinear coefficient $\gamma \bullet 1 - \delta R$ attains its minima where the linear refractive index attains maxima. Thus, the nonlinearity modulation and linear refractive index modulation are out-of-phase. The length of the arc between adjacent waveguides remains the same for any n since the radius of the array increases linearly with n. Equation (1) admits several conserved quantities, including the power or energy flow $U \bullet \int\int_{-\infty}^{\infty} |A|^2 dxdy$ and the Hamiltonian $H \bullet \int\int_{-\infty}^{\infty} \bullet \frac{1}{2}|\frac{\partial A}{\partial x}|^2 \bullet \frac{1}{2}|\frac{\partial A}{\partial y}|^2 - pR|A|^2 - \frac{1}{2}\gamma|A|^4 \bullet dxdy$.

We search for vortex solutions of Eq. (7) in the form of $A\bullet x, y, z\bullet \bullet \bullet q_r \bullet x, y\bullet \bullet iq_i\bullet x, y\bullet\bullet \exp\bullet ibz\bullet$, where $q_r\bullet x, y\bullet$ and $q_i\bullet x, y\bullet$ are real and imaginary parts of the solution profiles and b is a nonlinear propagation constant. Substituting the expression into Eq. (7) leads to two coupled ordinary differential equations in terms of q_r and q_i, which can be solved numerically by a two-dimensional relaxation algorithm. The stability characteristics of vortex solitons can be understood by numerically solving the eigenvalue of Eqs. (5).

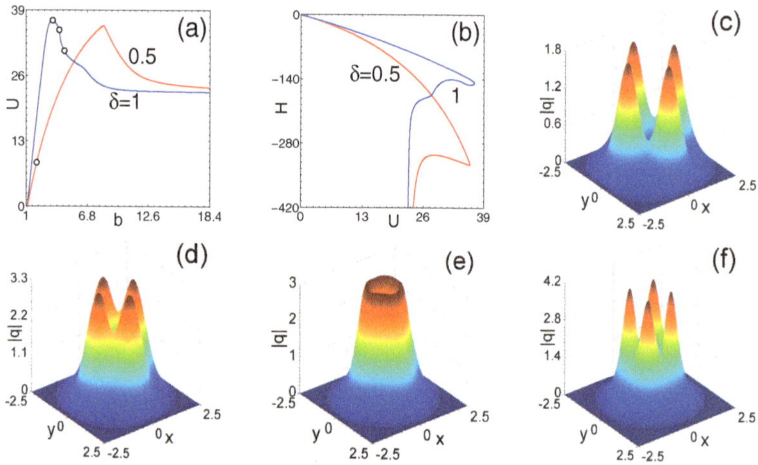

Figure 10. (a) Power of vortex solitons with unit charge versus b for different δ values. (b) Hamiltonian versus power corresponding to (a). (c-f) Soliton profiles marked in (a) at $b = 2.1$ (c) and $3.5, 4.12, 4.60$ in (d-f). In all panels $n = 4, p = 6$.

4.3. Single-charged vortices

First, we address the properties of vortex solitons supported by a circular waveguide array with $n = 4$. Typical vortex solitons with unit charge are illustrated in Fig. 10. Vortex solitons reflect their distinctive modulated profiles. The field modulus distribution of such state is a spatially localized ring modulated azimuthally, and the phase changes $2m\pi$ along a closed trajectory including the beam centre [Figs. 10(c)-10(f)]. Thus, they look like the so-called "azimuthons" in local and nonlocal nonlinear media [31, 53]. Vortex solitons are composed of n lobes located in different waveguides of the array with phase difference $2m\pi/n$ between the neighboring lobes. This phase difference determines the net force contributed by the interactions of vortex lobes. There are n local intensity minima connecting the n lobes for small and moderate propagation constant. The discreteness of solitons strengthens for $b \to b_{low}$ and $b \to \infty$.

If the modulation depth of the local nonlinearity is small ($\delta \leq 0.5$), the lobes of vortices stay around sites of linear waveguide array. However, the competition between the linear and nonlinear refraction may result in a remarkable shape transformation at $\delta > 0.5$, due to the concentration of the light energy in regions where the nonlinearity is stronger. Interestingly, there exists a transition point of propagation constant b_{tr} at which the vortex soliton features a ring-shaped profile [Fig. 10(e)], which is analogous to the cases in lattice-free media or in media with imprinted Bessel lattices [17]. The nonlinearity modulation gives rise to unusual power-controlled soliton shape transformations [Figs. 10(d)-10(f)]. Note that the four lobes of vortex solitons at $b > b_{tr}$ for $\delta = 1$ reside in the regions between neighboring waveguide sites of the circular array [Fig. 10(f)], while the lobes of vortex solitons at $b < b_{tr}$ reside in the waveguide sites [Fig. 10(d)]. Similar to the cases in harmonic lattices [56], vortex solitons may be termed as "on-site" vortices at $b < b_{tr}$ and "off-site" vortices at $b > b_{tr}$.

The power of vortex solutions is a nonmonotonic function of propagation constant, due to the competition between the linear and nonlinear refractive index contributions [Fig. 10(a)].

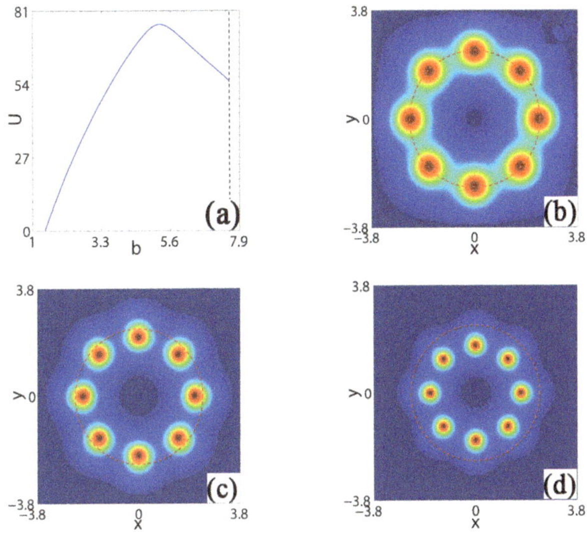

Figure 11. (a) Power of vortex solitons with unit charge versus b. Soliton profiles at $b = 2.50$ (b), 5.70 (c) and 7.57 (d). Rings in (b-d) denote circular array of waveguides. In all panels $n = 8, \delta = 0.7, p = 6$.

It increases firstly for small and moderate b and reaches a maximum, afterwards, the power decreases gradually and approaches to a constant ultimately. The initial increase of power indicates that the linear refractive index modulation plays a dominant role. However, the nonlinear contribution to the refractive index dominates when the power decreases with b. Thus, nonlinearity modulation imposes a restriction on the maximal power of vortex solitons. Hamiltonian of vortex solitons has inflexion points corresponding to the propagation constant value $b = b_{in}$ where $dU/db = 0$ [Fig. 10(b)]. Despite the fact that power is a decreasing function when $b > b_{in}$, the peaks of vortices still increase with the propagation constant, which implies that the width of vortex lobes decreases rapidly for solitons at $b > b_{in}$.

Next, we investigate the properties of vortex solitons with unit charge in circular waveguide array with $n = 8$. The vortex solitons bifurcate from the linear modes at $b = b_{low}$ ($U = 0$) and stop to exist at $b = b_{upp}$ [Fig. 11(a)]. The reason is that, for $b > b_{upp}$, where the nonlinearity modulation absolutely dominates, the strong focusing nonlinearity results in vortex solitons with high peaks and narrow widths, which in return destroy the global staircase-like phase connection between the neighboring lobes. In other words, the vortices become eight independent components for $b > b_{upp}$. Vortex solitons tend to shrink with the propagation constant [Figs. 11(b)-11(d)], which is quite different from the necklace solitons in the same regime, where necklace beams expand with the propagation constant [63]. It can be interpreted in physics that the phase difference between the neighboring lobes is $\pi/4$ which results in an attractive force between the neighboring components of vortices. The net attractive force strengthens with the propagation constant which leads to the radial shrinkage of vortices. In fact, the positions of four lobes of vortex solitons in waveguide array with $n = 4$ do not vary with the propagation constant because the phase difference between the neighboring lobes is $\pi/2$.

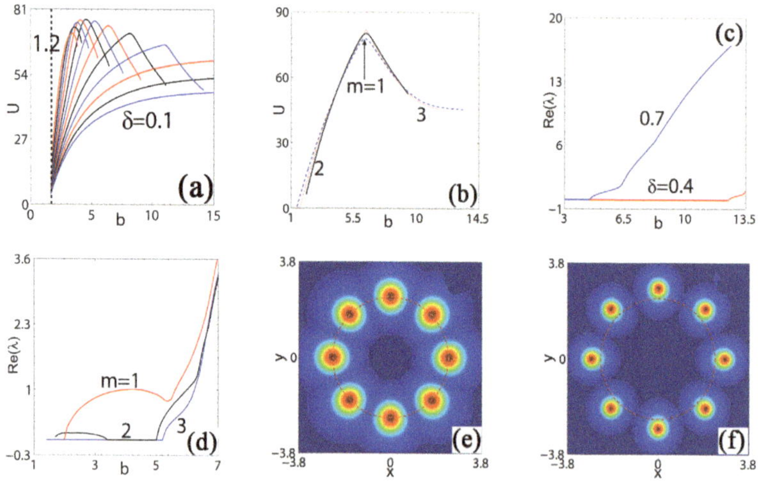

Figure 12. (a) Power of vortex solitons with $m = 1, p = 6, n = 8$ versus b at different δ values. (b) Power of vortex solitons with different charges. (c) Instability growth rate of vortices with $m = 1, n = 4$ versus b. (d) Instability growth rate of vortices with different m versus b. Field modulus distributions of vortex solitons with $m = 3$ at $b = 2.5$ (e) and 7.17 (f). $n = 8, p = 7$ in (b-f) and $\delta = 0.7$ except for (a, c).

4.4. Higher-charged vortices and their stability

The stability of vortex solitons, especially for higher-charged vortices, is always an important problem. The following discussions will focus on the stability of vortex solitons with different charges in circular waveguide arrays with different n and δ values. Figure 12(a) displays the power curves of vortex solitons with unit charge in the waveguide array with $n = 8$ for varying nonlinearity modulation depth δ. Vortex solitons cease to exist at $b = b_{upp}$ if the nonlinearity modulation depth exceeds a certain value, which is in contrast to the vortices in linear lattices. The power curves at different δ share a common lower propagation constant cutoff. It means that all vortices bifurcate from a common linear mode of linearized version of Eq. (7). The existence domains of vortex solitons become narrower with the increase of δ.

The power curves of vortex solitons with different charges at fixed δ are shown in Fig. 12(b). Due to the discrete symmetry group of the circular waveguide array, vortex solutions can only be found when the condition $0 < m < n/2$ is satisfied, just similar to the cases in azimuthally modulated Bessel lattices [25]. The existence domain of vortex solitons with $m = 2$ is narrower than those of $m = 1$ and 3 which may also be attributed to the group symmetry.

For vortex solitons with unit charge in a waveguide array with $n = 4$, the stability area shrinks quickly with the growth of nonlinearity modulation depth δ [Fig. 12(c)]. However, vortex solitons with unit charge in waveguide array with $n = 8$ are unstable in their entire existence domain. Interestingly enough, vortex solitons with higher topological charges can propagate stably in the same configuration [Fig. 12(d)]. The stability analysis results shown in Fig. 12(d) suggest that vortex solitons in a circular array can be stable only when the condition $n/4 \leq m < n/2$ is satisfied. Furthermore, the stability area of vortex solitons with $m = 3$ is broader than that of vortices with $m = 2$. That is to say, vortex solitons with higher charges are more stable than those with lower charges which constitutes one of our central findings. We

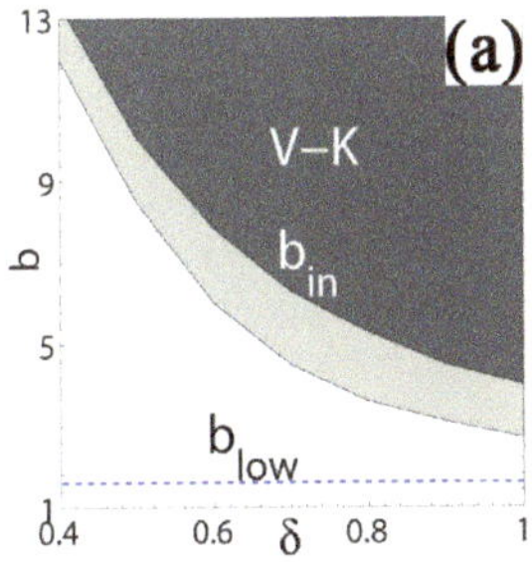

Figure 13. Stability (white) and instability (shaded) areas on the (δ, b) plane for vortex solitons at $m = 1, n = 4$ (a) and $m = 3, n = 8$ (b). Vortices in regions above b_{in} suffer V-K instability. In all panels $p = 7$.

conclude that waveguide array with larger n tends to stabilize the vortices with charges close to $n/2$.

Typical examples of vortices with $m = 3$ at different b are displayed in Figs. 12(e) and 12(f). Contrary to the vortices with $m = 1$ shown in Figs. 11(b)-11(d), vortices with $m = 3$ expand with the increase of b. As can also be explained by the phase difference between the neighboring lobes. The phase difference between the neighboring lobes of vortices with $m = 1, 2$ and 3 is $\pi/4, \pi/2$ and $3\pi/4$ respectively. The net force contributed by the eight lobes is attractive for vortices with $m = 1$, zero for vortices with $m = 2$ and repulsive for

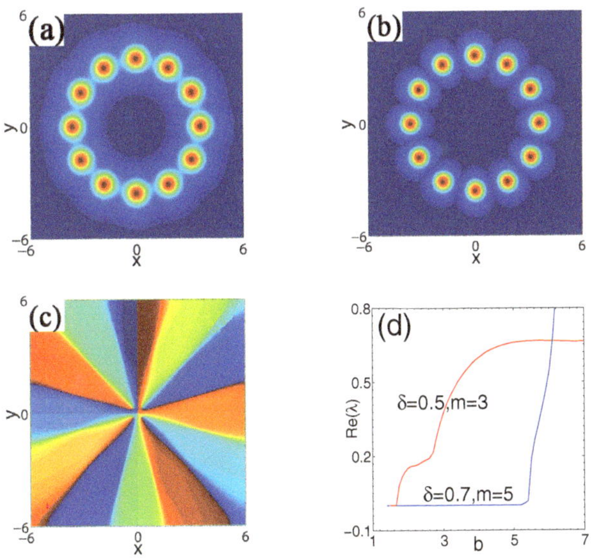

Figure 14. Field modulus distributions of vortex solitons at $m = 3, b = 2.1, \delta = 0.5$ (a) and $m = 5, b = 5, \delta = 0.7$ (b). (c) Phase structure of (b). (d) Instability growth rate of solitons with different m. In all panels $p = 7, n = 12$.

vortices with $m = 3$. Since the peaks of lobes grow with the propagation constant, the net force becomes stronger and the lobes cannot be trapped on the waveguide sites, which leads to the shrinkage of vortices with $m = 1$ and expansion of vortices with $m = 3$. The lobes of vortices with $m = 2$ do not change their positions with the variation of b, just similar to the vortices with $m = 1$ in waveguide array with $n = 4$ [Fig. 10]. Thus, we draw a conclusion that the expansion or shrinkage of vortices is solely determined by the radio between the topological charge and the number of waveguides.

To comprehensively understand the stability characteristics of vortex solitons, we conduct linear stability analysis on the stationary solutions with different m in circular waveguide array with different n for varying nonlinearity modulation depth δ. Representative stability and instability domains on the (δ, b) plane are illustrated in Fig. 13. At larger b where $dU/db < 0$ ($b > b_{in}$), vortices are expected to be linearly unstable according to the Vakhitov-Kolokolov (V-K) criterion [44]. This is in good agreement to the numerical analysis results. However, other types of instability (e.g. oscillatory instability with complex growth rates) may arise when $b < b_{in}$.

For vortex solitons with $m = 1, n = 4$ and $m = 3, n = 8$, the stability domains shrink with the growth of nonlinearity modulation depth. Note again vortex solitons with $m = 1, n = 8$ are completely unstable. Comparing Fig. 13(a) with Fig. 13(b), one can immediately find that the stability area of vortices with $m = 3, n = 8$ is broader than that of $m = 1, n = 4$. Thus, increasing waveguide number can significantly suppress the instability of vortex solitons with higher charges. There is an instability band near b_{low} for vortex solitons with $m = 2, n = 8$ [Fig. 12(d)]. We stress that vortex solitons with higher topological charges are stable in a wide parameter window which is hardly realized in other settings except in Bessel-like lattice modulated media [25, 26].

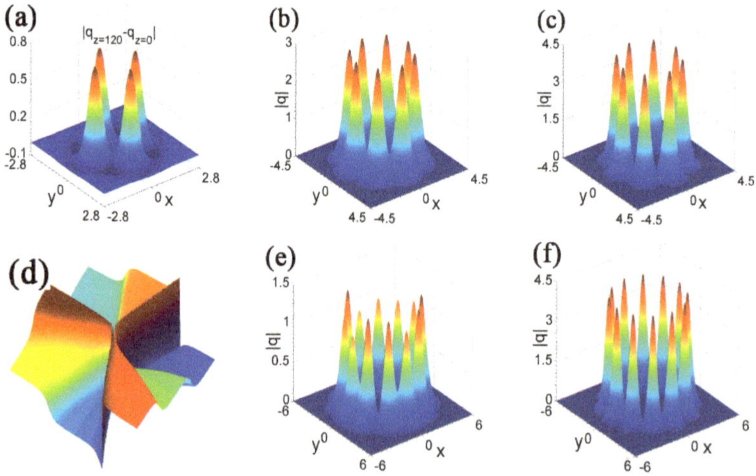

Figure 15. Propagation simulations of stable (b-d, f) and unstable (a, e) vortex solitons. (a) Difference between a vortex soliton at $z = 120$ and $z = 0$. $m = 1, b = 4.5, n = 4$. Propagation results of vortex solitons at $m = 2, b = 3.6, n = 8$ (b), $m = 3, b = 5.16, n = 8$ (c), $m = 3, b = 2.1, n = 12, \delta = 0.5, z = 69$ (e), and $m = 5, b = 5, n = 12$ (f). (d) Phase structure of (c). $\delta = 0.7$ except for (e) and $z = 512$ except for (a, e). In all cases $p = 7$.

To shed more light on the vortex solitons in circular waveguide array, we explore the stationary solutions, stability and propagation dynamics of vortices in a circular array with larger n (e. g. $n = 12, 18$ etc.). Figures 14(a) and 14(b) show the field modulus distributions of vortex solitons at $m = 3, \delta = 0.5, b = 2.1$ and $m = 5, \delta = 0.7, b = 5$ in waveguide array with $n = 12$ respectively. As mentioned above, the stability of vortex solitons weakens with the increase of nonlinearity modulation depth. Yet, vortex solitons with $m = 3, \delta = 0.5$ are stable in a very narrow area near b_{low} while vortex solitons with $m = 5, \delta = 0.7$ are stable in a wide parameter window [Fig. 14(d)]. It reveals again that vortex solitons tend to be more stable when m approaches to $n/2$.

Finally, we perform extensive propagation simulations of vortex solitons by the split-step Fourier method to verify the linear stability analysis results. Typical stable and unstable propagation examples are shown in Fig. 15. Figure 15(a) plots the difference of vortex profiles at $z = 120$ and 0. One can infer from the simultaneous increase of four lobes that the soliton experiences a V-K instability, that is, the unstable eigenvalues of λ are purely real, which do not destroy the phase distribution. On the other hand, vortex soliton at $m = 3, b = 2.1, n = 12, \delta = 0.5$ suffers oscillatory instability (growth rate with both real and imaginary parts) [Fig. 15(e)]. In the numerical simulations, we add white noises into the initial inputs for stable vortices and add no noises for unstable vortices.

Acknowledgements

This work is supported by the National Natural Science Foundation of China (Grant No. 11074221) and the Natural Science Foundation of Zhejiang Province, China (Grant No. Y6100381).

Author details

Liangwei Dong
Institute of Information Optics, Zhejiang Normal University, Jinhua, 321004, China

Huijun Li
Department of Physics, Zhejiang Normal University, Jinhua, 321004, China

5. References

[1] Pismen LM (1999) *Vortices in Nonlinear Fields*, Clarendon Press, Oxford.
[2] Kivshar YS, Agrawal GP (2003) *Optical Solitons: from Fibers to Photonic Crystals*, Academic, San Diego, Calif.
[3] Kivshar YS (1998) Dark optical solitons: physics and applications, *Phys. Rep.* 298(2-3): 81–197.
[4] Abo-Shaeer JR, Raman C, Vogels JM, Ketterle W (2001) Observation of vortex lattices in bose-einstein condensates, *Science* 292(5516): 476–479.
[5] Desyatnikov AS, Kivshar YS, Torner L (2005) Optical vortices and vortex solitons, *Prog. Opt.* 47: 291.
[6] Madison KW, Chevy F, Wohlleben W, Dalibard J (2000) Vortex formation in a stirred bose-einstein condensate, *Phys. Rev. Lett.* 84: 806–809.
[7] Raman C, Abo-Shaeer JR, Vogels JM, Xu K, Ketterle W (2001) Vortex nucleation in a stirred bose-einstein condensate, *Phys. Rev. Lett.* 87: 210402.

[8] Quiroga-Teixeiro M, Michinel H (1997) Stable azimuthal stationary state in quintic nonlinear optical media, *J. Opt. Soc. Am. B* 14(8): 2004–2009.
[9] Towers I, Buryak AV, Sammut RA, Malomed BA, Crasovan L-C, Mihalache D (2001) Stability of spinning ring solitons of the cubic-quintic nonlinear Schrödinger equation, *Phys. Lett. A* 288: 292–298.
[10] Mihalache D, Mazilu D, Crasovan L-C, Towers I, Buryak AV, Malomed BA, Torner L, Torres JP, Lederer F (2002) Stable spinning optical solitons in three dimensions, *Phys. Rev. Lett.* 88: 073902.
[11] Mihalache D, Mazilu D, Towers I, Malomed BA, Lederer F (2003) Stable spatiotemporal spinning solitons in a bimodal cubic-quintic medium, *Phys. Rev. E* 67: 056608.
[12] Raghavan S, Agrawal GP (2000) Spatiotemporal solitons in inhomogeneous nonlinear media, *Opt. Commun.* 180(4-6): 377 – 382.
[13] Ferrando A, Zacarés M, Fernández de Córdoba P, Binosi D, Monsoriu JA (2004) Vortex solitons in photonic crystal fibers, *Optics Express* 12: 817.
[14] Yang J, Musslimani ZH (2003) Fundamental and vortex solitons in a two-dimensional optical lattice, *Opt. Lett.* 28(21): 2094–2096.
[15] Alexander TJ, Sukhorukov AA, Kivshar YS (2004) Asymmetric vortex solitons in nonlinear periodic lattices, *Phys. Rev. Lett.* 93(6): 063901.
[16] Sakaguchi H, Malomed BA (2005) Higher-order vortex solitons, multipoles, and supervortices on a square optical lattice, *EPL (Europhysics Letters)* 72(5): 698.
[17] Kartashov YV, Vysloukh VA, Torner L (2005) Stable ring-profile vortex solitons in bessel optical lattices, *Phys. Rev. Lett.* 94(4): 043902.
[18] Kartashov YV, Vysloukh VA, Torner L (2008) Power-dependent shaping of vortex solitons in optical lattices with spatially modulated nonlinear refractive index, *Opt. Lett.* 33(19): 2173–2175.
[19] Kartashov YV, Malomed BA, Vysloukh VA, Torner L (2009) Two-dimensional solitons in nonlinear lattices, *Opt. Lett.* 34(6): 770–772.
[20] Dong L, Ye F (2010) Shaping solitons by lattice defects, *Phys. Rev. A* 82: 053829.
[21] Neshev DN, Alexander TJ, Ostrovskaya EA, Kivshar YS, Martin H, Makasyuk I, Chen Z (2004) Observation of discrete vortex solitons in optically induced photonic lattices, *Phys. Rev. Lett.* 92(12): 123903.
[22] Bartal G, Manela O, Cohen O, Fleischer JW, Segev M (2005) Observation of second-band vortex solitons in 2d photonic lattices, *Phys. Rev. Lett.* 95(5): 053904.
[23] Kartashov YV, Vysloukh VA, Torner L (2009) Solitons in complex optical lattices, *Eur. Phys. J. Special Topics* 173: 107–119.
[24] Ye F, Mihalache D, Hu B. (2009) Elliptic vortices in composite mathieu lattices, *Phys. Rev. A* 79: 053852.
[25] Kartashov YV, Ferrando A, Egorov AA, Torner L (2005b) Soliton topology versus discrete symmetry in optical lattices, *Phys. Rev. Lett.* 95(12): 123902.
[26] Dong L, Ye F, Wang H (2009) Suppression of azimuthal instability of ring vortex solitons, *New J. Phys.* 11(7): 073026.
[27] Fleischer JW, Carmon T, Segev M, Efremidis NK, Christodoulides DN (2003) Observation of discrete solitons in optically induced real time waveguide arrays, *Phys. Rev. Lett.* 90: 023902.
[28] Fleischer JW, Bartal G, Cohen O, Manela O, Segev M, Hudock J, Christodoulides, DN (2004) Observation of vortex-ring "discrete" solitons in 2d photonic lattices, *Phys. Rev. Lett.* 92(12): 123904.
[29] Lederer F, Stegeman GI, Christodoulides DN, Assanto G, Segev M, Silberberg Y (2008) Discrete solitons in optics, *Phys. Rep.* 463(1-3): 1 – 126.

[30] Kartashov YV, Vysloukh VA, Torner L (2009) Soliton shape and mobility control in optical lattices, Vol. 52 of *Progress in Optics*, Elsevier, pp. 63 – 148.
[31] Desyatnikov AS, Sukhorukov AA, Kivshar YS (2005) Azimuthons: Spatially modulated vortex solitons, *Phys. Rev. Lett.* 95(20): 203904.
[32] Sukhorukov AA, Kivshar YS (2001) Nonlinear localized waves in a periodic medium, *Phys. Rev. Lett.* 87: 083901.
[33] Makasyuk I, Chen Z, Yang J (2006) Band-gap guidance in optically induced photonic lattices with a negative defect, *Phys. Rev. Lett.* 96: 223903.
[34] Beličev PP, Ilić I, Stepić M, Maluckov A, Tan Y, Chen F (2010) Observation of linear and nonlinear strongly localized modes at phase-slip defects in one-dimensional photonic lattices, *Opt. Lett.* 35(18): 3099–3101.
[35] Szameit A, Molina MI, Heinrich M, Dreisow F, Keil R, Nolte S, Kivshar YS (2010) Observation of localized modes at phase slips in two-dimensional photonic lattices, *Opt. Lett.* 35(16): 2738–2740.
[36] Fedele F, Yang J, Chen Z (2005) Properties of defect modes in one-dimensional optically induced photonic lattices, *Studies in Applied Mathematics* 115(2): 279–301.
[37] Wang X, Young J, Chen Z, Weinstein D, Yang J (2006) Observation of lower to higher bandgap transition of one-dimensional defect modes, *Opt. Express* 14(16): 7362–7367.
[38] Szameit A, Kartashov YV, Heinrich M, Dreisow F, Pertsch T, Nolte S, Tünnermann A, Lederer F, Vysloukh VA, Torner L (2009) Observation of two-dimensional defect surface solitons, *Opt. Lett.* 34(6): 797–799.
[39] Chen WH, Zhu X, Wu TW, Li RH (2010) Defect solitons in two-dimensional optical lattices, *Opt. Express* 18(11): 10956–10962.
[40] Wang H, Wang J (2011) Defect solitons in parity-time periodic potentials, *Opt. Express* 19(5): 4030–4035.
[41] Yang J, Chen Z (2006) Defect solitons in photonic lattices, *Phys. Rev. E* 73: 026609.
[42] Wang J, Yang J, Chen Z (2007) Two-dimensional defect modes in optically induced photonic lattices, *Phys. Rev. A* 76: 013828.
[43] Ye F, Kartashov YV, Vysloukh VA, Torner L (2008) Bragg guiding of domainlike nonlinear modes and kink arrays in lower-index core structures, *Opt. Lett.* 33(12): 1288–1290.
[44] Vakhitov NG, Kolokolov AA (1973) Stationary solutions of the wave equation in a medium with nonlinearity saturation, *Izv. Vyssh. Uchebn. Zaved., Radiofiz.* 16: 783–789.
[45] Skryabin DV, Firth WJ (1998) Dynamics of self-trapped beams with phase dislocation in saturable kerr and quadratic nonlinear media, *Phys. Rev. E* 58: 3916–3930.
[46] Kartashov YV, Carretero-Gonzalez R, Malomed BA, Vysloukh VA, Torner L (2005) Multipole-mode solitons in bessel optical lattices, *Opt. Express* 13(26): 10703.
[47] Dong L, Wang H, Zhou W, Yang X, Lv X, Chen H (2008) Necklace solitons and ring solitons in bessel optical lattices, *Opt. Express* 16(8): 5649.
[48] Dong L, Wang J, Wang H, Yin G (2008) Broken ring solitons in bessel optical lattices, *Opt. Lett.* 33(24): 2989.
[49] Kartashov YV, Egorov AA, Vysloukh VA, Torner L (2004) Stable soliton complexes and azimuthal switching in modulated bessel optical lattices, *Phys. Rev. E* 70(6): 065602.
[50] Fischer R, Neshev DN, Lopez-Aguayo S, Desyatnikov AS, Sukhorukov AA, Krolikowski W, Kivshar YS (2007) Light localization in azimuthally modulated bessel photonic lattices, *J. MATER. SCI-MATER. EL* 18: 277.
[51] Fischer R, Neshev DN, Lopez-Aguayo S, Desyatnikov AS, Sukhorukov AA, Krolikowski W, Kivshar YS (2006) Observation of light localization in modulated bessel optical lattices, *Opt. Express* 14(7): 2825.

[52] Desyatnikov AS, Kivshar YS (2002) Rotating optical soliton clusters, *Phys. Rev. Lett.* 88(5): 053901.
[53] Lopez-Aguayo S, Desyatnikov AS, Kivshar YS (2006) Azimuthons in nonlocal nonlinear media, *Opt. Express* 14(17): 7903.
[54] Lopez-Aguayo S, Desyatnikov AS, Kivshar YS, Skupin S, Krolikowski W, Bang O (2006) Stable rotating dipole solitons in nonlocal optical media, *Opt. Lett.* 31(8): 1100.
[55] Law KJH, Kevrekidis PG, Alexander TJ, Królikowski W, Kivshar YS (2009) Stable higher-charge discrete vortices in hexagonal optical lattices, *Phys. Rev. A* 79(2): 025801.
[56] Yang J (2004) Stability of vortex solitons in a photorefractive optical lattice, *New J. Phys.* 6(1): 47.
[57] Terhalle B, Richter T, Law KJH, Göries D, Rose P, Alexander TJ, Kevrekidis PG, Desyatnikov AS, Krolikowski W, Kaiser F, Denz C, Kivshar YS (2009) Observation of double-charge discrete vortex solitons in hexagonal photonic lattices, *Phys. Rev. A* 79(4): 043821.
[58] Kartashov YV, Malomed BA, Torner L (2011) Solitons in nonlinear lattices, *Rev. Mod. Phys.* 83(1): 247–306.
[59] Rasmussen PD, Bennet FH, Neshev DN, Sukhorukov AA, Rosberg CR, Krolikowski W, Bang O, Kivshar YS (2009) Observation of two-dimensional nonlocal gap solitons, *Opt. Lett.* 34(3): 295–297.
[60] Chen F, Stepić M, Rüter C, Runde D, Kip D, Shandarov V, Manela O, Segev M (2005) Discrete diffraction and spatial gap solitons in photovoltaic linbo3 waveguide arrays, *Opt. Express* 13(11): 4314–4324.
[61] Blömer D, Szameit A, Dreisow F, Schreiber T, Nolte S, Tünnermann A (2006) Nonlinear refractive index of fs-laser-written waveguides in fused silica, *Opt. Express* 14(6): 2151–2157.
[62] Kartashov YV, Vysloukh VA, Torner L (2008) Soliton modes, stability, and drift in optical lattices with spatially modulated nonlinearity, *Opt. Lett.* 33(15): 1747–1749.
[63] Kartashov YV, Malomed BA, Vysloukh VA, Torner L (2009) Stabilization of multibeam necklace solitons in circular arrays with spatially modulated nonlinearity, *Phys. Rev. A* 80(5): 053816.
[64] Ye F, Kartashov YV, Hu B, Torner L (2009) Light bullets in bessel optical lattices with spatially modulated nonlinearity, *Opt. Express* 17(14): 11328–11334.
[65] Kartashov YV, Malomed BA, Vysloukh VA, Torner L (2009) Vector solitons in nonlinear lattices, *Opt. Lett.* 34(23): 3625–3627.
[66] Dong L, Li H (2010) Surface solitons in nonlinear lattices, *J. Opt. Soc. Am. B* 27(6): 1179–1183.
[67] Torner L, Kartashov YV (2009) Light bullets in optical tandems, *Opt. Lett.* 34(7): 1129–1131.
[68] Belmonte-Beitia J, Pérez-García VM, Vekslerchik V, Torres PJ (2007) Lie symmetries and solitons in nonlinear systems with spatially inhomogeneous nonlinearities, *Phys. Rev. Lett.* 98(6): 064102.
[69] Sakaguchi H, Malomed BA (2010) Solitons in combined linear and nonlinear lattice potentials, *Phys. Rev. A* 81(1): 013624.
[70] Sakaguchi H, Malomed BA (2006) Two-dimensional solitons in the gross-pitaevskii equation with spatially modulated nonlinearity, *Phys. Rev. E* 73(2): 026601.

Section 6

Optical Communication: MIMO Systems

Chapter 16

Performance Analysis of SISO and MIMO FSO Communication Systems Over Turbulent Channels

Kostas Peppas, Hector E. Nistazakis, Vasiliki D. Assimakopoulos
and George S. Tombras

Additional information is available at the end of the chapter

1. Introduction

Free-space optical (FSO) communications using intensity modulation and direct detection (IM/DD), is a cost-effective and high bandwidth access technique, which has recently received significant attention and commercial interest for a variety of applications [30, 44]. Optical wireless communication systems are rapidly gaining popularity as effective means of transferring data at high rates over short distances due to the necessity of a cost-effective, license-free, and high-bandwidth access communication technique [12, 30, 36, 44]. These systems facilitate rapidly deployable, lightweight, high-capacity communication without licensing fees and tariffs. Terrestrial FSO is not free of challenges though. A major impairment over FSO links is the atmospheric turbulence, caused by the variations in the refractive index because of inhomogeneities in temperature and pressure changes. In clear weather conditions, the atmospheric turbulence results in fluctuations at the intensity of the received signal, i.e., signal fading, also known as scintillation in optical communication terminology [6, 11]. Turbulence is caused by inhomogeneities of both temperature and pressure in the atmosphere and can severely degrade the link performance, particularly over link distances of 1 km or longer. The performance of this technology depends strongly on the atmospheric conditions between the transmitter and the receiver and the parameters of the link such as the length and the operation wavelength. Effects of fog, rain, atmospheric gases, and aerosols also result in beam attenuation due to photon absorption and scattering [25].

The performance of FSO systems over turbulence channels has been addressed in many previous works. Representative examples can be found in [47, 66, 70, 81] and the references therein. The results presented in these papers have demonstrated that the performance of single-input single-output (SISO) FSO links is severely degraded from turbulence. More specifically, the average bit error probability of such systems is far away from satisfying the typical targets for FSO applications within practical ranges of signal-to-noise ratio. To circumvent this problem, powerful fading mitigation techniques have to be deployed. In the open technical literature on FSO communication, the two most popular existing techniques for mitigation of the degrading effects of atmospheric turbulence are error control coding in conjunction with interleaving [70], [83] and maximum likelihood sequence detection (MLSD) [82]. However, for most scenarios the first one requires large-size interleavers to achieve the promised coding gains. On the other hand, MLSD requires complicated multidimensional integrations and suffers from excessive computational complexity. Some sub-optimal temporal-domain fading mitigation techniques are further explored in [83] and [84].

Another promising solution is the use of diversity techniques and the most popular scheme is the spatial diversity, i.e., the employment of multiple transmit/receive apertures, a well known diversity technique in Radio-Frequency (RF) systems [46, 48, 49, 61, 69, 72, 78, 80]. By using multiple apertures at the transmitter and/or the receiver, the inherent redundancy of spatial diversity has the potential to significantly enhance the performance. Moreover, the possibility for temporal blockage of the laser beams by obstructions is further reduced and longer distances can be covered through heavier weather conditions [46]. Concerning the performance analysis of FSO systems employing spatial diversity, the technical literature is rather rich. Representative past examples can be found in [14, 19, 21, 26, 27, 29, 46, 54–56, 62, 63, 69, 73–75].

On the other hand, various statistical models, e.g. the log normal, the gamma gamma (G – G), the I-K, the K, the negative exponential, and the Rician log normal distribution, have been used in order to describe the optical channel characteristics with respect to the atmospheric turbulence strength [4, 9, 10, 12, 23, 30, 32, 35, 38, 44, 47, 51–53, 59, 65, 71]. Recently, Al-Habash et al. proposed the G-G distribution [4] as a tractable mathematical model for atmospheric turbulence. This model is a two parameter distribution which is based on a doubly stochastic theory of scintillation and assumes that small-scale irradiance fluctuations are modulated by large-scale irradiance fluctuations of the propagating wave, both governed by independent gamma distributions. This distribution has become the dominant fading channel model for FSO links due to its excellent agreement with measurement data for a wide range of turbulence conditions [4].

For many practical FSO applications, however, irradiance is temporally correlated. Thus the derivation of a correlated G – G model is of significant theoretical and practical interest. It is noted that multivariate distributions have recently attracted the interest within the research community due to their importance in studying the performance of diversity systems operating over a multipath fading channels, see, e.g., [1, 2, 5, 15, 33, 45, 49, 58, 64, 68] and references therein. These distributions are particularly useful in the performance analysis of practical systems configurations where antenna branches are closely spaced and the correlation between diversity signals cannot be ignored [68, Chapt. 9]. In the past, several spatial correlation models have been proposed [5, 68] and used for the performance evaluation of wireless communication systems over correlated fading channels. Among them, the exponential correlation model has gained particular interest [2, 33, 45, 64]. This model corresponds to the scenario of multichannel reception by equispaced diversity antennas, in which the correlation between pairs of combined signals decays as the spacing between antenna branches increases [68].

In this chapter, we investigate the performance of multiple-input-multiple output (MIMO) FSO links over both independent and identically distributed (i.i.d.) G – G turbulence channels as well as for exponentially correlated G – G turbulent channels. A closed-form expression for the average bit error probability (ABEP) of SISO links is first derived. This result serves as benchmark for the performance analysis of FSO links employing multiple apertures and IM/DD. Rapidly convergent infinite series representations for the joint G – G probability density function (PDF), cumulative distribution function (CDF), and moment generating function (MGF) with exponential correlation are also derived. Based on these statistical results, the outage probability (OP) of selection diversity (SD) receivers as well as the ABEP of single-input-multiple-output (SIMO) FSO systems over exponentially correlated turbulent channels is investigated. Finally, we propose a simple yet highly accurate closed form approximation to the sum of arbitrary i.i.d. G – G variates. In the context of this chapter, the α-μ distribution [79] has been chosen as the convenient approximation, for which the parameters are adequately estimated from the sum of the G – G variates. Based on this result, simple accurate approximations for the OP and ABEP of MIMO FSO systems operating over i.i.d. G – G channels and employing equal gain combining (EGC) at the receiver are provided. Various numerically evaluated and computer simulation results demonstrate the accuracy of the proposed analysis. The validity of the presented analysis is testified by comparing numerically evaluated with equivalent computer simulations performance results.

2. Atmospheric turbulence

The refractive index of the atmosphere in the area between the transmitter and the receiver of a wireless optical link fluctuates randomly due to the atmospheric turbulence [12]. These fluctuations are induced mainly due to temperature oscillations among the atmosphere, the ground and the oceans [8, 12, 13, 24, 39]. More specifically atmospheric turbulence is a phenomenon belonging to different spatial and temporal scales. In the Planetary Boundary Layer (PBL), where human activities take place, it is generated by the wind's interaction with the earth's surface, which is said to be in a state of turbulent motion [18, 20, 37, 76, 77].

Turbulence is responsible for the transfer of heat, matter, and momentum within the PBL. However, it is random in nature and remains a complicated phenomenon with many unsolved aspects. The scientific community relies on the combination of experiments, theory, and computer models to understand it. Turbulence is created by thermal convection, wind shear and by the wind flowing over ground obstacles. Within the PBL, turbulence has a diurnal variation reaching a maximum about midday when the solar radiation is at a maximum. Solar radiation heats the surface which, in turn radiates heat to the air above it that becomes warmer and more buoyant and rises while cooler, denser air descends to displace it. The resulting vertical movement of air, together with flow disturbances around surface obstacles, makes low-level winds extremely irregular thus turbulent. The turbulence intensity depends primarily on the temperature lapse rate, i.e. $\lambda = dT/dH$, which is essentially the rate of temperature increase or decrease with increasing height. Under unstable conditions the temperature decreases with height and a hypothetical parcel of air which is warmer than its surrounding air would tend to rise [18, 20, 37, 76, 77].

Turbulence in the atmosphere is carried by rotational-like motions called eddies, which exist in different length scales characterized by different velocity and time scales. The larger eddies are unstable and break up into smaller ones with the subsequent transfer of the kinetic energy of the initial large eddy. The smaller eddies in turn break up into even smaller eddies and the energy is passed on to the new smaller ones. The energy is passed down from the larger scales to the smaller ones until the viscosity of the fluid (in this case air) can dissipate the kinetic energy into internal energy (or thermal energy). This is called the energy cascade and it is one of the main characteristics of turbulent motion. Other important features of turbulence are irregularity (or chaotic), diffusivity (mixing), turbulent diffusion (molecular diffusivity), rotationality (always three dimensional, the mechanism that aids the energy cascade), dissipation (the transfer of energy from larger to smaller eddies), length scales of turbulent eddies. The size of eddies spans from the order of a few millimetres to meters, namely the inner and outer scales, respectively [18, 20, 37, 76, 77]. This inner and outer scales is the main phenomenon for the choice of the appropriate distribution for turbulence's mathematical representation.

It is commonly known that in turbulent flow the actual flow velocity is broken into the mean velocity $\vec{U}$ plus the fluctuating turbulent velocity component $\vec{u}$, in the three directions, respectively. Thus, in the x-axis the instantaneous flow velocity is $U_x + u_x$. The values of the turbulent components namely u_x, u_y and u_z may not be expressed as functions of time, but a statistical description is only possible. Turbulence within the PBL can degrade the performance of free-space optical links, particularly over ranges of the order of 1 km or longer. This phenomenon is not caused by the turbulent eddies which possess different velocities but only by the parcels of air with different temperatures (and thus different densities) rising or descending that cross the path of the FSO links. They cause inhomogeneities in the temperature and pressure profiles of the atmosphere and lead to variations of the refractive index along the transmission path. Based on the fluctuations of the air density, the scientific community has developed the refractive index structure parameter C_n^2 (related to the temperature structure one, i.e. C_T^2) which takes different values depending on the strength of turbulence. The parameter C_n^2 may be measured experimentally of computed theoretically if one knows the outer scale of turbulence (i.e., the large eddy size scale) and the potential refractive index [18, 20, 37, 76, 77]. Additionally, as we will demonstrate below, there are many mathematical models for the estimation of the C_n^2 value [6, 7, 44, 52]

Thus, these variation of the refraction index in the free space area that the beam of the optical link propagates and causes deflections of the light beam into and mostly out of the transmit path [17, 52]. This random radiation of the laser beam results in fluctuations of the optical signal's irradiance at the receiver's side. This phenomenon is the so-called scintillation [4, 12, 17, 31, 42–44, 51]. The influence of scintillation in the performance of the wireless optical communication systems is very strong for the terrestrial links because induces fading of the signal arriving at the receiver in a random way. Thus, in order to estimate the optical signal arriving at the receiver it is necessary to study the appropriate statistical distribution which describes the fading statistics of each area.

Many statistical models have been proposed for the simulation of these fading statistics caused by the atmospheric turbulence effect. Some of them have been arising from experimental results, while, others, from theoretical studies. It is obvious that each location has irregularities, depending on the ground's morphology, the weather conditions, the time of the day [7, 24] and the turbulence strength. The proposed statistical models concern weak, moderate, strong or very strong turbulence conditions and the turbulence strength can be estimated through the turbulence parameter C_n^2, which depends on many parameters of the weather conditions.

One of the statistical parameters that we are using for the practical estimation of scintillation's influence at the wireless optical links' performance,is the scintillation index which is given by the following mathematical expression [44]:

$$\sigma_I^2 = \frac{< I^2 > - < I >^2}{< I >^2} \tag{1}$$

with I being the optical signal's irradiance at the receiver and $<>$ represents the ensemble average value. In the weak scintillation theory, under the assumption of plane wave propagation, the scintillation index is proportional to the Rytov variance and is given as [44]:

$$\sigma_{I,R}^2 = 1.23 C_n^2 k^{7/6} L^{11/6} \tag{2}$$

where C_n^2 is the parameter of turbulence, $k = 2\pi/\lambda$ is the optical wavenumber, while L is the link's length [44, 49, 52].

If we assume spherical wave propagation, $\sigma_{I,R}$ can be expressed as [44]:

$$\sigma_{I,R}^2 = 0.5 C_n^2 k^{7/6} L^{11/6} \tag{3}$$

In order to estimate the influence of the atmospheric turbulence conditions in the wireless optical communication system's performance, we assume a horizontal propagation path of up to a few kilometers (usually the FSO link is smaller than 4-5 km), where the turbulence strength value is considered as constant [50, 52]. Many models has been proposed for the estimation of the turbulence parameter C_n^2 [6, 12, 34, 44]. The most widely employed in research literature, are the so-called Hufnagel-Valley model and the Hufnagel and Stanley model, the SLC Day model and the SLC Night model [6, 7, 12, 34, 44, 49, 52].

The estimation of the parameter C_n^2 through the Hufnagel-Valley model (or HV5/7 model) [7, 44], depends on the wind speed, v, and the altitude, h, where the wireless optical link operates. The values 5 and 7 of the abbreviation of the Hufnagel-Valley model (i.e. HV5/7) correspond to the atmospheric coherence length (r_0) in cm and the isoplantic angle (θ_0) in μrad respectively, for $\lambda = 0.55\mu m$ while for $\lambda = 1.315\mu m$ the above values are 14cm and 20 μrad respectively [7, 44, 52]. The mathematical expression of the HV5/7 is the following [7, 44, 51],

$$C_n^2(h) = 0.00594(u/27)^2(10^{-5}h)^{10}e^{h/1000} + 2.7 \times 10^{-6}e^{-h/1500} + C_n^2(0)e^{-h/1000} \tag{4}$$

where $C_n^2(0)$ is the value of C_n^2 at the sea level in $m^{-2/3}$. In general, the turbulence parameter C_n^2 varies between the values 10^{-17} $m^{-2/3}$ and 10^{-13} $m^{-2/3}$ for weak up to very strong turbulence cases, respectively, [44, 51].

Another significant model for the estimation of the turbulence parameter is proposed by Hufnagel and Stanley [12, 34, 52]. Its mathematical expression is the following,

$$C_n^2(h) = K_0 h^{-1/3} e^{-h/h_0} \tag{5}$$

with K_0 being the turbulence strength parameter and depends on the characteristics of the specific location where the wireless optical link has been installed and h is the altitude.

Another model for the estimation of the atmospheric turbulence parameter, C_n^2, is the SLC Day model which depends only on the height where the FSO link operates [7, 52]. The mathematical expression for this model, which is accurate enough for daytime hours, is given as:

$$C_n^2(h) = \begin{cases} 1.7 \times 10^{-14} & \text{if } 0 < h < 18.5\text{m}; \\ \frac{3.13\times10^{-13}}{h^{1.05}} & \text{if } 18.5 < h < 240\text{m}; \\ 1.3 \times 10^{-15} & \text{if } 240 < h < 880\text{m}; \\ \frac{8.87\times10^{-7}}{h^3} & \text{if } 880 < h < 7200\text{m}; \\ \frac{2.0\times10^{-16}}{h^{0.5}} & \text{if } 7200 < h < 20000\text{m};. \end{cases} \tag{6}$$

while for accurate results for night hours are given by the SLC Night model [7, 52], depends only on the height where the optical link operates like the previous model are given by the following expression:

$$C_n^2(h) = \begin{cases} 8.4 \times 10^{-15} & \text{if } 0 < h < 18.5\text{m}; \\ \frac{2.87\times10^{-12}}{h^2} & \text{if } 18.5 < h < 110\text{m}; \\ 2.5 \times 10^{-16} & \text{if } 110 < h < 1500\text{m}; \\ \frac{8.87\times10^{-7}}{h^3} & \text{if } 1500 < h < 7200\text{m}; \\ \frac{2.0\times10^{-16}}{h^{0.5}} & \text{if } 7200 < h < 20000\text{m};. \end{cases} \tag{7}$$

The above mentioned models for the estimation of the turbulence parameter C_n^2, (i.e HV5/7, Hufnagel and Stanley, SLC Day and SLC Night), are valid mainly for wireless optical links which have been installed over terrestrial area. Thus, the estimation of the turbulence parameter for paths over maritime area can be done through the following accurate approximation which is valid for low altitude and for specific constant parameters c_1, c_2, c_3, c_4 and c_5. The values of the constant parameters are given in [41, 44] and the corresponding mathematical form is the following:

$$C_n^2(h) = c_1 + c_2 e^{-h/c_3} + c_4 e^{-h/c_5} \tag{8}$$

Obviously, as we mentioned above, for the estimation of the turbulence strength through the above Equations, we assumed that the value of C_n^2 remains constant for relatively long time interval and for the whole, horizontal, propagation path. This assumption is not always very accurate and thus, in some cases, it is necessary to handle this parameter as a random variable (RV), which follows a specific distribution [35, 52].

3. The Gamma-Gamma wireless optical channel model revisited

In the $G - G$ channel model for atmospheric turbulence channels the PDF of the irradiance I can be derived from the product of two independent Gamma-distributed RVs x and y with suitably defined parameters [4]. The PDF of the $G - G$ distribution is given by [44]

$$f_I(I) = \frac{2(km)^{(k+m)/2}}{\Gamma(k)\Gamma(m)\overline{I}} \left(\frac{I}{\overline{I}}\right)^{\frac{k+m}{2}-1} K_{k-m}\left[2\sqrt{km\left(\frac{I}{\overline{I}}\right)}\right] \tag{9}$$

where $K_\alpha(\cdot)$ is the modified Bessel function of the second kind and order α, $\Gamma(\cdot)$ is the gamma function and $\mathbb{E}\langle I\rangle$ with $\mathbb{E}\langle\cdot\rangle$. denoting expectation. The parameters $k > 0$ and $m > 0$ can be properly adjusted to provide good agreement between theoretical and experimental data.

Assuming spherical wave propagation, k, m can be directly related to atmospheric conditions through the following expressions [14]

$$k \bullet \left[\exp\left(\frac{0.49\sigma_{I,R}^2}{\left(1\bullet\ 0.18d^2 \bullet\ 0.56\sigma_{I,R}^{12/5}\right)^{7/6}}\right) - 1\right]^{-1} \tag{10}$$

$$m \bullet \left[\exp\left(\frac{0.51\sigma_{I,R}^2\left(1\bullet\ 0.69\sigma_{I,R}^{12/5}\right)^{-5/6}}{1\bullet\ 0.90d^2 \bullet\ 0.62d^2\sigma_{I,R}^{12/5}}\right) - 1\right]^{-1} \tag{11}$$

where $\sigma_{I,R}$ is the Rytov variance defined above and $d \bullet\ \sqrt{kD^2/4L}$, with D being the aperture diameter of the receiver.

The corresponding CDF can be expressed as [21]

$$F_I \bullet I \bullet \bullet\ \frac{1}{\Gamma\bullet k \bullet\ \Gamma \bullet m \bullet} G_{1,3}^{2,1}\left[\frac{km}{\overline{I}} I \middle|_{k,m,0}^{1}\right] \tag{12}$$

with $G_{p,q}^{m,n}\bullet\cdot\bullet$ being the Meijer-G function [60, Eq. (8.2.1.1)] [1].

The $\nu-$th moment of I defined as $\mathbb{E}\{X^\nu\} \bullet\ \int_0^\infty f_I \bullet I \bullet dI$ can be obtained using [28, Eq. (6.561/16)] as

$$\mathbb{E}\{I^\nu\} \bullet\ \left(\frac{km}{\overline{I}}\right)^{-\nu} \frac{\Gamma\bullet k \bullet\ \nu \bullet \Gamma \bullet m \bullet\ \nu \bullet}{\Gamma\bullet k \bullet \Gamma \bullet m \bullet} \tag{13}$$

For the correlated test case, to obtain an analytical expression for the multivariate G-G distribution, we first define **N** independent Gamma-distributed RVs W_n, $n \bullet\ 1, 2, \ldots, \mathbf{N}$, with marginal PDFs given by [70, Eq. (2)]

$$f_{W_n} \bullet w_n \bullet \bullet\ \frac{k^k w_n^{k-1}}{\Gamma \bullet k \bullet} \exp \bullet -k\, w_n \bullet\ H \bullet w_n \bullet \tag{14}$$

where $k \geq 1/2$ is the distribution's shaping parameter and $H \bullet\cdot\bullet$ the unit step function [28, p. xliv]. Also, let Y_n's be correlated Gamma-distributed RVs with correlation matrix given by $\Sigma_{i,j} \bullet\ \rho^{|i-j|}$, where $0 \leq \rho < 1$ is the power correlation coefficient [68, Eq. (9.195)]. Performing **N** RVs transformations in [33, Eq. (3)] and using [28, Eq. (8.445)], the joint PDF of $\mathbf{Y} \bullet \bullet Y_1\, Y_2 \cdots Y_{\mathbf{N}} \bullet$ is obtained as

$$f_{\mathbf{Y}} \bullet \mathbf{y} \bullet \bullet\ \frac{\left(1-\rho^2\right)^m}{\Gamma \bullet m \bullet} \exp\left[-\frac{m}{\overline{u} \bullet 1-\rho^2 \bullet} \bullet y_1 \bullet\ y_{\mathbf{N}} \bullet - \frac{m\left(1 \bullet\ \rho^2\right)}{\overline{u} \bullet 1-\rho^2 \bullet} \sum_{j\bullet\ 2}^{\mathbf{N}-1} y_j\right]$$
$$\times \sum_{i_1, i_2, \ldots, i_{\mathbf{N}-1} \bullet\ 0}^{\infty} \left[\frac{m}{\overline{u} \bullet 1-\rho^2 \bullet}\right]^{\mathbf{N}\, m \bullet\ 2\sum_{j\bullet\ 1}^{\mathbf{N}-1} i_j} y_1^{q_1-1} y_{\mathbf{N}}^{q_{\mathbf{N}}-1} \prod_{j \bullet\ 2}^{\mathbf{N}-1} y_j^{q_j-1} \prod_{n \bullet\ 1}^{\mathbf{N}-1} \left[\frac{1}{\Gamma \bullet m \bullet\ i_n \bullet\ i_n!}\right] \tag{15}$$

[1] The Meijer-G function is a standard built-in function available in the most popular mathematical software packages, such as Maple or Mathematica

where $\mathbf{y} = (y_1\, y_2\, \cdots\, y_{\mathbf{N}})$, with $y_n > 0, \forall\, n, m \geq 1/2$ is the distribution's shaping parameter, and

$$q_j = \begin{cases} m + i_1 & \text{if } j = 1; \\ m + i_{\mathbf{N}-1} & \text{if } j = \mathbf{N}; \\ m + i_{j-1} + i_j & \text{if } j = 2, 3, \ldots, \mathbf{N}-1. \end{cases} \tag{16}$$

Then, following the approach presented in [70] and [40], the multivariate G-G distribution with exponential correlation can be derived as [70, Eq. (5)]

$$f_{\mathbf{I}}(\mathbf{I}) = \int_0^\infty \int_0^\infty \cdots \int_0^\infty \prod_{n=1}^{\mathbf{N}} \left[y_n^{-1} f_{W_n}\left(\frac{I_n}{y_n}\right)\right] f_{\mathbf{Y}}(\mathbf{y})\, d\mathbf{y} \tag{17}$$

where $\mathbf{I} = (I_1\, I_2\, \cdots\, I_{\mathbf{N}})$, with $I_n = W_n Y_n, \forall\, n$, Substituting (14) and (15) to (17) and using [28, Eq. (3.471/9)], the joint PDF of $\mathbf{I}$ is obtained as

$$\begin{aligned} f_{\mathbf{I}}(\mathbf{I}) = {} & \frac{2^{\mathbf{N}}\left(1-\rho^2\right)^m}{(\Gamma(k))^{\mathbf{N}}\, \Gamma(m)} \sum_{i_1, i_2, \ldots, i_{\mathbf{N}-1}=0}^{\infty} \Xi^{\frac{\mathbf{N}(k+m)}{2} + \sum_{j=1}^{\mathbf{N}-1} i_j} \rho^{2\sum_{j=1}^{\mathbf{N}-1} i_j} \\ & \times I_1^{\omega_1} K_{q_1-k}\left[2\sqrt{\Xi I_1}\right] u_{\mathbf{N}}^{\omega_{\mathbf{N}}} K_{q_{\mathbf{N}}-k}\left[2\sqrt{\Xi I_{\mathbf{N}}}\right] \\ & \times \prod_{j=2}^{\mathbf{N}-1}\left\{\frac{I_j^{\omega_j}}{(1+\rho^2)^{\frac{q_j-k}{2}}} K_{q_j-k}\left[2\sqrt{\Xi(1+\rho^2) I_j}\right]\right\} \prod_{n=1}^{\mathbf{N}-1}\left[\frac{1}{\Gamma(m+i_n)\, i_n!}\right] \end{aligned} \tag{18}$$

with $\Xi = (km)/(\bar{I}\,(1-\rho^2))$, $\omega_n = (k+q_n)/2 - 1, \forall\, n = 1, 2, \ldots, \mathbf{N}$. Note that for $\rho = 0$, all infinite series terms that appear in (18) with $i_j \neq 0, \forall\, j$, vanish and (18) simplifies to the product of $\mathbf{N}$ independent G-G distributions [16, Eq. (2)].

Next, important statistical properties of the exponentially correlated multivariate G-G distribution, namely the joint G-G CDF and MGF will be presented.

3.1. Joint CDF

The joint CDF of $\mathbf{I}$ is given by

$$F_{\mathbf{I}}(\mathbf{I}) = \int_0^{I_1}\int_0^{I_2}\cdots\int_0^{I_{\mathbf{N}}} f_{\mathbf{I}}(\mathbf{t})\, d\mathbf{t} \tag{19}$$

with $\mathbf{t} = (t_1\, t_2\, \cdots\, t_{\mathbf{N}})$. In order to obtain an analytical expression for (19), the Bessel functions that appear in (18) are expressed in terms of Meijer-G functions [60, Eq. (8.4.3.1)]. Then, using [60, Eq. (8.4.23.1)], [60, Eq. (2.24.1.1)], and [60, Eq. (8.2.2.15)], an analytical expression for the joint G-G CDF with exponential correlation is derived in infinite series form as

$$\begin{aligned} F_{\mathbf{I}}(\mathbf{I}) = {} & \frac{\left(1-\rho^2\right)^m}{(\Gamma(k))^{\mathbf{N}}\, \Gamma(m)} \sum_{i_1, i_2, \ldots, i_{\mathbf{N}-1}=0}^{\infty} \prod_{n=1}^{\mathbf{N}-1}\left[\frac{\rho^{2i_n}}{\Gamma(m+i_n)\, i_n!}\right] \mathcal{G}(q_1, k, \Xi I_1) \\ & \times \mathcal{G}(q_{\mathbf{N}}, k, \Xi I_{\mathbf{N}}) \prod_{j=2}^{\mathbf{N}-1}\left\{\mathcal{G}\left[q_j, k, \Xi\left(1+\rho^2\right) I_j\right]\left(1+\rho^2\right)^{-q_j}\right\} \end{aligned} \tag{20}$$

where

$$\mathcal{G}(A, B, x) = G_{1,3}^{2,1}\left[x \,\middle|\, \begin{matrix} 1 \\ A, B, 0 \end{matrix}\right]. \tag{21}$$

3.2. Joint MGF

Using $f_{\mathbf{I}} \bullet \mathbf{I} \bullet$, the joint MGF of $\mathbf{I}$ can be obtained as

$$\mathcal{M}_{\mathbf{I}} \bullet \mathbf{s} \bullet \bullet \int_0^\infty \int_0^\infty \cdots \int_0^\infty \exp\left(-\sum_{n \bullet 1}^{\mathrm{N}} s_n I_n\right) f_{\mathbf{I}} \bullet \mathbf{I} \bullet \, d\mathbf{I} \tag{22}$$

with $\mathbf{s} \bullet \bullet s_1 \, s_2 \cdots s_{\mathrm{N}} \bullet$. Substituting (18) in (22) and using [60, Eq. (8.4.3.1)], [60, Eq. (8.4.23.1)], [60, Eq. (2.24.1.1)], and [60, Eq. (8.2.2.15)], an analytical expression for the joint G-G MGF is derived as

$$\begin{aligned} \mathcal{M}_{\mathbf{I}} \bullet \mathbf{s} \bullet \bullet & \frac{\left(1-\rho^2\right)^m}{\bullet \Gamma \bullet k \bullet \bullet^{\mathrm{N}} \, \Gamma \bullet m \bullet} \sum_{i_1, i_2, \ldots, i_{\mathrm{N}-1} \bullet \, 0}^{\infty} \prod_{j \bullet \, 1}^{\mathrm{N}-1} \left[\frac{\rho^{2 i_j}}{\Gamma\left(m \bullet \; i_j\right) i_j!}\right] \mathcal{H}\left(q_1, k, \frac{\Xi}{s_1}\right) \\ & \times \mathcal{H}\left(q_{\mathrm{N}}, k, \frac{\Xi}{s_{\mathrm{N}}}\right) \prod_{j \bullet \, 2}^{\mathrm{N}-1} \left\{\mathcal{H}\left[q_j, k, \frac{\Xi\left(1 \bullet \; \rho^2\right)}{s_j}\right] \left(1 \bullet \; \rho^2\right)^{-q_j}\right\} \end{aligned} \tag{23}$$

where

$$\mathcal{H} \bullet A, B, x \bullet \bullet \; G_{1,2}^{2,1}\left[x \,\middle|\, \begin{matrix} 1 \\ A, B \end{matrix}\right]. \tag{24}$$

4. System model

We consider a MIMO FSO system where the information signal is transmitted via M apertures and received by N apertures. The information bits are modulated using IM/DD with on-off keying (OOK) and transmitted through the M apertures using repetition coding [14]. A high-energy FSO system whose performance is limited by background radiation and thermal noise is assumed. Under this assumption, the use of the AWGN model as a good approximation of the Poisson photon counting detection model is applicable [46]. The received signal at the n-th receive aperture is expressed as

$$r_n \bullet \; \eta s \sum_{m \bullet \, 1}^{M} I_{mn} \bullet \; v_n, \; n \bullet \; 1, \ldots, N \tag{12}$$

where η is the optical-to-electrical conversion coefficient, $s \in \{0, 1\}$ represents the information bits and v_n is the AWGN with zero mean and variance $\sigma_v^2 \; \bullet \; N_0/2$. In the following analysis, we assume that the I_{mn}-s are either independent or exponentially correlated G-G random variables. It is noted that the assumption of independence can significantly simplify the underlying mathematical analysis and it is justified for link distances of the order of kilometers and for aperture separation distances of the order of centimeters [14]. Finally, we define the instantaneous electrical signal-to-noise ratio (SNR) as $\mu_{mn} \triangleq \bullet \eta I_{mn} \bullet^2 / N_0$ and the corresponding average electrical SNR as $\overline{\mu}_{mn} \triangleq \bullet \eta \mathbb{E}\langle I_{mn} \rangle \bullet^2 / N_0$.

5. Performance analysis of SISO links

The ABEP of the considered system in the presence of AWGN and under the assumption of perfect channel state information (CSI) at the receiver side is given by

$$P_e \bullet \; P \bullet 1 \bullet P \bullet e | 1 \bullet \bullet \; P \bullet 0 \bullet P \bullet e | 0 \bullet \tag{13}$$

where $P(1)$ and $P(0)$ are the probabilities of sending 1 and 0 bits, respectively, and $P(e|1)$, $P(e|0)$ are the conditional bit-error probabilities bit 1 or 0 has been transmitted, respectively. Without loss of generality, we assume $P(1) = P(0) = 0.5$ and $P(e|1) = P(e|0)$, a fact also justified by the symmetry of the problem. Using the analysis presented in [46] one obtains

$$P(e|I) = P(e|1, I) = P(e|0, I) = Q\left(\frac{\eta I}{\sqrt{2N_0}}\right) \tag{14}$$

where $Q(x) = \frac{1}{\sqrt{2\pi}} \int_x^\infty \exp\left(-\frac{t^2}{2}\right) dt$ is the Gauss-Q function. The ABEP can be obtaining by averaging $P(e|I)$ over the PDF of I, namely

$$\overline{P}_{be} = \int_0^\infty P(e|I) f_I(I) dI \tag{15}$$

By expressing $P(e|I)$ in terms of the electrical SNR μ, namely $Q\left(\frac{\eta I}{\sqrt{2N_0}}\right) = Q\left(\sqrt{\frac{\mu}{2}}\right) = 0.5\text{erfc}\left(\frac{\sqrt{\mu}}{2}\right)$, $\overline{P}_{be}$ can be obtained by performing a change of variables and averaging over the PDF of the electrical SNR, μ, instead of the PDF of I. Assuming that I follows a $\text{G} - \text{G}$ distribution with parameters k and m and by applying a simple power transformation of RVs, the PDF of μ can be expressed as

$$f_\mu(\mu) = \frac{(km)^{(k+m)/2}}{\Gamma(k)\Gamma(m)\sqrt{\overline{\mu}\mu}} \left(\sqrt{\frac{\mu}{\overline{\mu}}}\right)^{\frac{k+m}{2}-1} K_{k-m}\left[2\sqrt{km\sqrt{\frac{\mu}{\overline{\mu}}}}\right] \tag{16}$$

Finally $\overline{P}_{be}$ can be expressed in closed form as

$$\overline{P}_{be} = 0.5\mathcal{F}\left(k, m, \overline{\mu}, \frac{1}{2}\right) \tag{17}$$

where $\mathcal{F}(k, m, \overline{\mu}, s)$ is given by [57, Eq. (27)] [2]

$$\mathcal{F}(k, m, \overline{\mu}, s) = \frac{\Xi^{k+m} s^{-(k+m)/2}}{4\pi^{\frac{3}{2}}\Gamma(k)\Gamma(m)} G_{2,5}^{4,2}\left[\frac{k^2 m^2}{16\overline{\mu}s^2}\middle|_{b_q}^{a_p}\right] \tag{16}$$

with $a_p = \{1 - \frac{k+m}{4}, \frac{1}{2} - \frac{k+m}{4}\}$, $b_q = \{\frac{1}{2} + \frac{k-m}{4}, \frac{k-m}{4}, \frac{1}{2} + \frac{m-k}{4}, \frac{m-k}{4}, -\frac{k+m}{4}\}$ and $\Xi \triangleq \sqrt{\frac{km}{\sqrt{\overline{\mu}}}}$.

Another important performance metric is the outage probability. The outage probability is defined as the probability that the SNR of the signal at the output of the receiver, μ , falls below a specified threshold, μ_{th}. This metric is considered as an important parameter for FSO links to be operated as a part of a data network and is critical in the design of both transport and network layer [21]. With the help of (12) and assuming $\overline{I} = 1$, the outage probability is obtained in closed form as

$$P_{\text{out}} = \Pr\{\mu < \mu_{\text{th}}\} = \Pr\left\{\frac{I^2\eta^2}{N_0} < \mu_{\text{th}}\right\} = \Pr\left\{I < \sqrt{\frac{\mu_{\text{th}}}{\overline{\mu}}}\right\} = F_I\left(\sqrt{\frac{\mu_{\text{th}}}{\overline{\mu}}}\right) \tag{17}$$

[2] It is noted that Eq. (27) in [57] includes typos which we have corrected in (16)

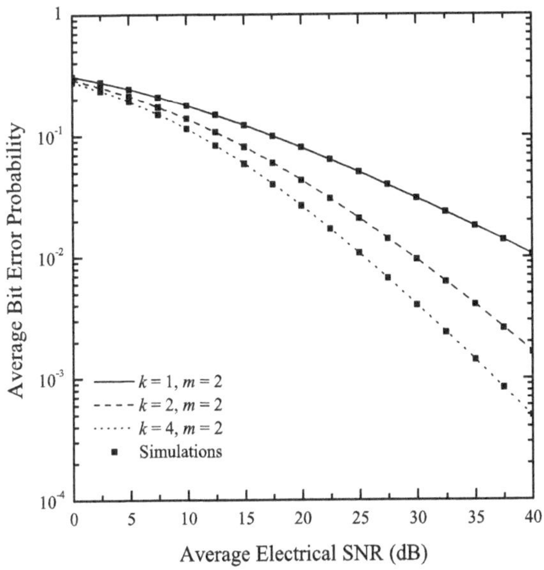

Figure 1. Average Bit Error Probability of SISO OOK receivers employing intensity modulation and direct detection and operating over G − G channels, for different values of parameters k and m, as a function of the Average Electrical SNR, $\overline{\mu}$.

Using (16), in Figure 1 the ABEP of SISO OOK receivers employing intensity modulation and direct detection and operating over G − G channels is depicted, for different values of parameters k and m, as a function of the Average Electrical SNR, $\overline{\mu}$. Moreover, in Figure 2 the OP of the system under consideration is depicted for the same set of parameters as a function of the inverse normalized outage threshold, $\frac{\mu_{\text{th}}}{\overline{\mu}}$.

As it can be observed, for a SISO FSO link, both the ABEP and OP performance is quite poor (i.e., higher than 10^{-3} in the SNR range of 30-40 dB , especially over strong atmospheric turbulence conditions (that corresponds to small values of k or m) and therefore the use of diversity techniques is absolutely justified. The use of spatial diversity can be implemented either at the transmitter (MISO) or at the receiver (SIMO case) or at both of them (MIMO case). In both figures, our numerically evaluated results are accompanied with semi-analytical Monte Carlo simulations. In our simulations, more than 10^6 square G − G samples are generated to guarantee statistical convergence. The following listings, written in Matlab 2008a demonstrate the evaluation of the ABEP and OP for the SISO case using Monte Carlo simulations.

```
function y = BER_OOK(SNR_dB, k, m, SIZE)
%SNR_dB: average electical SNR
%k, m: the distribution parameters
mu_bar = 10^(0.1*SNR_dB);
% The first gamma process:
g = gamrnd(k, sqrt(mu_bar)/k, 1, SIZE);
% The second gamma process:
```

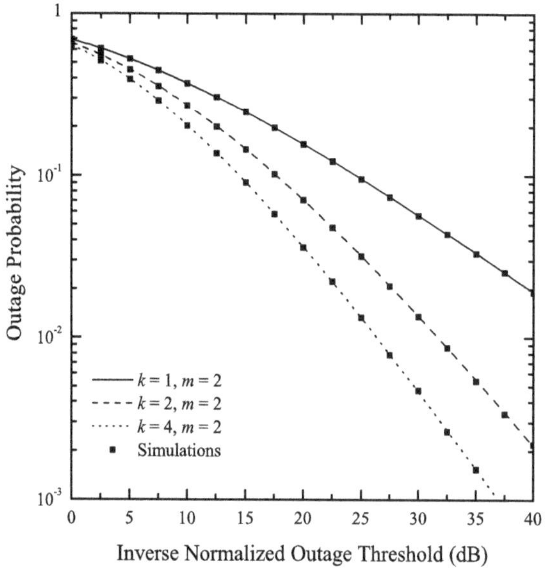

Figure 2. Outage Probability of SISO receivers employing intensity modulation and direct detection and operating over G – G channels, for different values of parameters k and m, as a function of the Inverse Normalized Outage Threshold, $\overline{\mu}/\mu_{th}$.

```
data = gamrnd(m, g/m, 1, SIZE);
% The final data
data = data.^2;
% ABEP of OOK
y = sum( 0.5*erfc(0.5*sqrt(data)))/SIZE;

function y = outage(k, m, inverse_th_dB , SIZE)
th = 10^(0.1*inverse_th_dB);
% first gamma process
first = gamrnd(k, 1/k, 1, SIZE);
% final process
data = (gamrnd(m, first/m, 1, SIZE));
y = length(find(data < 1/sqrt(th)))/SIZE;
```

In the following analysis, analytical results for both SIMO and MIMO cases will be presented.

5.1. SIMO case: Selection Diversity (SD)

The SD combining scheme is the least complicated among the considered combining schemes because of the fact that it processes the aperture with the maximum received irradiance (or electrical SNR). Consequently, the selection is made according to

$$I_{SD} = \max\{I_1, I_2 \cdots, I_N\} \tag{18}$$

If I_n are i.i.d random variables, then using a similar to the previous section analysis, the OP of SD receivers is readily obtained as

$$P_{\text{out}} = \left[F_I\left(\sqrt{\frac{\mu_{\text{th}}}{\overline{\mu}}}\right)\right]^N \tag{19}$$

If exponentially correlated irradiance is considered, the CDF of I_{SD} is readily obtained using (20) as

$$P_{\text{out}}\left(\gamma_{\text{th}}\right) = F_{\mathbf{I}}\left(\sqrt{\frac{\mu_{\text{th}}}{\overline{\mu}}}, \cdots, \sqrt{\frac{\mu_{\text{th}}}{\overline{\mu}}}\right). \tag{20}$$

In Figure 3 the OP of three- and four-branch SIMO receivers employing intensity modulation and direct detection and operating over exponentially correlated $G-G$ channels, for $\rho = 0.25$ and different values of parameters k and m, is depicted as a function of the Inverse Normalized Outage Threshold, $\overline{\mu}/\mu_{th}$. As it can be observed, the employment of spatial diversity significantly enhances the outage performance of the considered system. To double-check the correctness of the proposed analysis, the analytical results are accompanied with numerical ones obtained using Monte-Carlo simulations. Because of the long computational time, inherent to Monte-Carlo methods, simulation results of up to 10^{-6} are given. In order to generate exponentially correlated $G-G$ samples, we make use of the fact that I_n can be written as $I_n = W_n Y_n, \forall\, n$. where W_n and Y_n are samples of an uncorrelated and a correlated gamma process, respectively. Uncorrelated gamma samples with given parameters can be easily generated using the standard Matlab built-in function *gamrnd*(). The generation of correlated gamma random samples with given correlation matrix has been addressed in several past works. In the context of this work, we used the method presented in [67], which yields accurate results for the given system parameters, despite the fact that it imposes certain conditions or constraints on the PDF parameters. Finally, it is noted that although the analytical expression for the outage probability is given in terms of infinite series, it converges rapidly and steadily, requiring few terms to obtain sufficient numerical accuracy. As it was shown in [56], the number of terms depends on the values of the parameters k and m, the correlation coefficient as well as the SNR. In our terms, a truncation of the infinite series to 25 terms and 18 terms for the three- and four-branch case, respectively, resulted in an excellent match of the numerical results with the Monte-Carlo simulations.

5.2. SIMO case: Optimal combining

When receive diversity with optimal combining (OC) is applied, following a similar analysis as in [46], the ABEP is obtained as

$$\overline{P}_{be} = \int_0^\infty \int_0^\infty \cdots \int_0^\infty f_{\mathbf{I}}(\mathbf{I})\, Q\left(\frac{\eta}{\sqrt{2\,N\,N_0}}\sqrt{\sum_{n=1}^{N} I_n^2}\right) d\mathbf{I} \tag{21}$$

where the average irradiance is considered to be normalized to one.

In the following analysis, we assume that I_n are exponentially correlated G-G random variables. In general, the N-fold integral in (21) is difficult, if not impossible to be obtained in closed form. To circumvent this problem, we utilize a simple and accurate exponential

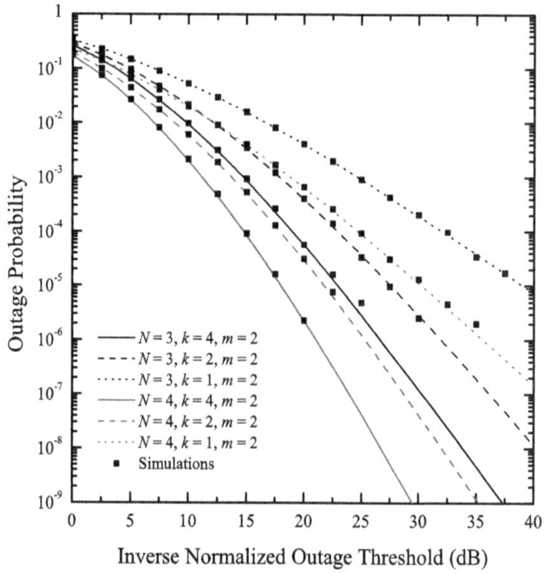

Figure 3. Outage Probability of triple and quadruple branch SIMO receivers employing intensity modulation and direct detection with selection diversity and operating over exponentially correlated G – G channels, for $\rho \bullet$ 0.25 and different values of parameters k and m, as a function of the Inverse Normalized Outage Threshold, $\overline{\mu}/\mu_{th}$.

approximation for the Gaussian-Q function using [22, eq. (2)] and [22, eq. (14)] namely

$$Q\bullet x\bullet \simeq \frac{1}{12}\exp\left(-\frac{x^2}{2}\right)\bullet \frac{1}{4}\exp\left(-\frac{2x^2}{3}\right). \tag{22}$$

Substituting (22) in (21) and using [60, Eq. (8.4.3.1)], [60, Eq. (8.4.23.1)], and [60, Eq. (2.24.1.1)], yields

$$\overline{P}_{be} \simeq \frac{1}{12}\mathcal{N}\left(k,m,\sqrt{\frac{\mu}{4N}}\right)\bullet \frac{1}{4}\mathcal{N}\left(k,m,\sqrt{\frac{\mu}{3N}}\right) \tag{23}$$

where μ is the average electrical SNR and

$$\begin{aligned}\mathcal{N}\bullet k,m,s\bullet\bullet &\frac{2^{N\bullet m\bullet k-2\bullet}\left(1-\rho^2\right)^m}{\pi^N\bullet\Gamma\bullet k\bullet\bullet^N\Gamma\bullet m\bullet}\sum_{i_1,i_2,\ldots,i_{N-1}\bullet 0}^{\infty}\bullet 2\rho\bullet^{2\sum_{j\bullet 1}^{N-1}i_j}\prod_{n\bullet 1}^{N}\left[\frac{1}{\Gamma\bullet m\bullet\ i_n\bullet i_n!}\right]\\ &\times\mathcal{I}\bullet q_1,k,\Xi,s\bullet\mathcal{I}\bullet q_N,k,\Xi,s\bullet\prod_{j\bullet 2}^{N-1}\left\{\mathcal{I}\left[q_j,k,\Xi\left(1\bullet\rho^2\right),s\right]\left(1\bullet\rho^2\right)^{-q_j}\right\}\end{aligned} \tag{24}$$

with

$$\mathcal{I}\bullet A,B,C,x\bullet\bullet\ G_{4,1}^{1,4}\left[\frac{16x^2}{C^2}\middle|\begin{matrix}\frac{1-A}{2},\frac{2-A}{2},\frac{1-B}{2},\frac{2-B}{2}\\0\end{matrix}\right]. \tag{25}$$

Figure 4 illustrates the ABEP performance of OC FSO links, operating at $\lambda \bullet$ 1550nm with $N \bullet$ 3 and 4 receive apertures, versus μ. Similarly to [70], exponentially correlated atmospheric turbulence channels are considered with $\rho = 0.175$. It is assumed that $C_n^2 = 1.7 \times 10^{-14}$, which is a typical value of refractive index for FSO links near the ground during daytime. Furthermore, it is assumed that $D \ll L$ leading to $d = 0$, and hence, no aperture averaging is possible. Three different link distances are considered: $L = 3, 4$, and 5 km. The resulting values for k and m are obtained via (10) and (11), respectively. As expected, the ABEP improves as N and/or μ increase and/or L decreases. Furthermore, as it is shown comparing numerically evaluated results for the ABEP with the equivalent exact ones obtained via Monte Carlo simulations, the proposed ABEP approximation given by (23) provides a tight upper bound for all test cases under consideration.

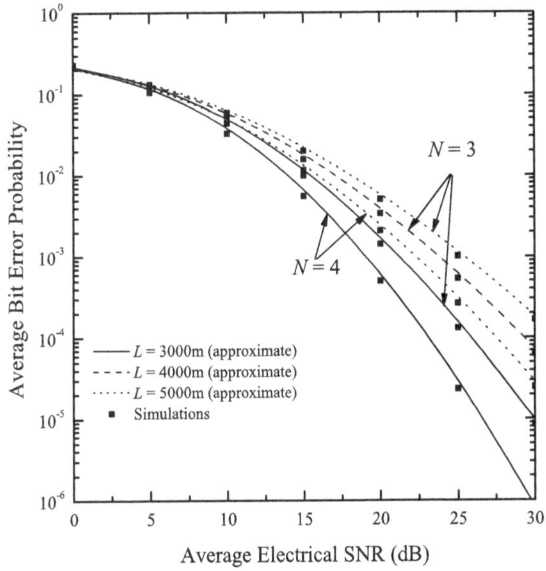

Figure 4. ABEP of triple and quadruple branch SIMO FSO links operating over exponentially correlated G – G atmospheric turbulent channels and employing optimal combining, versus average electrical SNR μ for $\rho = 0.175$ and different link distances.

5.3. MIMO case: Equal gain combining

Assuming perfect Channel State Information (CSI), the ABEP of the considered FSO system is given by [46]

$$\overline{P}_{be} = \int_0^\infty \int_0^\infty \cdots \int_0^\infty f_{\mathbf{I}}(\mathbf{I})\, Q\left(\frac{\eta}{N M \sqrt{2 N_0}} \sum_{n=1}^{N} I_n\right) \mathrm{d}\mathbf{I} \tag{26}$$

which can be further expressed as [21]

$$\overline{P}_{be} = \frac{1}{2} \int_0^\infty f_I(I)\, \mathrm{erfc}\left(\frac{\eta}{2 M N \sqrt{N_0}} I\right) \mathrm{d}I \tag{27}$$

where $I = \sum_{n=1}^{N} I_n$. This integral is very difficult to be obtained in closed form since an analytical expression for the statistical distribution of I is required. In the following it will be shown that when I_n are i.i.d G-G random variables, the distribution of I can be accurately approximated with the so-called $\alpha - \mu$ distribution [79].

5.3.1. $\alpha - \mu$ approximation to the sum of i.i.d. G − G variates

Let $I = \sum_{n=1}^{\mathbf{N}} I_n$ be a sum of **N** i.i.d. G − G variates. We propose to approximate the PDF $f_I(I)$ and CDF $F_I(I)$ of I by the α-μ PDF and CDF given in [79]

$$f_I(I) = \frac{\alpha \mu^{\mu} I^{\alpha\mu-1}}{\widehat{I}^{\alpha\mu}\Gamma(\mu)} \exp\left(-\mu \frac{I^{\alpha}}{\widehat{I}^{\alpha}}\right) \tag{28}$$

$$F_I(I) = 1 - \frac{\Gamma(\mu, \mu I^{\alpha}/\widehat{I}^{\alpha})}{\Gamma(\mu)} \tag{29}$$

In (28), (29), $\alpha, \mu > 0$ are the distribution parameters, $\widehat{I} = \mathbb{E}\{I^{\alpha}\}^{1/\alpha}$ is a scale parameter and $\Gamma(\cdot,\cdot)$ is the incomplete gamma function [28, Eq. (8.350/2)]. The motivation behind this approximation is twofold: First, in a recently published work [3] it was shown that the gamma distribution can be used to approximate the sum of independent G − G variates. We feel that the use of a more generic distribution, which includes as special case the gamma distribution (in the α-μ case by setting $\alpha = 1$), will result in a more accurate approximation. Second, as it will become evident, the estimation of the parameters of the resulting α-μ PDF requires the knowledge of the first, the second and the fourth moment of I, which can be easily evaluated given the moments of I_n. Therefore, the resulting PDF incorporates information regarding the mean, the variance *and* the kurtosis of I. In order to render (28) and (29) an accurate approximation, moment-based estimators for α, μ and $\widehat{I}$ are used. These estimators are obtained as [79]

$$\frac{\Gamma^2(\mu+1/\alpha)}{\Gamma(\mu)\Gamma(\mu+2/\alpha) - \Gamma^2(\mu+1/\alpha)} = \frac{\mathbb{E}^2\{I\}}{\mathbb{E}\{I^2\} - \mathbb{E}^2\{I\}} \tag{30}$$

$$\frac{\Gamma^2(\mu+2/\alpha)}{\Gamma(\mu)\Gamma(\mu+4/\alpha) - \Gamma^2(\mu+2/\alpha)} = \frac{\mathbb{E}^2\{I^2\}}{\mathbb{E}\{I^4\} - \mathbb{E}^2\{I^2\}} \tag{31}$$

$$\widehat{I} = \frac{\mu^{\frac{1}{\alpha}}\Gamma(\mu)\mathbb{E}\{I\}}{\Gamma\left(\mu + \frac{1}{\alpha}\right)} \tag{32}$$

The required moments $\mathbb{E}\{I\}$, $\mathbb{E}\{I^2\}$ and $\mathbb{E}\{I^4\}$ can be evaluated using (13) and the multinomial identity as

$$\begin{aligned}\mathbb{E}\{I^{\nu}\} = \sum_{j_1=0}^{\nu}\sum_{j_2=0}^{j_1}\cdots\sum_{j_{\mathbf{N}-1}=0}^{j_{\mathbf{N}-2}} \binom{\nu}{j_1}\binom{j_1}{j_2}\cdots\binom{j_{\mathbf{N}-2}}{j_{\mathbf{N}-1}} \\ \times \mathbb{E}\{I_1{}^{\nu-j_1}\}\mathbb{E}\{I_2{}^{j_1-j_2}\}\cdots\mathbb{E}\{I_{\mathbf{N}}{}^{j_{\mathbf{N}-1}}\}\end{aligned} \tag{33}$$

where ν is positive integer. Using Maple, the command lines given in (34), can be utilized to obtain α and μ in a computationally efficient manner. In this case, $In \triangleq \mathbb{E}\{I^n\}$, $n = 1,2,4$. The parameter $\widehat{I}$ can be finally obtained using (32). To demonstrate the accuracy of this analysis, Fig. 5 shows the exact and approximate CDFs of the sum of two and nine i.i.d. $G-G$ variates with $\overline{I}=1$ for different values of parameters k and m. As it can be observed, in all considered test cases, the proposed approximation is highly accurate and practically indistinguishable from the exact CDF curves. A comparison of the proposed method with the one proposed in [3] reveals that our method performs equally well for both small and large values of $\mathbf{N}$. Thus, a correcting factor, similar to the one introduced in [3] to obtain a sufficient approximation accuracy, is no longer required. Moreover, in [21], an approximate expression for the CDF of I in terms of Meijer-G functions [28, Eq. (9.301)] is provided. However, since the evaluation of Meijer-G functions can be sometimes laborious, (29) may be preferable to [21] in terms of computational complexity. Finally, our derived formulas are simpler than those presented in [14], since the latter are expressed as infinite series and require the computation of convolutional sums.

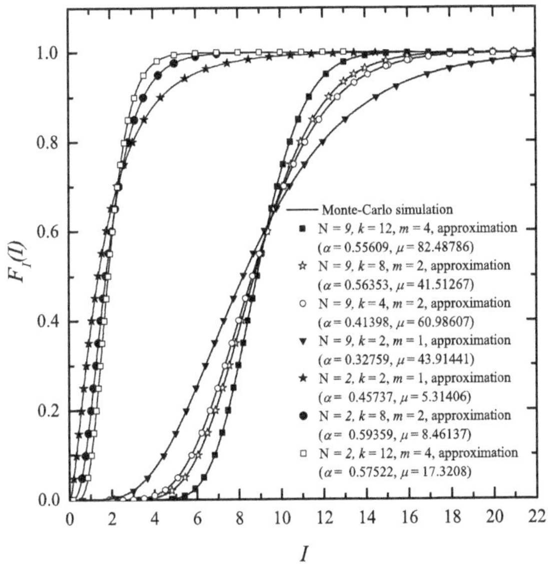

Figure 5. Exact and approximate CDF of the sum of $\mathbf{N}=2$ and $\mathbf{N}=9$ i.i.d $G-G$ variates

In Fig. 6 the OP of the considered system is plotted as a function of the inverse normalized outage threshold $\overline{\mu}/\mu_{th}$ for $L=2$km and $L=4$km. The parameters k and m are obtained using (10) and (11) assuming $\lambda=1550$nm, $C_n^2=1.7\times10^{-14}\mathrm{m}^{-2/3}$ and $D/L\rightarrow 0$. Both numerically evaluated and computer simulation results are depicted. From the above mentioned plot, it is clear that the derived approximative expressions are highly accurate for every considered MIMO deployment and for all considered link distances.

To evaluate ABEP, the PDF of I, $f_I(I)$, at the combiner output, will be approximated by the PDF of a single channel given in (28) where the parameters α and μ are estimated as functions of k and m. Having obtained these parameters, the ABEP is easily obtained by substituting (28) into (21) and performing symbolic or numerical integration. In Fig. 7 the ABEP of the considered MIMO system is depicted as a function of the average electrical SNR, using the same parameters considered in the OP case. From the observation of Fig. 7, one can

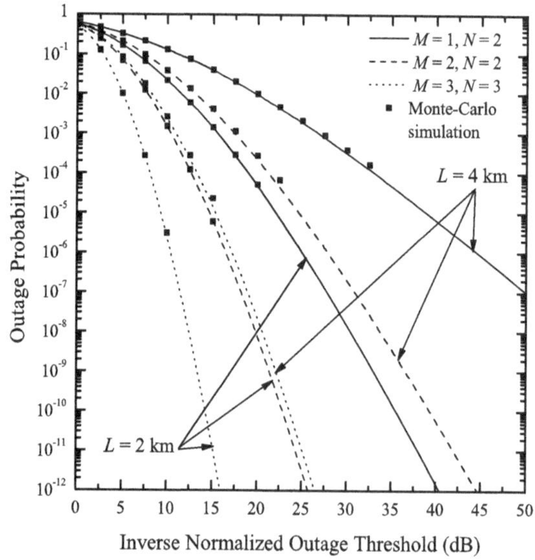

Figure 6. Outage Probability of MIMO FSO systems employing EGC and operating over i.i.d G – G fading channels as a function of the inverse normalized outage threshold, ($\lambda = 1550$nm, $C_n^2 = 1.7 \times 10^{-14}m^{-2/3}$ and $D/L \to 0$)

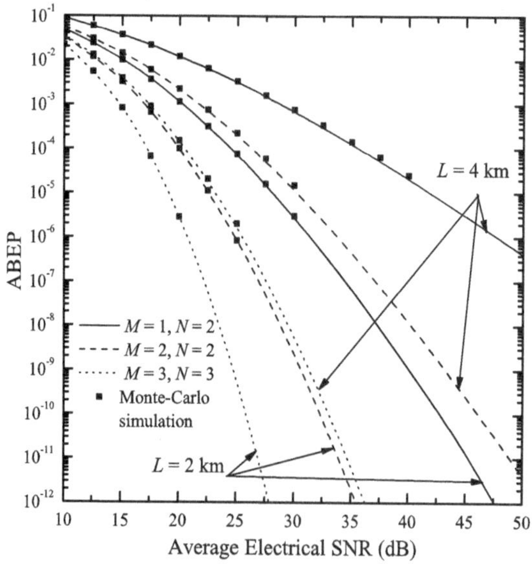

Figure 7. ABEP of MIMO FSO systems employing EGC and operating over i.i.d G – G fading channels as a function of the average electrical SNR, ($\lambda = 1550$nm, $C_n^2 = 1.7 \times 10^{-14}m^{-2/3}$ and $D/L \to 0$)

verify similar findings to that mentioned in Fig. 6 regarding the accuracy of the proposed approximation.

$$\begin{aligned}&\text{X1} := \text{I1\^{}2}/(\text{I2}-\text{I1\^{}2}); \text{X2} := \text{I2\^{}2}/(\text{I4}-\text{I2\^{}2});\\&\text{f1} := \text{X1} - (\text{GAMMA}(\text{mu}+1/\text{alpha}))\text{\^{}2}/(\text{GAMMA}(\text{mu})*\text{GAMMA}(\text{mu}+2/\text{alpha})\\&-(\text{GAMMA}(\text{mu}+1/\text{alpha}))\text{\^{}2}) = 0;\\&\text{f2} := \text{X2} - (\text{GAMMA}(\text{mu}+2/\text{alpha}))\text{\^{}2}/(\text{GAMMA}(\text{mu})*\text{GAMMA}(\text{mu}+4/\text{alpha})\\&-(\text{GAMMA}(\text{mu}+2/\text{alpha}))\text{\^{}2}) = 0;\\&\text{SOL} := \text{fsolve}(\{\text{f1},\text{f2}\},\{\text{alpha},\text{mu}\},\{\text{alpha}=0..1000,\text{mu}=0..1000\});\end{aligned} \tag{34}$$

6. Conclusions

In this chapter we presented a thorough performance analysis of FSO communication systems using spatial diversity over G – G distributed atmospheric turbulence channels. We obtained accurate approximated closed-form expressions and rapidly convergent infinite series representations for the average bit error probability and the outage probability of SIMO and MIMO FSO systems. Our results demonstrated that significant performance gains can be obtained when multiple apertures at the transmitter and/or receiver are used.

Acknowledgement

This work was partially supported by the Special Research Account of the National and Kapodistrian University of Athens.

Author details

Kostas Peppas
National Center For Scientific Research "Demokritos", Laboratory of Wireless Communications, Institute of Informatics and Telecommunications, 15310 Athens, Greece

Hector E. Nistazakis and George S. Tombras
Department of Electronics, Computers, Telecommunications and Control, Faculty of Physics, National and Kapodistrian University of Athens, 15784, Athens Greece

Vasiliki D. Assimakopoulos
Institute of Environmental Research and Sustainable Development of the National Observatory of Athens, I. Metaxa and Vas. Pavlou, Athens, 15236, Greece

7. References

[1] Aalo, V. A. [1995]. Performance of maximal-ratio diversity systems in a correlated Nakagami fading environment, *IEEE Trans. Commun.* 43: 2360–2369.

[2] Aalo, V. A. & Piboongungon, T. [2005]. On the multivariate generalized Gamma distribution with exponential correlation, *Proc. IEEE Global Telecommun. Conf. (GLOBECOM'05)*, Vol. 3, St. Louis, Missouri, USA.

[3] Al-Ahmadi, S. & Yanikomeroglu, H. [2010]. On the approximation of the generalized-K distribution by a gamma distribution for modeling composite fading channels, *IEEE Trans. Wireless Commun.* 9(2): 706–713.

[4] Al-Habash, M. A., Andrews, L. C. & Phillips, R. L. [2001]. Mathematical model for the irradiance pdf of a laser beam propagating through turbulent media, *Opt. Eng* 40(8): 1554–1562.

[5] Alexandropoulos, G. C., Sagias, N. C., Lazarakis, F. I. & Berberidis, K. [2009]. New results for the multivariate Nakagami-*m* fading model with arbitrary correlation matrix and applications, *IEEE Trans. Wireless Commun.* 8(1): 245–255.

[6] Andrews, L. [2004a]. Atmospheric optics, *SPIE Optical Engineering Press* .

[7] Andrews, L. [2004b]. Field guide to atmospheric optics, *SPIE Field Guides* FG02.

[8] Andrews, L., Al-Habash, M., Hopen, C. & Phillips, R. [2001]. Theory of optical scintillation: Gaussian beam wave model, *Waves in Random Media* 11: 271–291.

[9] Andrews, L. & Philips, R. [1985]. I-K distribution as a universal propagation model of laser beams in atmospheric turbulence, *Journal of the Optical Society of America A* 2(2): 160–163.

[10] Andrews, L. & Philips, R. [1986]. Mathematical genesis of the I-K distribution for random optical fields, *Journal of the Optical Society of America A* 3(11): 1912–1919.

[11] Andrews, L., Phillips, R. & Hopen, C. [2001]. Laser beam scintillation with applications, *SPIE Optical Engineering Press* .

[12] Arnon, S. [2003]. *Optical wireless communications*, Encyclopedia of Optical Engineering, New York: Marcel Dekker, pp. 1866–1886.

[13] Assimakopoulos, D., Lalas, D., Helmis, C. & Caroubalos, C. [1980]. An atmospheric turbulence probe, *IEEE Transactions on Geoscience and Remote Sensing* GE18: 347–353.

[14] Bayaki, E., Schober, R. & Mallik, R. [2009]. Performance analysis of MIMO free-space optical systems in gamma-gamma fading, *IEEE Trans. Commun.* 57(11): 3415–3424.

[15] Bithas, P. S., Sagias, N. C. & Mathiopoulos, P. T. [2009]. The bivariate generalized-K distribution and its application to diversity receivers, *IEEE Trans. Commun.* 57(9): 2655–2662.

[16] Bithas, P. S., Sagias, N. C., Mathiopoulos, P. T., Karagiannidis, G. K. & Rontogiannis, A. A. [2006]. On the performance analysis of digital communications over generalized-K fading channels, *IEEE Commun. Lett.* 10(5): 353–355.

[17] Bloom, S. [2001]. The physics of free-space optics, *AirFiber, Inc.* 2003.

[18] Bradley, E. & Antonia, R. [1979]. Structure parameters in the atmospheric surface layer, *Quarterly Journal of the Royal Meteorological Society* 105(445): 695–705.

[19] Castillo-Vazquez, C., Garcia-Zambrana, A. & Castillo-Vazquez, B. [2009]. Closed-form BER expression for FSO links with transmit laser selection over exponential atmospheric turbulence channels, *Electr. Lett.* 23(45).

[20] Caughey, S. & Palmer, S. [1979]. Some aspects of turbulence structure through the depth of the convective boundary layer, *Quarterly Journal of the Royal Meteorological Society* 105(446): 811–827.

[21] Chatzidiamantis, N. & Karagiannidis, G. [2011]. On the distribution of the sum of gamma-gamma variates and applications in RF and optical wireless communications, *IEEE Trans. Commun.* 59(5): 1298–1308.

[22] Chiani, M., Dardari, D. & Simon, M. K. [2003]. New exponential bounds and approximations for the computation of error probability in fading channels, *IEEE Trans. Wireless Commun.* 2: 840–845.

[23] Churnside, J. & Clifford, S. [1987]. Log-normal ricean probability density function of optical scintillations in the turbulent atmosphere, *Journal of the Optical Society of America A* 4(10): 1923–1930.

[24] Filho, F. C. M., Jayasuriya, D., Cole, R., Helmis, C. & Assimakopoulos, D. [1988]. Correlated humidity and temperature measurements in the urban atmospheric boundary layer, *Meteorol. Atmos. Phys.* 39: 197–202.
[25] Gagliardi, R. & Karp, S. [1995]. *Optical Communications*, John Wiley, New York.
[26] Garcia-Zambrana, A. [2007]. Error rate performance for STBC in free-space optical communications through strong atmospheric turbulence, *IEEE Commun. Lett.* 11(5): 390–392.
[27] Garcia-Zambrana, A., Castillo-Vazquez, C., Castillo-Vazquez, B. & Hiniesta-Gomez, A. [2009]. Selection transmit diversity for FSO links over strong atmospheric turbulence channels, *IEEE Photonics Technol. Lett.* 21(14): 1017–1019.
[28] Gradshteyn, I. & Ryzhik, I. M. [2000]. *Tables of Integrals, Series, and Products*, 6 edn, Academic Press, New York.
[29] Haas, S. M. & Shapiro, J. H. [2003]. Capacity of wireless optical communications, *IEEE J. Select. Areas Commun.* 21(8): 1346–1357.
[30] Henniger, H. & Wilfert, O. [2010]. An introduction to free-space optical communications, *Radioengineering* 19(2): 203–212.
[31] Ishimaru, A. [1978]. Wave propagation and scattering in random media, *Academic Press* 2.
[32] Kamalakis, T., Sphicopoulos, T., Muhammad, S. & Leitgeb, E. [2006]. Estimation of the power scintillation probability density function in free-space optical links by use of multicanonical monte carlo sampling, *Optics Letters* 31(21): 3077–3079.
[33] Karagiannidis, G. K., Zogas, D. A. & Kotsopoulos, S. A. [2003]. On the multivariate Nakagami-*m* distribution with exponential correlation, *IEEE Trans. Commun.* 51(8): 1240–1244.
[34] Karp, S., Gagliardi, R. M., Moran, S. E. & Stotts, L. [1988]. *Optical channels: fibers, clouds, water and the atmosphere*, New York: Plenum Press.
[35] Katsis, A., Nistazakis, H. E. & Tombras, G. S. [2009]. Bayesian and frequentist estimation of the performance of free space optical channels under weak turbulence conditions, *Journal of the Franklin Institute* 346: 315–327.
[36] Keddar, D. & Arnon, S. [2004]. Urban optical wireless communication networks: the main challenges and possible solutions, *IEEE Opt. Commun.* 42(5): 51–57.
[37] Kunkel, K. & Walters, D. [1982]. Intermittent turbulence in measurements of the temperature structure parameter under very stable conditions, *Boundary-layer meteorology* 22(1): 49–60.
[38] Laourine, A., Stephenne, A. & Affes, S. [2007]. Estimating the ergodic capacity of log-normal channels, *IEEE Communications Letters* 11(7): 568–570.
[39] Lawrence, R. & Strohben, J. [1970]. A survey of clear air propagation effects relevant to optical communications, *Proceedings of the IEEE* 58: 1523–1545.
[40] Lombardo, P. & Farina, A. [1996]. Coherent radar detection against K-distributed clutter with partially correlated texture, *Signal Process.* 48: 1–15.
[41] MacGovern, A., Nahrstedt, D. & Johnson, M. [2000]. Atmospheric propagation for tactical directed energy applications, *Proceedings SPIE* 4034: 128–139.
[42] Majumdar, A. [1984a]. Higher-order statistics of laser irradiance fluctuations due to turbulence, *Journal of the Optical Society of America* 1: 1067–1074.
[43] Majumdar, A. [1984b]. Uniqueness of statistics derived from moments of irradiance fluctuations in atmospheric optical propagation, *Optics Communications* 50(1): 1–7.

[44] Majumdar, A. K. [2005]. Free-space laser communication performance in the atmospheric channel, *J. Opt. Fiber Commun. Rep.* 2: 345–396.

[45] Mallik, R. K. [2003]. On multivariate Rayleigh and exponentials distributions, *IEEE Trans. Inf. Theory* 49(6): 1499–1515.

[46] Navidpour, S. M., Uysal, M. & Kavehrad, M. [2007]. Performance of free-space optical transmission with spatial diversity, *IEEE Trans. Wireless Commun.* 6(8): 2813–2819.

[47] Nistazakis, H., Assimakopoulos, V. & Tombras, G. [2011]. Performance estimation of free space optical links over negative exponential atmospheric turbulence channels, *Optik* 122(24): 2191–2194.

[48] Nistazakis, H. E. [in press, 2012]. A time-diversity scheme for wireless optical links over exponentially modeled turbulence channels, *OPTIK* .

[49] Nistazakis, H. E. & Tombras, G. [in press, 2012]. On the use of wavelength and time diversity in optical wireless communication systems over gamma-gamma turbulence channels, *Journal of Optics and Laser Technology* .

[50] Nistazakis, H. E., Tsiftsis, T. A. & Tombras, G. S. [2009]. Performance analysis of free space optical communication systems over atmospheric turbulence channels, *IET Communications* 3(8): 1402–1409.

[51] Nistazakis, H., Karagianni, E., Tsigopoulos, A., Fafalios, M. & Tombras, G. [2009]. Average capacity of optical wireless communication systems over atmospheric turbulence channels, *J. Lightw. Technol* 27(8): 974–979.

[52] Nistazakis, H., Katsis, A. & Tombras, G. [2012]. *On the Reliability and Performance of FSO and Hybrid FSO Communication Systems over Turbulent Channels*, Nova Publishers, Turbulence: Theory, Types and Simulation.

[53] Nistazakis, H., Tsigopoulos, A., Hanias, M., Psychogios, C., Marinos, D., Aidinis, C. & Tombras, G. [2011]. Estimation of outage capacity for free space optical links over I-K and K turbulent channels, *Radioengineering* 20(2): 493–498.

[54] Niu, M., Cheng, J. & Holzman, J. F. [2010]. Exact error rate analysis of equal gain and selection diversity for coherent free-space optical systems on strong turbulence channels, *Optics Express* 18(13): 13915–13926.

[55] Peppas, K. [2011]. A simple, accurate approximation to the sum of gamma-gamma variates and applications in MIMO free-space optical systems, *IEEE Photon. Technol. Lett.* 23(13): 839–841.

[56] Peppas, K. P., Alexsandropoulos, G. C., Datsikas, C. K. & Lazarakis, F. I. [2011]. Multivariate gamma-gamma distribution with exponential correlation and its applications in radio frequency and optical wireless communictions, *IET Antennas and Propag.* 5: 364–371.

[57] Peppas, K. P. & Datsikas, C. K. [2010]. Average symbol error probability of general-order rectangular quadrature amplitude modulation of optical wireless communication systems over atmospheric turbulence channels, *J. Opt. Commun. Netw.* 2(2): 102–110.

[58] Peppas, K. & Sagias, N. C. [2009]. A trivariate Nakagami-*m* distribution with arbitrary covariance matrix and applications to generalized selection diversity receivers, *IEEE Trans. Commun.* 57(7): 1896–1902.

[59] Popoola, W., Ghassemlooy, Z., Lee, C. & Boucouvalas, A. [2010]. Scintillation effect on intensity modulated laser communication systems - laboratory demonstration, *Optics and Laser Technology* 42: 682–692.

[60] Prudnikov, A. P., Brychkov, Y. A. & Marichev, O. I. [1986]. *Integrals and Series Volume 3: More Special Functions*, 1 edn, Gordon and Breach Science Publishers.

[61] Rachmani, R. & Arnon, S. [2010]. Wavelength diversity in turbulence channels for sensor networks, *IEEE 26th Convention of Electrical and Electronics Engineers in Israel, IEEEI* pp. 915–918.

[62] Razavi, M. & Shapiro, J. [2005]. Wireless optical communications via diversity reception and optical preamplification, *IEEE Trans. Wireless Commun.* 4: 975–983.

[63] Safari, M. & Uysal, M. [2008]. Do we really need space-time coding for free-space optical communication with direct detection, *IEEE Trans. Wireless Commun.* 7(11): 4445–4448.

[64] Sagias, N. C. & Karagiannidis, G. K. [2005]. Gaussian class multivariate Weibull distributions: Theory and applications in fading channels, *IEEE Trans. Inf. Theory* 51(10): 3608–3619.

[65] Sandalidis, H. G. & Tsiftsis, T. A. [2008]. Outage probability and ergodic capacity of free-space optical links over strong turbulence, *Electronics Letters* 44(1): 46–47.

[66] Sandalidis, H. G., Tsiftsis, T. A., Karagiannidis, G. K. & Uysal, M. [2008]. BER performance of FSO links over strong atmospheric turbulence channels with pointing errors, *IEEE Commun. Lett.* 12(1): 44–46.

[67] Sim, C. H. [1993]. Generation of poisson and gamma random vectors with given marginals and covariance matrix, *Journal of Statistical Computation and Simulation* 47(1): 1–10.

[68] Simon, M. K. & Alouini, M. S. [2005]. *Digital Communication over Fading Channels*, 2 edn, Wiley, New York.

[69] Tsiftsis, T. A., Sandalidis, H. G., Karagiannidis, G. K. & Uysal, M. [2009]. Optical wireless links with spatial diversity over strong atmospheric turbulence channels, *IEEE Trans. Wireless Commun.* 8(2): 951–957.

[70] Uysal, M., Navidpour, S. M. & Li, J. [2004]. Error rate performance of coded free-space optical links over strong turbulence channels, *IEEE Commun. Lett.* 8(10): 635–637.

[71] Vetelino, F., Young, S. & Andrews, L. [2007]. Fade statistics and aperture averaging for gaussian beam waves in moderate to strong turbulence, *Applied Optics* 46(18): 3780–3789.

[72] Wainright, E., Refai, H. H. & Jr., J. J. S. [2005]. Wavelength diversity in free-space optics to alleviate fog effects, *Proceedings of SPIE* 5712(16): 110–118.

[73] Wang, Z., Zhong, W.-D., Fu, S. & Lin, C. [2009]. Performance comparison of different modulation formats over free-space optical (FSO) turbulence links with space diversity reception technique, *IEEE Photnics Journal* 1(6): 277–285.

[74] Wilson, S. G., Brandt-Pearce, M., Cao, Q. & Baedke, M. [2005]. Optical repetition MIMO transmission with multipulse PPM, *IEEE J. Sel. Areas Commun.* 23: 1901–1910.

[75] Wilson, S. G., Brandt-Pearce, M., Cao, Q. & Leveque, J. H. [2005]. Free-space optical MIMO transmission with Q-ary PPM, *IEEE Trans. Commun.* 53(8): 1402–1412.

[76] Wyngaard, J. [1992]. Atmospheric turbulence, annual review of fluid mechanics, *Annual Review of Fluid Mechanics* 24: 205–234.

[77] Wyngaard, J., Izumi, Y. & Jr., S. C. [1971]. Behavior of the refractive-index-structure parameter near the ground, *Journal of the Optical Society of America A* 61(12): 1646–1650.

[78] Xu, F., Khalighi, A., Causse, P. & Bourennane, S. [2009]. Channel coding and time-diversity for optical wireless links, *Optics Express* 17(2): 872–887.

[79] Yacoub, M. D. [2007]. The α-μ distribution: A physical fading model for the Stacy distribution, *IEEE Trans. Veh. Technol.* 56(1): 27–34.

[80] Z. Wang, W.-D. Zhong, S. F. & Lin, C. [2009]. Performance comparison of different modulation formats over free-space optical (fso) turbulence links with space diversity reception technique, *IEEE Photonics Journal* 1(6): 277–285.

[81] Zhu, X. & Kahn, J. M. [2002]. Free-space optical communications through atmospheric turbulence channels, *IEEE Transaction on Communications* 50(8): 1293–1300.

[82] Zhu, X. & Kahn, J. M. [2003a]. Markov chain model in maximum-likelihood sequence detection for free-space optical communication through atmospheric turbulence channels, *IEEE Trans. Commun.* 51(3): 509–516.

[83] Zhu, X. & Kahn, J. M. [2003b]. Performance bounds for coded free-space optical communications through atmospheric turbulence channels, *IEEE Trans. Commun.* 51(8): 1233–1239.

[84] Zhu, X., Kahn, J. M. & Jin, W. [2003]. Mitigation of turbulence-induced scintillation noise in free-space optical links using temporal-domain detection techniques, *IEEE Photon. Technol. Lett.* 15(4): 623–625.

Chapter 17

Multiple-Input Multiple-Output (MIMO) Optical Wireless Communications

Oswaldo González

Additional information is available at the end of the chapter

1. Introduction

People's ideas about communications have changed completely, nowadays when this subject is mentioned almost everyone thinks of wireless communications. The demand for broadband wireless communications offering greater and greater data rates is endless, and the radio-technology community is trying harder and harder to satisfy this demand. Recently, there was the worldwide launch of 4th generation (4G) systems promising 100 Mbit/s for high mobility communications and up to 1 Gbit/s for stationary or low mobility communications. The key to this technology is the combination of orthogonal frequency-division multiple access (OFDMA) applied to multiple-input multiple-output (MIMO) systems.

On the other hand, researchers in wireless optical communications (WOC) are trying to find a way to gain the interest of communication companies by providing new and attractive alternatives to radio communications, as we must not lose sight of the fact that most wireless communications are established inside rooms. Thus, WOC systems offer some advantages over their radio-frequency (RF) counterparts [15]: they are, theoretically, unregulated and have unlimited[1] bandwidth. There is also an inherent security capability, as light (communication) is confined to the room, and there is immunity to multipath fading. However, they are not exempt from drawbacks: strict power limitations due to eye-safety constraints, severe path losses and multipath dispersion and, last but not least, limited maximum achievable signal-to-noise ratio (SNR) due to unavoidable natural and artificial noise sources are the main problems.

Over the last few years, OFDM has begun to be proposed for both fibre and wireless optical communications [4] as an effective solution to mitigating inter-symbol interference (ISI) caused by dispersive channels. Furthermore, the frequency-domain channel equalization provided by an OFDM system does not undergo severe complexity penalty when data rates and dispersion increase as opposed to serial time-domain equalizers, and MIMO techniques can be applied to these systems with relative ease. Finally, the complexity of transmitters

[1] This is not strictly true for new visible-light communication (VLC) systems, where LEDs present a transmission bandwidth limited to several MHz

and receivers is transferred from an analogue to a digital domain by employing Fast Fourier Transform (FFT) and Inverse FFT (IFFT) blocks as demodulators and modulators, respectively. Therefore, all these aspects favour the implementation of OFDM systems in the current digital era.

This chapter describes the characteristics of MIMO-OFDM systems applied to WOC, discussing their benefits, but also their drawbacks, as compared with other techniques used in order to obtain high-capacity optical data networks.

2. The optical channel model

In wireless optical communications, the optical link is typically established by means of *optical modulation* (IM), in which the desired waveform is modulated onto the instantaneous power of the carrier, in conjunction with *direct detection* (DD) as a down-conversion technique at the receiver end. Therefore, the transmitted waveform $x(t)$ is the instantaneous optical power of the emitter, and the received waveform $y(t)$ is the instantaneous current in the receiving photodetector. In this way, the optical channel with IM/DD can be modelled as a *baseband* linear system with impulse response $h(t)$ or, alternatively, this can be described in terms of the frequency response:

$$H(f) = \int_{-\infty}^{\infty} h(t)e^{-j2\pi ft}dt \tag{1}$$

which is the Fourier transform of $h(t)$. This channel model $h(t)$ is practically stationary because it only varies when emitter, receiver or objects in the room are moved by tens of centimetres. In many applications, optical links are operated in the presence of intense infrared and visible background light. The received background light adds shot noise, which is usually the limiting noise source in a well-designed receiver. Due to its high intensity, this shot noise can be modelled as white, Gaussian, and independent of $x(t)$. When little or no ambient light is present, the dominant noise source is receiver preamplifier noise, which is also signal-independent and Gaussian (though often nonwhite) [15]. Thus, the noise $n(t)$ is usually modelled as Gaussian and signal-independent, and the instantaneous output current at the receiver can be represented as:

$$y(t) = Rx(t) \otimes h(t) + n(t) \tag{2}$$

where the "$\otimes$" symbol denotes convolution and R is the detector responsivity (A/W). According to (2), the optical link can be completely characterized by means of the impulse response $h(t)$ and the noise sources $n(t)$. The knowledge of $h(t)$ allows us to determine the multipath penalty, which limits the maximum baud rate. The second term is related to the signal-to-noise ratio (SNR), which determines the performance of the digital link.

In order to evaluate the impulse response on indoor wireless optical channels, several deterministic methods were first proposed [5]. However, these methods can only be implemented to determine the impulse response until the third reflection due to their computational complexity. Later on, modified Monte Carlo-based ray-tracing algorithms were introduced, which present a lower computational cost and without limit to the number of reflections being considered [10].

In these algorithms, ray directions are randomly generated according to the radiation pattern from the emitter. The contribution of each ray from the source or after a bounce to the receiver is computed deterministically. Consequently, the discretisation error is due to the number of random rays. The line-of-sight (LOS) and multiple-bounce impulse responses are considered when calculating the total impulse response.

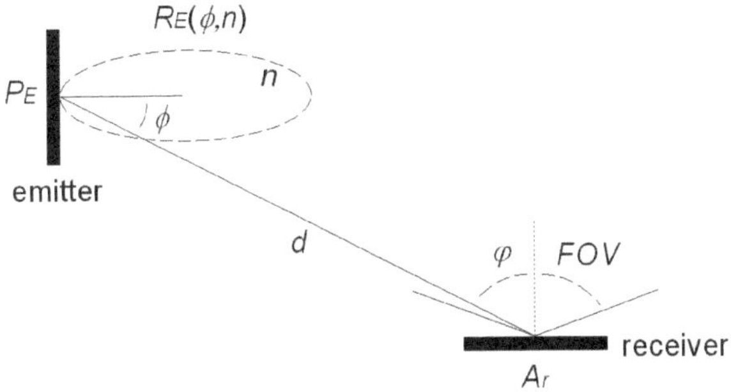

Figure 1. LOS contribution to the total received power

2.1. LOS impulse response

Given an emitter E and receiver R in an environment without reflectors (see Fig. 1), with a large distance d between both, the received power is approximately:

$$P_R = \frac{1}{d^2} R_E(\phi, n)\, A_{\text{eff}}(\varphi) \tag{3}$$

where the emitter is modelled using a generalized Lambertian radiation pattern $R_E(\phi, n)$. $A_{\text{eff}}(\varphi)$ represents the effective signal collection area of the receiver.

$$R_E(\phi, n) = \tfrac{n+1}{2\pi} P_E \cos^n(\phi), \; 0 \le \phi \le \frac{\pi}{2} \tag{4}$$

$$A_{\text{eff}} = A_r \cos\varphi \; \text{rect}\left(\tfrac{\varphi}{\text{FOV}}\right) \tag{5}$$

$$\text{rect}(x) = \begin{cases} 1, & |x| \le 1 \\ 0, & |x| > 1 \end{cases} \tag{6}$$

Here n is the *order* or *mode number* of the radiation lobe that specifies the directionality of the emitter, P_E the power radiated by the emitter, A_r the physical area of the receiver, and FOV the receiver field of view (half-angle from the surface normal).

2.2. Multiple-bounce impulse response

If we now consider an emitter E and receiver R in a room with reflectors, the radiation from the emitter can reach the receiver after any number of reflections (see Fig. 2). In the algorithm, many rays are generated at the emitter position with a probability distribution equal to its normalized radiation pattern $R_E(\phi, n)/P_E$. The power of each generated ray is initially P_E/N, where N is the number of rays used to discretise the source. When a ray impinges on a surface, the reflection point is converted into a new optical source, thus, a new ray is generated with a probability distribution provided by the reflection pattern of that surface, $R_S(\theta, \theta')$. The process continues during the simulation time t_{max}. After each reflection, the power is reduced

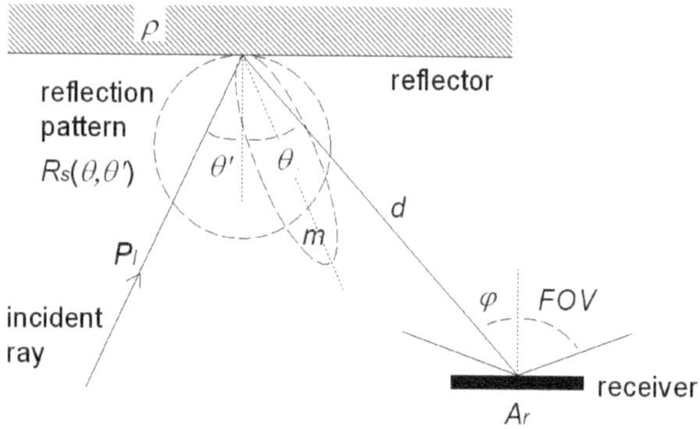

Figure 2. Geometry of emitter and receiver with reflectors. Reflection pattern of the surface is described by Phong's model

by the reflection coefficient of the surface ($P_{\text{new ray}} = \rho P_I$), and the reflected power reaching the receiver is computed by:

$$P_R = \frac{1}{d^2} R_S\left(\theta,\theta'\right) A_{\text{eff}}\left(\varphi\right) \tag{7}$$

where d is the distance between the reflection point and receiver, and $R_s\left(\theta,\theta'\right)$ is Phong's model, used to describe the reflection pattern of the surface [21]. This model is able to approximate the behaviour of those surfaces that present a strong specular component. It considers the reflection pattern as a sum of both diffuse and specular components. In this way, surface characteristics are defined by two new parameters, the percentage of incident signal that is reflected diffusely r_d and the directivity of the specular component of the reflection m. This model is described by:

$$R_S\left(\theta,\theta'\right) = \rho P_I \left[r_d \frac{\cos\theta}{\pi} + \left(1 - r_d\right) \frac{m+1}{2\pi} \cos^m\left(\theta - \theta'\right) \right] \tag{8}$$

where ρ is the surface reflection coefficient, P_I represents the optical power of the incident ray, θ is the observation angle, and θ' represents the incidence angle (see Fig. 2).

2.3. The MIMO optical channel

In a typical multi-user application, several emitters can be placed in the room and an *angle-diversity* receiver, composed of multiple receiving elements oriented in different directions, could be used. By using (7), the contribution of the ith ray emitted by the lth user to the received power reaching each jth branch of the receiver during a certain time interval ($p^{(l)}_{j,i,k}$, k^{th} time interval) can be computed. The total received power at the jth branch of the optical detector in the kth time interval (width Δt) is computed as the sum of the power of the $N^{(l)}_{j,k}$ rays that contribute in that interval:

$$p^{(l)}_{j,k} = \sum_{i=1}^{N^{(l)}_{j,k}} p^{(l)}_{j,i,k} \tag{9}$$

If we consider a normalized receiver responsivity of 1 A/W, the impulse responses $h_j^{(l)}(t)$ at all branches ($j = 1, \ldots, P$) due to the lth user ($l = 1, \ldots, L$) are given by:

$$h_j^{(l)}(t) = \sum_{k=0}^{K-1} p_{j,k}^{(l)} \delta(t - k\Delta t) \tag{10}$$

where $K = t_{\max}/\Delta t$, and where we have assumed as the time origin the instant when the rays are generated from the emitter. This process must be repeated in order to obtain the different impulse responses between each emitter and each receiving branch in the multi-user scenario.

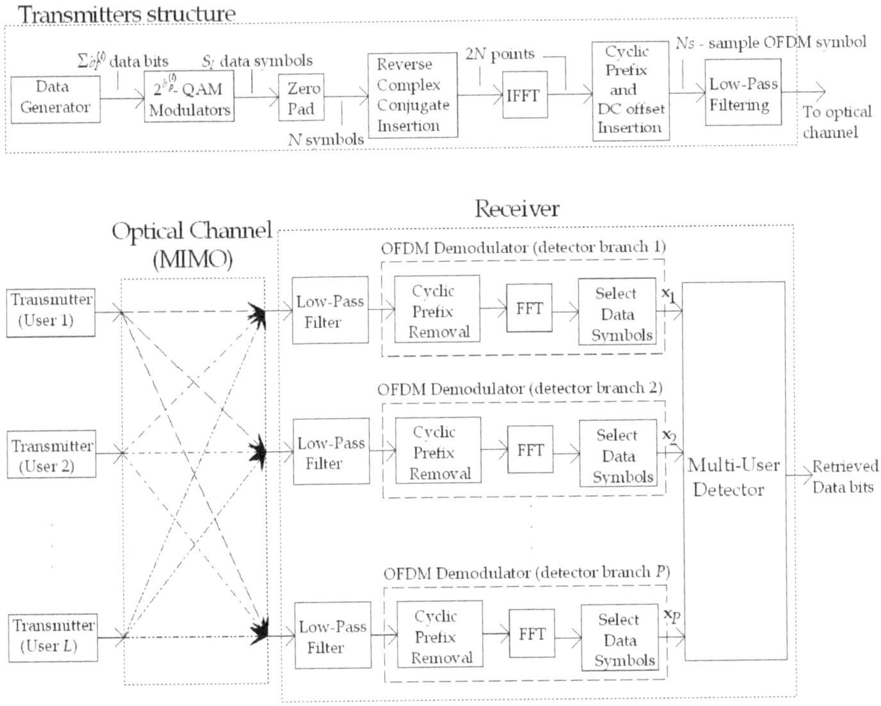

Figure 3. Multi-user MIMO-OFDM system for optical wireless communications

3. The MIMO-OFDM system

Fig. 3 shows the block diagram of the MIMO-OFDM system for multi-user communications over an indoor wireless optical channel. As we can see in the transmitter structure, $\sum_p b_p^{(l)}$ data bits are generated by l^{th} user and modulated by using an appropriate $2^{b_p^{(l)}}$-QAM (quadrature amplitude modulation) modulator, $b_p^{(l)}$ being the number of bits conveyed by the pth subcarrier ($1 \leq p \leq S_l$). In general, we will assume $b_p^{(l)} = B$ for every subcarrier p and user l, but we could accommodate different number of bits to each subcarrier and user, adapting the throughput to the frequency characteristics of the channel [9]. Then, $N - S_I$ zeros are padded to this set of S_I symbols generated by the modulators, the first zero being added before, corresponding to the direct-current (DC) value, and the remaining ones after. Finally, the complex conjugate of the mirror of the word of N symbols is added before computing the IFFT. The outgoing IFFT symbol is a real sequence of $2N$ points, which can be made non-negative by adding an appropriate DC level. Moreover, a cyclic prefix (N_e-sample extension) is inserted to combat ISI, leading to a transmitted signal with a total of $N_S = 2N + N_e$ samples and duration T_S. This positive and real signal can modulate the intensity of light emitted by the optical source. Note that the cyclic extension prefix will be ISI-corrupted due to the low-pass channel response and will be discarded at receiver. Next, the FFT of the remaining $2N$ points will be computed at each receiving branch, and only the first S_I points (after the DC value) will be considered by the multi-user (MU) detector.

As we have L transmitters and the receiver possesses P receiving branches, the optical channel constitutes a MIMO system, whose individual impulse responses are given by (10). We can consider that, after the FFT processing carried out by each branch of a new received OFDM symbol, the P-branch detector array provides the MU detector with a vector of complex symbols $\mathbf{x}_p$ at p^{th} subcarrier, which is the superposition of the independently distorted signals associated with the L users sharing the same space-frequency resource and also corrupted by noise at the detector array elements:

$$\mathbf{x}_p = \mathbf{H}_p \mathbf{s}_p + \mathbf{n}_p \tag{11}$$

where the vector $\mathbf{x}_p \in \mathbb{C}^{P \times 1}$ of received signals at the pth subcarrier, the vector $\mathbf{s}_p \in \mathbb{C}^{L \times 1}$ of transmitted signals and the array noise vector $\mathbf{n}_p \in \mathbb{C}^{P \times 1}$, respectively, are given by:

$$\begin{aligned} \mathbf{x}_p &= (x_{1p}, x_{2p}, \ldots, x_{Pp})^{\mathrm{T}} \\ \mathbf{s}_p &= (s_p^{(1)}, s_p^{(2)}, \ldots, s_p^{(L)})^{\mathrm{T}} \\ \mathbf{n}_p &= (n_{1p}, n_{2p}, \ldots, n_{Pp})^{\mathrm{T}} \end{aligned} \tag{12}$$

where $(\,)^{\mathrm{T}}$ means the transpose of $(\,)$, then $\mathbf{x}_p$, $\mathbf{s}_p$ and $\mathbf{n}_p$ are all column vectors. The frequency domain channel transfer function factor matrix $\mathbf{H}_p \in \mathbb{C}^{P \times L}$ is composed of the set of channel transfer vectors $\mathbf{H}_p^{(l)} \in \mathbb{C}^{P \times 1}, l = 1, \ldots, L$ of the L users:

$$\mathbf{H}_p = (\mathbf{H}_p^{(1)}, \mathbf{H}_p^{(2)}, \ldots, \mathbf{H}_p^{(L)}) \tag{13}$$

each of which hosts the frequency domain channel transfer factor between the single emitter source associated with a particular user l and the receiving branches $j = 1, \ldots, P$ at the pth subcarrier band:

$$\mathbf{H}_p^{(l)} = (H_{1p}^{(l)}, H_{2p}^{(l)}, \ldots, H_{Pp}^{(l)})^{\mathrm{T}} \tag{14}$$

with $l = 1, \ldots, L$. Note that the frequency domain channel transfer factors $H_{jp}^{(l)}, \forall p$ between the lth emitter and the jth receiving branch are obtained by computing the Fourier transform (1) at the subcarrier frequency $f = pf_0$ over the corresponding impulse response (10), where $f_0 = 1/T$. Here, T is the time duration of the part of the OFDM symbol which actually conveys information, i.e. the OFDM symbol duration excluding the cyclic prefix extension, $T = 2NT_S/N_S = 2NT_S/(2N + N_e)$.

3.1. Optimum multi-user detection

The optimum detector, when a *maximum a posteriori probability* (MAP) criterion is considered and all the symbols are assumed, *a priori*, to have equal probability of being transmitted, is the *maximum likelihood* (ML) detector [26]. This detector finds the specific L-dimensional vector of M-ary symbols ($M = 2^B$), which is more likely to have been transmitted. In formal terms, ML detection is based on the idea of maximising the *a posteriori* probability $P(\check{\mathbf{s}}_p \,|\mathbf{x}_p, \mathbf{H}_p)$ that a specific vector $\check{\mathbf{s}}_p \in \mathbb{C}^{L\times 1}$ of the different users' symbols was transmitted over the MIMO channel characterised by the channel transfer factor matrix $\mathbf{H}_p \in \mathbb{C}^{P\times L}$ under the condition that the vector $\mathbf{x}_p \in \mathbb{C}^{P\times 1}$ was received at the MU detector end. Note that each evaluated vector $\check{\mathbf{s}}_p$ is an element of a set $\mathcal{M}_p^L$ with a total of M^L possible trial-vectors. The aforementioned maximisation can be expressed as:

$$\check{\mathbf{s}}_{p,\text{ML}} = \arg\max_{\check{\mathbf{s}}_p \in \mathcal{M}_p^L} P(\check{\mathbf{s}}_p \,|\mathbf{x}_p, \mathbf{H}_p) \tag{15}$$

For the optical AWGN channel, maximising $P(\check{\mathbf{s}}_p \,|\mathbf{x}_p, \mathbf{H}_p)$ is equivalent to minimising the Euclidean distance metric $\|\mathbf{x}_p - \mathbf{H}_p\check{\mathbf{s}}_p\|_2^2$, $\forall \check{\mathbf{s}}_p \in \mathcal{M}_p^L$ [12], and hence we have:

$$\check{\mathbf{s}}_{p,\text{ML}} = \arg\max_{\check{\mathbf{s}}_p \in \mathcal{M}_p^L} P(\check{\mathbf{s}}_p \,|\mathbf{x}_p, \mathbf{H}_p) \Leftrightarrow \check{\mathbf{s}}_{p,\text{ML}} = \arg\min_{\check{\mathbf{s}}_p \in \mathcal{M}_p^L} \|\mathbf{x}_p - \mathbf{H}_p\check{\mathbf{s}}_p\|_2^2 \tag{16}$$

where $\| \, \|_2^2$ denotes the squared Euclidean norm. The ML detection is excessively costly for typical values of M-ary modulation and number of transmitting users L, since the receiver must evaluate a total of M^L Euclidean distance calculations (16) for each pth subcarrier, with S_I being the total number of subcarriers. For example, for 16-QAM modulation and $L = 3$ users, the total number of operations by each subcarrier is 4 096, which is impractical. Therefore, sub-optimum detection techniques have been devised in order to reduce the dimensionality of the classification problem associated with selecting the specific constellation point, which is most likely to have been transmitted by each user. Linear detection techniques are based on obtaining linear estimates of the different users' transmitted signals with the aid of the weighted combining of the signals received by the different detector elements. This is followed by separately demodulating each of the L different users' combiner output signals. Thus, the original M^L-dimensional classification problem associated with the optimum ML detection is reduced to L individual classification steps, each having a dimensionality of M. In fact, the corresponding receiver associated with the lth transmitting user only has to carry out one of these L classifications to retrieve the data that concerns it. However, when using optimal ML detection, the receiver has to carry out M^L evaluations of (16) in spite of being concerned with only the data transmitted by its corresponding l^{th} user.

3.2. Least squares error detector

Using a linear detector, an estimate $\hat{\mathbf{s}}_p \in \mathbb{C}^{L\times 1}$ at the pth subcarrier band of the vector of transmitted signals $\mathbf{s}_p$ of the L simultaneous users is generated by linear combining the signals received by the P different receiving branches with the aid of a weight matrix $\mathbf{W}_p \in \mathbb{C}^{P\times L}$:

$$\hat{\mathbf{s}}_p = \mathbf{W}_p^H \mathbf{x}_p \tag{17}$$

where $\mathbf{W}_p^H$ denotes the complex conjugate matrix of the matrix $\mathbf{W}_p$. When least-squares (LS) error detector is considered, also called zero-forcing (ZF) combiner, the weight matrix $\mathbf{W}_{p,\mathrm{LS}} \in \mathbb{C}^{P\times L}$ is given by [12]:

$$\mathbf{W}_{p,\mathrm{LS}} = \mathbf{H}_p \left(\mathbf{H}_p^H \mathbf{H}_p\right)^{-1} \tag{18}$$

By substituting the received signal's model of (11) and the LS estimation based weight matrix (18) into (17), we obtain:

$$\hat{\mathbf{s}}_{p,\mathrm{LS}} = \mathbf{s}_p + \mathbf{W}_{p,\mathrm{LS}}^H \mathbf{n}_p \tag{19}$$

which indicates that the LS-estimate $\hat{\mathbf{s}}_{p,\mathrm{LS}}$ of the transmitted signal vector $\mathbf{s}_p$ of the L simultaneous users is an unbiased noise-contaminated version of $\mathbf{s}_p$, so that $E\left\{\hat{\mathbf{s}}_{p,\mathrm{LS}}\right\} = \mathbf{s}_p$. When using the LS combiner, the lth user's associated vector component of estimate (17) of the vector of transmitted signals at the pth subcarrier can be expressed as:

$$\hat{s}_{p,\mathrm{LS}}^{(l)} = \mathbf{W}_{p,\mathrm{LS}}^{(l)H} \mathbf{x}_p \tag{20}$$

where the lth user's associated weight vector $\mathbf{W}_{p,\mathrm{LS}}^{(l)} \in \mathbb{C}^{P\times 1}$ coincides with the lth column vector of the matrix $\mathbf{W}_{p,\mathrm{LS}}$. The complex symbol that is most likely to have been transmitted by the lth user can be determined upon minimising the Euclidean distance between estimate (20) of the transmitted signal obtained at the lth user's combiner output and all the constellation points associated with the specific modulation scheme employed. This only requires M evaluations as compared to the M^L ones of the optimum ML detector.

3.3. Frequency-domain channel transfer factor matrix estimation

From (18) and (20), we can observe that the LS detector requires the knowledge of the transfer factor matrix $\mathbf{H}_p$. The method described in [8], where TS known training sequences are used to estimate the channel transfer function between a single user and the receiver, can be used to estimate the matrix $\mathbf{H}_p$, assuming that only one user is transmitting in a specific instant during the training periods. If a certain known training sequence X_p is transmitted by the lth user over the slowly time-varying wireless optical channel, the channel response of each sub-band $H_{jp}^{(l)}$ for this user at the jth receiving branch can be estimated from the S_I received symbols ($Y_{jp}^{(l)}$) as follows:

$$\hat{H}_{jp}^{(l)} = Y_{jp}^{(l)} / X_p \tag{21}$$

Note that the $\hat{H}_{jp}^{(l)}$ values are distorted by noise and therefore several training sequences should be transmitted, and then the mean values of the estimates can be used to obtain a better channel characterisation. In fact, by using ten training sequences, the channel response estimation given by (21) is very close to the actual channel response, and we will obtain a system performance very similar to that of the ideal case when a perfect knowledge of the channel response is available [8].

4. Simulation results

In this section, the previously described multi-user OFDM system is evaluated by means of simulations with a maximum number of $L = 6$ simultaneous users. The simulated room is 7.5 m long ×5.5 m wide ×3.5 m high [5] (see Table 1 for additional information about the room's characteristics and other simulation parameters: the north wall is the first you come across if you enter the room along the x-direction). The emitters are attached to the ceiling at different positions and their main lobes are pointed straight down (see Fig. 4). The emitters have three lobes of radiation modelled as second order Lambertian (half-power beam width, HPBW = 45°): their side lobes are titled at 45°pointing in opposite directions with respect to the ceiling normal and one of them points towards the centre of the room floor, where the receiver is located. The receiver consists of $P = 6$ branches of equal field of view (FOV), $\phi_b = 40°$. The central branch is oriented towards the ceiling, whereas the five side branches are tilted at 45°and distributed uniformly at 72°in azimuth, as depicted in Fig. 4. Table 2 summarises the main characteristics of emitters and receiver (in the table, γ_i is the angle subtended between the x-axis and an imaginary line drawn from the position of the ith emitter towards the centre of the ceiling).

Room parameters	
length (x):	7.5 m
width (y):	5.5 m
height (z):	3.5 m
ρ_{NORTH}:	0.30
ρ_{SOUTH}:	0.56
ρ_{EAST}:	0.30
ρ_{WEST}:	0.12
$\rho_{CEILING}$:	0.69
ρ_{FLOOR}:	0.09
Simulation parameters	
Number of rays (N):	1 000 000
Number of maximum reflections (K):	20
Resolution (Δt):	0.2 ns
Simulation time (t_{max}):	100 ns

Table 1. Parameters for simulation

Emitters Locations: ($x\ y\ z$) m	**Radiation pattern**	**Receiver**
		Position ($x\ y\ z$) m: (3.75 2.75 0.75)
Emitter 1: (0.75 0.75 3.5)	Mode(n): 2	Area: 1 cm^2
Emitter 2: (6.75 0.75 3.5)	Number of radiation lobes: 3	FOV: 40°
Emitter 3: (3.75 2.75 3.5)	Orientation (elevation,	Orientation of detector branches
Emitter 4: (0.75 4.75 3.5)	azimuth):	(elevation, azimuth):
Emitter 5: (3.75 4.75 3.5)	(135° γ_i), (180° 0°),	(0° 0°), (45° 0°), (45° 72°)
Emitter 6: (6.75 4.75 3.5)	(135° $\gamma_i + 180°$)	(45° 144°), (45° 216°), (45° 288°)

Table 2. Parameters of emitters and receiver

The channel impulse responses from each emitter (user) to each receiving branch have been computed by using the modified Monte Carlo based ray-tracing algorithm described in section 2. We can easily anticipate that the impulse responses of the main channel between a certain user and the receiver will have a predominant LOS contribution because there is always at least one branch of the receiver directly observing every point on the ceiling (see observed impulses responses at each receiving branch in Fig. 5). This configuration could be considered as not sufficiently exploiting the angle-diversity provided by the receiver and

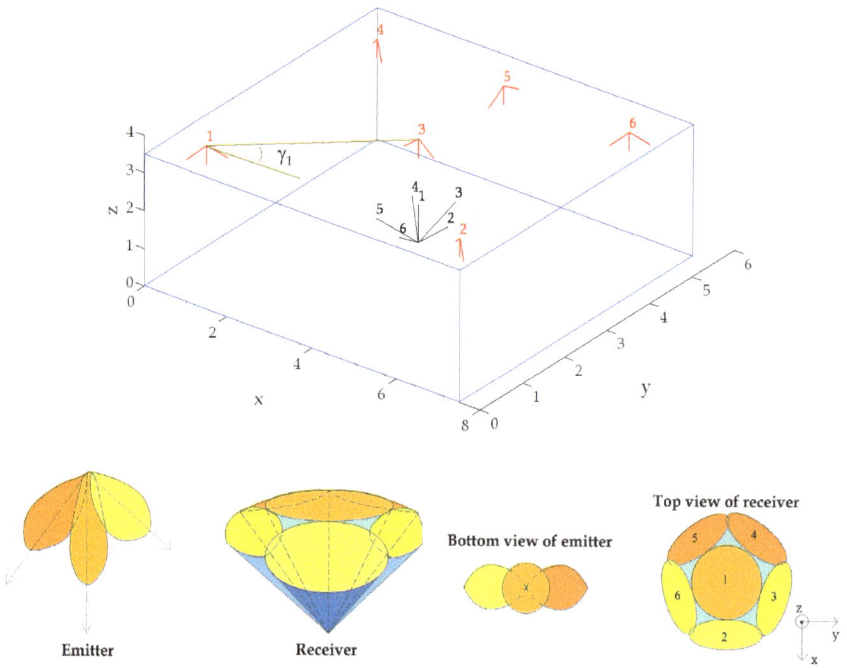

Figure 4. Location and structure of the six transmitters and the receiver in the simulation room

that non-LOS links should also be taken into account. However, the combination of the angle-diversity receiver with the MU detector enables multi-user communication whenever the emitters are moved apart, so the receiver does not see them as quasi-identical light sources. In addition, the LS combiner also requires that the maximum number L of simultaneous users does not outweigh the number P of receiving branches [12]. Therefore, we have considered LOS links as a more realistic situation, especially for a potential system based on visible LED lighting (VLC: *Visible Light Communications*). The main advantage of LOS links is the remarkably lower transmission power required to obtain the same performance as their non-LOS counterparts. However, it is vulnerable to shadowing: the LOS link can be blocked momentarily by the presence of an obstacle or a human between emitter and receiver. This can be solved by using redundant emitters or providing the system with more directive emitters (larger lobe order n) and receivers (narrower FOVs) where shorter quasi-vertical links prevail over longer distance smaller-angle horizontally-tilted ones, which are prone to blocking. The shorter quasi-vertical links allow us to establish different communication networks inside a room as their high directivity avoids interferences with neighbouring zones beyond a few metres. This is the underlying idea in VLC, where illumination lamps are used to enable transmission mainly with the receiving devices just underneath them. However, despite having chosen to evaluate LOS links in this chapter as a more convenient study of future applications, the results presented here can be easily extrapolated to non-LOS configurations. The only difference is that, in the latter, the emissions from one user usually affect several branches (three or more) at the receiver, which demands a greater "effort" from the MU detector to "separate" the different signals received. Hence, a greater signal-to-noise ratio

(SNR) will be required, as compared to the LOS counterparts, to obtain a certain level of performance. Moreover, we must not forget that greater transmission powers are also needed to reach those required SNR values. Typical non-LOS configurations would require up to ten decibels more in the SNR at the receiver than LOS links [9], which would make them impractical in many applications.

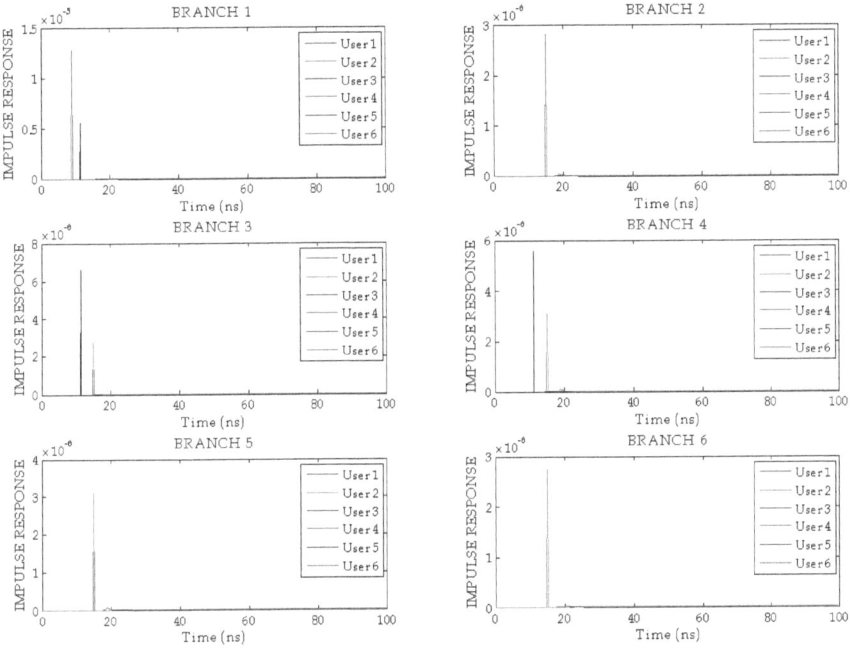

Figure 5. Impulses responses between the transmitting users and each receiving branch

In all the simulation results presented below, the number of subcarriers is $N = 64$ (128-point FFT), but only $S_I = 52$ convey data (2^B-QAM modulated). A cyclic prefix extension of $N_e = 32$ samples was used and the transmission symbol rate was 1 Msymbol/s, which leads to a total system throughput of $B \times S_I \times L$ Mbit/s (e.g. 1.248 Gbit/s for $L = 6$ users with every subcarrier 16-QAM modulated). This OFDM symbol extension ($N_e/(2N + N_e) = 200$ ns) is large enough to compensate for ISI in most of the non-LOS diffuse links scenarios, hence its length could be reduced up to four times if only LOS configurations are considered (in this case, the efficiency of the OFDM transmission $2N/(2N + N_e)$ would increase from 80% to roughly 95%). However, here we have considered a more general system, valid for practically any kind of optical link. In this case, the separation between subcarriers is $\Delta f = f_0 =$ 1.25 MHz, leading to a transmission bandwidth of approximately $B_T = S_I \Delta f = 65$ MHz plus a convenient out-of-band guard [8], which is extremely practical for typical optical devices. Finally, we used ten training sequences to estimate (using the method described in section 3.3) the frequency-domain channel transfer function factors of matrix $\mathbf{H}_p$ required by LS detector to obtain the weight matrix (18) or for evaluating (16) in the case of ML detector.

4.1. Comparison of maximum-likelihood and least-squares detection

In this subsection, the bit error rate (BER) performance of the multi-user detector based on the maximum-likelihood criterion is compared with that of least-squares error detector for $L = 3$ simultaneous users. Fig. 6 shows the BER performance as a function of the number B of transmitted bits per subcarrier (BPSK, QPSK and 16-QAM are considered). For reference purposes, the figure also depicts the results for the case of single-user transmission ($L = 1$) in contrast to those of the multi-user ($L = 3$) scenario. As mentioned in subsection 3.1, ML detection would require $2^{BL} = 4\,096$ classifications for each subcarrier when 16-QAM is considered for $L = 3$ users or when addressing QPSK for $L = 6$ users, both situations being impractical in a real application. Therefore, we are not able to assess the $L_{\max} = P = 6$ users scenario in this comparison because it would require $2^{4\times 6} = 16\,777\,216$ classifications per subcarrier for 16-QAM, a prohibitive number even in a simulation. Thus, the results shown have been obtained for the fifth user of the communication scenario illustrated in Fig. 4 when users 4 and 6 are also transmitting. This constitutes one of the worst-case situations, since emissions from user 5 are strongly interfered at receiving branches 1, 3 and 4 by other users (see Fig. 5), especially by users 1, 4 and 6.

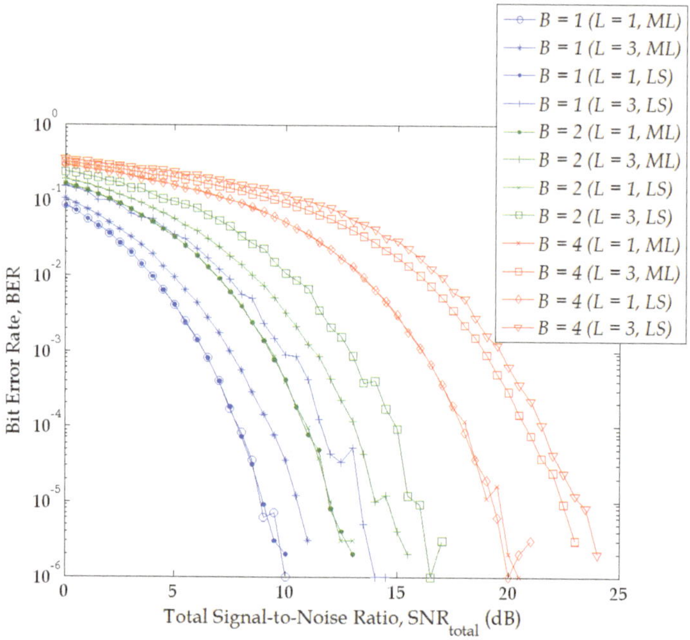

Figure 6. Performance comparison of ML and LS detection for the fifth user (worst case) when the subcarriers are 2^B-QAM modulated

For the single-user scenario, $\mathbf{H}_p \in \mathbb{C}^{P\times 1}$ is a column-vector which leads to the LS estimation based weight matrix $\mathbf{W}_{p,\mathrm{LS}}$ being reduced to:

$$\mathbf{W}_{p,\mathrm{LS}}^{(L=1)} = \mathbf{H}_p \left(\mathbf{H}_p^H \mathbf{H}_p\right)^{-1} = \frac{\mathbf{H}_p}{\mathbf{H}_p^H \mathbf{H}_p} = \frac{\mathbf{H}_p}{\|\mathbf{H}_p\|^2} \tag{22}$$

where $\|\mathbf{H}_p\|^2 = H_{1p}^H H_{1p} + H_{2p}^H H_{2p} + \ldots + H_{Pp}^H H_{Pp}$ is a real number. Therefore, for the single-user case, the estimation of the transmitted symbol s_p at the pth subcarrier given by (17) yields:

$$\hat{s}_p^{(L=1)} = \mathbf{W}_{p,\text{LS}}^{(L=1)H} \mathbf{x}_p = \frac{\mathbf{H}_p^H \mathbf{x}_p}{\|\mathbf{H}_p\|^2} \tag{23}$$

Minimising the Euclidean distance between the previous estimate $\hat{s}_p^{(L=1)}$ and all the constellation points associated with the specific modulation scheme employed is equivalent to the evaluation of (16) carried out by the ML detector. Hence, both ML and LS methods perform equally and optimally in the single-user scenario. This is why we can observe identical results for a given $2^B - QAM$ modulation scheme in the $L = 1$ cases irrespective of the detection method used. However, when receivers deal with a multi-user scenario, the ML detector outperforms the LS-based one by several decibels. Nonetheless, the greater the modulation scheme employed, the smaller the SNR gain obtained by the optimum receiver with respect to that of the suboptimal LS detector. Thus, we can see how SNR gains of about 3 dB for BPSK are reduced to less than 1 dB when 16-QAM is considered. In addition, taking into account that the LS detector only requires 4 and 16 classifications associated with a specific user for QPSK and 16-QAM, respectively, in contrast to the 64 and 4 096 needed by the ML detector for $L = 3$ users, we can conclude that the simplicity and fairly close to optimum performance of LS detection makes it a practical and better choice for multi-user communication scenarios.

4.2. Performance of multi-user LS detector

Fig. 7 draws together the simulation results obtained for each lth user depicted in Fig. 4 for two situations: when only the desirable user is transmitting (ideal case, specified as $L = 1$ in the legend) and for the full-user scenario, i.e. when all the $L = L_{\text{max}} = 6$ users are transmitting simultaneously.

In Fig. 7(a), the BER is depicted versus the total signal-to-noise ratio, the latter obtained as $\sum_{j=1}^{P} \text{SNR}_j^{(l)}$ ($\text{SNR}_j^{(l)}$ is the SNR at j^{th} receiving branch for the lth user). It can be seen how all the users present the same performance under the single-user scenario, which coincides with that of 16-QAM OFDM systems over ideal additive white Gaussian noise (AWGN) channels [8]. We can also observe how the fifth user ($l = 5$) performs the worst under the multi-user ($L = 6$) scenario, requiring about ten decibels more than the single-user case. In contrast, the first user ($l = 1$) presents a SNR loss of hardly 1 dB between multi-user and single-user scenarios. The previous results are logical, since the fifth user is located in the worst position of the room, sharing several receiving branches at the detector with other users as we mentioned previously, whereas the first emitter is alone in a corner and the only one that illuminates the fifth receiving branch of the detector (see Fig. 4 and 5).

Despite Fig. 7(a) showing a performance loss of the multi-user system with respect to the single-user scenario, it does not allow us to discover the required SNR at each branch to obtain the observed performance. Fig. 7(b) shows the same results as before, but representing the BER versus the maximum SNR at the receiver for the lth user, i.e. $\max(\text{SNR}_j^{(l)}), j = 1, \ldots, P$. In this graph, we can see how users $l = 3, 4$ and 5 present a very similar performance when facing the multi-user scenario, i.e. all of them require approximately the same SNR at the branch which receives the most light power from the corresponding user in order to obtain a given

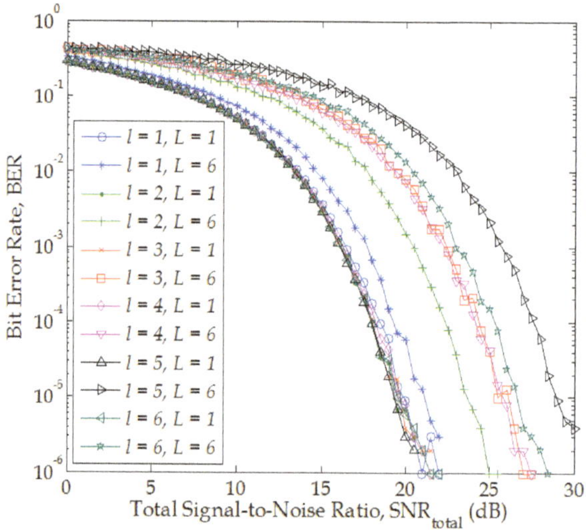

(a) BER vs. Total SNR

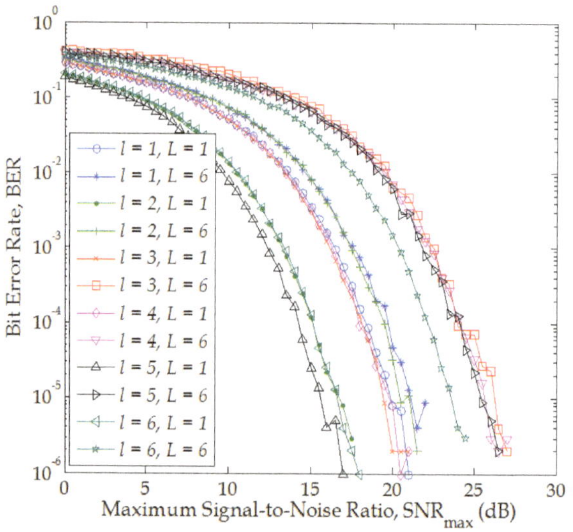

(b) BER vs. Maximum SNR

Figure 7. BER vs. total and maximum SNR for each lth user when all the subcarriers are 16-QAM modulated

BER. That is, they require similar "effective" emission powers to obtain a certain performance. However, user 5 directly illuminates three branches (j = 1, 3 and 4, see Fig. 5), although the electrical power of the induced-photocurrent at the third receiving branch practically doubles those at the other two branches. To the contrary, users 3 and 4 only directly illuminate one branch, the first and fourth, respectively, the received power at the remaining branches being negligible. Therefore, given a similar "effective" emitted power, the total electrical power at the detector due to user 5 is approximately twice that due to users 3 and 4, leading to the different performances observed in Fig. 7(a) of approximately 3 dB (actually, slightly more) worse for user 5. Additionally, with this new representation criterion, the loss in dB between the multi-user and single-user scenario performances remains unchanged for each user (with respect to that observed in Fig. 7(a)), as expected. However, single-user performances are not all identical as before: the users that directly illuminate more than one branch improve by three or more dB their single-user performance, with respect to those which only illuminate one branch directly. Finally, users 1 and 2 present the best multi-user performances.

Therefore, we can conclude that Fig. 7(b) offers a better performance comparison than Fig. 7(a), since it shows us the true SNR required at receiver branches to obtain a specific BER, regardless of how many branches are simultaneously illuminated by each user.

4.3. Complexity reduction of least-squares detector

The main limitation of linear multi-user detectors is the fact that a weight matrix $\mathbf{W}_p$, $\forall p$ has to be determined. For the simplest LS detector, obtaining each of these weight matrices involves, according to (18), multiplying two complex matrices ($\mathbb{C}^{L\times P}$ and $\mathbb{C}^{P\times L}$, respectively), the inversion of the resulting $\mathbb{C}^{L\times L}$ matrix and a final $\mathbb{C}^{P\times L} \times \mathbb{C}^{L\times L}$ multiplication. However, this very complicated operation only has to be carried out once after a specific training period has provided new estimates of the frequency-domain channel transfer factor matrices $\mathbf{H}_p$. For slowly time-varying wireless optical channel, this operation could occur with a very low periodicity in time.

In contrast, obtaining the estimates $\hat{\mathbf{s}}_p$ of the different users' transmitted signals for each subcarrier must be done every OFDM symbol period T_S. According to (17), this operation requires a $\mathbb{C}^{L\times P} \times \mathbb{C}^{P\times 1}$ matrix multiplication by each subcarrier. When we are only concerned with the data due to a specific user, operation (20) must be evaluated, which only requires the multiplication of two P-element complex vectors, i.e. P complex multiplications and sums by each subcarrier. In any case, the total number of operations can be reduced considerably if we take into account the actual interrelation between users at the receiver, that is, how emissions from other users interfere with that of our desired user. For example, from Fig. 4 and 5, it can be easily observed how the first user is the only one that directly illuminates the fifth branch of the receiver: the received power on it due to the remaining users is negligible. Therefore, demodulating the first user's data could be carried out by only considering this branch, then turning the evaluation of (20) into a single-user single-branch (SISO, *single-input single-output*) problem:

$$\hat{s}_{p,\text{LS}}^{(l=1)} = \mathbf{W}_{p,\text{LS}}^{(l=1,j=5)H} \mathbf{x}_p^{(j=5)} = \left(\frac{H_{5p}^{(l=1)}}{H_{5p}^{(l=1)H} H_{5p}^{(l=1)}} \right)^H x_{5p} = \frac{x_{5p}}{H_{5p}^{(l=1)}} \tag{24}$$

Obviously, the multi-user detection problem could be more complex for other users. For example, user 3 only directly illuminates the first branch. But this branch is also illuminated by user 5, which also illuminates branches 3 and 4. However, users 6 and 4, respectively, also illuminate these branches. Therefore, the demodulation of the third user's data involves considering the received signals at, at least, branches 1, 3 and 4. In this case, the number of users ($L = 4$) would outweigh the number of branches concerned ($P = 3$). Then, an additional branch must be considered in order to fulfil the condition $P \geq L$ required by LS detectors. Thus, branch 2 must be chosen, since this is another one illuminated by user 6, in addition to branch 3. However, branch 2 is also affected by emissions from user 2, which additionally illuminates branch 6. Finally, the detection of the data coming from only user 3 requires processing the signals received at all the branches except the fifth.

A simple algorithm can be developed to find the minimum number of branches required for the detection of a certain user's data. Let u_j be a vector that includes the users' indexes that significantly illuminate the jth receiving branch, n_j the total number of users illuminating that branch, and b_l a vector that includes the indexes of the branches illuminated by a certain user l. The algorithm for obtaining the minimum set of branches required to demodulate the data from the l'^{th} user can be described as follows:

1. *Find j so that minimises n_j: $l' \in u_j$*
2. *Set $x = j$*
3. *Set $x' = x$*
4. *$\forall j \in x$, repeat 5*
5. *$\forall l \in u_j, x = x \cup b_l$*
6. *If $x \neq x'$ go back to 3*
7. *x contains the branches to consider for demodulating l'^{th} user's data*

Fig. 8 compares the performance of the original LS detector, which uses all the $P_{\text{max}} = 6$ branches to demodulate the data corresponding to the different users, and a reduced version of this which only uses the minimum set of P_{min} branches obtained by using the previous algorithm. The modulation mode was 16-QAM for every subcarrier. The results have been shown versus the total SNR because they are easier to visualise as compared with maximum SNR results where the greater overlapping of the curves hinders its analysis enormously. However, here we are concerned with comparing BER performance of reduced and original LS detectors for each lth user and not comparing the different performances between users. In the graphs, total SNR refers, as in section 4.2, to the sum of the SNRs at each branch for the corresponding user: $\sum_{j=1}^{P} \text{SNR}_j^{(l)}$ where $P = P_{\text{max}} = 6$ irrespective of whether complexity reduction is applied by using fewer branches during demodulation or not. Therefore, reduced and original LS detectors are compared fairly, contrasting their performances with the same total SNR. Surprisingly, we can see how the reduced LS detector outperforms, although with less than 1 dB gain, the original one for all the users (the unique exception being $l = 2$ for $\text{SNR}_{\text{total}} > 23$ dB). This can be explained by the fact that, when using all the receiving branches, the residual interference due to the remaining users at each branch impairs the multi-user detector performance, especially when some branches do not receive any signal contribution from the corresponding user. Thus, for example, in the case of the first user, it seems evident that a better performance would be obtained by "switching off" all the receiving branches except the fifth. Therefore, the results show that not only better performance is achieved with the reduced LS detector but also a significant reduction in the complexity of the detection process. Except for users 3 and 4 with slight receiver reductions from 6 to 5 branches, the remaining users can operate perfectly by using only half or fewer of the total receiving branches.

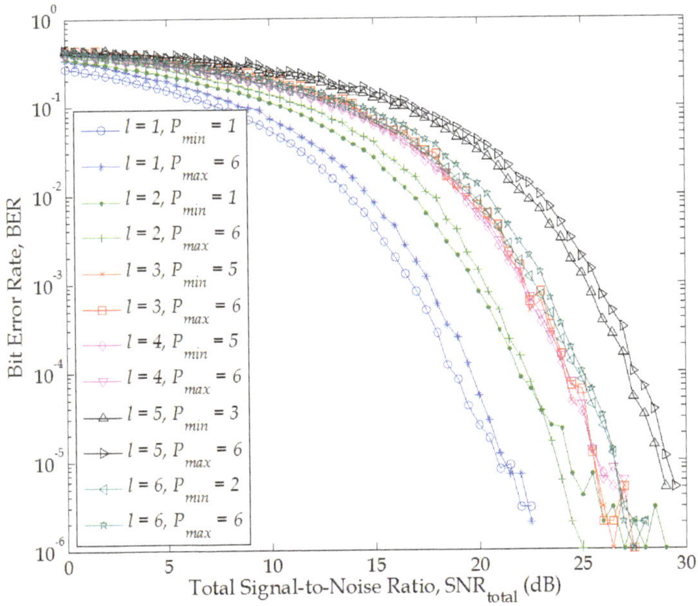

Figure 8. Performance comparison of reduced and original LS detectors for each lth user when all the subcarriers are 16-QAM modulated

5. Other MIMO techniques for WOC

As we have seen, the main limitation of LS-based multi-user detection is its high complexity, since it is implemented on a subcarrier-to-subcarrier basis. This complexity can be reduced significantly as described in section 4.3, without forgetting the beneficial capacity of OFDM for accommodating the system throughput to the channel characteristics and number of simultaneous users [9]. However, despite the complexity of the detection process for a particular user being only proportional to the number P of receiving branches, obtaining the weight matrix grows exponentially with this number. Therefore, this multi-user technique is only appropriate for scenarios with a reduced number L of users (since P must always meet the requirement $P \geq L$).

When we are interested in scenarios with large numbers of users, multi-carrier code-division multiple access (MC-CDMA) is a more practical solution [2]. The lower complexity of MC-CDMA to accommodate larger numbers of users implies a compensation payment. Now, the total throughput provided by the OFDM system is shared among all the simultaneous users. This is accomplished by their corresponding orthogonal codes, which avoid any collision between the user's transmissions at a certain subcarrier in a specific instant in time. That is, never there are two users occupying the same frequency-time resource. This can be compared with the multi-user detection techniques presented in this chapter, where the system takes advantage of the spatial diversity provided by the MIMO channel to separate all the users' signals that are continuously being transmitted using all the available bandwidth.

Outside the OFDM context, there are other multiple works addressing MIMO techniques for indoor wireless optical communications, which are generally applied in conjunction with conventional optical modulation schemes (on-off keying, pulse-position modulation, etc.). The main idea of an important group of them relies on creating many nearly-ideal and independent channels between a specific user and receiver by using multibeam transmitters and angle-diversity receivers [1, 14]. Imaging receivers have also been proposed to greatly increase the number of receiving channels at a reduced cost [6, 16], hence providing higher data rates [3, 29]. Sometimes, the distinctive spatial nature of the channel, which is unique, between a specific transmitter and the receiver is exploited to carry additional information as in optical spatial modulation [22].

6. Visible light communications: the new challenge

Visible light communications (VLC), first proposed by researchers at Keio University in Tokyo [18, 19] have prompted the interest of the scientific community in the last few years [7, 11, 17, 25, 29]. There have also been regulatory efforts made on this technology that have led to the appearance of a standard [13]. These new VLC systems, using visible LED lamps to simultaneously transmit information together with their normal use as illumination devices, share the same advantages as their infrared counterparts [15]. They are also eye-safe (visible light is not harmful to the human eye), which enables the use of higher transmission powers. However, the main drawback is the limited transmission bandwidth of current LED devices, typically several MHz, and whose enhancement has been one of the main issues addressed by researchers [23, 24]. Zeng et al. [29] have proposed the use of MIMO schemes based on imaging receivers in order to obtain high capacity VLC networks. Additionally, the OFDM technique has been proved to be a feasible candidate to obtain these high-speed networks [20], demonstrating impressive experimental data rates for short-range communications [27, 28].

Therefore, it appears to be clear that combining OFDM technique and imaging reception could be an interesting research field for the future. Table 3 shows the main parameters of a simulation scenario in which MIMO-OFDM, based on imaging reception, is evaluated. Fig. 9 illustrates the images of the LED arrays on the pixelated imaging receiver at two different positions in the room at a height of 0.75 m, which have been obtained by using a paraxial optic approach, as in the work by Zeng et al. [29]. The performance results of the multi-user LS receiver for these two positions of the detector array are compared in Fig. 10. In order to carry out a fair comparison, the BER performance is shown versus the maximum SNR observed at the receiving pixels for the emissions from the lamp $l = 6$, which is located the furthest from the detector, when this is positioned close to the corner (Fig. 9(b)), i.e. $\max(\mathrm{SNR}_j^{(l=6,\mathrm{pos=b})})$, $j = 1, \ldots, P$. Here, the case labelled as $L = 6$ is referred to the aggregate system BER (considering those of all the users $l = 1, \ldots, L$ jointly). Regarding those labelled as $L = 1$, they represent the single-user performances for the user $l = 6$ (the worst). Finally, $P_{\min}$ denotes the number of receiving channels (pixels) required for a correct joint demodulation, which is always 8 (those illuminated by the lamps, see Fig. 9). We can observe that the detector, when located close to a corner, requires more than 40 dB in the SNR to make the BER drop below 10^{-6} when considering the aggregate performance. This represents a SNR loss of roughly 20 dB with respect to its corresponding single-user scenario. In contrast, evaluated under the same illumination conditions (versus the maximum $\mathrm{SNR}_j^{(l=6,\mathrm{pos=b})}$), the receiver at the centre of the room requires more than 15 dB less electrical power to obtain the same aggregate performance. Although these SNR values could be practical in a VLC

Parameter	Value
Room size (length×width×height):	7.5 m ×5.5 m ×3.5 m
Number of LED arrays:	6 (3 × 2)
Number of LEDs per array:	3600 (60 × 60)
Dimensions of each LEDs array:	1.2 m ×1.2 m
Positions of LED arrays	array 1: (1.50,1.50,3.50)
(central point) (x, y, z):	array 2: (3.75,1.50,3.50)
	array 3: (6.00,1.50,3.50)
	array 4: (1.50,4.00,3.50)
	array 5: (3.75,4.00,3.50)
	array 6: (6.00,4.00,3.50)
LED lambertian order (n)	1
LED transmission bandwidth	∼ 15 MHz
Total number of subcarriers (N)	64
Number of information subcarriers (S_I)	48
Number of bits per subcarrier (B)	4 (16-QAM)
OFDM symbol period (T_S)	4 μs
Cyclic prefix extension (N_e)	16
Aggregate throughput	288 Mbit/s
Imaging lens f-number ($f_{\#}$)	1
Lens diameter (D)	4 cm
Number of pixels or receiving channels (P)	16 (4 × 4)
Detector size (silicon area)	1.6 cm ×1.6 cm

Table 3. Parameters for simulation

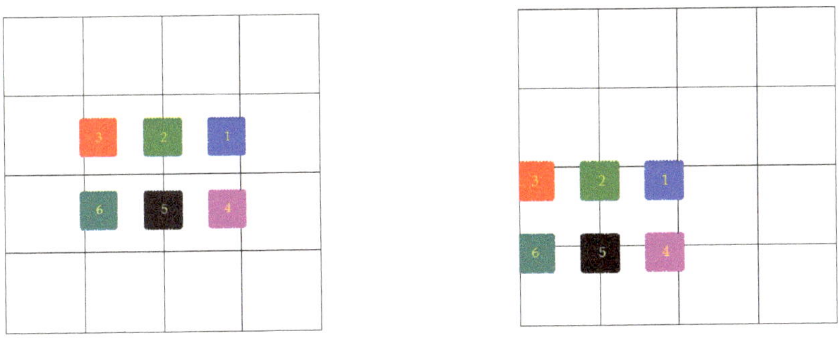

(a) Detector at the room centre

(b) Detector close to a room corner

Figure 9. Images of the LED arrays on the detector at two positions

environment, it is evident that moving towards corners degrades enormously the aggregate performance due to the long distances and low inclination of the rays coming from the furthest lamps. Hence, it would be a more efficient solution to employ higher-order modulation modes (greater values of B) for the lamps closer to the receiver and lower ones for those further away, even deactivating them as necessary, by using adaptive OFDM schemes [9]. In any case, what seems apparent, according to these preliminary results, is that MIMO-OFDM with imaging reception is a promising technique, which is worthy of further work.

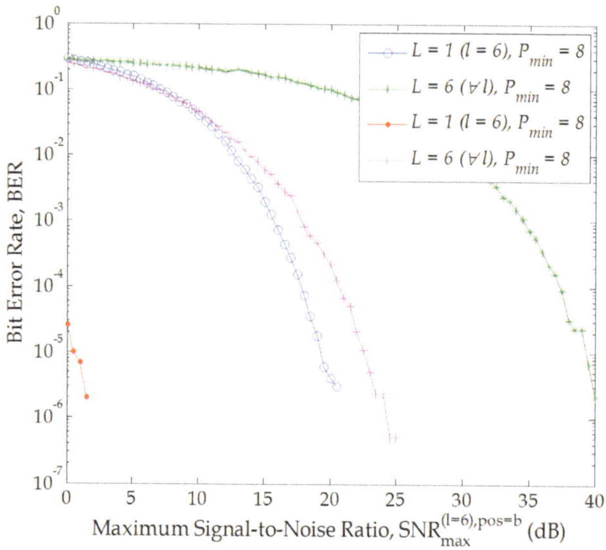

Figure 10. Performance comparison of multi-user LS detection based on imaging reception for two positions of the detector array

7. Conclusions

In this chapter, the employment of multi-user LS detection in conjunction with angle-diversity receivers and OFDM modulation technique for wireless optical communication has been evaluated. The MIMO optical channel model, which can be determined by using Monte Carlo-based ray-tracing algorithms, has been described in detail. This algorithm allows us to determine LOS and multiple-bounce reflection contributions to the received optical power at the photodetector, which enables a more accurate analysis of the proposed system performance. The multi-user detector, which is based on the linear combining of the incoming signals at its receiving branches, is shown to have a performance very close to that of the optimum ML detector. Additionally, it has been observed that strategies to find the truly significant receiving branches, during the data retrieving for a specific user, can be applied to reduce the complexity of the demodulation problem while maintaining and, even improving, the system performance. The results have also shown that aggregate high data rates can be obtained for indoor wireless optical communications at practical signal-to-noise ratios. Finally, the proposed multi-user OFDM system has been assessed for a visible-light communication scenario, where imaging reception is assumed. The preliminary results show that the previous scheme is an attractive candidate for developing high-capacity VLC networks, but further research still needs to be carried out.

Acknowledgments

This work has been funded in part by the Spanish Research Administration (TEC2009-14059-C03-02), by the Canary Government (SolSubC200801000306) and the Plan E (Spanish Economy and Employment Stimulation Plan).

Author details

Oswaldo González
Dept. Fundamental and Experimental Physics, Electronics and Systems
University of La Laguna, Spain

8. References

[1] Al-Ghamdi, A. G. & Elmirghani, J. M. H. [2004]. Analysis of diffuse optical wireless channels employing spot-diffusing techniques, diversity receivers, and combining schemes, *IEEE Transactions on Communications* 52(10): 1622–1631.
[2] Alsaadi, F. E. & Elmirghani, J. M. H. [2009a]. Adaptive mobile line strip multibeam MC-CDMA optical wireless system employing imaging detection in a real indoor environment, *IEEE Journal on Selected Areas in Communications* 27(9): 1663–1675.
[3] Alsaadi, F. E. & Elmirghani, J. M. H. [2009b]. Performance evaluation of 2.5 Gbit/s and 5 Gbit/s optical wireless systems employing a two dimensional adaptive beam clustering method and imaging diversity detection, *IEEE Journal on Selected Areas in Communications* 27(8): 1507–1519.
[4] Armstrong, J. [2009]. OFDM for optical communications, *Journal of Lightwave Technology* 27(3): 189–204.
[5] Barry, J. R., Kahn, J. M., Krause, W. J., Lee, E. A., & Messerschmitt, D. G. [1993]. Simulation of multipath impulse response for wireless optical channels, *IEEE Journal on Selected Areas in Communications* 11(3): 367–379.
[6] Djahani, P. & Kahn, J. M. [2000]. Analysis of infrared wireless links employing multibeam transmitter and imaging diversity receivers, *IEEE Transactions on Communications* 48(12): 2077–2088.
[7] Elgala, H., Mesleh, R. & Haas, H. [2011]. Indoor optical wireless communication: potential and state-of-the-art, *IEEE Communications Magazine* 49(9): 56–62.
[8] González, O., Pérez-Jiménez, R., Rodríguez, S., Rabadán, J. & Ayala, A. [2005]. OFDM over indoor wireless optical channel, *IEE Proceedings-Optoelectronics* 152(4): 199–204.
[9] González, O., Rodríguez, S., Pérez-Jiménez, R., Delgado, F. & Ayala, A. [2007]. Muti-user adaptive orthogonal frequency-division multiplexing system for indoor wireless optical communications, *IET Optoelectronics* 1(2): 68–76.
[10] González, O., Rodríguez, S., Pérez-Jiménez, R., Mendoza, B. R. & Aayala, A. [2011]. Comparison of Monte Carlo ray-tracing and photon-tracing methods for calculation of the impulse response on indoor wireless optical channels, *Optics Express* 19(3): 1997–2005.
[11] Grubor, J., Randel, S., Langer, K.-D. & Walewski, J. W. [2008]. Broadband information broadcasting using LED-based interior lighting, *Journal of Lightwave Technology* 26(24): 3883–3892.
[12] Hanzo, L., Münster, M., Choi, B.-J. & Keller, T. [2003]. *OFDM and MC-CDMA for broadband multi-user communications, WLANs and broadcasting*, John Wiley & Sons, West Sussex, England.
[13] IEEE Std 802.15.7 [2011]. IEEE standard for local and metropolitan area networks–part 15.7: Short-range wireless optical communication using visible light, pp. 1–309.
[14] Jivkova, S. & Kavehrad, M. [2001]. Receiver designs and channel characterization for multi-spot high-bit-rate wireless infrared communications, *IEEE Transactions on Communications* 49(12): 2145–2153.
[15] Kahn, J. M. & Barry, J. R. [1997]. Wireless infrared communications, *Proceedings of IEEE* 85(2): 265–298.

[16] Kahn, J. M., You, R., Djahani, P., Weisbin, A. G., Teik, B. K. & Tang, A. [1998]. Imaging diversity receivers for high-speed infrared wireless communication, *IEEE Communications Magazine* 36(12): 88–94.

[17] Komine, T., Lee, J. H., Haruyama, S. & Nakagawa, M. [2009]. Adaptive equalization system for visible light wireless communication utilizing multiple white LED lighting equipment, *IEEE Transactions on Wireless Communications* 8(6): 2892–2900.

[18] Komine, T. & Nakagawa, M. [2003]. Integrated system of white LED visible-light communication and power-line communication, *IEEE Transactions on Consumer Electronics* 49(1): 71–79.

[19] Komine, T. & Nakagawa, M. [2004]. Fundamental analysis for visible-light communication system using LED lights, *IEEE Transactions on Consumer Electronics* 50(1): 100–107.

[20] Li, X., Vučić, J., Jungnickel, V. & Armstrong, J. [2012]. On the capacity of intensity-modulated direct-detection systems and the information rate of ACO-OFDM for indoor optical wireless applications, *IEEE Transactions on Communications* 60(3): 799–809.

[21] Lomba, C. R., Valadas, R. T. & de Oliveira Duarte, A. M. [1998]. Experimental characterisation and modelling of the reflection of infrared signals on indoor surfaces, *IEE Proceedings-Optoelectronics* 145(3): 191–197.

[22] Mesleh, R., Elgala, H. & Haas, H. [2011]. Optical spatial modulation, *IEEE/OSA Journal of Optical Communications and Networking* 3(3): 234–244.

[23] Minh, H. L., O'Brien, D. C., Faulkner, G. E., Zeng, L., Lee, K., Jung, D., Oh, J. & Won, E. T. [2009]. 100-Mb/s NRZ visible light communications using a postequalized white LED, *IEEE Photonics Technology Letters* 21(15): 1063–1065.

[24] O'Brien, D. C., Minh, H. L., Faulkner, G. E., Zeng, L., Lee, K., Jung, D. & Oh, J. [2008]. High-speed visible light communications using multiple-resonant equalization, *IEEE Photonics Technology Letters* 20(15): 1243–1245.

[25] O'Brien, D. C., Zeng, L., Minh, H. L., Faulkner, G. E., Walewski, J. W. & Randel, S. [2008]. Visible light communications: challenges and possibilities, *Proceedings of IEEE 19th International Symposium on Personal, Indoor and Mobile Radio Communications, PIMRC 2008*, pp. 1–5.

[26] Proakis, J. G. [2001]. *Digital Communications*, McGraw-Hill International Edition, Singapore.

[27] Vučić, J., Kottke, C., Habel, K. & Langer, K.-D. [2011]. 803 Mbit/s visible light WDM link based on DMT modulation of a single RGB LED luminary, *Proceedings of Optical Fiber Communication Conference and Exposition, and the National Fiber Optic Engineers Conference (OFC/NFOEC) 2011*, pp. 1–3.

[28] Vučić, J., Kottke, C., Nerreter, S., Langer, K.-D. & Walewski, J. W. [2010]. 513 Mbit/s visible light communications link based on DMT-modulation of a white LED, *Journal of Lightwave Technology* 28(4): 3512–3518.

[29] Zeng, L., O'Brien, D. C., Minh, H. L., Faulkner, G. E., Lee, K., Jung, D., Oh, J. & Won, E. T. [2009]. High data rate multiple iinput multiple output (MIMO) optical wireless communications using white LED lighting, *IEEE Journal on Selected Areas in Communications* 27(9): 1654–1662.

Permissions

The contributors of this book come from diverse backgrounds, making this book a truly international effort. This book will bring forth new frontiers with its revolutionizing research information and detailed analysis of the nascent developments around the world.

We would like to thank Narottam Das, for lending his expertise to make the book truly unique. He has played a crucial role in the development of this book. Without his invaluable contribution this book wouldn't have been possible. He has made vital efforts to compile up to date information on the varied aspects of this subject to make this book a valuable addition to the collection of many professionals and students.

This book was conceptualized with the vision of imparting up-to-date information and advanced data in this field. To ensure the same, a matchless editorial board was set up. Every individual on the board went through rigorous rounds of assessment to prove their worth. After which they invested a large part of their time researching and compiling the most relevant data for our readers. Conferences and sessions were held from time to time between the editorial board and the contributing authors to present the data in the most comprehensible form. The editorial team has worked tirelessly to provide valuable and valid information to help people across the globe.

Every chapter published in this book has been scrutinized by our experts. Their significance has been extensively debated. The topics covered herein carry significant findings which will fuel the growth of the discipline. They may even be implemented as practical applications or may be referred to as a beginning point for another development. Chapters in this book were first published by InTech; hereby published with permission under the Creative Commons Attribution License or equivalent.

The editorial board has been involved in producing this book since its inception. They have spent rigorous hours researching and exploring the diverse topics which have resulted in the successful publishing of this book. They have passed on their knowledge of decades through this book. To expedite this challenging task, the publisher supported the team at every step. A small team of assistant editors was also appointed to further simplify the editing procedure and attain best results for the readers.

Our editorial team has been hand-picked from every corner of the world. Their multi-ethnicity adds dynamic inputs to the discussions which result in innovative outcomes. These outcomes are then further discussed with the researchers and contributors who give their valuable feedback and opinion regarding the same. The feedback is then collaborated with the researches and they are edited in a comprehensive manner to aid the understanding of the subject.

Apart from the editorial board, the designing team has also invested a significant amount of their time in understanding the subject and creating the most relevant covers. They scrutinized every image to scout for the most suitable representation of the subject and create an appropriate cover for the book.

The publishing team has been involved in this book since its early stages. They were actively engaged in every process, be it collecting the data, connecting with the contributors or procuring relevant information. The team has been an ardent support to the editorial, designing and production team. Their endless efforts to recruit the best for this project, has resulted in the accomplishment of this book. They are a veteran in the field of academics and their pool of knowledge is as vast as their experience in printing. Their expertise and guidance has proved useful at every step. Their uncompromising quality standards have made this book an exceptional effort. Their encouragement from time to time has been an inspiration for everyone.

The publisher and the editorial board hope that this book will prove to be a valuable piece of knowledge for researchers, students, practitioners and scholars across the globe.

List of Contributors

K. Harasawa
Hitachi Information & Communication Engineering, Ltd., Japan

Hiroshi Maeda
Dept. of Information and Communication Eng., Fukuoka Institute of Technology, Japan

Jeong Kim
Chungnam National University, Department of Electric, Electronic and Communication Engineering Education, Daejeon, South Korea

Edward G. Kostsov
Institute of Automation and Electrometry, Russian Academy of Sciences, Russia

Sergey V. Piskunov and Mike B. Ostapkevich
Institute of Computational Mathematics and Mathematical Geophysics, Russian Academy of Sciences, Russia

Narottam Das
Department of Electrical and Computer Engineering, Curtin University, Australia
School of Engineering, Edith Cowan University, Australia

Hitoshi Kawaguchi
Graduate School of Materials Science, Nara Institute of Science and Technology, Japan

Mohammad Razaghi
Department of Electrical and Computer Engineering, University of Kurdistan, Iran

Alejandro García Juárez
University of Sonora, Department of Physics Research, Mexico

Ignacio Enrique Zaldívar Huerta
National Institute of Astrophysics, Optics and Electronics. Department of Electronics, Mexico

Jorge Rodríguez Asomoza
University of the Americas-Puebla, Department of Electronics, Mexico

Muzahim I. Azawe
Mosul University, Mosul, Iraq

Mingbo Niu, Julian Cheng and Jonathan F. Holzman
The University of British Columbia (UBC), School of Engineering, Faculty of Applied Science,
Kelowna, British Columbia, Canada

Vjaceslavs Bobrovs and Girts Ivanovs
Riga Technical University, Institute of Telecommunications, Latvia

Koichi Takahashi
Future Creation Laboratory, Olympus Corporation, Japan

Xiaobo Xing
Education Ministry's Key Laboratory of Laser Life Science & Institute of Laser LifeScience, College of Biophotonics, South China Normal University,Guangzhou, Guangdong, China
Department of Chemistry, Jinan University, Guangdong, Guangzhou, 510632, China

Huaqing Yu
School of Physics and Electronic-information Engineering, Hubei Engineering University, Xiaogan, China

Debin Zhu
Education Ministry's Key Laboratory of Laser Life Science & Institute of Laser LifeScience, College of Biophotonics, South China Normal University,Guangzhou, Guangdong, China

Jiapeng Zheng
School of Physics and Optoelectronic Science and Engineering, South China Normal University,
Guangzhou, Guangdong, China

Huang Chen
School of Information and Optoelectronic Science and Engineering, South China Normal University,
Guangzhou, Guangdong, China

Wei Chen
School of Information and Optoelectronic Science and Engineering, South China Normal University,
Guangzhou, Guangdong, China

Ruibin Xie
School of Information and Optoelectronic Science and Engineering, South China Normal University,
Guangzhou, Guangdong, China

Jiye Cai
Department of Chemistry, Jinan University, Guangdong, Guangzhou, 510632, China

Bo Yuan
University of Minnesota-Twin Cities, USA

Li Li
Nanjing University, China

Zhongfeng Wang
Broadcom Corporation, USA

Antonio García-Zambrana and Beatriz Castillo-Vázquez
Department of Communications Engineering, University of Málaga, Spain

Carmen Castillo-Vázquez
Department of Statistics and Operations Research, University of Málaga, E-29071 Málaga, Spain

Y. Ben Ezra
Holon Institute of Technology (HIT), Holon, Israel; Optiway Integrated Solutions LTD, Rosh Haayin, Israel

B.I. Lembrikov
Holon Institute of Technology (HIT), Holon, Israel

Avi Zadok, Ran Halifa
School of Engineering and Institute for Nano-Technology and Advanced Materials, Bar-Ilan
University, Israel

D. Brodeski
Optiway Integrated Solutions LTD, Rosh Haayin, Israel

Liangwei Dong
Institute of Information Optics, Zhejiang Normal University, Jinhua, 321004, China

Huijun Li
Department of Physics, Zhejiang Normal University, Jinhua, 321004, China

Kostas Peppas
National Center For Scientific Research "Demokritos", Laboratory of Wireless Communications,
Institute of Informatics and Telecommunications, 15310 Athens, Greece

Hector E. Nistazakis and George S. Tombras
Department of Electronics, Computers, Telecommunications and Control, Faculty of Physics, National
and Kapodistrian University of Athens, 15784, Athens Greece

Vasiliki D. Assimakopoulos
Institute of Environmental Research and Sustainable Development of the National Observatory of
Athens, I. Metaxa and Vas. Pavlou, Athens, 15236, Greece

Oswaldo González
Dept. Fundamental and Experimental Physics, Electronics and Systems
University of La Laguna, Spain